# ADVANCES IN PROTEIN CHEMISTRY

Volume 60

Copper-Containing Proteins

# ADVANCES IN PROTEIN CHEMISTRY

EDITED BY

FREDERIC M. RICHARDS
Department of Molecular Biophysics
and Biochemistry
Yale University
New Haven, Connecticut

DAVID S. EISENBERG
Department of Chemistry
and Biochemistry
University of California, Los Angeles
Los Angeles, California

JOHN KURIYAN
Department of Molecular Biophysics
Howard Hughes Medical Institute
Rockefeller University
New York, New York

VOLUME 60

## Copper-Containing Proteins

EDITED BY

JOAN SELVERSTONE VALENTINE
UCLA, Los Angeles, California

EDITH BUTLER GRALLA
UCLA, Los Angeles, California

ACADEMIC PRESS
An imprint of Elsevier Science

Amsterdam Boston London New York Oxford Paris
San Diego San Francisco Singapore Sydney Tokyo

This book is printed on acid-free paper. ∞



Academic Press
*An imprint of Elsevier Science*
525 B Street, Suite 1900, San Diego, California 92101-4495, USA
http://www.academicpress.com

Academic Press
84 Theobolds Road, London WC1X 8RR, UK
http://www.academicpress.com

Library of Congress Catalog Card Number: 0065-3233

International Standard Book Number: 0-12-034260-X

PRINTED IN THE UNITED STATES OF AMERICA
02 03 04 05 06 07 SB 9 8 7 6 5 4 3 2 1

# CONTENTS

## Bacterial Copper Transport

Zen Huat Lu and Marc Solioz



## Understanding the Mechanism and Function of Copper P-type ATPases

Ilia Voskoboinik, James Camakaris, and Julian F. B. Mercer



## Copper Chaperones

Jennifer Stine Elam, Susan T. Thomas, Stephen P. Holloway, Alexander B. Taylor, and P. John hart

## Fet3p, Ceruloplasmin, and the Role of Copper in Iron Metabolism

DANIEL J. KOSMAN



## Blue Copper-Binding Domains

ARAM M. NERSISSIAN AND ERIC L. SHIPP



## Cytochrome *c* Oxidase

SHINYA YOSHIKAWA

## Nuclear Magnetic Resonance Spectroscopy Studies on Copper Proteins

Lucia Banci, Roberta Pierattelli, and Alejandro J. Vila

# PREFACE

This is an auspicious time to publish a volume on copper proteins. The number of known proteins with metallic cofactors continues to increase steadily, and the availability of structural and sequence data is enabling much more specific characterizations of the interactions between the metal ions and proteins as well as of their functions and mechanisms. Numerous investigators are choosing copper proteins and copper metabolism as their model systems for such studies. While copper-containing proteins play essential roles, their numbers are few enough that a comprehensive understanding is a reasonable goal.

In planning this volume we chose to emphasize some of the areas of copper proteins in which research has been moving particularly rapidly in recent years and that we felt would particularly benefit from a timely review. We have not included some topics for which several such reviews have already appeared elsewhere, such as copper-zinc superoxide dismutase, a copper enzyme particularly close to our own hearts.

In the area of copper metabolism, four topics are covered: bacterial copper transport reviewed by Huat Lu and Solioz; copper P-type ATPases reviewed by Voskoboinik, Camakaris, and Mercer; copper chaperones reviewed by Stine Elam *et al.*; and copper metalloregulation of gene expression reviewed by Winge. An important related topic is the link between copper and iron metabolism. In this area, Kosman has reviewed the multicopper oxidase enzymes, such as Fet3p and ceruloplasmin, which catalyze the conversion of iron(II) to iron(III) in preparation for its specific transport by partner transporter proteins.

Increasing knowledge of copper protein structures is allowing a much more detailed understanding of copper protein reactivities and biophysical properties. Two very different examples are found in galactose oxidase reviewed by Whittaker and cytochrome *c* oxidase reviewed by Yoshikawa. Increased availability of sequence data across many species is leading to the discovery of large classes of copper proteins containing blue copper binding domains as described in the review by Nersissian and Shipp. Copper is a wonderful metal ion for biophysical studies of metalloproteins, and a technique that has seen a particularly high level of activity in recent years is NMR, and an area reviewed by Banci, Pierattelli, and Vila.

We thank the authors for their excellent contributions to this volume.

Joan Selverstone Valentine
Edith Butler Gralla

# GALACTOSE OXIDASE

BY JAMES W. WHITTAKER

Department of Biochemistry and Molecular Biology, OGI School of Science and Engineering at OHSU, Beaverton, Oregon 97006



## I. Introduction

Galactose oxidase (GAOX) is an extraordinary copper metalloenzyme combining the reactivity of a free radical ligand with a redox-active metal center in a unique catalytic complex, the free radical-coupled copper active site (Whittaker and Whittaker, 1988; Whittaker, 1994). This novel structure is the basis for the distinctive chemistry and spectroscopy of a new family of enzymes, the radical copper oxidases (Whittaker and Whittaker, 1998), that includes galactose oxidase (from *Dactylium dendroides*) (Avigad *et al.*, 1962; Hamilton, 1981; Kosman, 1985) and glyoxal oxidase (from *Phanerochaete chrysosporium*) (Kersten and Kirk, 1987). The extensive characterization of these enzymes has made them prototypes for biological radical chemistry in the growing field of free radical enzymology (Stubbe, 1989; Pederson and Finazzi-Agro, 1993; Marsh, 1995; Frey, 1997; Stubbe and van der Donk, 1998). This survey will principally



ADVANCES IN PROTEIN CHEMISTRY, Vol. 60



0065-3233/02 $35.00

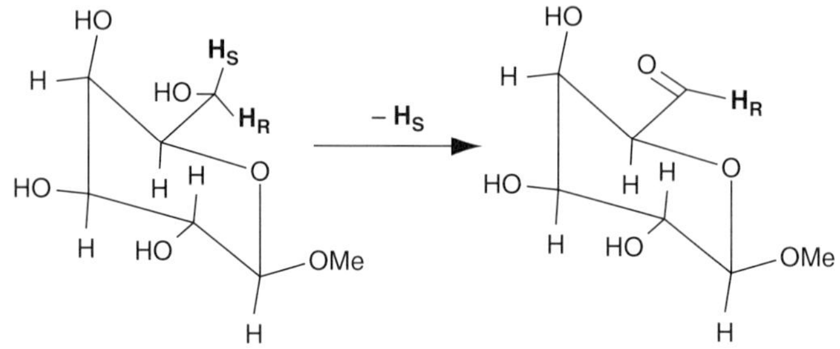


FIG. 1. Stereospecific dehydrogenation catalyzed by galactose oxidase. Prochiral hydrogens (*pro*-R, $H_R$; *pro*-S, $H_S$) of the 6-hydroxymethyl group are indicated.

focus on the structure and properties of galactose oxidase, referring to the other members only for comparison.

GAOX catalyzes the oxidation of primary alcohols to the corresponding aldehydes (Fig. 1), coupling substrate oxidation to the reduction of $O_2$ to hydrogen peroxide (Avigad *et al.*, 1962). The enzymatic reaction is strictly stereospecific, with irreversible abstraction of the *pro*-S hydrogen from the 6-hydroxymethyl group of the substrate (Maradufu *et al.*, 1971). The aldehyde product formed in this step may then be further oxidized by the same enzyme to the carboxylic acid, but at a much slower rate (Kelleher and Bhavanandan, 1986). Alcohol oxidation is strictly regioselective, and no secondary alcohols are oxidized. However, the enzyme accepts a wide variety of primary alcohols as reducing substrates, including simple sugars, aliphatic and aromatic alcohols, and even protein glycoconjugates (Table I) (Avigad *et al.*, 1962). Interestingly, GAOX strongly differentiates between the epimeric sugars galactose and glucose, a property that makes it useful in a variety of bioanalytical applications (Loken, 1966; Johnson *et al.*, 1982). Table I shows that, despite the enzyme's name, the best substrate for galactose oxidase appears to be dihydroxyacetone, supporting nearly four times the turnover rate of the canonical substrate (galactose) under comparable conditions. However, emphasis on the metabolism of these organic substrates by GAOX is probably misleading, and the physiologically important reaction is almost certainly the formation of hydrogen peroxide. The precise biochemical role of the peroxide product is not completely clear, but it may be a bacteristatic agent, as proposed for galactose oxidase (Whittaker, 1994), fuel extracellular peroxidases, as is the case for glyoxal oxidase (Kersten, 1990), or even serve as an intercellular signal.

TABLE I
*Reducing Substrates for Galactose Oxidase*

| Substrate | Relative rate |
|---|---|
| Carbohydrates | |
| D-Galactose | 1.0 |
| D-Galactosamine | 0.75 |
| 2-Deoxy-D-galactose | 0.32 |
| L-Galactose | 0.0 |
| D-Glucose | 0.0 |
| Aliphatic alcohols | |
| Glycerol | 0.01 |
| Dihydroxyacetone | 3.8 |
| Methanol | 0.00015 |
| Aromatic alcohols | |
| Benzyl alcohol | 0.05 |

## II. PROTEIN STRUCTURE

The 639 residues of the mature GAOX polypeptide derive from a 680-residue precursor protein containing N-terminal features typical of fungal prepro signal sequences (McPherson *et al.*, 1992). The prepro leader peptide is functionally important, as it directs secretion of the enzyme into the extracellular space. Posttranslational processing of the signal sequence involves proteolytic cleavage at two sites. The first cut, removing the 24-amino-acid presequence leader peptide (residues −41 to −17), is presumably performed by a signal peptidase in the secretory pathway. Subsequent tryptic-like cleavage on the C-terminal side of an arginine (residue −1 in the leader sequence) removes the 17-residue propeptide, generating the N-terminal of the mature protein. Additional maturation steps are required for the posttranslational addition of a novel covalent cross-link between Tyr-272 and Cys-228 that has been identified by X-ray crystallography (Ito *et al.*, 1991, 1994). As described in Section IX, the cross-link appears to form spontaneously in the proenzyme in the presence of dioxygen and copper ions. The mature, active enzyme is secreted into the extracellular medium, although it has also been detected intracellularly (Mendonça and Zancan, 1987), possibly representing protein that was blocked from export by incorrect processing or premature cross-linking. Under metal-deprivation stress, fungi may secrete both the prosequence (lacking the Tyr–Cys crosslink) and the mature protein (Rogers *et al.*, 2000). The mature native, folded protein, a monomer of 68 kDa, has been reported to contain a small amount of covalently bound carbohydrate (less than 10% of the total mass) (Mendonça and Zancan, 1987), although none has been identified by crystallography (see below).

High-resolution crystal structures have been solved for the native enzyme and several complexes (Knowles and Ito, 1993), providing valuable information on the metal environment as well as more global features of protein structure. Three distinct domains, composed almost exclusively of beta structure with short turns, can be structurally and functionally distinguished (Fig. 2). The N-terminal domain (residues 1–154, Domain I) forms a globular unit consisting of eight strands of antiparallel beta sheet folded into a sandwich. This segment of the protein contains a structural metal-binding site formed by the side chains of Asp-32, Asn-34, Thr-37, and Glu-142 together with two peptide carbonyls (from Lys-29 and Ala-141). These groups create a roughly octahedral site coordinating a metal cation that has been identified crystallographically as a monovalent $Na^+$ ion. However, a monovalent cation would only partly cancel the −2 charge on the two acidic side chains, leaving a net charge on the site that would be unusual for a structural metal center. This suggests that the structural metal might be a divalent ion (e.g., $Ca^{2+}$). In addition to this metal center, a carbohydrate-binding site has been detected in Domain I that may function in targeting the enzyme to extracellular carbohydrates. GAOX binding to Sepharose polymers (Tressel and Kosman, 1982) or mellobiose–polyacrylamide (Kelleher *et al.*, 1988) is, in fact, used for affinity chromatography purification of the enzyme from culture medium, a process that may be mediated by this site in Domain I. A hydrophobic patch anchors Domain I on the circumference of Domain II.

The second domain, representing the bulk of the protein (residues 155–552, Domain II), contains the catalytic active site (Fig. 3a). This domain has an unusually complex structure, variously described as a flower or a propeller, and belongs to the *kelch* superfamily of protein folds (Bork and Doolittle, 1994; Adams *et al.*, 2000). The family name derives from the name of a *Drosophila* structural protein involved in oogenesis, the first member to be identified. The characteristic feature of this structural group is a modular organization based on a four-stranded antiparallel beta sheet repeat element (Fig. 3b). Each of these modules is typically 45 to ~55 residues long with 4–5 residues per strand and includes a hexapeptide consensus feature (*hhhhGG*) consisting of 4 hydrophobic residues followed by a pair of glycines. In galactose oxidase, this consensus element occurs in the second strand from the inside of the beta module. Its composition reflects the special requirements of chain

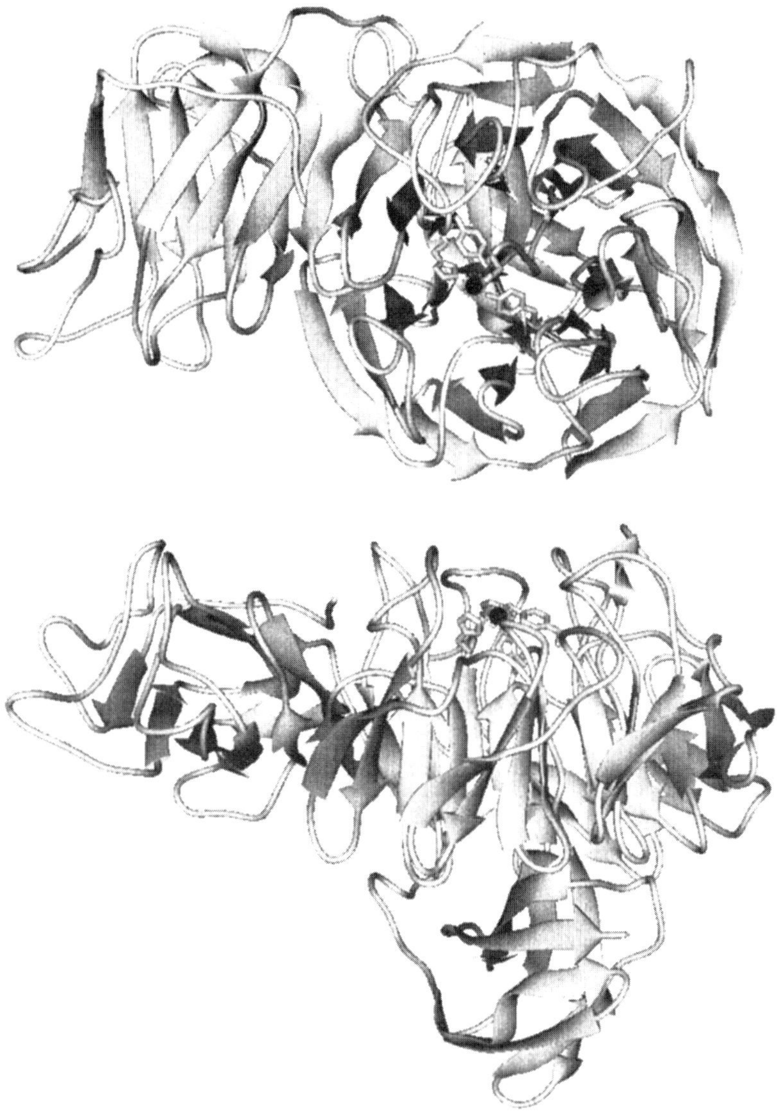

FIG. 2. Structure of galactose oxidase. (Top) View along axis of Domain II. (Bottom) View perpendicular to the wheel axis. The location of the active site copper is indicated by a black dot (•).

packing in the protein interior, as the initial hydrophobic *hhhh* tetrapeptide is entirely buried in the protein core. The glycine pair contributes steric compactness and flexibility, being able to participate in the tight

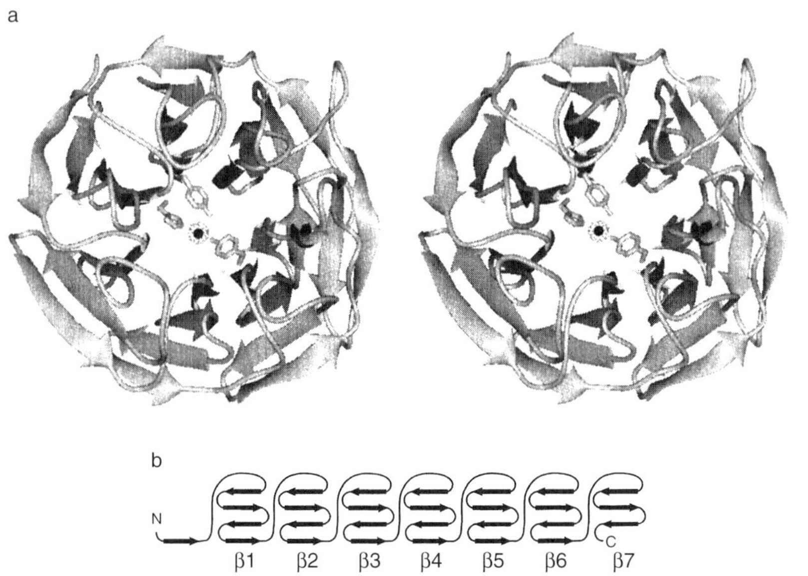


FIG. 3. Architecture of the *kelch* domain. (a) Stereoview of Domain II with the active site metal ion and protein ligands superimposed on a ribbon diagram of the polypeptide chain. (b) The modular organization of the sevenfold propeller domain is based on four-stranded antiparallel beta sheet subdomains.

turn at the end of the strand. Each module contributes a wedge-shaped segment, with smaller side chains lying along the innermost strand and the bulk of the side chains increasing toward the outer strands. Because the sheet motif in each module has the same intrinsic curvature, the segments can nest together, and the string of beta-wedges closes to form a ring or wheel. The last segment (β7) contains three beta strands from the C-terminal end of the central domain sequence and one from the beginning, forming a clasp that closes the circle and holds the structure together. This central domain contributes three of the four metal ligands (Tyr-272, Tyr-495, and His-496), all arising from turns associated with the innermost strands of the beta modules.

The third, C-terminal domain (residues 553–639, Domain III) forms a hub lying on one side of the protein wheel, capping Domain II from below. This essentially globular unit is formed from seven antiparallel beta sheets. Between sheets 3 and 4, there is a long unstructured strand (residues 572–590) that extends away from the globular hub, threading all the way through the core of the middle domain to the other side, where it contributes the fourth ligand (His-581) to the metal-binding site.

The copper is bound near the surface of the protein in a slight concavity on the wheel axis opposite Domain III.

## III. SEQUENCE CORRELATIONS

The pattern of repeated four-stranded beta modules that form the *kelch* propeller domain might be expected to be associated with a well-defined signature in the amino acid sequence permitting detailed structural correlations to be developed within the family. However, the presence of large gaps between the rather short consensus sequence repeats complicates a simple BLAST comparison (Tatusova and Madden, 1999), and GAPPED-BLAST, PSI-BLAST (Altschul *et al.*, 1997), or even more complex algorithms based on secondary structure prediction (Bork and Doolittle, 1994) may be required for alignment in some cases.

The *kelch* superfamily of proteins is a functionally diverse group in which metal ion binding and catalysis appear to be relatively rare. However, this apparent rarity may simply reflect the fact that only a few of these proteins have been purified and characterized in detail. The majority are actually virtual proteins or open reading frames predicted from nucleotide sequence data. Genomic research has added a large number of such virtual proteins to bioinformatics databases, providing an expanded set of sequences for comparative structural studies. A recent search of GenBank reveals a number of protein sequences closely related to galactose oxidase (Fig. 4). Alignment of these homologous sequences not only reproduces the diglycine motif characteristic of the beta-propeller architecture, but also conserves the metal ligands in the active site. The sequence homologues included in this alignment arise from organisms spanning the phylogenetic map from prokaryotes (*Streptomyces coelicolor* and *Stigmatella aurantiaca*) to fungi (*Ph. chrysosporium*) and green plants (*Arabidopsis thaliana*). The extent of sequence overlaps is illustrated in Fig. 4, which identifies the key conserved elements among these structures. The fact that the metal ligand residues (marked by asterisks) and their context are both conserved over these sequences supports the identification of these homologues as structural variants within a larger radical copper oxidase family of enzymes.

The only homologue from this set other than galactose oxidase that has been biochemically characterized is glyoxal oxidase (GLOX) from *Ph. chrysosporium*, which was originally identified and established as a radical-copper oxidase independent of sequence correlations (Kersten and Kirk, 1987; Kersten, 1990; Kersten and Cullen, 1993; Whittaker *et al.*, 1996b). The enzyme is a distinct protein, sharing less than 20% overall sequence similarity with GAOX. The GLOX polypeptide is slightly

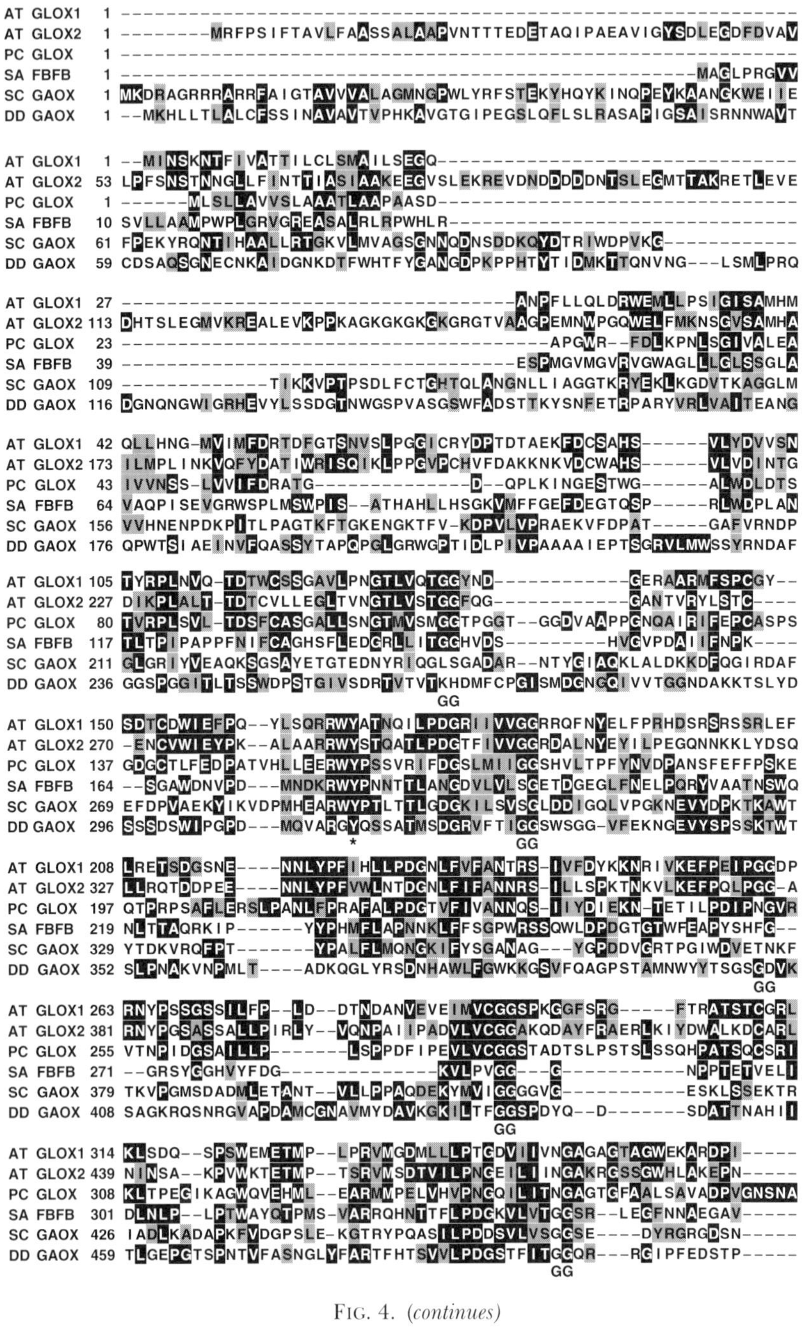


FIG. 4. (*continues*)

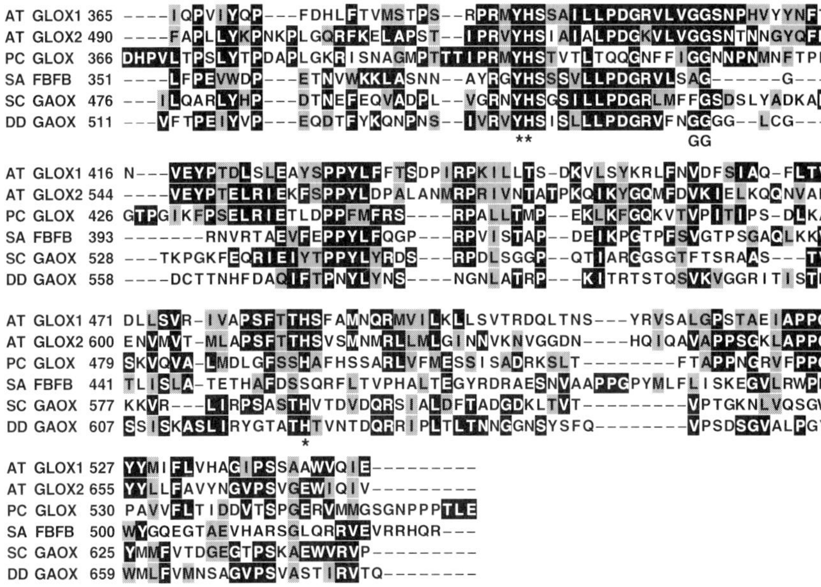


FIG. 4. Sequence correlations among copper oxidase homologues. Regions of identity are highlighted in black; regions of conservative similarity are highlighted in gray. Active site residues are marked with an asterisk (*) and the positions of the repeating glycine-pair (GG) motif within the *kelch* domain are indicated. (AT GLOX1, *Arabidopsis thaliana* glyoxal oxidase homologue, GenBank Accession No. 4678266; AT GLOX2, *A. thaliana* glyoxal oxidase homologue, GenBank Accession No. AC002130; PC GLOX, *Phanerochaete chrysosporium* glyoxal oxidase, GenBank Accession No. 1050301; SA FBFB, *Stigmatella aurantiaca* FbfB protein, GenBank Accession No. 1360138; SC GAOX, *Streptomyces coelicolor* galactose oxidase homologue, GenBank Accession No. 6689159; DD GAOX, *Dactylium dendroides* galactose oxidase, GenBank Accession No. 167225.)

smaller than GAOX, lacking the N-terminal putative targeting domain, but is highly glycosylated, with covalently bound carbohydrate representing more than 15% of the total molecular mass. GLOX is also functionally distinct from GAOX, preferentially catalyzing the oxidation of simple aldehydes to carboxylic acids. In nature, GLOX serves as a peroxide factory fueling extracellular peroxidases (lignin peroxidase and manganese peroxidase) secreted by the wood rot fungus under nutrient limitation (Kersten and Kirk, 1987; Hammel *et al.*, 1994). Despite the differences in catalytic specificity and significantly different primary structures, spectroscopic comparison indicates that GAOX and GLOX have nearly identical active sites (see below) (Whittaker *et al.*, 1996b). The extensive

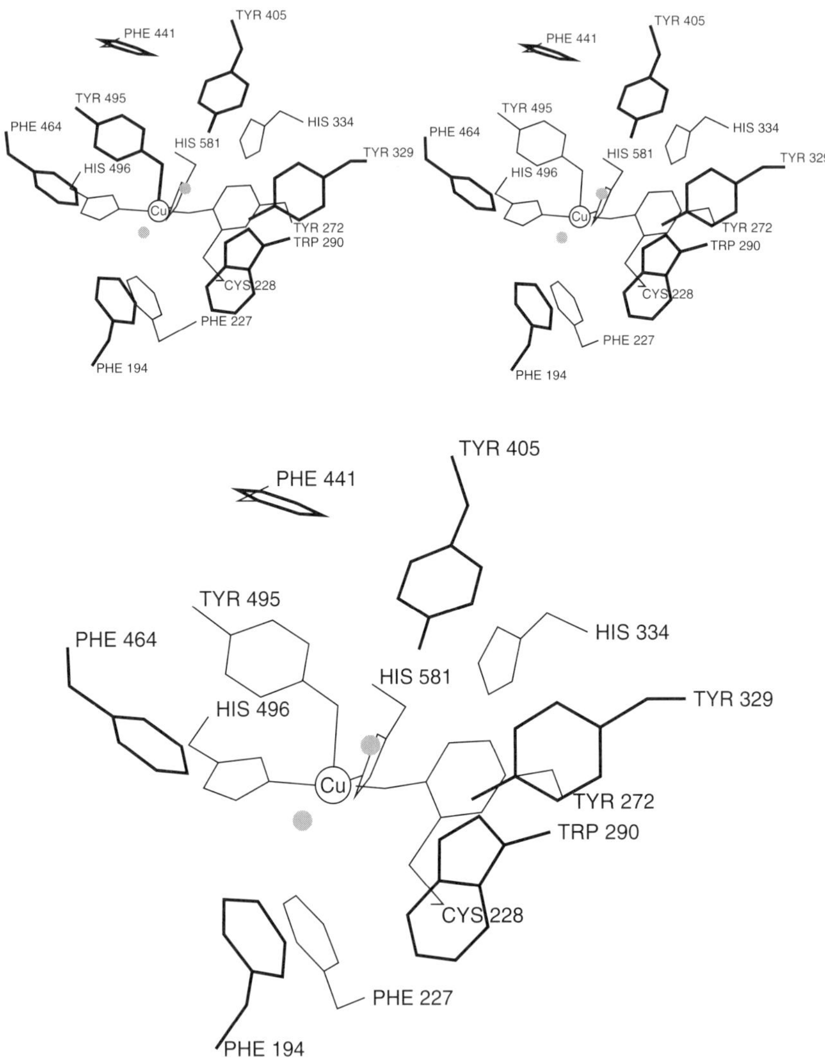


FIG. 5. The active site of galactose oxidase. (Top) Stereoview of the metal environment including catalytic residues. (Bottom) Expanded view of active site. Two crystallographic solvent molecules located in the active site are indicated by gray disks. (Based on protein coordinates PDB ID 1GOG.)

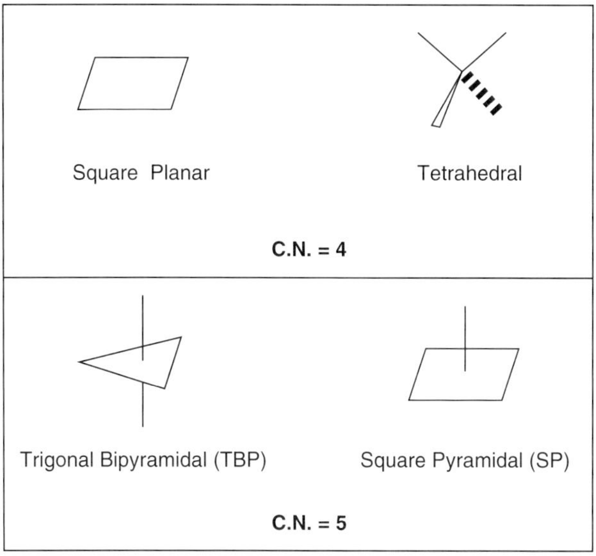


FIG. 6. Idealized coordination modes for metal complexes. Limiting geometries for four-coordinate ($ML_4$, top) and five-coordinate ($ML_5$, bottom) complexes.

the complex (tilted 11° above the plane), while the ring of His-581 is more nearly perpendicular (approximately 72°), aligned with the Cu–Tyr-272 bond vector. The tyrosine coordination makes the copper site in galactose oxidase particularly unusual for biological copper, as

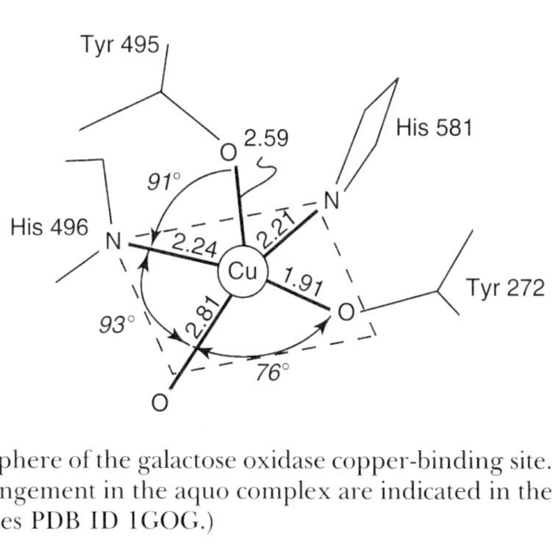


FIG. 7. Inner sphere of the galactose oxidase copper-binding site. Geometric details of the ligand arrangement in the aquo complex are indicated in the figure. (Based on protein coordinates PDB ID 1GOG.)

most biological copper complexes involve ligation by histidine and/or thiolate ligands. The only other example of tyrosine-coordinated copper is that of copper (mis)incorporated into the iron transport protein transferrin (Smith *et al.*, 1991). The effects of tyrosine ligation depend on electronic overlaps that are sensitive to the phenolate bond angle ($\theta$) and ring torsion angle ($\tau$) as previously defined (Whittaker and Whittaker, 1998; Whittaker *et al.*, 2000). Each of the two tyrosine ligands is distinct in this regard, with Tyr-272 being oriented more nearly perpendicular to the Cu complex ($\tau = -72°$, $\theta = 127°$) while Tyr-495 has a more in-plane orientation ($\tau = 53°$, $\theta = 105°$) (Fig. 8). These geometric differences are likely to contribute significantly to the distinctly different reactivities for these two phenolate ligands (see below).

The entire active site environment seems quite rigid, based on comparison of the structure of the metallated enzyme (PDB ID 1GOG) and the metal-free apoprotein (PDB ID 1GOH). The four protein ligands retain their positions between holo and apo structures with only minor reorganization. The largest changes in the site resulting from removal of the metal ion are associated with Tyr-272 and Tyr-495, for which the root mean square (rms) deviations between holo and apo forms are 0.21 and 0.32 Å, respectively. Both of the histidine side chains are relatively fixed, with rms variation over the side chain atoms of 0.12 Å for both His-496 and His-581. Tight packing of the protein around the active site and hydrogen bonding to the remote nitrogens of the His-496 and His-581 imidazoles may contribute to the stability of the ligand array. However, the apparent

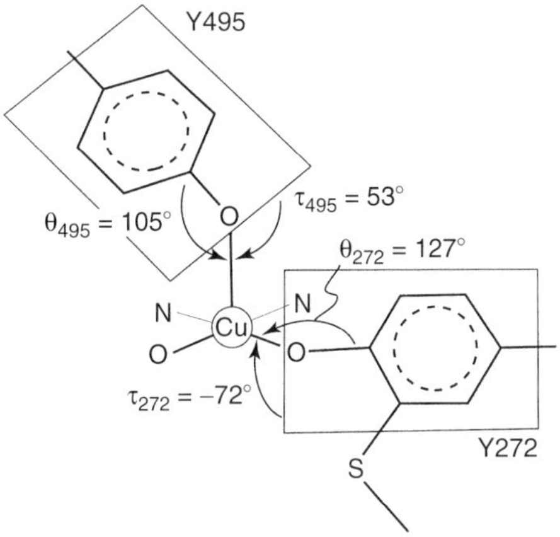


FIG. 8. Tyrosine coordination modes in galactose oxidase–copper complex. Tyrosine phenolate bond angles ($\theta$) and ring torsion angles ($\tau$) are indicated. (Based on protein coordinates PDB ID 1GOG.)

rigidity of the site may be actually be an artifact of crystallography, since the apoprotein was prepared by extracting the metal from the preformed crystal (Knowles and Ito, 1993) and lattice forces may have constrained its relaxation.

### *B. Outer Sphere and Extended Environment*

Beyond the inner sphere of the copper complex, the feature having the greatest effect on the chemistry of the site is a surprising covalent linkage between the coordinated Tyr-272 and a cysteine residue (Cys-228) (Ito *et al.*, 1991). The cross-link occurs between one of the *ortho* ring carbons and the cysteine $S_\gamma$, the result of a novel posttranslational protein modification forming a new, cross-linked amino acid, cysteinyltyrosine. As a result of the covalent bond, the sulfur becomes a thioether substituent on the tyrosine ring. The $C_\beta$ of the cysteinyl side chain lies within 7° of the tyrosine ring plane, adopting a *syn* orientation. This arrangement places the sulfur beyond bonding distance at approximately 3.5 Å from the copper, but without any intervening atoms. The cysteinyltyrosine cross-link affects both the structure and the reactivity of the amino acid side chains involved in the adduct. The most direct structural consequence is the increased rigidity of the Tyr-272 side chain, which completely loses its rotational flexibility as a result of the *ortho* ring coupling.

The reactivity of the side chain is dramatically altered by the formation of the cross-link. Biochemical and spectroscopic studies have shown that the Tyr–Cys feature is a specialized redox site in GAOX, forming a protein side-chain free radical under mild conditions (Whittaker and Whittaker, 1990; Babcock *et al.*, 1992; Gerfen *et al.*, 1996). Other proteins are known to stabilize free radicals in their structures, localized on tyrosine, tryptophan, cysteine, or glycine residues, as part of their biological function. For example, tyrosine free radicals are involved in the function of the oxygen-evolving active site of Photosystem II of green plants (Tommos *et al.*, 1998), in mammalian ribonucleotide reductase (Sjöberg *et al.*, 1978; Eklund *et al.*, 1997), and in prostaglandin H synthase (Lassman *et al.*, 1993). In each of these cases, the redox-active group is an unmodified tyrosine residue (Fig. 9a). Oxidation of an unmodified tyrosine is relatively difficult, requiring high-potential oxidant (>0.8 V vs $H^+/H_2$) for the formation of the tyrosyl phenoxyl radical, making the radical generally unstable in biological samples. Modified tyrosines may be more easily oxidized and may stabilize the radical products. A hydroxylated tyrosine side chain, trihydroxyphenylalanine (TOPA) (Fig. 9b), occurs as a redox cofactor in the quinoprotein, amine oxidase (Janes *et al.*, 1990). This group is redox-active; the trihydroxy substitution pattern allows TOPA to undergo two-electron oxidation to form the 2,5-quinone.

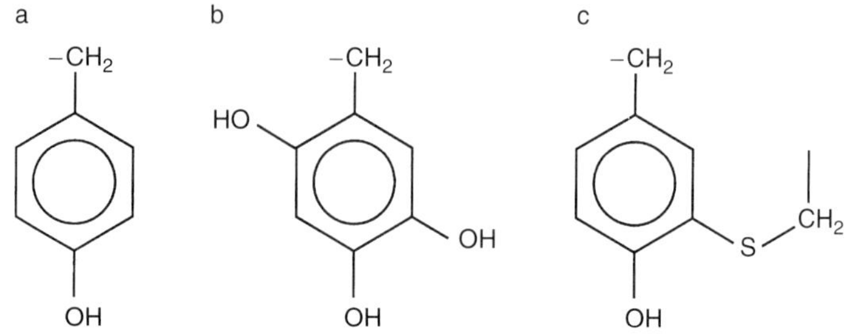


FIG. 9. Redox-active amino acid residues related to tyrosine. (a) Tyrosine, the redox center in ribonucleotide reductase, prostaglandin H synthase, and the photosynthetic oxygen evolving complex. (b) 2,4,5-Trihydroxyphenylalanine, the redox cofactor of the quinoprotein amine oxidase. (c) Tyrosine–cysteine (Tyr–Cys), the redox cofactor of galactose oxidase.

During enzyme turnover, the latter electrophile appears to react directly with amine substrates to form a covalent Schiff base adduct that is an intermediate in amine oxidation. While radicals associated with a one-electron oxidized form of the TOPA cofactor have been reported (Dooley *et al.*, 1991), their mechanistic significance is unclear. Like hydroxylation, thioether substitution (Fig. 9c) makes the tyrosine side chain easier to oxidize, but restricts the Tyr–Cys cofactor to one-electron oxidation, leading to stabilization of a protein radical in GAOX (Whittaker and Whittaker, 1988). Cross-linked amino acids have also been found in several other copper proteins. Tyrosylhistidine has been identified crystallographically in the heme–copper dioxygen-reduction site of cytochrome *c* oxidase (Ostermeier *et al.*, 1997; Yoshikawa *et al.*, 1998; Buse *et al.*, 1999), where it may be responsible for the formation of a phenoxyl free radical during turnover (MacMillan *et al.*, 1999). Cysteinylhistidine has been found in the copper-binding sites of tyrosinase (Lerch, 1982) and molluscan hemocyanins (Gielens *et al.*, 1997), where it appears to have a structural, nonredox role.

The cysteinyltyrosine group in GAOX is shielded from the solvent by Trp-290, which stacks over the thioether side chain. The ring planes of Trp-290 and Tyr-272 are nearly parallel with a separation of 3.5 Å, with $C_\delta$ of Trp-290 approximately eclipsing the *para* ring carbon of Tyr-272 and the indole nitrogen NE eclipsing the phenolate oxygen. The close contact between these two conjugated systems suggests the possibility of electronic $\pi$-orbital overlaps, but at present there is no evidence for such interactions. The role of this residue is unclear, but the orientation of Trp-290 places the indole nitrogen in a position where it might form a hydrogen bond to coordinated substrate or product. Mutation of

Trp-290 affects the stability of the active enzyme and ligand interactions (Reynolds *et al.*, 1997).

The active site region is rich in aromatic residues, particularly tyrosine and phenylalanine, contributing to the hydrophobic character of the active site. Water is also present in the outer sphere region, including a well-ordered solvent molecule that appears to hydrogen bond to both Tyr-495 phenol oxygen and the coordinated solvent and anchors a hydrogen-bond chain extending through an outer sphere tyrosine (Tyr-405) to a more remote histidine (His-334). The active site is surrounded by a radiating web of hydrogen bonds extending through the protein and forming a specialized environment for the redox center.

## V. Spectroscopic Probes of Metal Interactions

The X-ray crystallographic studies on GAOX described in the previous sections have defined the metal environment in the protein at atomic resolution, providing a detailed structural description of the active site. However, crystallographic studies are limited in the information they can provide. First, crystallization conditions impose a fundamental restriction on which complexes may be studied. At present, crystal structures are available for only one of the three possible oxidation states of GAOX, because of the intrinsic instability of the other forms under the crystallization conditions. Further, the crystallization process may make use of ionic buffers and precipitants that can generally be assumed to be innocent, but may potentially bind and perturb the metal center. Most importantly, though, the information available from protein crystallography is restricted to the atomic level of resolution and can only indirectly provide information on the electronic interactions that are fundamental to chemistry. Spectroscopic methods, on the other hand, are sensitive to precisely these electronic factors. Spectroscopy and crystallography therefore contribute essential and complementary information on the metalloenzyme complex.

For GAOX, structural analysis is particularly complicated, because of the existence of multiple states of the enzyme differing essentially only in the number of electrons, i.e., the oxidation state of the metalloprotein complex. Three distinct oxidation states can be prepared, each with properties and reactivities dramatically different from the others, as indicated in Fig. 10 (Whittaker and Whittaker, 1988). When isolated from culture medium, GAOX is a mixture of two of these states: a blue, one-electron reduced, catalytically inactive form (IAGO) that contains a Cu(II) ion and no radical and a green form that is catalytically active (AGO) and contains both Cu(II) and a free radical. The enzyme may be converted to

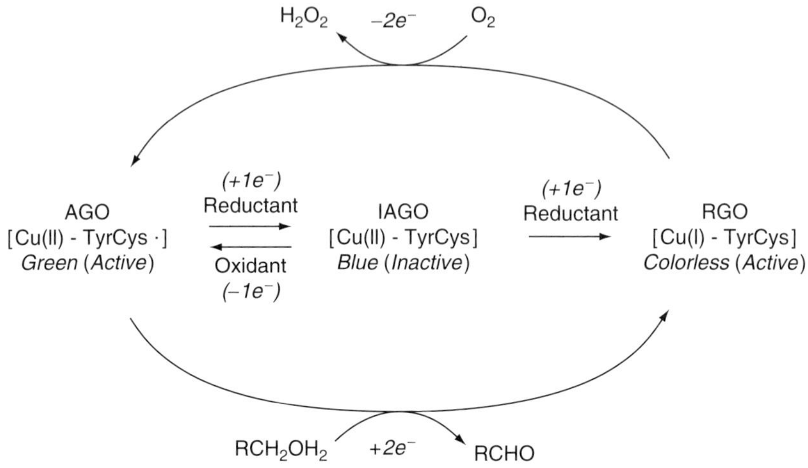


FIG. 10. Interconversion of redox states for galactose oxidase. Three distinct states (AGO, IAGO, and RGO) may be prepared and interconverted by either one-electron or two-electron redox steps.

these limiting oxidized and reduced forms under mild conditions by treatment with the appropriate redox buffer [e.g., $K_3Fe(CN)_6$ for oxidation; $K_4Fe(CN)_6$ or ascorbate for reduction] (Fig. 10). During the reaction cycle, reduction by substrate leads to the formation of a colorless, two-electron reduced enzyme complex (RGO) that contains Cu(I) and no radical. This species is active as a dioxygen reduction catalyst, and reaction with $O_2$ converts it to AGO. These three enzyme forms may be reversibly interconverted in one-electron steps with inorganic redox agents or in two-electron steps involving oxidation and reduction of substrates.

A large number of spectroscopic techniques are available for probing specific features of the enzyme. Each technique has inherent advantages and limitations, and a combination of several approaches is usually required to form an accurate description of the protein complex. The most important spectroscopic methods for metalloenzyme applications include optical absorption (including circular dichroism); resonance Raman; electron paramagnetic resonance; magnetic susceptibility (which, while strictly speaking not a spectroscopic method, complements the other techniques); and X-ray absorption spectroscopies. Each of these approaches has contributed valuable information on the GAOX active site.

### *A. Optical Absorption*

Optical spectroscopy involves electronic excitations in molecules, making it well suited to extending the atomic structural information available from crystallography to the higher resolution of electronic structure and bonding. Optical absorption spectroscopy is especially useful for characterizing GAOX samples because each of the three redox forms of the resting enzyme has a distinct spectrum (Whittaker and Whittaker, 1988), allowing the state of the enzyme to be defined by a simple absorption measurement. These spectra are also sensitive to perturbations by ligand binding and changes in protonation state that reflect intrinsic properties of the active site. Optical absorption measurements may be used qualitatively, in detecting the effects of ligand perturbations and identifying structural changes, as well as quantitatively, resolving mixtures of enzyme states into their pure components.

The fully reduced enzyme (RGO) has a very simple absorption spectrum lacking any significant features in the near UV–visible spectral range (Fig. 11, line C) below the protein absorption cut-off in the UV. The absence of low-energy absorption for this species is a consequence of the full-shell character of both the reduced $d^{10}$ Cu(I) metal ion and the Tyr–Cys redox cofactor. For both of these species, allowed electronic excitations all lie at very high energy because there are no low-lying empty orbitals available for a transition. On the other hand, the open-shell $d^9$ Cu(II) metal ion in the inactive enzyme (IAGO) has a vacancy in a metal valence orbital, leading to the appearance of broad-band

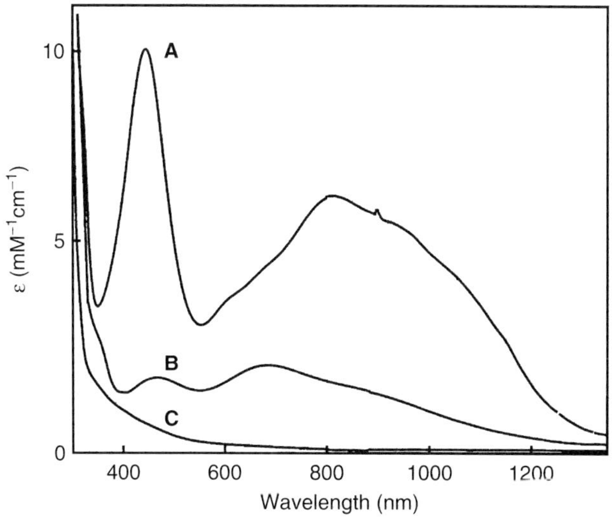


FIG. 11. Optical absorption spectra for galactose oxidase. (A) Redox-activated (AGO) complex. (B) Reductively inactivated (IAGO) complex. (C) Fully reduced (RGO) complex.

absorption spectra at low energy, arising from metal-centered (ligand field or $d \rightarrow d$) and ligand-to-metal charge transfer (LMCT) excitations (Fig. 11, line B). These low-energy transitions involve spatial redistribution of metal valence electrons ($d \rightarrow d$) or optical electron transfer from the ligand valence shell to Cu (LMCT). The spectra are generally broad, reflecting the sensitivity of these transitions to geometric distortions. The sensitivity to ligand environment is also reflected in the average energy of the $d \rightarrow d$ spectra, which correlates with the strength and arrangement of the metal ligands. Similarly, the energy and intensity of CT features in the spectra are characteristic of ligand type and may be used to further define the metal site.

The active enzyme (AGO) is distinguished by an unusual spectrum (Fig. 11, line A) unlike spectra observed for either RGO or IAGO forms, with extremely strong absorption that spans the entire visible region and extends deep into the near infrared. The principal absorption in this form is associated with two intense features centered at 445 and 850 nm. This spectrum is, in fact, unlike spectra observed for any other metalloprotein complex, emphasizing the unique character of the GAOX active site. The origin of this spectrum is discussed in more detail below (Section VII).

Circular dichroism spectra for these complexes resolve additional structure in the broad absorption bands (Fig. 12, Table II). Significantly, the CD spectra of AGO and IAGO cross at several points, called *isodichroic points*, that are the counterpart of isosbestic points in absorption spectra. These isodichroic points compensate for the absence of isosbestic points in the absorption spectra for IAGO and AGO species and allow partly active mixtures of the two oxidation states to be resolved into these two limiting forms. This spectroscopic evidence gave the first clear demonstration that redox activation involved interconversion between discrete IAGO and AGO forms and provided a direct method for quantitating the extent of conversion (Whittaker and Whittaker, 1988).

The optical spectrum of reductively inactivated enzyme (IAGO) is perturbed by ligand binding, resulting in decreased intensity and a shift of the absorption maximum to higher energy (Fig. 13A). These spectral changes imply a change in the effective geometry at the Cu(II) center and are consistent with a distortion toward square planar coordination of the metal ion (Fig. 6), with the exogenous ligand replacing solvent in the plane. The structure of GAOX crystals prepared in acetate buffer supports the identification of the coordinated solvent as a labile exchange site. In the crystals, acetate substitutes for water in the active site, resulting in a modest decrease in metal–ligand bond distance to the nonprotein ligand in the anion complex (from 2.81 to 2.26 Å) and a slight increase in the Cu–Tyr-495 bond length (from 2.59 to 2.69 Å).

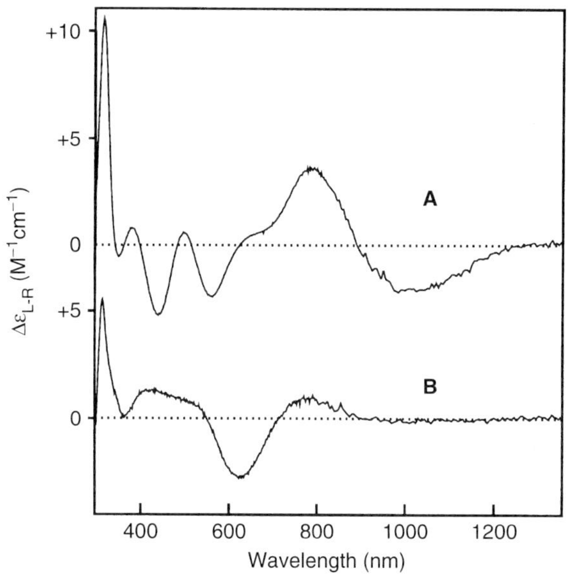


FIG. 12. Circular dichroism spectra for galactose oxidase. (A) Redox-activated (AGO) complex. (B) Reductively inactivated (IAGO) complex.

Acetate binding produces the same type of spectral changes as other anions, demonstrating that exogenous ligand binding occurs by substitution for the solvent.

TABLE II
*Spectroscopic Parameters for Galactose Oxidase Complexes*

| | Absorption | | Circular dichroism | |
|---|---|---|---|---|
| Complex | $\lambda_{max}$ (nm) | $\varepsilon$ ($M^{-1}cm^{-1}$)[a] | $\lambda_{max}$ (nm) | $\Delta\varepsilon$ ($M^{-1}cm^{-1}$)[a] |
| AGO | 445 | 5500 | 320 | +10.7 |
| | 850 | 3400 | 400 | +1.0 |
| | | | 445 | −3.5 |
| | | | 500 | +1 |
| | | | 550 | −2.5 |
| | | | 800 | +3.8 |
| | | | 1040 | −1.6 |
| IAGO | 450 | 865 | 320 | +5 |
| | 620 | 1050 | 420 | +1 |
| | | | 610 | −3 |
| | | | 800 | +1.5 |

[a]Per active site.

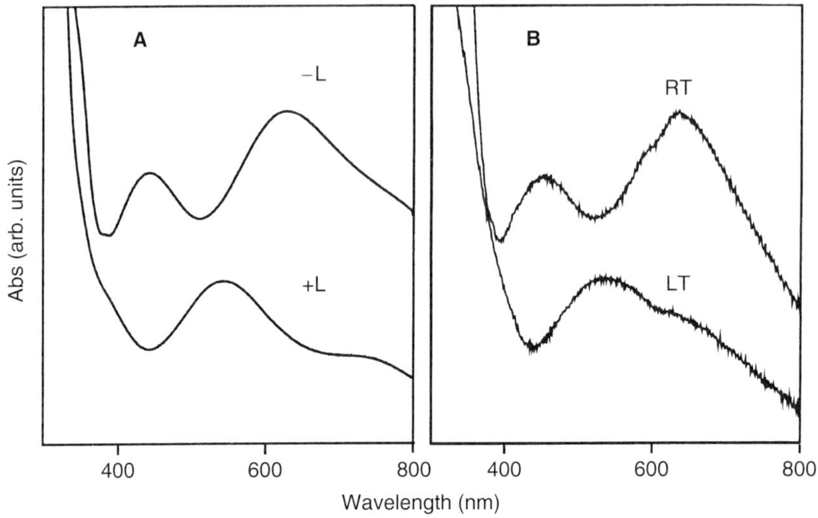


FIG. 13. Response of the active site copper complex to chemical and physical perturbations. (A) Absorption spectra for the IAGO Cu(II) complex in the absence (−L) and presence (+L) of a coordinating anion, cyanate ($OCN^-$). (B) Absorption spectra for IAGO Cu(II) complex at ambient (300 K, RT) and cryogenic temperatures (200 K, LT).

This reorganization of the copper site would be expected to lead to decreased covalency of the Tyr-495 phenolate–Cu(II) bonding and result in an increase in the basicity of the phenolate oxygen. Proton uptake experiments designed to test this hypothesis confirm that anion binding to the metal center is coupled to uptake of a single proton per active site, associated with a base having a $pK_a > 9$ (Whittaker and Whittaker, 1993). This experiment indicates that Tyr-495 is displaced and protonated when exogenous ligands coordinate to the active site copper. Based on this interpretation, the ligand-free GAOX and ligand-bound GAOX (Fig. 13A) correspond to $Tyr_{ON}$ and $Tyr_{OFF}$ forms of the enzyme that differ in coordination of the copper by Tyr-495.

Nearly identical spectral changes are observed when GAOX, prepared in a glassing solvent (e.g., 50% glycerol) that prevents formation of microcrystals and preserves the optical transparency of the sample, is cooled to cryogenic temperatures (Fig. 13B) (Whittaker and Whittaker, 1993; Whittaker *et al.*, 2000). A color change from blue (RT) to red (LT) reflects a thermochromic transition in the protein structure. The similarity of the optical spectrum of the low-temperature complex to the spectra observed for anion adducts suggests that the RT aquo complex ($Tyr_{ON}$

form) is converted to a hydroxide complex ($Tyr_{OFF}$ form) on cooling, as a result of transfer of one solvent proton to the Tyr-495 phenolate. Ligand binding and thermal effects thus similarly perturb the active site.

Both the coupling of proton uptake to anion binding and the surprising low-temperature structural instability of the active site complex may mimic abstraction of the hydroxylic proton from a coordinated substrate molecule during turnover. Ionization of a coordinated alcohol by a base in the active site would be an important step for substrate activation in the catalytic mechanism. The nonturnover reactions serve as models for this process, giving us clearer insight into the catalytic reaction by mapping out an intrinsic proton transfer coordinate in the active site.

The underlying chemistry involved in this process is illustrated in Fig. 14. For the native complex, the weakest interaction is with the coordinated solvent, and this determines the unique ligand field axis for the Cu(II) center. Replacing the solvent with an anion (or ionization of water to hydroxide) strengthens the interactions with this ligand (Fig. 14, 1) at the expense of interactions with Tyr-495 (Fig. 14, 2), which is consequently displaced and protonated, becoming the new direction of weak interaction in the resultant complex. This reorganization of the ligands may be described as a pseudorotation of the metal complex, as the weakest ligand interaction shifts direction without a physical rotation of the ligand set.

As predicted from this analysis, Tyr-495 is absolutely required for catalytic activity. A mutant in which this residue is replaced by phenylalanine (Y495F GAOX) is inactive, even though it contains both copper

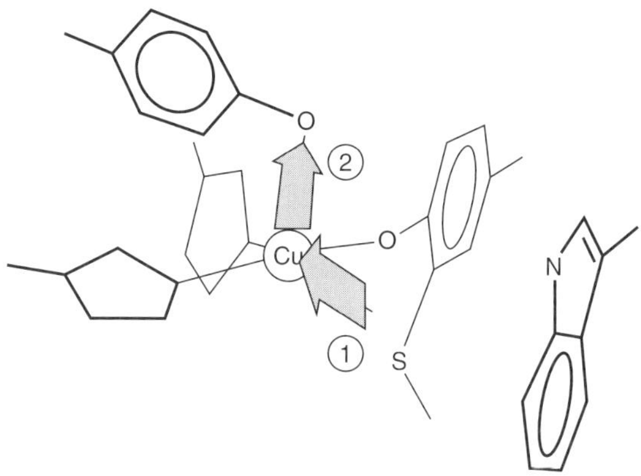


FIG. 14. Proposed modulation of copper–protein interactions by exogenous ligand binding. Replacement of the coordinated solvent by an exogenous ligand results in stronger interactions along the metal–ligand axis (1) at the expense of the interactions with the Tyr-495 phenolate, which is displaced in the complex (2).

and the Tyr–Cys cofactor (Reynolds *et al.*, 1995; Rogers *et al.*, 1998). Anion binding is also uncoupled from proton uptake in the mutant, reinforcing the assignment of Tyr-495 as a general base in the active site. The corresponding Y377F mutant of GLOX has also been prepared and is also found to be inactive, despite being able to form a stable oxidized free radical–copper active site complex (Whittaker *et al.*, 1999).

The optical spectrum of redox-activated AGO also undergoes exogenous ligand and temperature perturbations related to those seen for IAGO. For AGO, the principal change is in the intensity of the broad near-infrared absorption band. This "red band" can therefore be used to monitor whether the enzyme is in the $Tyr_{ON}$ or $Tyr_{OFF}$ state of the oxidized enzyme complex. Based on the crystal structure, exogenous ligand perturbations, and model correlations, it has been possible to propose a detailed assignment of many features in these spectra. These assignments provide a basis for interpretation of the spectra in structural terms and lead to deeper insight into understanding the electronic structural origins of catalytic reactivity in the radical copper oxidase active site.

### *B. Resonance Raman*

Resonance Raman spectroscopy experimentally connects a specific absorption band to the vibrational spectrum of the chromophore giving rise to it, providing important information on the structure of the chromophore. This link between spectra and molecular structure also makes it possible to propose detailed spectroscopic assignments. In the resonance Raman experiment, laser radiation in resonance with an optical absorption band is used to excite a molecule. Interaction between the light and the absorbing molecule leads to enhanced scattering of photons at a discrete set of frequencies, differing from that of the exciting radiation by precisely the spectrum of molecular vibrational frequencies, yielding a resonance Raman (rR) spectrum.

The rR spectrum for AGO excited within the red band (875 nm) contains two distinct sets of vibrations, one corresponding to modes of a simple tyrosinate (1170, 1246, 1499, and 1603 $cm^{-1}$) and a second set corresponding to a tyrosine residue having slightly perturbed vibrational frequencies (1185, 1246, 1487, and 1595 $cm^{-1}$) (Whittaker *et al.*, 1989; McGlashen *et al.*, 1995). For the azide adduct of AGO, a single set of frequencies that closely matches the perturbed set of the unliganded enzyme is observed (1185, 1246, 1490, and 1595 $cm^{-1}$). Since the intense absorption of the active enzyme is associated with the presence of a free radical-copper active site, the spectra directly identify these two tyrosines as being involved in the free radical complex. The ligand-binding experiments on IAGO establishing the $Tyr_{ON}$ or $Tyr_{OFF}$ character of the different ligation states suggest assignment of the normal and perturbed spectra to the coordinated Tyr–495 and Tyr–272, respectively. Resonance Raman analysis of distinct ligation states of glyoxal oxidase gives very similar results, emphasizing the close structural similarity of the two active sites (Whittaker *et al.*, 1996b).

### C. *Electron Paramagnetic Resonance*

Electron paramagnetic resonance (EPR) spectroscopy is specifically sensitive to the presence of unpaired electrons in a sample that may be associated with transition metal ions and free radicals. For IAGO, EPR spectroscopy gives information on the environment of the Cu(II) metal ion, which has a single unpaired electron in its valence shell. Although divalent copper is present in both IAGO and AGO, Cu(II) EPR signals are observed for only the former complex, as a result of electronic coupling between the copper and the free radical spins in the active enzyme complex (see below).

The EPR signals observed for IAGO ($g_x \sim g_y = 2.055$, $g_z = 2.28$) (Fig. 15, line b) are typical of Type II copper sites in proteins, reflecting an axial electronic symmetry for the complex that agrees with the tetragonal or square pyramidal geometry expected on the basis of the crystal structure. In this environment, orbital paramagnetism shifts one resonance ($g_z$) away from the free electron $g$-value ($g_e = 2.0023$), providing information on the nature of the metal $d$-orbital containing the unpaired electron and effectively defining the redox orbital in the complex. The large Cu electron-nuclear hyperfine splitting ($a_z = 175$ G) reflects lower covalency and a stronger tetragonal distortion compared to the trigonal or pseudotetrahedral Cu(II) of a "blue" (Type I) redox active site. Ligand nuclear hyperfine splittings in the $g_\perp$ region of the spectrum derive from perturbation by the two coordinated nitrogens ($^{14}$N, $I = 1$) from His-496 and His-581 in the $xy$ ligand plane. Analysis of the superhyperfine splitting superimposed on the second ($M_I = +1/2$) feature of the $a_z$ hyperfine quartet resolves all five components associated with a pair of equivalent $I = 1$ ligand nuclei ($2nI_L + 1 = 5$; $n = 2$), with a uniform splitting of 15 G. The near-equivalence of the two nitrogens implies that the Cu(II) site is accurately described as a square pyramidal complex under the conditions of the experiment. Spin quantitation by double integration of the derivative EPR spectrum shows that all of the copper in the protein contributes to the observed signal.

Like the optical spectra described above, the EPR spectra are temperature-dependent (Whittaker and Whittaker, 1993), and the square complex reflected in the ground state parameters in frozen solution at low temperature may not accurately describe the complex under more

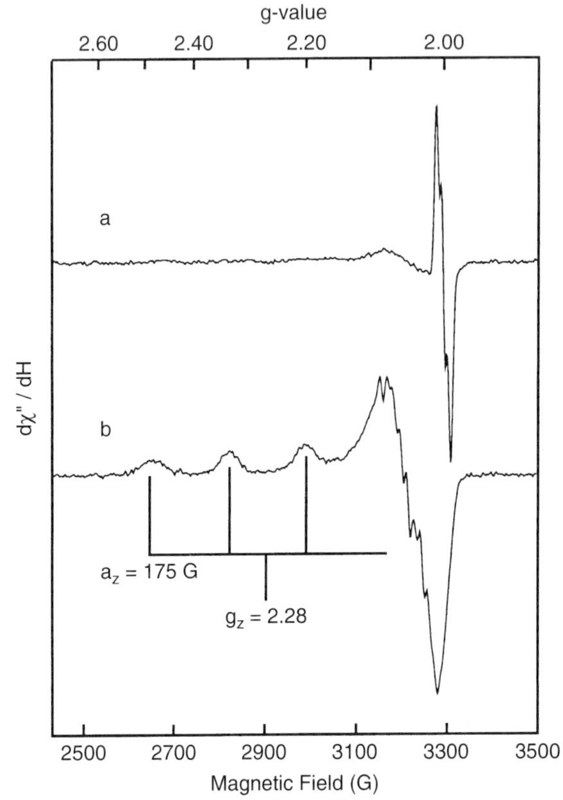


FIG. 15. EPR spectra for galactose oxidase complexes. (a) Oxidized enzyme (AGO) prepared by treating native galactose oxidase with $K_3Fe(CN)_6$. (b) Radical-free IAGO complex, prepared by treating native galactose oxidase with $K_4Fe(CN)_6$. Instrumental parameters: microwave power, 10 μW; microwave frequency, 9.223 GHz; modulation amplitude, 5 G; temperature, 30 K.

physiological conditions (e.g., liquid sample at 25°C). There are clear changes in the EPR spectrum as the temperature is raised, reflecting a change in the Cu(II) environment between cryogenic and ambient temperatures, where $g_z = 2.27$ and $a_z = 127$ G. This trend toward a smaller value of $g_z$ and reduced $a_z$ hyperfine splitting may be interpreted in terms of a lowering of the site symmetry as the temperature is raised. This in turn would be consistent with increased interaction with Tyr-495, effectively redefining the axial direction in the complex.

Oxidation of the enzyme, required for activation, eliminates the Cu(II) EPR signals characteristic of IAGO (Fig. 15, line a) (Whittaker and Whittaker, 1988). For AGO, a new signal is observed, near the free-electron $g$-value. This sharp signal quantitates to less than ~0.1 spins/protein and represents a small fraction of apoenzyme in the sample that is able to form a stable free radical (see below). In addition, there is a minority Cu(II) EPR signal (approximately 0.2 spins/protein) but the majority of the oxidized enzyme lacks any detectable EPR signal.

## D. *Magnetic Susceptibility*

Like EPR, bulk magnetic susceptibility ($\chi_{sus}$) is sensitive to paramagnetic species in a sample, but susceptibility measurements tend to be less selective than resonance methods. However, $\chi_{sus}$ has the advantage of being able to detect all paramagnetic species, regardless of spin state. Thus, while conventional EPR spectroscopy is essentially limited to half-integer spin ground states, magnetic susceptibility can provide information on integer-spin ground states, as well (Day, 1993). This aspect of susceptometry is important for investigating the EPR-silent active enzyme complex. Variable temperature saturation magnetization profiles recorded for AGO over a temperature range from 2 to 200 K and magnetic fields up to 5.5 T demonstrate that the electronic ground state of the active enzyme is diamagnetic (Whittaker *et al.*, 2000). Small paramagnetic contributions that can be detected in these samples at low temperatures are quantitatively explained by the presence of a small amount of IAGO complex that can be independently estimated by EPR measurements. Aside from this minority impurity species, the only other paramagnetic contribution appears at the higher temperatures and exhibits a temperature- and magnetic field-dependence consistent with a low-lying non-Kramers (integer electronic spin) excited state. The singlet–triplet splitting estimated from multifield magnetization experiments is $J > 200\,\mathrm{cm}^{-1}$ (evaluated for a Heisenberg exchange hamiltonian $H = JS_1 \cdot S_2$), in the same range as has been reported for a crystallographically characterized Cu(II)–nitroxide spin label complex (Lim and Drago, 1972; Dickman and Doedens, 1981).

## E. *X-Ray Absorption*

X-ray absorption spectroscopy [including edge excitation (XANES) and extended fine structure (EXAFS) methods] provides important information on metal ion oxidation state and coordination environment in the protein. XANES analysis has been useful in defining the oxidation state of the metal center in redox-activated GAOX, by comparison of the Cu X-ray absorption edge structures for AGO with results for the IAGO form that is known to contain a Cu(II) metal center. The X-ray absorption edge essentially measures the binding energy of a core 1s electron, which systematically increases by approximately 5–10 eV as the oxidation level of the metal ion is increased. For GAOX, the edge shifts to slightly *lower* energy on redox activation, in the opposite direction expected for metal-centered oxidation (Clark *et al.*, 1990). Model studies have confirmed that oxidation of Cu(II) to Cu(III) results in an *increase* in edge energy of 2–6 eV, so the XANES results require a divalent [Cu(II)] oxidation state for the metal in the AGO complex and indicate that redox activation involves oxidation of some other group in the protein (see below).

EXAFS provides information on the immediate environment of a metal center (ligand type, coordination number) from the analysis of oscillations in the X-ray absorption above the edge. These oscillations arise from interference effects resulting from ligand back-scattering of the excited photoelectron. Ligand scattering depends on several factors, including the atomic number (Z), the number of ligands of the same type, and the metal–ligand bond distance, making EXAFS particularly sensitive to these parameters. EXAFS has been able to provide structural information on the metal environment in the reduced, Cu(I)-containing GAOX, which is inaccessible by optical absorption and EPR methods, and for which no crystallographic data are available. The results are consistent with coordination of the reduced metal center by 2–3 low-Z atoms (e.g., N, O) (Clark *et al.*, 1994). EXAFS does not resolve atoms from the same row of the periodic table and therefore cannot distinguish between nitrogen and oxygen ligands, but suggests that the two histidines (His-496 and His-581) remain bound in the Cu(I) state. Further interaction with Tyr-272 (possibly protonated to form a phenol) would complete a T-shaped arrangement of ligands that would be consistent with the results and favorable for stabilizing low-valent copper in the protein.

## VI. Probes of the Radical Site

The involvement of a free radical in the active site of galactose oxidase is not obvious and was, in fact, overlooked for many years. As mentioned earlier (Section V, C), the Cu(II) ion strongly interacts with the radical in the active enzyme (AGO), and the normal spectroscopic signatures of a free radical are absent. The strong interactions between the radical and the metal ion result in an EPR-silent complex and an unusual absorption spectrum that is clearly not a simple superposition of spectra for Cu(II) and a free radical (see below). However, a distinctive and unusual free radical EPR signal is consistently present as a minority species in EPR spectrum of oxidant-treated GAOX. This same feature is produced when the metal-free apoenzyme is treated with mild oxidant, showing that the protein itself is redox-active. Depending on conditions and method of preparation, as many as 40% of the molecules in the sample may contain the radical, as judged by quantitative EPR analysis. The EPR spectra of these samples are clean, and a single signal is observed (Fig. 16b), showing that there is only one redox-active group in the protein. The free radical can also be detected in the optical spectrum of oxidized GAOX through the appearance of a new near-UV absorption band in the radical-containing protein (Fig. 16a). A variety of approaches are thus available for investigating and identifying the radical, but paramagnetic resonance methods (EPR, electron-nuclear double resonance (ENDOR) spectroscopy, and high-field EPR) have proven to be been especially valuable in developing a detailed structural description of this novel protein free radical.

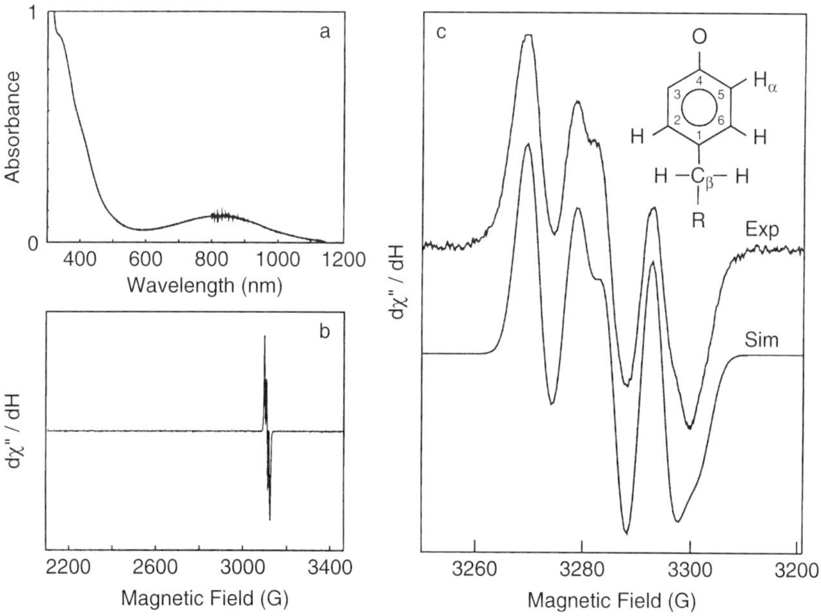


FIG. 16. Spectroscopic characterization of the oxidized apogalactose oxidase free radical. (a) Optical absorption spectrum for the radical-containing apoprotein. (b) X-band EPR spectrum of the metal-free protein following Ir(IV) oxidation. (c) Expansion of the region near $g = 2$ comparing experimental data (Exp) with a theoretical simulation (Sim) based on coupling of the unpaired electron spin with one $H_\alpha$ and one $H_\beta$ proton of a tyrosine phenoxyl. Simulation parameters: $g_1 = 2.0017$, $g_2 = 2.0073$; $A_1(H_\alpha) = 8.4$ G, $A_2(H_\alpha) = 8.8$ G; $A_1(H_\beta) = 12.7$ G, $A_2(H_\beta) = 13.8$ G.

### *A. X-Band EPR Spectroscopy*

In conventional X-band (9 GHz) EPR spectroscopy, oxidized apo-GAOX (Figs. 16b and 16c) exhibits a sharp resonance with an average *g*-value near 2.0055 and resolved hyperfine structure (Fig. 16c, *Exp*), as would be expected for a tyrosine phenoxyl free radical such as that formed in ribonucleotide reductase. However, theoretical simulation of this signal (Fig. 16c, *Sim*) yields estimates of hyperfine parameters that are consistent with interaction of the unpaired electron spin with only two hydrogen nuclei, rather than three (two *ortho* α ring protons and one β-methylene proton) typically observed for simple tyrosine phenoxyl radicals. As described above (Section IV, B), protein crystallography has revealed a novel feature in GAOX, the Tyr–Cys site, that can account for these unusual spectra. Covalent attachment of cysteine at one of the *ortho* ring carbons of Tyr-272 eliminates one of the α hydrogens, accounting for the unusual hyperfine behavior of the GAOX radical. These spectroscopic observations provided the initial basis for assignment of a redox function to the Tyr–Cys side chain.

Although this assignment is quite convincing in itself, positive identification of a Tyr–Cys radical as the origin of the free radical EPR signal in oxidized apoGAOX requires further evidence. One line of evidence draws on the isotope sensitivity of EPR spectroscopy. Isotope perturbations can be very important in making structural assignments for EPR spectra, defining contributions from specific atoms in a sample. Even before the crystallographic data were available, the similarity of the oxidized apoGAOX signal to the spectrum of a tyrosine free radical suggested that a tyrosine residue was involved, and, to test this prediction, the enzyme was uniformly labeled with perdeuterated tyrosine. Isotopic substitution on tyrosine has a dramatic effect on the hyperfine splitting pattern for the GAOX radical spectrum, confirming that the signal derives from an oxidized tyrosine residue (Whittaker and Whittaker, 1990). However, the anomalous hyperfine parameters demonstrate that it is not a simple tyrosine free radical, like that found in ribonucleotide reductase.

### *B. Electron-Nuclear Double Resonance Spectroscopy*

Further analysis of the radical signal has required more advanced spectroscopic techniques, including ENDOR spectroscopy. ENDOR is used to extend the structural information available from EPR, allowing more precise determination of the resonance parameters. For GAOX, ENDOR's sensitivity to hyperfine coupling in the ground state has allowed two inequivalent protons coupled to the unpaired electron spin to be distinguished for the apoGAOX radical (Babcock *et al.*, 1992). One proton is characterized by a relatively small and anisotropic $A$-tensor ($|A_y| = 8.4\,\text{MHz}$; $|A_z| = 21.6\,\text{MHz}$), values typical of an α-hydrogen attached to an aromatic ring interacting via spin polarization coupling. Unlike an ordinary tyrosine phenoxyl free radical, however, a single *ortho* ring proton is found to be strongly coupled. The other proton, with a larger and more isotropic $A$-tensor ($|A_\perp| = 39.8\text{MHz}$; $|A_\parallel| = 43.4\,\text{MHz}$), may be assigned to a single β-methylene hydrogen in the tyrosine side chain. Exocyclic hydrogens of this type are coupled mainly via a hyperconjugation mechanism, resulting in a strong dependence of the hyperfine coupling on the head group torsion angle (θ). A McConnell-type relation allows estimation of the side chain torsion for the phenoxyl radical

$$A^{\text{C-H}} = \rho_{\text{C1}} B \cos^2 \theta, \qquad (1)$$

where $A^{\text{C-H}}$ is the experimentally determined hyperfine coupling to the β-methylene proton, $\rho_{\text{C1}}$ is the unpaired electron spin density in the $p_z$ orbital on the adjacent ring carbon, $B$ is a constant equal to 162 MHz, and θ is the dihedral angle between the ring carbon $p_z$ orbital and the $C_1$–$C_\beta$–H plane. EPR and ENDOR data for a model radical, *O*-methylhiocresyl phenoxyl free radical, indicate that thioether substitution reduces the C1 spin density by about 25% relative to the unsubstituted cresyl phenoxyl, for which $\rho_{\text{C1}} = 0.49$. Using a scaled value of $\rho_{\text{C1}} = 0.37$ for the oxidized apoGAOX radical, this analysis leads to a predicted methylene hydrogen torsion angle $\theta = 34°$ for Tyr-272 in apoGAOX. This value is significantly larger than is predicted from the head group $C_\alpha$ dihedral determined by X-ray crystallography for both the metallated IAGO complex ($\theta = 13°$) and the metal-free apoprotein ($\theta = 8°$). In the absence of a crystal structure for the radical-containing apoenzyme, this may indicate a significant ( ~50°) twist of the Tyr-272 side chain on oxidation. More likely, based on density functional theory calculations (see below), this estimate of the *C*1 spin density may be high and a lower estimate ( ~0.25) would result in good agreement between the computational model, crystallography, and spectroscopy.

### *C. High-Field EPR Spectroscopy*

High-field EPR spectroscopy is a powerful new technique that is especially well suited to characterizing and identifying biological free radicals (Bennati *et al.*, 1999). At conventional EPR frequencies (near 10 GHz), spectra tend to be dominated by complex and overlapping nuclear hyperfine splittings that obscure the underlying electronic contributions to the spectrum. However, while the electronic orbital Zeeman splitting, which determines the spectral resolution of the *g*-shifts, is field-dependent

( $= \mathbf{g}_L \beta \mathbf{H}$), the magnitude of the electron-nuclear hyperfine splittings is a field-independent quantity. Thus, at sufficiently high magnetic field, electronic rather than hyperfine contributions dominate the spectra. High-field spectra contain important information on the electronic structure of the radical that is not easily accessible in low-field spectra, since the orbital $g$-shifts revealed at high field directly relate to valence electronic structure of the radical. For oxidized apoGAOX, high-frequency (139.5 GHz) EPR leads to a simple first-order EPR spectrum in which the electronic (orbital) $g$-shifts dominate (Fig. 17, line a) (Gerfen *et al.*, 1996). The powder spectrum has a clear axial form, with $g_1 = 2.0074 \cong g_2 = 2.0064$, $g_3 = 2.0021$. This is markedly different from the behavior of a simple tyrosine phenoxyl, such as that found in ribonucleotide reductase, whose spectrum exhibits a strong rhombic splitting (Fig. 17, line c) but precisely the same as observed for the *O*-methylthiocresyl model radical (Fig. 17, line b). This clearly identifies the Tyr–Cys side chain as the site of the oxidized apoGAOX radical and demonstrates that the electronic structure of the thioether-substituted phenoxyl is distinct from that of a simple phenoxyl radical.

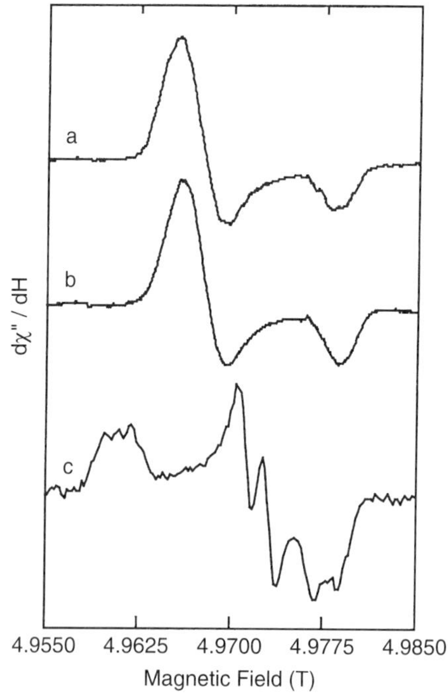


FIG. 17. High-frequency (139.5 GHz) free radical EPR spectra. (a) Oxidized apogalactose oxidase free radical. (b) Photochemically generated *O*-(methylthio)cresyl (**mtc**) phenoxyl radical. (c) Ribonucleotide reductase tyrosyl radical.

### *D. Computational Approaches*

Computational methods give further insight into the origins of these spectra and relate the spectroscopic data to electronic structure and therefore the chemistry of the radical site. *Ab initio* density functional theory electronic structure calculations have been performed on cresyl and *o*-methylthiocresyl radical ground states as models for the oxidized apoGAOX radical site (Gerfen *et al.*, 1996). These aromatic radicals have $C_S$ molecular symmetry, and the electronic wavefunctions transform under the a′ (and a″) irreducible representations of the group, depending on whether they are symmetric (or antisymmetric) with respect to the ring plane. Isosurface contours for the valence molecular orbitals spanning the Fermi level are shown in Fig. 18. For the cresyl radical (**cre**), the spin-occupied molecular orbital (SOMO) lies at relatively deep binding energy (−6.37 eV) compared to that of the thioether substituted radical (**mtc**) (−4.89 eV). The nodal character of the **cre** SOMO wavefunction defines the odd alternant unpaired spin distribution in the ground state, with the unpaired electron localized on the phenoxyl oxygen and the $p_z$ orbitals from *ortho* and *para* ring carbons.

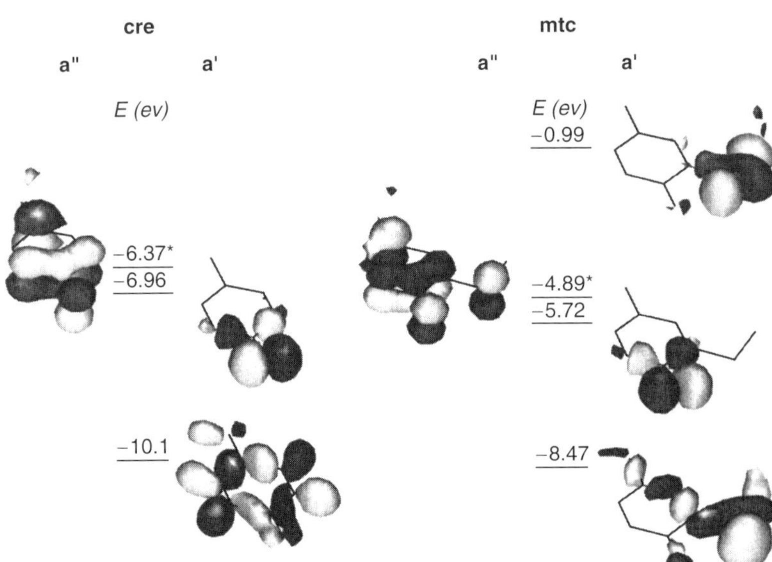


FIG. 18. Ground state electronic structures for cresyl and *o*-(methylthio)cresyl phenoxyl radicals. Isosurface representations of molecular orbitals solved by *ab initio* density functional theory methods for cresyl (**cre**) and *o*-(methylthio)cresyl (**mtc**) phenoxyl radicals. Eigenvalues are listed and for each the SOMO is identified with an asterisk (*).

The SOMO/LUMO gap is large for **cre**, but there are filled valence levels lying at slightly deeper binding energy.

The presence of the thioether side chain significantly perturbs the electronic structure of the **mtc** radical (Fig. 18, **mtc**). The SOMO is an antisymmetric wavefunction delocalized over the aromatic ring, as found for **cre**, but is perturbed by participation of the exocyclic sulfur in the SOMO. The involvement of this sulfur atom leads to a shift in the electron distribution in the SOMO, increasing the contribution from the adjacent ring carbon. The calculation shows that for the **mtc** radical, both oxygen and sulfur support unpaired electron density. This additional delocalization in the SOMO provided by the thioether side chain may thus contribute to the nearly 0.4-V stabilization of the Tyr–Cys radical relative to a simple phenoxyl protein free radical. For the **mtc** phenoxyl, there are near-lying orbitals at both lower and higher binding energies. The unpaired electron distribution for the **mtc** radical *in vacuo* (without external interactions with, for example, hydrogen bond donors) is shown in Fig. 19 and indicates that the oxygen and sulfur atoms bear the majority of the unpaired electron in the ground state. More recent calculations within a variety of frameworks confirm these basic results, although the extent of sulfur contribution to the SOMO, as estimated by the spin density on the sulfur atom in the **mtc** phenoxyl, appears to be sensitive to the computational method used [PM3 (0.17) (Itoh *et al.*, 1993); B3LYP/6–31G(d) (0.11) (Wise *et al.*, 1999); PWP86 (0.15) (Himo *et al.*, 1999)].

The predicted electronic delocalization onto the thioether side chain in the **mtc** phenoxyl free radical ground state has been confirmed experimentally by EPR measurements on **mtc** specifically deuterated in the side

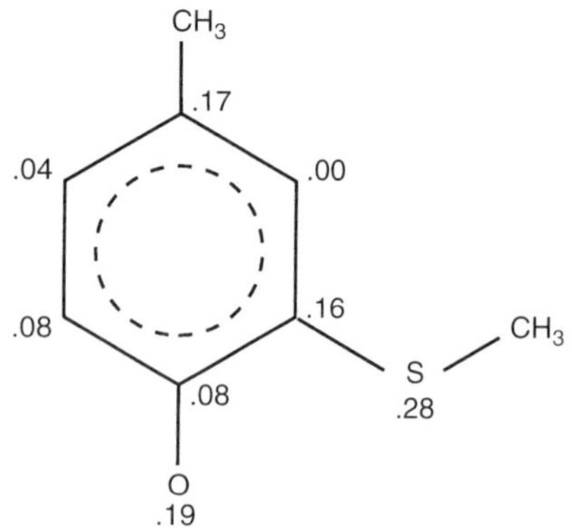


FIG. 19. Unpaired spin distribution in the *o*-(methylthio)cresyl phenoxyl ground state. The geometry of the *o*-(methylthio)cresyl model is shown together with unpaired spin associated with ring carbons and the exocyclic oxygen and sulfur atoms.

chain methyl group (Gerfen *et al.*, 1996). A significant sharpening of the radical EPR spectrum is observed for the deuterated radical, and simulations indicate a decrease in the isotropic hyperfine coupling to the side chain methyl group from 0.5 mT (protons) to 0.077 mT (deuterons). Based on the magnitude of the α-proton hyperfine coupling measured for a thiyl radical (for which the spin density on sulfur approaches 1.0), these observations are consistent with a significant localization of unpaired electron spin on sulfur, although somewhat lower than calculated (~0.20 vs 0.28).

The effect of sulfur participation on the orbital *g*-shifts in the EPR spectra, illustrated in Fig. 20, accounts for the qualitatively different spectra observed for tyrosyl phenoxyl and Tyr–Cys phenoxyl radicals (Gerfen *et al.*, 1996). The rhombicity of the simple tyrosyl radical EPR spectrum is a consequence of the splitting between $g_x$ and $g_y$ principal *g*-values. These *g*-shifts deviate from the free electron *g*-value ($g_e = 2.00023$) as a result of orbital angular momentum contributions. While a nondegenerate electronic state (such as the A″ ground state for **cre**) contains no first-order unquenched orbital momentum, second-order spin-orbit mixing between close-lying a′ and a″ functions results

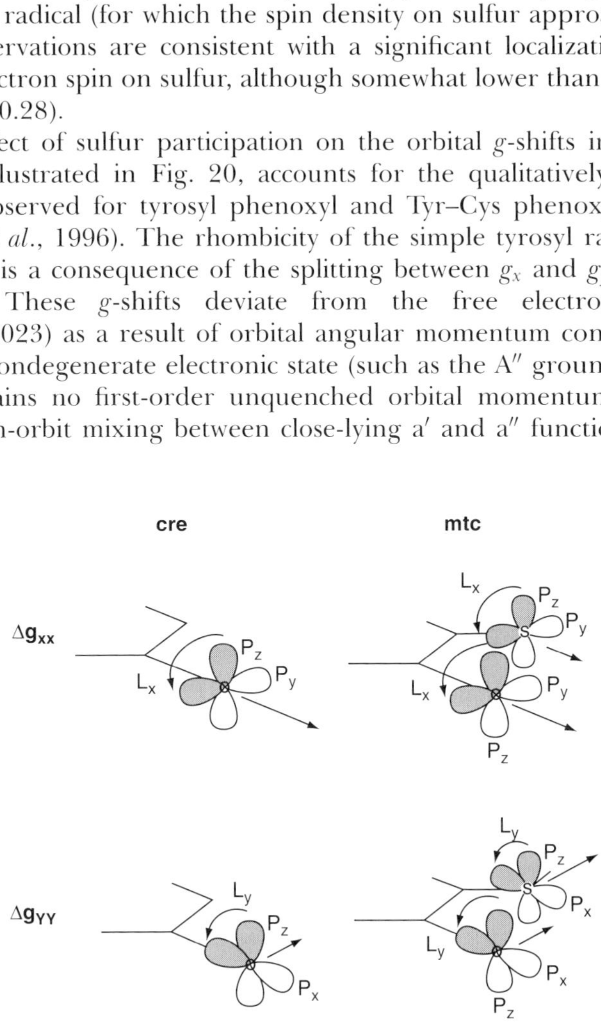


FIG. 20. Spin-orbit mixing mechanism for orbital *g*-shifts in substituted phenoxyl radicals. The electronic *g*-tensor is perturbed by spin-orbit effects which can be viewed as orbital rotation elements. The perturbation of the $g_{xx}$ term ($\Delta g_{xx}$) arises from mixing perpendicularly oriented valence orbitals on the same atom under the $L_x$ orbital operator and summing these individual contributions over all atoms to produce the resultant molecular *g*-shift.

in the appearance of a small orbital contribution that produces an electronic orbital Zeeman resonance shift. The effect is very small for lighter atoms (hydrogen and carbon) but becomes significant when heavier atoms (like oxygen and sulfur) are involved in the wavefunction. The magnitude of the effect also depends on the relative energies of the SOMO and admixed levels; thus orbitals that lie at deeper binding energy (as a consequence of stabilization in covalent bonds) make relatively minor contributions. The largest contributions to the $g_x$ and $g_y$ $g$-shifts for the **cre** radical result from mixing the oxygen O $p_y$ and O $p_x$ orbitals, respectively, with O $p_z$. Since the O $p_x$ orbital is strongly stabilized through covalent bonding with the C-4 ring carbon atom, the $g_x$ and $g_y$ $g$-shifts have very different magnitudes, and consequently a rhombic splitting is observed. The relatively axial spectrum observed for the thioether-substituted radicals can be similarly explained. While both sulfur and oxygen contributions to the $g$-shifts would be expected to be anisotropic, for the same reason as the **cre** phenoxyl, the individual anisotropies from sulfur and oxygen contributions would partly cancel in the resultants that determine the overall magnitude of the two $g$-shift terms. Thus, the spectrum of the **mtc** radical exhibits relatively small differences between the principal $g$-values in the $xy$ plane ($g_x$ and $g_y$ $g$-values), giving rise to the characteristic axial high-field EPR signature for the thioether-substituted phenoxyl radical (Fig. 17). The computed $g$-tensor for a thioether-substituted phenoxyl based on calculations that predict small sulfur contributions to the SOMO fails to reproduce this experimental axial trend (Engström *et al.*, 2000), which is clearly evident for the model predicting more significant sulfur contributions.

In addition to giving a theoretical interpretation of the resonance spectra, these calculations provide insight into the chemistry of the Tyr–Cys radical. The calculated ground state structure indicates that the phenoxyl oxygen is a site of reduced electron density, making it an electrophilic site in the radical. A phenoxyl thus has potential for hydrogen atom abstraction from the substrate, with the O–H bond enthalpy of the resulting phenol [$\sim$80 kcal mol$^{-1}$ (Lucarini *et al.*, 1996)] providing the driving force for the reaction. This chemistry is fundamental to the proposed catalytic mechanism and provides a convincing rationale for the presence of a free radical in galactose oxidase.

## VII. The Free Radical-Coupled Copper Active Site

The free radical–copper complex in active GAOX combines two distinct reactive sites to form a two-electron redox unit in the protein with new properties, different from those of the individual components. The spectrum of this novel complex is more than a simple superposition of spectra for the isolated copper center and protein free radical as would be expected for weakly interacting centers and is evidence of the strong interactions between the Cu(II) metal ion and the coordinated free radical ligand.

The second, nonredox-active tyrosine (Tyr-495) also appears to be involved in defining the properties of the radical–metal ion pair by extending the covalent pathways for electronic delocalization in the ground state. Delocalization of radical character over both tyrosine ligands is implied by the intense, low-energy transition of the $Tyr_{ON}$ form of AGO that is assigned to a ligand-to-ligand charge transfer absorption in this complex (McGlashin *et al.*, 1995). From this analysis, it is clear that the site is organized around a pair of tyrosines (one phenoxyl, one phenolate) bridged by a metal ion that electronically couples the two ligands, analogous to a mixed-valent atom-bridged binclear metal complex, but with the ligands and metal ions playing opposite roles. Both complexes have delocalized electronic structures and the interacting elements are strongly coupled. For the radical-copper-active site, extended delocalization of the radical in the ground state may contribute to the stability of the oxidized complex. As mentioned in earlier sections, the GAOX radical forms under relatively mild conditions and is unusually stable compared to most protein free radicals. Unlike a typical radical, which has at best a transitory existence, the radical site in AGO has a half-life on the order of a week when protected from stray reductants (Whittaker *et al.*, 1998).

In addition to acting as a charge accumulation site that stores electrons between the two half-reactions, the radical-coupled copper complex also binds protons (hydrogens) removed from the organic substrate during turnover. The pair of tyrosine ligands is also believed to be responsible for the proton reactivity of the site. One protonation step involves Tyr-495 as a general base whose reactivity depends on the ligation state of the active site metal ion. As shown by proton uptake experiments, coordinating ligands drive a structural transition in the metal complex, unmasking the basic Tyr-495 phenolate oxygen. Tyr-272 may also be protonated in the fully reduced complex, forming a weak phenol coordination to the Cu(I) center. The special reactivity of this biological metal complex appears to be a consequence of its ability to accommodate both Cu(I) and Cu(II) oxidation states and to structurally control the reactivity of the ligands.

## VIII. Catalytic Mechanism

An impressive range of approaches have been applied over the years to investigating the molecular mechanism of catalysis by the radical copper oxidases. The foundations for these studies have been the identification and accurate characterization of the distinct states in which the enzyme may be prepared. Early studies were complicated by apparently conflicting data that resulted from studying enzyme preparations that included mixtures of the various enzyme forms.

The stereo- and regiospecific oxidation of primary alcohols (Fig. 1) that comprises the reductive chemistry of the GAOX reaction can be mechanistically resolved from the second half-reaction, reoxidation by dioxygen. This is convincingly shown by the preparation of a stable, $O_2$-reactive reduced enzyme intermediate by anaerobic reaction with substrate (Fig. 11, line C), indicating that each of the two half-reactions can proceed independently of the other. While no crystal structure is available for the galactose oxidase substrate complex, the lability of the coordinated water makes it likely that the substrate binds to the copper by replacing the solvent in the complex. In fact, for this enzyme, water might be regarded as a substrate analog (ROH with R=H). The broad substrate specificity of GAOX (Table I) and the relatively large $K_m$ values for substrates suggest that the ES complex may be more or less collisional, without the lock-and-key array of precise contacts characteristic of enzymes with restricted substrate ranges. The orientation of the side chain of the substrate coordinated to the metal ion in the ES complex may be constrained as shown in Fig. 21 to account for the *pro*-S stereoselectivity. Binding to Cu(II) is expected to dramatically perturb the acidity of the coordinated substrate hydroxyl, lowering the $pK_a$ of the hydroxylic proton by as much as 10 pH units (Kimura *et al.*, 1994). Tyr-495 is expected to serve as a general base at this point, abstracting the acidified hydroxyl proton along the proton transfer coordinate. Ionization activates the coordinated substrate, and the alkoxide complex is relatively easy to oxidize. The rate-limiting step for substrate oxidation is known to involve hydrogen atom abstraction from the substrate methylene carbon, based on the unusually large kinetic isotope effect associated with oxidation of methylene-deuterated substrate (Maradufu *et al.*, 1971; Villafranca *et al.*, 1993; Whittaker *et al.*, 1998). The strong temperature-dependence of the kinetic isotope effect (Whittaker *et al.*, 1998) suggests that hydrogen atom abstraction may occur through a tunneling mechanism (Cha *et al.*, 1989; Bahnson and Klinman, 1995). This step probably involves the Tyr–Cys phenoxyl free radical, explaining the requirement for an organic redox cofactor in the active site, since it is energetically more favorable to transfer hydrogen to phenoxyl rather than to the metal to form a hydride. Because there is no convincing evidence for occurrence of a stable ketyl radical intermediate during turnover, strong coupling between hydrogen transfer and Cu(II) reduction may make the two steps

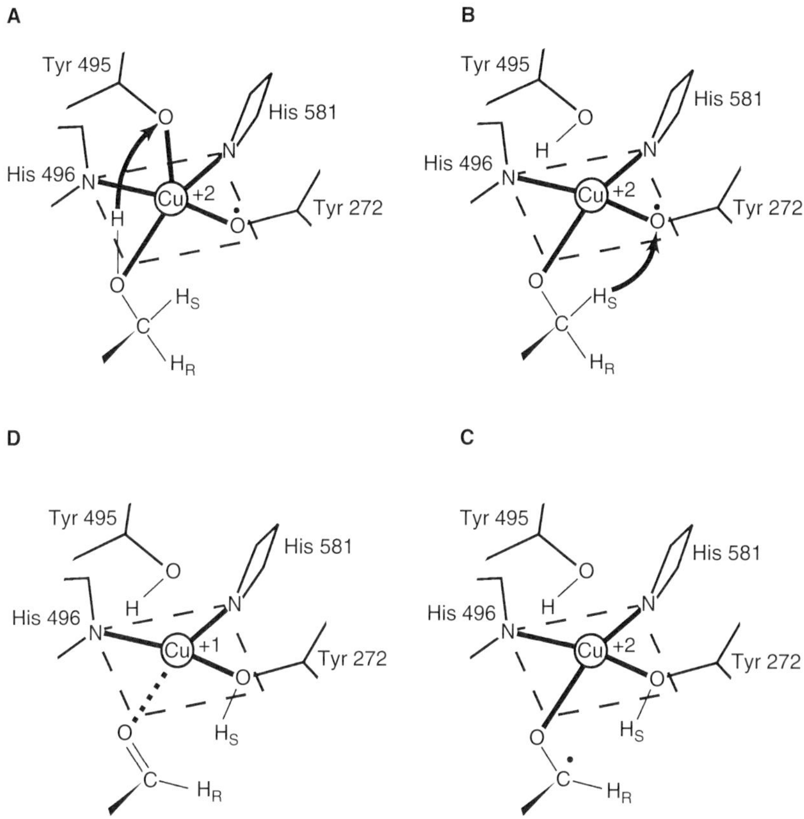


FIG. 21. Proposed catalytic mechanism for substrate oxidation by galactose oxidase. (A) Substrate binding displaces Tyr-495 phenolate which serves as a general base for abstracting the hydroxylic proton. (B) Stererospecific *pro*-S hydrogen abstraction by the Tyr–Cys phenoxyl radical. (C) Inner sphere electron transfer reducing Cu(II) to Cu(I). (D) Dissociation of the aldehyde product.

essentially concerted despite the different character of the two redox sites in the protein.

Dissociation of the aldehyde product would leave a low-coordinate, Cu(I) redox center associated with two protonated tyrosine phenols in the active site. This complex is known to be very reactive toward dioxygen, the second-order kinetic constant for reoxidation of the reduced enzyme by $O_2$ being nearly $8 \times 10^6\ M^{-1}\ s^{-1}$ (Borman *et al.*, 1997; Whittaker *et al.*, 1998). This extremely fast reaction makes it difficult to investigate by conventional rapid reaction techniques, and the nature of the oxygenated complex remains speculative. One clue to the nature of the oxy complex may be derived from a detailed inspection of the structure of the IAGO active site. Two water molecules can be seen to lie in close proximity to the metal center in this complex (Fig. 22). One, labeled HO 703 (PDB AGOG numbering), is the directly coordinated solvent molecule, while the second, labeled HOH 294, is approximately 3.2 Å away and may be hydrogen bonded to both the coordinated solvent and the phenolic oxygens of Tyr-495 and Tyr-272. This pair of water molecules may be imagined to represent the "ghost" of a bound peroxide complex, preserving the hydrogen-bonding and metal interactions involved in stabilizing the catalytic peroxide intermediate. This type of analysis has previously been applied to the copper–quinoenzyme amine oxidase (Whittaker, 1999), and the predicted structure of the oxygenated adduct has been confirmed by cryogenic X-ray crystallography (Wilmot *et al.*, 1999). For GAOX, the structure of the predicted peroxide intermediate would imply that dioxygen reacts at the open face of the reduced Cu(I) complex. Following initial reduction by inner sphere electron transfer to dioxygen, the distal, noncoordinated oxygen atom of the bound superoxide could abstract a hydrogen atom from the adjacent Tyr-272 phenol, producing a hydroperoxide adduct of the radical–copper complex. Protonation of the proximal oxygen of the hydroperoxide would release the product and permit rebinding of Tyr-495 phenolate. This process would essentially mirror the substrate oxidation reaction, with the two tyrosines once again performing proton transfer and redox steps in the mechanism.

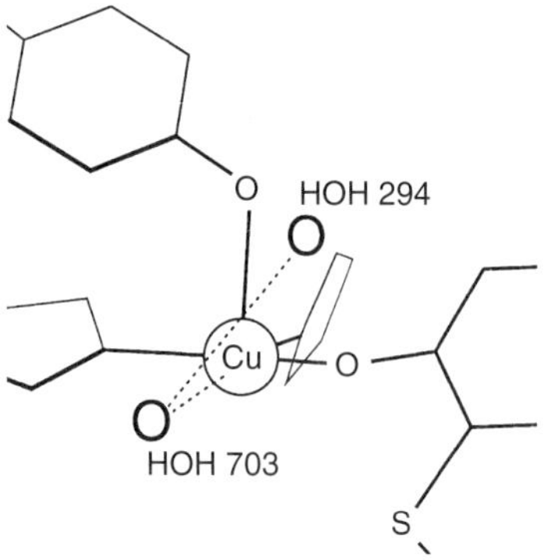


FIG. 22. A possible peroxide "ghost" in the galactose oxidase active site. Two crystallographically well-defined solvent molecules (HOH 294 and HOH 703, PDB 1GOG numbering) lie along the face of the active site metal complex at the base of the substrate access channel in the resting enzyme.

## IX. Cofactor Biogenesis

The origin of the Tyr–Cys redox cofactor in GAOX has been the subject of lively discussion ever since its discovery (Ito *et al.*, 1991; Dooley, 1999). Biochemical studies demonstrate that it forms posttranslationally as a result of coupling between Cys-228 and Tyr-272 side chains (Baron *et al.*, 1994; Rogers *et al.*, 2000). The reaction appears to be spontaneous in the presence of copper and dioxygen and does not involve other enzymes. Conversion of the proenzyme to the cofactor-containing mature protein product can be monitored either spectrophotometrically, through the appearance of the optical spectra of the mature protein, or electrophoretically, through the shift in mobility of the protein that occurs on cross-linking.

The mechanism for cofactor biogenesis in this enzyme has not yet been completely resolved. In fact, in contrast to extensive studies that have been aimed at elucidating the biogenesis reactions for the TOPA cofactor in the quinoprotein amine oxidases (Cai and Klinman, 1994; Choi *et al.*, 1995; Matsuzaki *et al.*, 1995; Tanizawa, 1995; Ruggiero *et al.*, 1997), investigations of the Tyr–Cys cofactor formation are just beginning. The *ortho* substitution pattern of the cysteine sulfur relative to the phenolic group in the side chain is reminiscent of products of oxidative phenol radical coupling and is consistent with either (A) nucleophilic addition of a thiol to a phenoxyl radical intermediate or (B) addition of an electrophilic thiyl radical to a phenol. In the former case, the *ortho* ring position would direct the reaction through the significant unpaired electron density on that carbon, and in the latter case the reactivity of the ring carbon would reflect the *ortho*-directing release of electron density by the phenolic oxygen.

A mechanism for cofactor biogenesis involving a phenoxyl intermediate has previously been proposed (Fig. 23A) (Rogers *et al.*, 2000). This mechanism requires that initial Cu(II) binding to the enzyme stabilizes a Cu(I)–Tyr 272 phenoxyl resonance contribution in the ground state of the complex. This phenoxyl would be highly reactive and expected to readily form a covalent bond with an adjacent thiol, and subsequent reaction with dioxygen would drive the aromatization of the Tyr–Cys ring system. However, this mechanism suffers from serious difficulties. First, the "resonance contribution" of a phenoxyl–Cu(I) structure corresponds to a charge transfer excited state of the phenolate–Cu(II) complex and its ground state contribution will be small (proportional to $1/\Delta E_{CT}$, the LMCT excitation energy). Second, the overall reaction described in this scheme is a three-electron process, which does not lead to a stable dioxygen reduction product, in contrast to a two- or four-electron reaction.

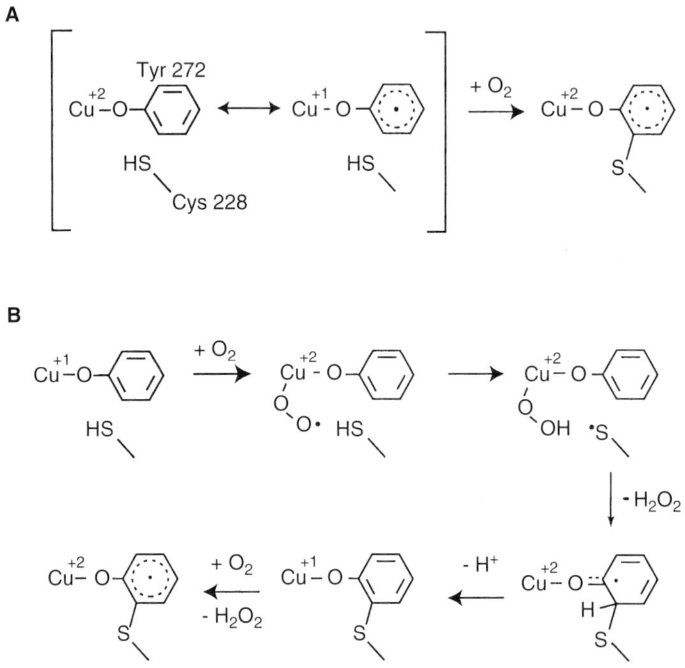


FIG. 23. Proposed mechanisms for Tyr–Cys cofactor biogenesis. (A) Initial Cu(II) binding leads to the appearance of resonance contributions from a reactive phenoxyl that electrophilically attacks the neighboring Cys-228 thiol. (B) Cu(I) binding produces an oxygen-reactive complex that drives hydrogen abstraction from Cys-228 to form a thiyl free radical which subsequently attacks the neighboring Tyr-272 ring with formation of a carbon–sulfur covalent bond.

An alternative mechanism is outlined in Fig. 23B. In this scheme, an initially formed Cu(I) complex reacts with dioxygen to form a superoxide adduct. The superoxide radical is capable of abstracting hydrogen from the nearby Cys-228 thiol to form a thiyl free radical, at the same time reducing oxygen to the level of peroxide. Dissociation of peroxide followed by electrophilic addition of the electron-deficient radical to the Tyr-272 ring would result in formation of a carbon–sulfur covalent bond. Tautomerization of the radical ring system and reduction of Cu(II) to Cu(I) would generate a species equivalent to the reduced enzyme complex (RGO) of the mature protein. Subsequent reaction of this complex with a second molecule of dioxygen would result in conversion to the AGO complex. Clearly, further experiments will be required to determine the mechanism of cofactor formation.

## X. Biomimetic Model Studies

The intriguing structure of the GAOX active site has raised a challenge to the field of bioinorganic chemistry and inspired the synthesis of an array of molecular models. Models for the isolated Tyr–Cys side chain have yielded important information on the chemistry and spectroscopy of the dissected cofactor, as described earlier (Section VI) (Whittaker *et al.*, 1993; Itoh *et al.*, 1993, 1997; Gerfen *et al.*, 1996). More recently, attention has been directed at mimicking the complex structure, spectroscopy, and even the catalytic reactivity of the intact radical–copper complex in model chemistry.

The first ligand system to incorporate the essential elements of the enzyme complex is shown in Fig. 24 (Whittaker *et al.*, 1996a). This unsymmetrical duncamine (**dnc**) chelate supports tripodal coordination of copper by two nitrogen donors and two phenolates, one of the phenolates being modified as an *ortho*-thioether derivative, reproducing the $CuN_2O_2$ core of the resting enzyme. Although the ligand binds Cu(II) and forms a topologically correct structural model of the enzyme active site, this and related chelate systems constrain the phenolate groups to an in-plane orientation ($\tau \approx 0°$) quite different than that found for the corresponding groups in GAOX. This has been experimentally shown to affect the ground state magnetism in phenoxyl–Cu(II) complexes (Müller *et al.*, 1998). More seriously, the [Cu(II)**dnc**] complex does not convert to a stable oxidized, radical-containing species. Subsequent studies have identified the C–H bonds in the methylene linkers as weak points in the structure that may be responsible for this instability resulting in the decomposition of the radical complex. However, further work has

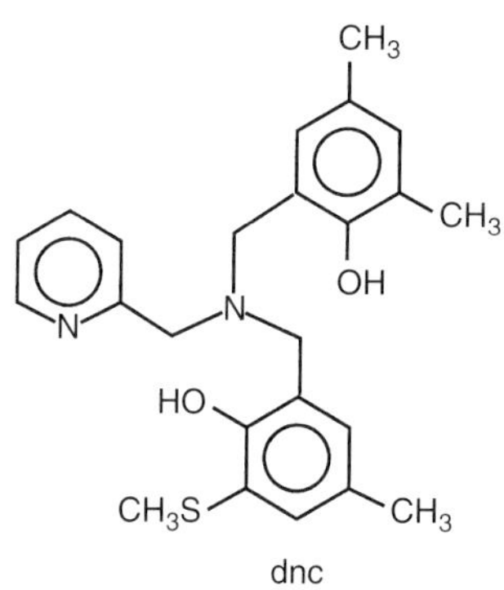


Fig. 24. The duncamine chelate (**dnc**), a structural model for the galactose oxidase active site.

demonstrated that under suitable conditions, this type of complex can be converted to a radical form (Zurita *et al.*, 1997) that is capable of performing a hydrogen atom abstraction (Halfen *et al.*, 1997; Halcrow *et al.*, 1998; Saint-Aman *et al.*, 1998; Taki *et al.*, 2000).

Other models have successfully avoided the instability of ligands related to **dnc**, allowing stoichiometric substrate oxidation. One model has even been crystallized with benzyl alcohol (substrate) bound to the reduced Cu(II) nonradical complex (Halfen *et al.*, 1996). This structure is particularly interesting in that it implies a hydrogen-bonding interaction between a coordinated phenolate oxygen and a methylene C–H of the alcohol, interactions that could be important for organizing the substrate complex and directing hydrogen atom transfer between these groups. Models for the fully reduced Cu(I) complex have also been prepared (Jazdzewski *et al.*, 1998; Holland and Tolman, 1999). Mechanism-based designs using alternative ligands less closely related to the protein have produced a catalytically active, functional mimic of the active site that is capable of catalytic turnover in the presence of $O_2$ (Wang *et al.*, 1998; Mahadevan *et al.*, 2000). This complex incorporates the radical-coupled copper catalytic motif of GAOX but lacks an endogenous base for substrate activation and therefore requires a catalytic amount of base in solution to support turnover. Physical characterization of the oxidized complex has confirmed a Cu(II)–radical assignment for the reactive species, as previously found for the AGO enzyme complex (Wang *et al.*, 1998). Car-Parinello *ab initio* molecular dynamics calculations have been used to investigate the electronic structure and energetics of reactive intermediates for both the catalytic model and the enzyme active site, emphasizing the importance of geometric factors for radical stabilization and hydrogen abstraction (Rothlisberger *et al.*, 2000). The field of galactose oxidase model chemistry has recently been comprehensively reviewed (Itoh *et al.*, 2000).

## XI. Biomedical Applications

The reaction performed by galactose oxidase links oxidation of an organic substrate to $O_2$ reduction, forming hydrogen peroxide. Reactions of this type are ideally suited to bioanalytical applications, since the stoichiometric relation between substrate oxidation and dioxygen reduction allows well-established polarographic oxygen detection to be used for quantitation of biological compounds (Johnson *et al.*, 1982; Karube *et al.*, 1990). Formation of hydrogen peroxide by GAOX also permits the reaction to be coupled to the peroxidase-catalyzed oxidation of dyes for colorimetric detection. The standard clinical determination of galactose makes use of this type of chemistry (Loken, 1966).

The unique reactivity of GAOX has also been taken advantage of in synthetic organic chemistry (Wong and Whitesides, 1994). Enzymatic semisynthesis using GAOX catalysis allows complex aldehydes and carboxylic acids to be formed from compounds containing a single primary alcohol functional group (Mazur, 1991). Applied in this way, GAOX has permitted the large-scale preparation of a number of novel sugars and nucleotides (Basu *et al.*, 2000).

Modification of cell surface carbohydrates is another important application for GAOX. The enzyme is widely used in cell labeling studies and histochemical staining, taking advantage of the broad substrate specificity that allows it to metabolize even macromolecular substrates like glycoproteins (Roberts and Gupta, 1965; Schulte and Spicer, 1983). Oxidation of carbohydrate side chains in glycoconjugates followed by reduction with $NaB^3H_4$ presents a facile method for tagging surface-exposed proteins. Carbonyl groups formed by GAOX oxidation may also be reacted with other nucleophiles to generate chromophoric products.

One of the recent developments in cancer screening makes use of the selectivity of GAOX for galactose among hexoses to modify D-galactose-β[1,3]-*N*-acetylgalactosamine [Gal-β[1,3]-GalNac] (Fig. 25), also known as the Thomas–Freidrich or T-antigen, an important tumor marker (Springer, 1997). The occurrence of this protein-bound disaccharide is strongly correlated with certain cancers, particularly colon cancer. Oxidation of accessible 6-hydroxymethyl groups in Gal-β[1,3]-GalNAc produces reactive carbonyls that combine with simple amines to form highly chromophoric Schiff base adducts. Formation of these conjugates provides a rapid visual cytochemical staining method and has been applied as a simple screening assay for early detection of colon cancer (Carter *et al.*, 1997; Said *et al.*, 1999).

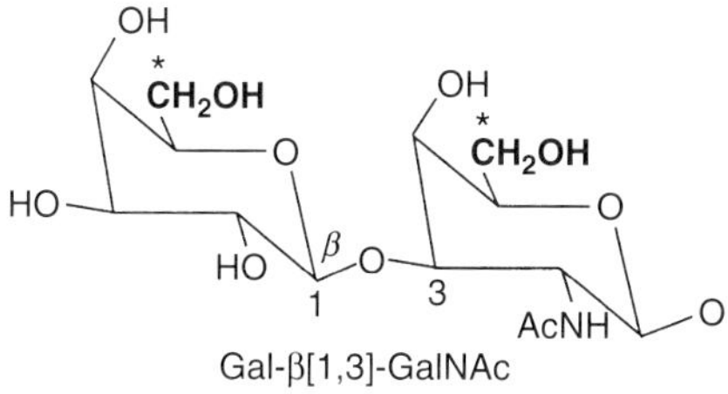


FIG. 25. The structure of the disaccharide tumor marker D-galactose-β[1,3]-*N*-acetyl galactosamine.

## XII. Summary and Conclusions

The free radical-coupled copper catalytic motif has emerged as the unifying feature of a new family of enzymes, the radical copper oxidases. Their highly evolved active sites include a novel amino acid modification, the Tyr–Cys dimer, that forms spontaneously through self-processing of the protein during its maturation. The active site is remarkable in the extent to which metal ligands participate in the catalytic process. Rather than simply coordinating the metal ion, the ligands perform essential redox and proton-transfer functions in the chemistry of the active site, directed by their interactions with the copper center in the protein. The wide phylogenetic distribution and range of functions represented within the family hint of a fundamental role for these enzymes in the biology of oxygen. The roles for these enzymes are further expanding through a variety of biotechnological applications.

## Acknowledgments

I thank Mei Whittaker for discussions and for valuable assistance in preparation of the manuscript. Support from National Institutes of Health Grant GM 46749 is also gratefully acknowledged.

## References

Adams, J., Kelso, R., and Cooley, L. (2000). *Trends Cell Biol.* **10**, 17–24.

Altschul, S. F., Madden, T. L., Schaffer, A. A., Zhang, J., Zhang, Z., Miller, W., and Lipman, D. J. (1997). *Nucleic Acids Res.* **25**, 3389–3402.

Avigad, G., Amaral, D., Asensio, C., and Horecker, B. L. (1962). *J. Biol. Chem.* **237**, 2736–2743.

Babcock, G. T., El-Deeb, M. K., Sandusky, P. O., Whittaker, M. M., and Whittaker, J. W. (1992). *J. Am. Chem. Soc.* **114**, 3727–3734.

Bahnson, B. J., and Klinman, J. P. (1995). *Methods Enzymol.* **249**, 373–397.

Baron, A. J., Stevens, C., Wilmot, C., Seneviratne, K. D., Blakely, V., Dooley, D. M., Phillips, S. E. V., Knowles, P. F., and McPherson, M. J. (1994). *J. Biol. Chem.* **269**, 25095–25105.

Basu, S. S., Dotson, G. D., and Raetz, R. H. (2000). *Anal. Biochem.* **280**, 173–177.

Bennati, M., Farrar, C. T., Bryant, J. A., Inati, S. J., Weiss, V., Gerfen, G. J., Riggs-Gelasco, P., Stubbe, J., and Griffin, R. G. (1999). *J. Magn. Res.* **138**, 232–243.

Bork, P., and Doolittle, R. F. (1994). *J. Mol. Biol.* **236**, 1277–1282.

Borman, C. D., Saysell, C. G., and Sykes, A. G. (1997). *J. Biol. Inorg. Chem.* **2**, 480–487.

Brown, K., Tegoni, M., Prudencio, M., Pereira, A. S., Besson, S., Moura, J. J., Moura, I., and Cambillau, C. (2000). *Nat. Struct. Biol.* **7**, 191–195.

Buse, G., Soulimane, T., Dewor, M., Meyer, H. E., and Bluggel, M. (1999). *Protein Sci.* **8**, 985–990.

Cai, D., and Klinman, J. P. (1994). *J. Biol. Chem.* **269**, 32039–32042.

Carter, J. H., Deddens, J. A., Pullman, J. L., Colligan, B. M., Whiteley, L. O., and Carter, H. W. (1997). *Clin. Cancer Res.* **3**, 1479–1489.

Cha, Y., Murray, C. J., and Klinman, J. P. (1989). *Science* **243**, 1325–1330.

Choi, Y. H., Matsuzaki, R., Fukui, T., Shimizu, E., Yorifuji, T., Sato, H., Ozaki, Y., and Tanizawa, K. (1995). *J. Biol. Chem.* **270**, 4712–4720.

Clark, K., Penner-Hahn, J. E., Whittaker, M. M., and Whittaker, J. W. (1990). *J. Am. Chem. Soc.* **112**, 6433–6434.

Clark, K., Penner-Hahn, J. E., Whittaker, M., and Whittaker, J. W. (1994). *Biochemistry* **33**, 12553–12557.

Day, E. P. (1993). *Methods Enzymol.* **227**, 437–463.

Dickman, M. H., and Doedens, R. J. (1981). *Inorg. Chem.* **20**, 2677–2681.

Dooley, D. M. (1999). *J. Inorg. Biol. Chem.* **4**, 1–11.

Dooley, D. M., McGuirl, M. A., Brown, D. E., Turowski, P. N., McIntire, W. S., and Knowles, P. F. (1991). *Nature* **349**, 262–264.

Eklund, H., Eriksson, M., Uhlin, U., Nordlund, P., and Logan, D. (1997). *Biol. Chem.* **378**, 821–825.

Engström, M., Himo, F., and Ågren, H. (2000). *Chem. Phys. Lett.* **319**, 191–196.

Frey, P. A. (1997). *Curr. Opin. Chem. Biol.* **1**, 347–356.

Gerfen, G. A., Bellew, B., Griffin, R., Singel, D., Ekberg, C. A., and Whittaker, J. W. (1996). *J. Phys. Chem.* **100**, 16739–16748.

Gielens, C., De Geest, N., Xin, X. Q., Devreese, B., Van Beeumen, J., and Preaux, G. (1997). *Eur. J. Biochem.* **248**, 879–888.

Halcrow, M. A., Chia, L. M. L., Liu, X., McInnes, E. J. L., Yellowlees, L. J., Mabbs, F. E., and Davies, J. E. (1998). *Chem. Commun.* **22**, 2465–2466.

Halfen, J. A., Young, V. G., Jr., and Tolman, W. B. (1996). *Angew. Chem. Int. Ed.* **35**, 1687.

Halfen, J. A., Jazdzewski, B. A., Mahapatra, S., Berreau, L. M., Wilkinson, E. C., Que, L., Jr., and Tolman, W. B. (1997). *J. Am. Chem. Soc.* **119**, 8217–8227.

Hamilton, G. A. (1981). *In* "Copper Proteins" (Spiro, T. G., Ed.), pp. 193–218. Wiley, New York.

Hammel, K. E., Mozuch, M. D., Jensen, K. A., Jr., and Kersten, P. J. (1994). *Biochemistry* **33**, 13349–13354.

Himo, F., Babcock, G. T., and Eriksson, L. A. (1999). *Chem. Phys. Lett.* **313**, 374–378.

Holland, P. L., and Tolman, W. B. (1999). *J. Am. Chem. Soc.* **121**, 7270–7271.

Ito, N., Phillips, S. E. V., Stevens, C., Ogel, Z. B., McPherson, M. J., Keen, J. N., Yadav, K. D. S., and Knowles, P. F. (1991). *Nature* **350**, 87–90.

Ito, N., Phillips, S. E. V., Yadav, K. D. S., and Knowles, P. F. (1994). *J. Mol. Biol.* **238**, 794–814.

Itoh, S., Hirano, K., Furuta, A., Komatsu, M., Ohshiro, Y., Ishida, A., Takamuku, S., Kohzuma, T., Nakamura, N., and Suzuki, S. (1993). *Chem. Lett.* **1993**, 2099–2102.

Itoh, S., Takayama, S., Arakawa, R., Furuta, A., Komatsu, M., Ishida, A., Takamuku, S., and Fukuzumi, S. (1997). *Inorg. Chem.* **36**, 1407–1416.

Itoh, S., Taki, M., and Fukuzumi, S. (2000). *Coord. Chem. Rev.* **198**, 3–20.

Janes, S. M., Mu, D., Wemmer, D., Smith, A. J., Kaur, S., Maltby, D., Burlingame, A. L., and Klinman, J. P. (1990). *Science* **248**, 981–987.

Jazdzewski, B. A., Young, V. G., Jr., and Tolman, W. B. (1998). *Chem. Commun.* **22**, 2521–2522.

Johnson, J. M., Halsall, H. B., and Heineman, W. R. (1982). *Anal. Chem.* **54**, 1394–1399.

Karube, I., Kimura, J., Yokoyama, K., and Tamiya, E. (1990). *Ann. N.Y. Acad. Sci.* **613**, 385–389.

Kelleher, F. M., and Bhavanandan, V. P. (1986). *J. Biol. Chem.* **261**, 11045–11048.

Kelleher, F. M., Dubbs, S. B., and Bhavanandan, V. P. (1988). *Arch. Biochem. Biophys.* **263**, 349–354.

Kersten, P. J. (1990). *Proc. Natl. Acad. Sci. USA* **87**, 2936–2940.

Kersten, P. J., and Cullen, D. (1993). *Proc. Natl. Acad. Sci. USA* **90**, 7411–7413.

Kersten, P. J., and Kirk, T. K. (1987). *J. Bacteriol.* **169**, 2195–2201.

Kimura, E., Nakamura, I., Koike, T., Shionaya, M., Kodama, Y., Ikeda, T., and Shiro, M. (1994). *J. Am. Chem. Soc.* **116**, 4764–4771.

Knowles, P. F., and Ito, N. (1993). *Perspect. Bioinorg. Chem.* **2**, 207–244.

Kosman, D. J. (1985). *In* "Copper Proteins and Copper Enzymes" (Lontie, R., Ed.), Vol. 2, pp. 1–26. CRC Press, Boca Raton, FL.

Lassmann, G., Odenwaller, R., Curtis, J. F., Degray, J. A., Mason, R. P., Marnett, L., and Eling, T. E. (1993). *Dev. Oncol.* **71**, 51–53.

Lerch, K. (1982). *J. Biol. Chem.* **257**, 6414–6419.

Lim, Y. Y., and Drago, R. S. (1972). *Inorg. Chem.* **11**, 1334–1338.

Loken, H. F. (1966). *Scand. J. Clin. Lab. Invest. Suppl.* **92**, 99–100.

Lucarini, M., Pedrielli, P., Pedulli, G. F., Cabiddu, S., and Furruoni, C. (1996). *J. Org. Chem.* **61**, 9259–9263.

MacMillan, F., Kannt, A., Behr, J., Prisner, T., and Michel, H. (1999). *Biochemistry* **38**, 9179–9184.

Mahadevan, V., Klein-Gebbink, R. J. M., and Stack, T. D. P. (2000). *Curr. Opin. Chem. Biol.* **4**, 228–234.

Maradufu, A., Cree, G. M., and Perlin, A. S. (1971). *Can. J. Chem.* **49**, 3429–3437.

Marsh, E. N. (1995). *BioEssays* **17**, 431–441.

Matsuzaki, R., Suzuki, S., Yamaguchi, K., Fukui, T., and Tanizawa, K. (1995). *Biochemistry* **34**, 4524–4530.

Mazur, A. W. (1991). *In* "Enzymes in Carbohydrate Synthesis" (Bednarski, M. O., and Simon, E. S., Eds.), pp. 99–110. Am. Chem. Soc., Washington, DC.

McGlashin, M. L., Eads, D. D., Spiro, T. G., and Whittaker, J. W. (1995). *J. Phys. Chem.* **99**, 4918–4922.

McPherson, M. J., Ogel, Z. B., Stevens, C., Yadev, K. D., Keen, J. N., and Knowles, P. F. (1992). *J. Biol. Chem.* **267**, 8146–8152.

McPherson, M. J., Stevens, C., Baron, A. J., Ogel, Z. B., Senevriatne, K., Wilmot, C., Ito, N., Brocklebank, I., Phillips, S. E. V., and Knowles, P. F. (1993). *Biochem. Soc. Trans.* **21**, 752–756.

Mendonça, M. H., and Zancan, G. T. (1987). *Arch. Biochem. Biophys.* **252**, 507–514.

Müller, J., Weyhermüller, T., Bill, E., Hildebrandt, P., Ould-Moussa, L., Glaser, T., and Wieghardt, K. (1998). *Angew. Chem. Int. Ed.* **37**, 616–619.

Ostermeier, C., Harrenga, A., Ermler, U., and Michel, H. (1997). *Proc. Nat. Acad. Sci. USA* **94**, 10547–10553.

Pedersen, J. Z., and Finazzi-Agro, A. (1993). *FEBS Lett.* **325**, 53–58.

Reynolds, M. P., Baron, A. J., Wilmot, C. M., Phillips, S. E. V., Knowles, P. F., and McPherson, M. J. (1995). *Biochem. Soc. Trans.* **23**, 510S.

Reynolds, M. P., Baron, A. J., Wilmot, C. M., Vinecombe, E., Stevens, C., Phillips, S. E. V., Knowles, P. F., and McPherson, M. J. (1997). *J. Biol. Inorg. Chem.* **2**, 327–335.

Roberts, G. P., and Gupta, S. K. (1965). *Nature* **207**, 425–426.

Rogers, M. S., Knowles, P. F., Baron, A. J., McPherson, M. J., and Dooley, D. M. (1998). *Inorg. Chim. Acta* **275/276**, 175–181.

Rogers, M. S., Baron, A. J., McPherson, M. J., Knowles, P. F., and Dooley, D. M. (2000). *J. Am. Chem. Soc.* **122**, 990–991.

Rothlisberger, U., Carloni, P., Doclo, K., and Parrinello, M. (2000). *J. Biol. Inorg. Chem.* **5**, 236–250.

Ruggiero, C. E., Smith, J. A., Tanizawa, K., and Dooley, D. M. (1997). *Biochemistry* **36**, 1953–1959.

Said, I. T., Shamsuddin, A. M., Sherief, M. A., Taleb, S. G., Aref, W. F., and Kumar, D. (1999). *Histol. Histopathol.* **14**, 351–357.

Saint-Aman, E., Ménage, S., Pierre, J.-L., Defrancq, E., and Gellon, G. (1998). *New J. Chem.* **22**, 393–394.
Schulte, B. A., and Spicer, S. S. (1983). *J. Histochem. Cytochem.* **31**, 391–403.
Silakowski, B., Ehret, H., and Schairer, U. (1998). *J. Bacteriol.* **180**, 1241–1247.
Sjöberg, B.-M., Reichard, P. Gräslund, A., and Ehrenberg, A. (1978). *J. Biol. Chem.* **253**, 6863–6865.
Smith, C. A., Baker, H. M., and Baker, E. N. (1991). *J. Mol. Biol.* **219**, 155–159.
Springer, G. F. (1997). *J. Mol. Med.* **75**, 595–602.
Stubbe, J. (1989). *Annu. Rev. Biochem.* **58**, 257–285.
Stubbe, J., and van der Donk, W. A. (1998). *Chem. Rev.* **98**, 705–762.
Taki, M., Kumei, H., Itoh, S., and Fukuzumi, S. (2000). *J. Inorg. Chem.* **78**, 1–5.
Tanizawa, K. (1995). *J. Biochem.* **118**, 671–678.
Tatusova, T. A., and Madden, T. L. (1999). *FEMS Microbiol. Lett.* **174**, 247–250.
Tommos, C., Hoganson, C. W., Valentin, M. D., Lydakis-Simantiris, N., Dorlet, P., Westphal, K., Chu, H. A., McCracken, J., and Babcock, G. T. (1998). *Curr. Opin. Chem. Biol.* **2**, 244–252.
Tressel, P. S., and Kosman, D. J. (1982). *Methods Enzymol.* **89**, 163–171.
Villafranca, J. J., Freeman, J. C., and Kotchevar, A. (1993). *In* "Bioinorganic Chemistry of Copper" (Karlin, K. D., and Tyeklar, Z., Eds.), pp. 439–446, Chapman and Hall, New York.
Wang, Y., DuBoise, J. L., Hedman, B., Hodgeson, K. O., and Stack, T. D. P. (1998). *Science* **279**, 537–540.
Whittaker, J. W. (1994). *In* "Metal Ions in Biological Systems" (Sigel, H., Ed.), Vol. 30, pp. 315–360. Dekker, New York.
Whittaker, J. W. (1999). *Essays Biochem.* **34**, 155–172.
Whittaker, J. W., and Whittaker, M. M. (1998). *Pure Appl. Chem.* **70**, 903–910.
Whittaker, M. M., and Whittaker, J. W. (1988). *J. Biol. Chem.* **263**, 6074–6080.
Whittaker, M. M., and Whittaker, J. W. (1990). *J. Biol. Chem.* **265**, 9610–9613.
Whittaker, M. M., and Whittaker, J. W. (1993). *Biophys. J.* **64**, 762–772.
Whittaker, M. M., DeVito, V. L., Asher, S. A., and Whittaker, J. W. (1989). *J. Biol. Chem.* **264**, 7104–7106.
Whittaker, M. M., Chuang, Y. Y., and Whittaker, J. W. (1993). *J. Am. Chem. Soc.* **115**, 10029–10035.
Whittaker, M. M., Duncan, W. R., and Whittaker, J. W. (1996a). *Inorg. Chem.* **35**, 382–386.
Whittaker, M. M., Kersten, P. J., Nakamura, N., Sanders-Loehr, J., Schweizer, E. S., and Whittaker, J. W. (1996b). *J. Biol. Chem.* **271**, 681–687.
Whittaker, M. M., Ballou, D. P., and Whittaker, J. W. (1998). *Biochemistry* **37**, 8426–8436.
Whittaker, M. M., Kersten, P. J., Cullen, D., and Whittaker, J. W. (1999). *J. Biol. Chem.* **274**, 36226–36232.
Whittaker, M. M., Ekberg, C. A., Peterson, J., Sendova, M. S., Day, E. P., and Whittaker, J. W. (2000). *J. Mol. Catal. B Enzym.* **8**, 3–15.
Wilmot, C. M., Hajdu, J., McPherson, M. J., Knowles, P. F., and Phillips, S. E. V. (1999). *Science* **286**, 1724–1728.
Wise, K. E., Pate, J. B., and Wheeler, R. A. (1999). *J. Phys. Chem. B* **103**, 4764–4772.
Wong, C. H., and Whitesides, G. M. (1994). "Enzymes in Synthetic Organic Chemistry." Pergamon, New York.
Yoshikawa, S., Shinzawa-Itoh, K., Nakashima, R., Yaono, R., Yamashita, E., Inoue, N., Yao, M., Fei, M. J., Libeu, C. P., Mizushima, T., Yamaguchi, H., Tomizaki, T., and Tsukihara, T. (1998). *Science* **280**, 1723–1729.
Zurita, D., Gautier-Luneau, I., Ménage, S., Pierre, J.-L., and Saint-Aman, E. (1997). *J. Biol. Inorg. Chem.* **2**, 46–55.

# COPPER METALLOREGULATION OF GENE EXPRESSION

BY DENNIS R. WINGE

University of Utah Health Sciences Center,
Salt Lake City, Utah 84132



## I. Copper Homeostasis

Copper is an essential nutrient in the normal physiology of most cells. Copper ions are required cofactors in a myriad of enzymes including oxidases, monooxygenases, electron transfer proteins, and redox reaction enzymes. It is possible that copper is not essential in all cells. Curiously, the Lyme disease pathogen lacks all common copper enzymes (Posey and Gherardini, 2000). Cells regulate the intracellular concentration of essential metal ions to ensure that adequate, but not excessive, levels of metal ions exist. Homeostatic mechanisms maintain metal ion concentrations within an optimal range. Deviations from the optimal range for copper occur in humans leading to either Cu deficiency or Cu toxicosis. Children afflicted with chronic diarrhea or fed cow's milk exclusively may experience Cu deficiency (Lonnerdal *et al.*, 1985). In addition, excessive ingestion of zinc may lead to a Cu-deficient state (Danks, 1988). Copper deficiency is rare in Western countries since diets provide 2–4 mg Cu/day. The average dietary copper is more than the minimal daily requirement of between 0.3 and 1 mg assuming no fluid loss (Shulman, 1989). In contrast, copper toxicity can occur in humans by consumption of water supplied by copper pipes, accidental or abusive ingestion of Cu-contaminated materials, or environmental exposure (Scheinberg and Sternlieb, 1976). Hypercupremia occurs in certain inherited disorders such as Bedlington terrier's toxicosis and Wilson's disease in humans. In Wilson's disease, hepatic copper levels



*ADVANCES IN PROTEIN CHEMISTRY, Vol. 60*


0065-3233/02 $35.00

are 50-fold elevated over controls, resulting in hepatic, neurologic, or psychiatric dysfunction (Brewer *et al.*, 1987).

All species exhibit a dynamic range in copper levels that separates negative copper balance from toxicity. Hepatic copper levels can reach a 10- to 30-fold elevation in presymptomatic Wilson's patients without apparent liver dysfunction (Scheinberg and Sternlieb, 1968). The dynamic range is species specific (Winge and Mehra, 1990). Average hepatic copper levels in sheep and goats are nearly 10 times higher than in the normal human adult. The tissue copper concentration separating copper balance from toxicosis is dicated largely by the available homeostatic mechanisms and detoxification systems in a given species. The variety of these systems and their regulation determine the threshold range of cytotoxicity.

Copper homeostasis involves regulation of absorption, tissue distribution, and excretion of the metal ion. In addition, all species have detoxification systems that minimize copper-induced toxicosis. Two common detoxification mechanisms involve the sequestration of copper ions by metallothionein-type proteins and facilitated efflux by cation-exporting P-type ATPases. Some species, such as *Saccharomyces cerevisiae*, achieve Cu tolerance primarily through regulated biosynthesis of metallothioneins (Fogel and Welch, 1982; Thiele, 1988). In contrast, regulated Cu efflux confers Cu tolerance to species ranging from bacteria, to pathogenic fungi (*Candida albicans*), to animals (Petris *et al.*, 1996; Solioz and Odermatt, 1995; Weissman *et al.*, 2000).

Sensory mechanisms to detect deviations from the optimal Cu range exist, although the mechanism of Cu ion sensing is unresolved. The sensory mechanism may be one of the following possibilities. First, sensory systems may detect variations in a free or reactive Cu ion pool as the control mechanism. One argument against the model of a reactive Cu pool is the calculations showing the lack of any significant quantities of free Cu in yeast cells (Rae *et al.*, 1999). These studies do not preclude the possibility of a highly transient and localized labile Cu pool forming after Cu uptake across the plasma membrane. Sensing of a labile Cu pool requires precise control of metal ion binding by a sensory molecule to minimize signaling by competing metal ions. It is important that regulation of Cu levels does not affect the nutritional status of other essential metal ions. Second, the Cu population of a binding site on a copper metalloenzyme may be sensory. Variations in the pool size of the enzyme's substrate or product through changes in the fraction of active enzyme may be a sensory cue. Third, the flux of Cu ions transported across a membrane barrier may be a homeostatic signal. High- and low-affinity Cu ion permeases exist, so one possibility is that cells sense permeases engaged in transport. The signal may translate into a physiological response through either switching or signal transduction.

## II. Copper Metalloregulation in Prokaryotes

Copper ion homeostasis in prokaryotes involves Cu ion efflux and sequestration. The proteins involved in these processes are regulated in their biosynthesis by the cellular Cu ion status. The best studied bacterial Cu metalloregulation system is found in the gram-positive bacterium *Enterococcus hirae*. Cellular Cu levels in this bacterium control the expression of two P-type ATPases critical for Cu homeostasis (Odermatt and Solioz, 1995). The CopA ATPase functions in Cu ion uptake, whereas the CopB ATPase is a Cu(I) efflux pump (Solioz and Odermatt, 1995). The biosynthesis of both ATPases is regulated by a Cu-responsive transcription factor, CopY (Harrison *et al.*, 2000). In low ambient Cu levels Cop Y represses transcription of the two ATPase genes. On exposure to Cu(I), CopY dissociates from promoter/operator sites on DNA with a $K_d$ for Cu of 20 $\mu$M (Strausak and Solioz, 1997). Transcription of *copA* and *copB* proceeds after dissociation of CuCopY. The only other metal ions that induce CopY dissociation from DNA *in vitro* are Ag(I) and Cd(II), although the *in vivo* activation of *copA* and *copB* is specific to Cu salts. The CuCopY complex is dimeric with two Cu(I) ions binding per monomer (C. T. Dameron, personal communication). The structural basis for the Cu-induced dissociation of CopY is unknown. Curiously, CopY is also activated in Cu-deficient cells, but the mechanism is distinct from the described Cu-induced dissociation from DNA (Wunderli-Ye and Solioz, 1999).

Cu signaling to CopY requires the function of CopZ, a 69-residue protein homologous to the Atx1 family of Cu metallochaperones (Lin *et al.*, 1997; Pufahl *et al.*, 1997). Atx1 is a Cu metallochaperone that shuttles Cu(I) ions to a P-type ATPase in post-Golgi vesicles in yeast (O'Halloran and Culotta, 2000). The Cu(I)-binding CopZ routes Cu(I) to CopY in Cu-treated *En. hirae* cells (Cobine *et al.*, 1999). In the absence of CopZ, transcription of *copB* and *copA* is repressed due to persistent binding of the CopY transcriptional repressor. The CopZ/Atx1 metallochaperones have a conserved structure with a single digonally bent Cu(I) site with two thiolate ligands arising from a conserved Cys-x-x-Cys sequence motif that is also found in most Cu-specific ATPase effluxers (Rosenzweig *et al.*, 1999; Wimmer *et al.*, 1999). Figure 1 shows a structural comparison of the apo-CopZ molecule and the yeast Atx1 conformer. The apo-CopZ conformer has the two Cys ligands oriented away from each other, indicating that a limited structural rearrangement of the Cys-x-x-Cys loop is necessary for Cu(I) binding. The linear binding arrangement of the Hg(II) to the two Cys residues in Atx1 is clearly seen in Fig. 1. Both structures have a four-stranded antiparallel $\beta$ sheet covered by two $\alpha$ helices (Rosenzweig *et al.*, 1999; Wimmer *et al.*, 1999).

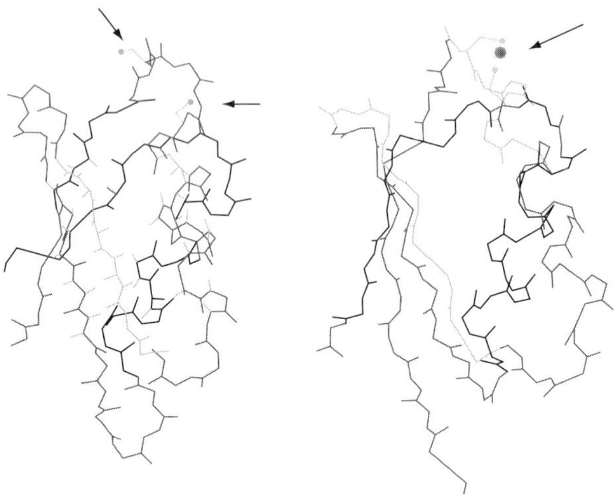


FIG. 1. Comparison of structures of the apo-CopZ and Hg-Atx1 metallochaperones from *Enterococcus hirae* and *Saccharomyces cerevisiae*, respectively (Rosenzweig *et al.*, 1999; Wimmer *et al.*, 1999). In the HgAtx1 structure the Hg(II) atom (shown as a dark ball) is ligated by two cysteines (the sulfurs in the side chains are shown as smaller balls). The coordination of the Hg(II) is linear; a similar coordination geometry is expected for Cu(I). In the CopZ structure the two corresponding cysteinyl residues shown by arrows are not in the proper orientation to ligate Cu(I). A limited structural rearrangement is expected in the loop to permit linear coordination as seen in Hg-Atx1.

Intriguing questions persist with regard to the *En. hirae* system. First, it is unclear whether CopZ senses a transient rise in a cytoplasmic Cu pool or is metallated directly by interaction with a permease. Presumably, CopZ is metallated only when some cellular threshold Cu level is exceeded. At lower cellular Cu levels, other Cu metallochaperones presumably target Cu ions to sites of Cu metalloenzyme biosynthesis and folding. Second, it is unclear whether CopZ functions as a metallochaperone only for CopY or, alternatively, participates in Cu(I) ion delivery to the CopB Cu(I) efflux pump. Third, it is perplexing why copper activates expression of both an efflux pump and a permease. Additional levels of regulation may exist to modulate the function of the two ATPase pumps. Fourth, little is known of how signal transduction pathways are turned off. If the CuCopZ and CuCopY complexes are of either low stability of low affinity, *copA/copB* transcription would cease when the signaling Cu pool was reduced below some threshold level. Efflux of Cu(I) by the CopB efflux pump may lower the cytosolic Cu ion concentration below the $K_d$ of CopY.

A series of chromosomal and plasmid genes in *Escherichia coli* encode proteins involved in cellular resistance to copper ions. *Es. coli* limits intracellular Cu levels through energy-dependent Cu export in log-phase cells, but sequesters Cu ions in stationary-phase cells (Brown *et al.*, 1995). Both plasmid and chromosomal genes control Cu efflux. The chromosomal CopA is a Cu-inducible P-type ATPase that functions in Cu efflux (Rensing *et al.*, 2000). Disruption of the *copA* locus results in hypersensitivity of cells to copper salts. Cu-induced expression of CopA is regulated by CueR (Outten *et al.*, 2000). CueR is a homologue of the Hg-responsive transcription factor merR. Deletion of the *Es. coli* gene encoding CueR, *ybbI*, abrogated Cu induction of *copA* (Outten *et al.*, 2000). The candidate DNA-binding domain of CueR contains a helix-turn-helix motif. The C-terminal segment of CueR, which may function as a Cu(I)-binding domain, contains two Cys residues separated by seven residues. CueR also mediates Cu induction of *yacK*, which encodes a putative multicopper oxidase. If CueR is a Cu(I) sensor, an important question is whether a metallochaperone exists to shuttle Cu(I) to CueR analogous to the CopZ of *En. hirae*.

The bacterial P-type ATPases resemble well-known cation pumps that translocate cations against their electrochemical potential gradient by using the energy from hydrolysis of ATP (Lutsenko and Kaplan, 1995). During the reaction cycle a conserved Asp residue is phosphorylated. A prototype of the P-type ATPase is the sarcoplasmic reticulum Ca(II) ATPase that functions in translocating Ca(II) ions from the cytoplasm into the sarcoplasmic reticulum (MacLennan *et al.*, 1997). Clues to the mechanism of action of copper-ATPases come from inspection of the known Ca(II) ATPase. The structure of the Ca(II) ATPase was recently solved at 2.6 Å (Toyoshima *et al.*, 2000). The 994-residue protein exists with a cytoplasmic headpiece consisting of three separate domains and a transmembrane segment consisting of 10 helices (Toyoshima *et al.*, 2000). Two Ca(II) ions are bound within the transmembrane region by six oxygen atoms per Ca(II) ion (Fig. 2). One cytoplasmic headpiece domain (domain N) is the ATP-binding domain. A second domain (domain P) contains the Asp residue phosphorylated during the reaction cycle. The third headpiece domain (domain A) is a small jellyroll structure projected to move during active transport. Two of the helices in the transmembrane region are unwound in a region where Ca(II) binds. The coordination geometry of the Ca(II) site requires unwinding of the helix (Toyoshima *et al.*, 2000). Since the phosphorylated Asp is $>$ 25 Å away from the site of ATP binding, domain closure must occur during ATP hydrolysis. Ca(II) binding to the two sites within the transmembrane region appears to initiate the domain closure that is completed on phosphorylation, and domain closure is coupled to Ca(II) translocation (MacLennan *et al.*,

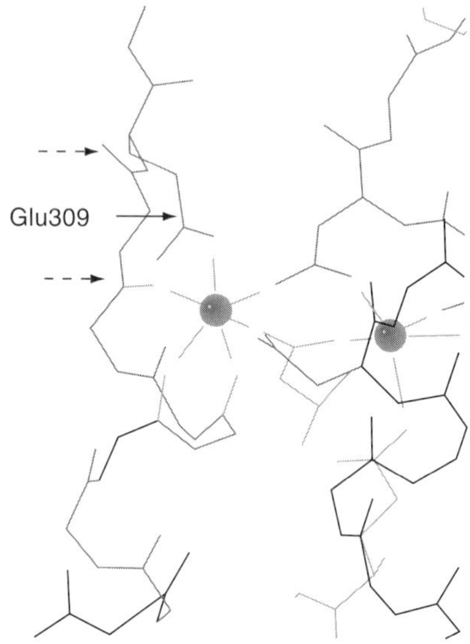


FIG. 2. View of the two Ca(II) ions bound to transmembrane helices in the Ca(II) P-type ATPase (Toyoshima *et al.*, 2000). The Ca(II) on the left in the picture is site II. The transmembrane helix M4 is visible on the left. Ca(II) in site II is coordinated by Glu-309 (solid arrow) and carbonyl oxygens of residues 304, 305, and 307. These carbonyl oxygens are shown in light gray. In order for Ca(II) to be bound in this site using carbonyl oxygens the helix must be partially unwound. Notice how the carbonyl oxygen dipoles point up toward the bottom of the helix and this pattern changes at the carbonyl oxygens near Glu-309 (dotted arrows). In the Cu P-type ATPases, a CysProCys motif exists at an equivalent position to Glu-309. These two Cys residues are expected to coordinate Cu(I) ions prior to extrusion across the bilayer.

1997). The residues within the unwound segment of transmembrane helix M4 form part of the Ca(II) site 2 (Fig. 2). The Cu P-type ATPases have a conserved CysProCys sequence in the corresponding segment. Thus, one clear prediction is that Cu(I) binding by the CysProCys motif initiates the reaction cycle of ATP hydrolysis and cytoplasmic channel closure. Cu(I) binding by Cys-x-Cys sequence motifs usually occurs in combination with other Cys-x-Cys motifs to create a polycopper cluster. In the Cu P-type ATPases Cu(I) binding by the Cys-x-Cys motif is only transient so a full ligand field is not expected. Other features of the catalytic cycle are likely to resemble those of the Ca(II) ATPase. One distinction between the two types of ATPases is the presence of Cu(I)-binding modules at the N-terminal region of Cu ATPases. The *Es. coli* Cu ATPase contains two N-terminal Cu(I)-binding modules with a conserved Cys-x-x-Cys sequence. These modules structurally resemble the Atx1 metallochaperone motif shown in Fig. 1 (Rosenzweig *et al.*, 1999). A single Cu(I) ion binds to the Atx1 motif (Rosenzweig and O'Halloran, 2000). The Atx1 motifs in the P-type ATPases appear to function as docking sites for a specific metallochaperone prior to transfer of Cu(I) ions to the ATPase (O'Halloran and Culotta, 2000). Cu(I) binding within the Atx1 motifs of an ATPase may be an initial metallation event prior to transfer of the Cu(I) ion to the conserved CysProCys motif within the transmembrane region (Huffman and O'Halloran, 2000). Other bacterial P-type ATPases that function in Cu homeostasis have been identified in *Listeria* (Francis and Thomas, 1997).

## III. Copper Metalloregulation in Eukaryotes

### *A. Metalloregulation of Genes Involved in Nutritional Responses*

Cells in the natural world encounter a changing environment with fluctuations in nutrient concentrations. Cell survival requires physiological responses to such changes. Nutritional responses to Cu limitation are known in several eukaryotic organisms. Cu-dependent regulation of alternate metalloproteins exists in green algae. The availability of copper in the unicellular alga *Chlamydomonas reinhardtii* is the determining factor of whether cells synthesize the Cu-containing plastocyanin or the heme-containing cytochrome *c*6 to mediate electron transfer in the reaction center of photosystem I (Merchant *et al.*, 1991). If Cu ions are available, plastocyanin is the preferred molecule synthesized. Cu-deficient conditions result in the turnover of apoplastocyanin and transcriptional activation of cytochrome *c*6. In addition, Cu-deficient conditions lead to the up-regulation of coprophyrinogen oxidase transcription and genes encoding a Cu uptake system (Hill *et al.*, 1996). The Cu uptake system induced by Cu-deficient conditions is a high-affinity system that appears to contain a cupric reductase (Hill *et al.*, 1996). Cu-responsive promoter elements exist in the 5′ sequences of cytochrome *c*6 and coproporphyrinogen oxidase, forming binding sites for an undefined transcriptional activator (Quinn *et al.*, 2000; Quinn and Merchant, 1995). The sequences containing the copper-responsive element do not resemble known copper-responsive elements found in yeast. The response to added Cu is rapid, with a halftime response of less than 10 min. The evidence is consistent with a homeostatic mechanism in *Ch. reinhardtii* involving Cu metalloregulation of a transcriptional activator.

The yeast *Sa. cerevisiae* presents the most complete picture of copper nutritional regulation in a eukaryote. Copper ions are required for at least three key enzymes in the yeast. The ability of cells to grow on nonfermentable carbon sources is dependent on having an active cytochrome oxidase complex that requires Cu ions as cofactors. Oxidative growth requires defense molecules against reactive oxygen intermediates. Superoxide dismutase is a Cu metalloenzyme that dismutes superoxide anions. A third, key Cu metalloenzyme is Fet3, which is a ferrooxidase critical for uptake of Fe(II) (Askwith *et al.*, 1996). A myriad of other oxidases and oxygenases require Cu(II) as a functional cofactor in other species, so additional Cu metalloenzymes may exist in *Sa. cerevisiae*. The essentialness of Cu for normal physiology is consistent with the fact that yeast possess mechanisms to ensure a positive copper balance.

Copper homeostasis is achieved, in part, through Cu-regulated expression of genes involved in copper ion uptake. Conditions of copper deficiency in *Sa. cerevisiae* result in the derepression of three genes whose products are involved in cellular uptake of copper ions (Dancis *et al.*, 1994; Georgatsou *et al.*, 1997; Hassett and Kosman, 1995; Labbe *et al.*, 1997; Yamaguchi Iwai *et al.*, 1997). Two of the three genes encode high-affinity, plasma-membrane, Cu permeases, Ctr1 and Ctr3 (Dancis *et al.*, 1994; Labbe *et al.*, 1997; Pena *et al.*, 2000) (Fig. 3). Ctr1 and Ctr3 are functionally redundant in that a Cu-deficient state arises only when both

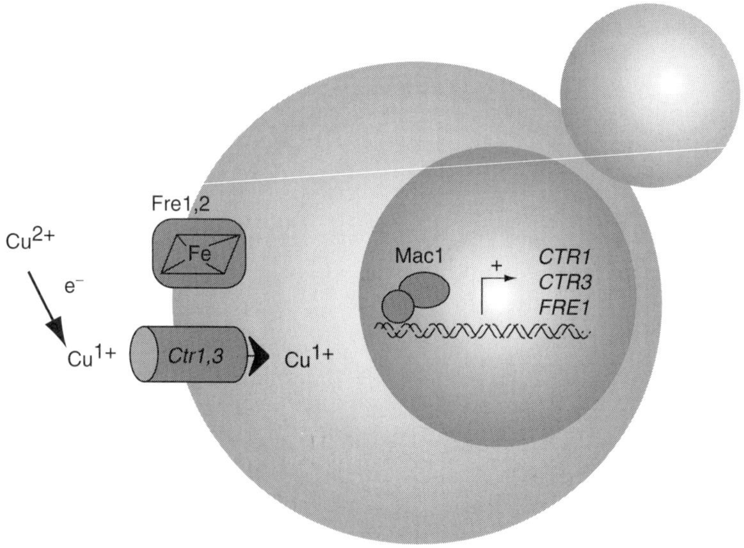


FIG. 3. Mac1 activates the expression of three gene products involved in high-affinity copper ion uptake in *Saccharomyces cerevisiae*. Two genes encode Cu ion permeases Ctr1 and Ctr3. The third is one of two metalloreductases that reduce Cu(II) ions prior to uptake. Mac1 is a transcriptional activator in Cu-deficient yeast cells.

permeases are nonfunctional (Knight *et al.*, 1996). Ctr1 is oligomeric with three candidate transmembrane segments per subunit (Dancis *et al.*, 1994). The putative ectodomain has multiple Met-x-x-Met sequence motifs that may function as low-affinity Cu(II) or Cu(I) scavenging modules gathering extracellular Cu ions for subsequent transport. Uptake across the lipid bilayer may be facilitated, in part, by binding of Cu ions to two Cys-x-Cys sequence motifs within the candidate cytoplasmic domain. Cu uptake in yeast is energy dependent, although the mechanism of energy-coupled transport is unresolved (Lin and Kosman, 1990). *CTR3* is not expressed in most laboratory yeast strains due to the insertion of a transposable element (Knight *et al.*, 1996). Ctr3 (241 residues) is smaller than Ctr1 (406 residues) and exhibits only limited similarity to Ctr1. Ctr3 has three transmembrane segments per monomer and exists in an oligomeric state, most likely as a trimeric complex (Pena *et al.*, 2000).

Cu ion uptake by the high-affinity Ctr1/Ctr3 system is facilitated by the metalloreductase Fre1. Fre1 is a flavocytochrome-containing NADPH oxidase that pumps a diffusible reductant into the growth medium to mobilize oxidized Cu(II) complexes (Finegold *et al.*, 1996; Lesuisse *et al.*, 1996; Shatwell *et al.*, 1996). The regulation of *FRE1* is more complex than that of *CTR1*. *FRE1* is up-regulated in Cu-deficient cells (Georgatsou *et al.*, 1997; Hassett and Kosman, 1995; Lesuisse *et al.*, 1996), but is also expressed in iron-deficient cells through a distinct mechanism (Dancis *et al.*, 1990; Georgatsou and Alexandraki, 1994). A second metalloreductase, Fre2, functions in both copper and iron ion uptake, but *FRE2* is actively expressed only under iron-limiting conditions (Dancis *et al.*, 1990; Georgatsou and Alexandraki, 1994; Georgatsou *et al.*, 1997; Shatwell *et al.*, 1996). The significance of cell surface metalloreductases in copper and iron homeostasis is that Cu and especially Fe ions in the environment are largely present in insoluble, oxidized valent states. The bioavailability of these ions is increased via reduction.

Three additional genes are up-regulated in Cu-deficient yeast, but the functions of these molecules remain to be elucidated. These genes include *FRE7* and two ORFs, YFR055w and YJL217w (Gross *et al.*, 2000). Fre7 exhibits sequence similarity to Fre1 in regions expected to bind FAD, NADPH, and the two hemes (Martins *et al.*, 1997). Thus, Fre7 is expected to be a NADPH oxidase, although its cellular localization has not been defined. YFR055w and YJL217w were recently identified by genomics transcript profiling using microarray technology (Gross *et al.*, 2000). YFR055w belongs to a family of transsulfuration enzymes which includes yeast and rat cystathionine γ-lyase, yeast homocysteine synthase, and *Es. coli* cystathionine γ synthase. YFR055w is 28% identical to Cys3, which generates cysteine from cystathionine, and 26% identical to Met17, which converts *O*-acetylhomoserine into homocysteine. Cys3 is the major cystathionine γ-lyase important in cysteine biosynthesis in yeast (Ono *et al.*, 1999). YFR055w may be one of several cystathionine γ-lyase isozymes in *Sa. cerevisiae* that generates cysteine from cystathionine under Cu-limiting conditions. YJL217w, on the other hand, has no known function and exhibits no homology to known genes that may provide a clue as to its physiological function. Since the three known genes activated in Cu-deficient cells function in Cu ion uptake, the prediction is that the newly identified *FRE7*, YFR055w, and YJL217w gene products will likewise be important in cellular Cu ion acquisition or utilization under Cu-deficient conditions.

The expression of these six genes is a cellular response to inadequate, intracellular Cu levels. Cells shifted to Cu-limiting conditions [$< 1$ nM Cu(II)] exhibit a rapid derepression of *CTR3* expression with maximal response occurring within 10 min (Labbe et al., 1997). In contrast, expression of these genes is attenuated in cells cultured in medium containing greater than 1 nM Cu(II) (Dancis *et al.*, 1994; Hassett and Kosman, 1995; Labbe *et al.*, 1997). Cells lacking the two permeases gain greater resistance to copper toxicity (Dancis *et al.*, 1994), suggesting that the permeases remain partially functional in Cu-replete cells. Cu ion uptake can also occur through low-affinity permeases including Fet4, Ctr2, and Smf1 (Dix *et al.*, 1997; Kampfenkel *et al.*, 1995; Liu *et al.*, 1997).

One cellular response to sufficient intracellular copper is the mentioned repression of high-affinity Cu uptake. A second response is the Cu-dependent removal of the Ctr1 transporter in the plasma membrane (Ooi *et al.*, 1996). Immunofluorescence analyses of epitope-tagged Ctr1 revealed Cu-induced internalization of Ctr1 from the cell surface (Ooi *et al.*, 1996; Pena *et al.*, 2000). In addition, Western analysis revealed a Cu-induced degradation of Ctr1 (Ooi *et al.*, 1996). Cu-induced Ctr1 degradation occurs independent of internalization and requires a higher exogenous Cu level [10 μM Cu(II)] than that required for inhibition of *CTR1* expression [1 nM Cu(II)] (Ooi *et al.*, 1996). The mechanism of the apparent Cu metalloregulation of Ctr1 degradation remains unresolved. Ctr3 differs from Ctr1 in not undergoing the same Cu-induced degradation or change in localization (Pena *et al.*, 2000).

### *1. Transcriptional Regulation of the Copper Nutritional Response*

Genes transcriptionally expressed under Cu-deficient conditions in *Sa. cerevisiae* are regulated by the transcription factor, Mac1 (metal-binding activator 1) (Georgatsou *et al.*, 1997; Hassett and Kosman, 1995; Jungmann *et al.*, 1993; Labbe *et al.*, 1997; Yamaguchi-Iwai *et al.*, 1997). *MAC1* was originally identified as a partially dominant *MAC1* mutation, $MAC1^{up1}$ (Jungmann *et al.*, 1993). Cells harboring the $MAC1^{up1}$ allele are immune to Cu-induced repression of the copper regulon and exhibit elevated metalloreductase activity and Cu uptake rates (Hassett and Kosman, 1995; Yamaguchi-Iwai *et al.*, 1997). As a result, *MAC1*$^{up1}$ cells are hypersensitive to copper salts added to the growth medium (Jungmann *et al.*, 1993). The copper sensitivity of *MAC1*$^{up1}$ cells demonstrates the importance of down-regulating the high-affinity uptake system in Cu-replete cells. In contrast, a frameshift mutation in *MAC1*, designated *mac1-1*, results in substantial loss of both Cu(II) and Fe(III) reduction and loss of Cu ion uptake (Hassett and Kosman, 1995; Jungmann *et al.*, 1993). These cells have very low transcript levels of *CTR1*, *CTR3*, and *FRE1* (Labbe *et al.*, 1997; Yamaguchi-Iwai *et al.*, 1997). The phenotypes of *mac1-1* cells, including respiratory deficiency and sensitivity to a myriad of stresses, are reversed on addition of exogenous copper salts. These phenotypes and the reduced Cu uptake rate are consistent with a Cu-deficient state in *mac1-1* cells (Jungmann *et al.*, 1993).

Mac1-activated genes contain a conserved element [TTTGC(T,G)CA] repeated in the 5′ promoter regions (Labbe *et al.*, 1997; Martins *et al.*, 1997; Yamaguchi-Iwai *et al.*, 1997). Mac1-mediated transcription requires two functional elements (Labbe *et al.*, 1997; Martins *et al.*, 1997; Yamaguchi-Iwai *et al.*, 1997). The binding sites, designated copper-regulatory elements or CuRE (Labbe *et al.*, 1997), can exist as inverted or direct repeats with variable spacing between the elements ranging from 7 to 40 bp (Labbe *et al.*, 1997; Martins *et al.*, 1997; Yamaguchi-Iwai *et al.*, 1997). Expansion of the spacing between the two elements upstream of *CTR1* revealed attenuated expression (Martins *et al.*, 1997). The Mac1-responsive ORF YFR055w has two inverted candidate Mac1 sites separated by only 3 bp. YJL217w has one perfect consensus site, but additional nonconsensus sites may be functional (Gross *et al.*, 2000).

A number of genes differentially expressed in *MAC1*$^{up1}$ cells appear not to be direct targets of Mac1. Induction or repression of these genes is likely a secondary effect of physiological changes in cells containing a constitutively active Mac1. Genes up-regulated in *MAC1*$^{up1}$ cells include *CUP1* and phosphate regulon genes (Gross *et al.*, 2000). *CUP1* expression is likely enhanced by virtue of constitutive expression of the high-affinity copper transport system, since elevated copper transport will stimulate *CUP1* expression through Cu activation of a second transcription factor, designated Ace 1. Phosphate regulon, including *PHO5*, *PHO11*, *PHO12*, and *PHO84*, is derepressed presumably by virtue of phosphate deprivation in *MAC1*$^{up1}$ cells. The apparent phosphate deprivation occurring in *MAC1*$^{up1}$ cells is not obviously explained based on the function of known Mac1-regulated genes. Genes down-regulated in *MAC1*$^{up1}$ cells include *ZRT1* and *ZRT2*, which encode Zn ion plasma membrane transporters (Gross *et al.*, 2000). These genes are down-regulated in Zn-replete cells, suggesting that $MAC1^{up1}$ cells may either preferentially accumulate Zn(II) or contain an elevated nonsequestered Zn(II) pool. Clearly, Cu-induced changes occur in Zn pools in $MAC1^{up1}$ cells.

*2. Structural Dissection of Mac 1*

Mac 1 is a typical fungal transcriptional activator protein with two separate functional domains, an N-terminal DNA-binding domain and a C-terminal transactivator consisting of two Cys-rich sequence motifs (Fig. 4). The minimal DNA-binding domain maps to the N-terminal 159 residues, which binds two Zn(II) ions (Jensen *et al.*, 1998). A segment of the Mac1 DNA-binding domain (residues 1–40) is homologous to a conserved Zn module found in the Cu-activated transcription factors Ace 1 and Amt 1 found in *Sa. cerevisiae* and *Ca. glabrata*, respectively (Jungmann *et al.*, 1993; Thiele, 1988; Welch *et al.*, 1989; Zhou and Thiele, 1991) (Fig. 5). The Zn module from Amt1 is distinct from previously characterized zinc finger motifs and consists of a three-stranded antiparallel β sheet with two short helical turns that project from one end of the sheet followed by a conserved short unstructured motif with a consensus sequence of (R/K) GRP (Turner *et al.*, 1998) (Fig. 6). A single Zn(II) ion is tetrahedrally coordinated in the motif by three thiolates and a single histidyl residue with the ligands spaced in a C-$x_2$-C-$x_8$-C-x-H sequence (x is any residue) (Posewitz *et al.*, 1996). The (R/K) GRP motif adjacent to the Zn module is homologous to a minor groove DNA-binding motif

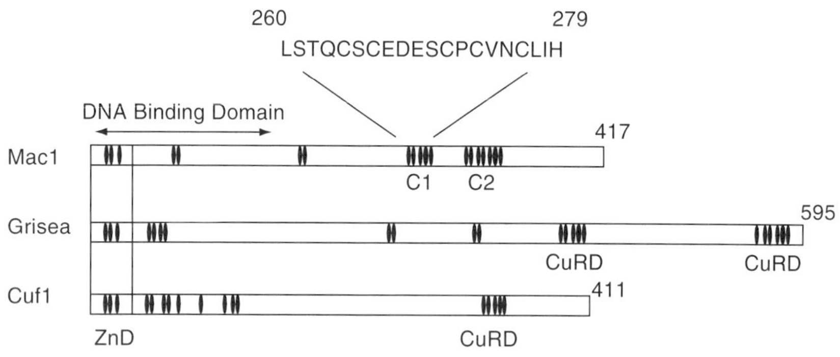


FIG. 4. Comparison of Mac1 and Mac1 orthologues from *Podospora anserina* (Grisea) and *Schizosaccharomyces pombe* (Cuf1). All contain a conserved N-terminal 40-residue Zn(II) module that constitutes part of the DNA-binding interface. Two Cys-rich motifs in the C-terminal segment of Mac1, designated C1 and C2, are conserved in Grisea and Cuf1. The C1 and C2 motifs are part of transactivation domains. The C1 motif is a functional Cu-regulatory domain in Mac1. The sequence of the C1 motif is shown at the top. The dark ovals represent the positions of cysteinyl residues in each molecule. Each Cys-rich motif in Mac1 binds four Cu(I) ions.

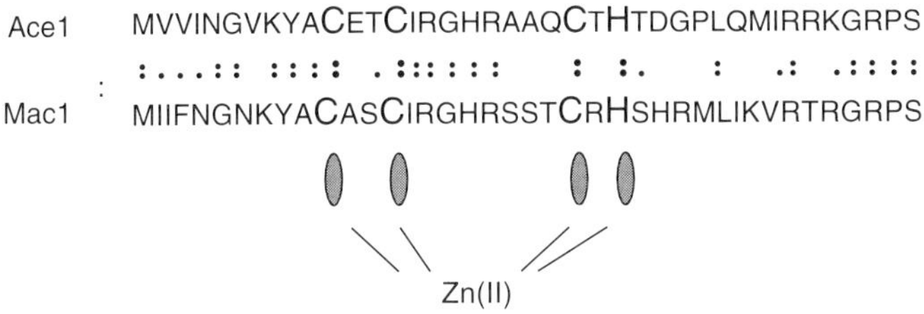


FIG. 5. Sequence comparison of the Zn modules in Ace 1 and Mac 1. The four residues involved in Zn(II) ligation are shown in large font and are highlighted by filled ovals below the sequence.

found in various nuclear proteins from animals, plants, insects, yeast, and bacteria (Bustin and Reeves, 1996; Geierstanger *et al.*, 1994). One well-characterized RGRP motif involved in minor groove DNA binding is the high mobility group (HMG) protein I(Y) (Bustin and Reeves, 1996). The motif, designated the AT-hook, binds A/T-rich DNA sequences. The Zn module in Mac1 is one of two domains involved in DNA binding. DNA binding requires an additional 110 residues that also bind a single Zn(II) ion, although the ligands for the second site have not been identified.

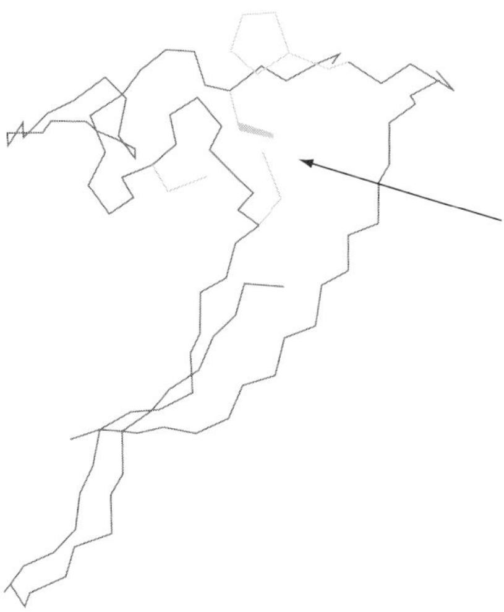

FIG. 6. Structure of the Zn module of Amt1 (Turner *et al.*, 1998). Side chains of the four residues that coordinate Zn(II) are shown in light gray. The arrow points to their position. The module forms an L-shaped domain. The bottom segment of the L consists of three β strands. The perpendicular segment contains the bound Zn(II), and the RGHR motif predicted to be important in major groove DNA binding in Ace1 and Mac1 projects off this face.

Mac1 binds to the CuREs in a modular fashion with the RGRP AT-hook motif interacting in the minor groove of the TTT sequence and a second Mac1 domain making major groove base contacts in the GC(T/G)C sequence (Jamison McDaniels *et al.*, 1999). The RGRP motif found in HMG-I(Y) inserts as an extended conformation deep within the minor groove of A/T-rich sequences (Aravind and Landsman, 1998; Huth *et al.*, 1997). When bound within the minor groove, the RGR tripeptide adopts a concave surface packed tightly against the bases within the groove (Fig. 7). The Arg side chains are oriented parallel to the minor groove and extend away from the central basepair (Huth *et al.*, 1997). Interactions made by the arginines are important in conferring sequence-specific binding. The proline in the motif directs the peptide backbone away from the minor groove. The RGRP motif in Mac1 is predicted to make contacts within the TTT minor groove of the CuRE similar to those of an AT-hook motif. One curious aspect of the Mac1 contact with the CuRE is

FIG. 7. The conserved Zn module in Ace 1 and Mac 1 contains an RGRP motif that resembles the AT-hook domain of HMG-I(Y). In the HMG-I protein the RGR sequence adopts an extended structure that is buried deeply within the A/T minor groove (Huth *et al.*, 1997). The two Arg residues of the motif are shown in spacefill as light gray and the Gly residue is shown as a dark spacefill.

the preferential strong protection Mac1 imparts to the TTT strand DNA backbone relative to the AAA strand (Jamison McDaniels *et al.*, 1999). The prediction is that in addition to the RGRP minor groove contacts and the major groove contacts within the GC(T/G)C sequence, a domain of Mac1 makes extensive backbone contacts on the TTT strand. The affinity of Mac1 for a CuRE is enhanced by an additional TA dinucleotide sequence upstream of the TTT motif (Joshi *et al.*, 1999).

The transactivator motif in Mac1 lies within two repeated C-terminal Cys-rich motifs with a Cys-x-Cys-$x_4$-Cys-x-Cys-$x_2$-Cys-$x_2$-His consensus sequence (Graden and Winge, 1997; Zhu *et al.*, 1998) (Fig. 4). Transactivation domains function in the assembly of the transcription preinitiation complex involving the TATA element-binding complex, the mediator complex, and the RNA polymerase. The gain-of-function mutation in $MAC1^{up1}$ is a single T-A transversion resulting in a $His_{279}Gln$ substitution in the first of the two cysteine-rich motifs, designated C1. The Cys-rich motifs are related to Cu(I)-binding sequences found in metallothioneins and the Cu-activated transcription factors Ace1 and Amt1.

Cu inhibition of Mac1 function may arise from Cu-mediated proteolysis of Mac1, Cu-dependent nuclear export of Mac1, or inhibition of DNA-binding or transactivation activities. Cu-mediated proteolysis does occur with Mac1, but a higher Cu concentration is necessary to induce proteolysis than is necessary to inhibit Mac1 function (Zhu *et al.*, 1998). The nuclear localization of Mac1 does not change depending on the copper status of yeast cells (Jensen and Winge, 1998; Jungmann *et al.*, 1993). Cu inhibition of Mac1 function involves both Cu-dependent loss of *in vivo* DNA-binding activity (Labbe *et al.*, 1997) and Cu-dependent inhibition of transactivation function (Georgatsou *et al.*, 1997; Graden and Winge, 1997). *In vivo* footprinting studies of *CTR3* revealed that the CuRE is bound by Mac1 in Cu starvation but not Cu-replete cells (Labbe *et al.*, 1997). Thus, DNA binding by Mac1 is inhibited in Cu-replete cells. However, no copper dependency is observed in DNA binding by the minimal DNA-binding motif in Mac1 (Jensen *et al.*, 1998). Cu inhibition of N-terminal DNA binding requires C-terminal sequences. Transactivation activity of Mac1 is also inhibited in Cu-replete cells (Georgatsou *et al.*, 1997; Graden and Winge, 1997). Cu modulation of transactivation activity requires both the C-terminal transactivation domains and a portion of the N-terminal DNA-binding domain (Graden and Winge, 1997). The prediction is that the repressed conformation of Mac1 is an intramolecular complex involving both the N-terminal and the C-terminal domains (Fig. 8). Cu(I) binding to the C-terminal Cys-rich motifs was shown to induce an intramolecular interaction with the N-terminal DNA-binding domain (Jensen and Winge, 1998). The intramolecular interaction appears to inhibit both DNA binding and transactivation.

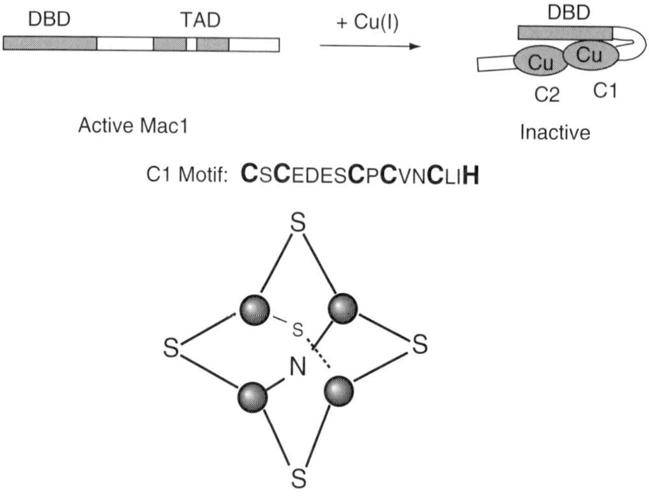


FIG. 8. Model for the Cu-induced inactivation of Mac 1. Experiments suggest that Cu(I) binding to the C1 motif induces an intramolecular interaction between the N-terminal DNA-binding domain and the C-terminal Cu-binding modules, resulting in an inactive Mac1. The Cu-regulatory domain of Mac1 is shown at the bottom with the candidate ligands shown in large font. This motif binds four Cu(I) ions, and a postulated tetracopper cluster is shown.

Mac1 was recently shown to form a dimer stabilized by a C-terminal dimerization motif (residues 388–406) (Serpe *et al.*, 1999). A deletion mutation in the dimerization element abolished Mac1 function (Serpe *et al.*, 1999). Cu inhibition of Mac1 may alter the dimeric state of Mac1. However, this cannot entirely account for the Cu inhibition of Mac1 since Cu regulation of Mac1 occurs with a minimal Mac1 molecule lacking the C-terminal dimerization motif (Jensen and Winge, 1998).

The two carboxyl-terminal Cys-rich repeats, designated C1 and C2, bind a total of eight Cu(I) ions (Jensen and Winge, 1998). Several mutagenesis studies mapped residues important in the copper inhibition of Mac1 activity. All mutations exhibiting Cu-independent Mac1 transcriptional activation cluster in the C1 Cys-rich motif consisting of residues 264–279 (Yamaguchi Iwai *et al.*, 1997; Zhu *et al.*, 1998). All but one of the constitutive Mac1 mutations occur in one of the conserved six residues in the $C_{264}$-x-C-$(x)_4$-C-x-C-$(x)_2$-C-$(x)_2$-$H_{279}$C1 motif (Fig. 8) (Keller *et al.*, 2000; Yamaguchi Iwai *et al.*, 1997; Zhu *et al.*, 1998). The lone exception is a $L_{260}$S substitution (Keller *et al.*, 2000). Two additional *MAC1* mutations exhibiting constitutive activity are in-frame deletions encompassing portions of the C1 motif (Keller *et al.*, 2000).

Engineered mutations in the second Cys-rich motif did not yield a constitutively active Mac1. These results are consistent with the C1 motif being the major Cu-regulatory switch. Both Cys-rich motifs exhibited transactivation activity, although the C1 activator was weak relative to the C2 activator (Keller *et al.*, 2000). Limited copper metalloregulation of Mac1 was observed with only the C1 activator fused to the N-terminal DNA-binding domain. Thus, the two Cys-rich motifs appear to function independently. The C1 Cys-rich motif appears to be a functional copper-regulatory domain, whereas the C2 motif is the major transactivator.

Six candidate ligands exist within the C1 motif including five Cys residues and a single His residue (Fig. 8). The prediction is, therefore, that a $Cu_4S_5N_1$ polycopper cluster forms. X-ray absorption spectroscopy of the CuC1 domain of Mac1 revealed the presence of a polycopper cluster (Winge and George, unpublished observation). The spectroscopic data are consistent with Cu(I) ions binding within the C1 domain in apparent trigonal geometry with predominantly sulfur ligands. The mean Cu–S distance of 2.26 Å is clearly consistent with trigonal geometry (Pickering *et al.*, 1993). No clear indication of histidyl Cu(I) coordination is apparent. An outer shell interaction at 2.7 Å was best fitted as a Cu–Cu scatter peak. The magnitude of the Cu–Cu scatter peak was indistinguishable from that of CuAce1, which forms a tetracopper-thiolate cluster. Thus, the prediction is that a $Cu_4$ center forms in the Cu-inhibited C1 domain. The function of metallation of the C1 motif appears to be the conformational switch that favors an intramolecular interaction between the N-terminal DNA-binding domain and the C-terminal polycopper domain. Cu(I) binding to the C1 motif is abrogated in $Mac1^{up1}$. Since the C2 domain contains a similar Cys-rich repeat, a polycopper cluster likely forms in this domain as well. It is unclear whether or how metallation of the C2 motif modulates Mac1 function.

*3. Mechanism of Mac1 Copper Sensing*

Mac1 is the nutritional Cu sensor that regulates the expression of the high-affinity copper uptake system in fungi. The function of Mac1 as a transcriptional activator is inhibited in copper-replete cells. The ability of Mac1 to bind Cu(I) with high affinity suggests that Mac1 is a direct Cu sensor within the nucleus. Since the nuclear localization of Mac1 does not change with the perturbations in the copper status of cells, the regulation of Mac1 function appears to arise from differential metallation of Mac1. Based on the CopZ/CopY story in *En. hirae*, a prediction is that a metallochaperone exists to transit Cu to the nucleus for Mac1 binding. Yeast have metallochaperones for Cu delivery to post-Golgi vesicles, to mitochondria, and to superoxide dismutase (Harrison *et al.*, 2000; O'Halloran and Culotta, 2000; Rae *et al.*, 1999). No nuclear Cu metallochaperones have been identified to date. Mac1 is activated only under conditions of transient copper deprivation. Yeast cultured in standard laboratory medium are predicted to contain an inactive Mac1 in which both its DNA-binding and transactivation activities are Cu inhibited. The polycopper clusters formed within the repressed conformer of Mac1 are expected to form in an all-or-nothing manner analogous to the cooperative polycopper cluster formation in CuCup1. Alternatively, tetracopper cluster formation in the C1 domain may occur by stepwise population of sites yielding molecules with increasing Cu nuclearities from 1 to 4. Since Cu inhibition of Mac1 likely requires formation of the full tetracopper center, a Cu ion concentration insufficient to fully inhibit all Mac1 molecules may have one of two results. If cluster formation is all-or-nothing, then the low Cu concentration would result in a subpopulation of fully inhibited Mac1 molecules within a population of Cu-free Mac1 molecules. In contrast, stepwise formation of the cluster may not result in Mac1 inhibition until a sufficient Cu pool exists to form a cluster in most Mac1 molecules. Therefore, only cooperative cluster formation would yield a graded response in Mac1 function with respect to the cellular copper status. The extent of Mac1 inhibition may correlate with the Cu concentration ferried to the nucleus by a candidate metallochaperone.

The Cu signaling pathway in yeast remains unclear. Mac1 is activated in yeast cultured in standard laboratory medium within 10 min of the addition of a Cu(I) selective chelator, bathocuproine sulfonate (BCS) (Labbe *et al.*, 1997; Pena *et al.*, 1999). Short incubations of cells with BCS do not generate Cu-deficient yeast. The total Cu content of cells remains unchanged, yet the activation of Mac1 results in up-regulation of the high-affinity Cu uptake system. It is unclear what Mac1 is sensing. If activation occurs by dissociation of Mac1-bound Cu(I) ions, the prediction is that the inactive CuMac1 must exhibit a high Cu exchange rate to account for the rapid kinetics of Mac1 activation. One model of Cu signaling is that a specific metallochaperone senses changes in a highly transient and localized Cu pool formed subsequent to Cu uptake across the plasma membrane. Cu metallation of the metallochaperone results in translocation of the complex to the nucleus and subsequent metallation of Mac1. The CuMac1 complex is transcriptionally inactive. If a copper-deficient environment is encountered, the Cu flux across the membrane results in a transient diminution in the labile Cu pool that metallates the nuclear metallochaperone. Failure to metallate the metallochaperone results in interrupted nuclear Cu translocation. If the CuMac1 complex has only a transient stability, dissociation of Cu(I) ions from Mac1 would activate the molecule. The proportion of apochaperone molecules would correlate with the extent of active Mac1 molecules. In this manner, expression of the high-affinity Cu uptake system may be tied to the Cu flux into the cells.

### B. *Copper Nutritional Responses in Other Species*

Copper ion nutritional regulation occurs in other fungi and in plants. Mac1 orthologues exist in *Schizosaccharomyces pombe* and *Podospora anserina*. The Mac1 orthologue in *Sc. pombe* is Cuf1 (Labbe *et al.*, 1999). Cuf1 resembles Mac1 in containing an N-terminal Zn module and a single C-terminal Cys-rich motif (Fig. 4). A functional Cuf1 is necessary in *Sc. pombe* to provide adequate cellular copper levels for metalloenzymes (Labbe *et al.*, 1999). Cuf1 activates the expression of *CTR4*, which encodes a copper permease (Labbe *et al.*, 1999). Ctr4 appears to be a hybrid of the *Sa. cerevisiae* Ctr1 and Ctr3 permeases (Labbe *et al.*, 1999). Cuf1 is localized within the nucleus regardless of the copper status of the cells. One curious function of Cuf1 is the repression activity on two genes in the iron uptake pathway, the fio1$^+$ Fe(II) oxidase and fip1$^+$ permease (Askwith and Kaplan, 1997; Labbe *et al.*, 1999). These two molecules share significant similarity to Fet3 and Ftr1 from *Sa. cerevisiae*. Cuf1 inhibits the expression of the *Sc. pombe* Fe uptake genes under conditions of copper deprivation (Labbe *et al.*, 1999). Deletion of *cuf*1$^+$ elevated the basal expression of the two iron uptake genes. In addition, mutations in the candidate Cuf1 sites in *fip*1$^+$ within a *fip*1$^+$/*lacZ* fusion increased elevated basal expression of the reporter gene (Labbe *et al.*, 1999). The inhibition of expression of Fe uptake genes by copper deprivation is one of several interconnections between iron and copper metabolism (Kaplan and O'Halloran, 1996).

The Mac1 orthologue from the filamentous fungus *P. anserina* is Grisea (Borghouts and Osiewacz, 1998; Osiewacz and Nuber, 1996). Grisea contains an N-terminal DNA-binding domain analogous to Mac1 and two C-terminal Cys-rich motifs (Borghouts and Osiewacz, 1998) (Fig. 4). Grisea can functionally replace Mac1 in *Sa. cerevisiae* (Borghouts and Osiewacz, 1998). The transactivation domains of Grisea appear to be attenuated by separate segments of the DNA-binding domain, consistent with one or more intramolecular interactions analogous to the Cu-induced repressive interactions observed in Mac1 (Borghouts and Osiewacz, 1998).

Cells with a nonfunctional Grisea exhibit an enhanced life span compared to wild-type cells due to copper deprivation (Marbach *et al.*, 1994; Osiewacz and Nuber, 1996). The enhanced life span of *P. anserina* cells harboring a nonfunctional Grisea correlates with an enhanced stabilization of mitochondrial DNA. The addition of Cu salts to mutant cells reverses the life span phenotype and increases the instability of mitochondrial DNA (Borghouts *et al.*, 2000). Copper appears to facilitate amplification and recombination events in mitochondrial DNA, resulting in rearrangements seen in senescent cells (Borghouts *et al.*,

2000). Mutations in Grisea limit Cu ion uptake in *P. anserina*, thereby minimizing the Cu-induced mitochondrial rearrangements. The mechanism of the Cu-induced deleterious changes in the mitochondrial genome is unclear, but may relate to Cu-induced accumulation of reactive oxygen intermediates (Borghouts *et al.*, 2000).

Gene activation by conditions of copper deficiency occurs in the plant *Arabidopsis*. A series of metalloreductase genes has been identified in *Arabidopsis*. One gene, *FRO2*, rescues a mutant defective in ferric reductase activity (Robinson *et al.*, 1999). *FRO2* accumulated under iron limitation, whereas expression of two FRO genes is enhanced in Cu-deficient plants (Nigel Robinson, unpublished observation). The transcription factor mediating Cu-deficient expression of the FRO genes is unknown, but it is likely to be distinct from Mac1 as the *Arabidopsis* genome does not contain a sequence homologous to the highly conserved Mac1 Zn module.

It is unclear whether any mammalian genes are activated under Cu deficiency. The likely candidates for mammalian genes induced by Cu deficiency are those involved in intestinal Cu ion absorption, since stringent control of absorption and excretion of dietary copper across the mucosal epithelium is known (Linder, 1991; Linder and Hazegh-Azam, 1996). Cu ion absorption occurs primarily within the small intestine, although some absorption of Cu also occurs in the stomach (Lonnerdal, 1996; Wapnir and Stiel, 1987). Mucosal absorption of Cu is regulated in accordance with metal status; hence absorption is enhanced when body stores of copper are low and vice versa (Arredondo *et al.*, 2000; Johnson, 1989; Linder and Hazegh-Azam, 1996; Turnlund *et al.*, 1989). Absorption of dietary copper is achieved through transfer across the apical surface of the mucosal brush border into the gut epithelium. The basolateral transfer of metal ions into the circulation is also regulated (Arredondo *et al.*, 2000; Linder, 1991). Cu ion transport in intestinal monolayers from the apical surface to the basolateral medium is strongly influenced by the Cu status of the monolayer cells (Arredondo *et al.*, 2000). Basolateral transport of Cu ions was enhanced in cells preconditioned in low-copper medium (Arredondo *et al.*, 2000). The only gene known to function in intestinal copper absorption is the Menkes MNK protein, which likely is important for the basolateral transfer of copper (Bull and Cox, 1994; Linder and Hazegh-Azam, 1996). Menkes disease patients are effectively copper deficient by virtue of inadequate copper transfer into the circulation (Vulpe and Packman, 1995). A candidate apical copper permease has been identified, hCtr1 (Zhou and Gitschier, 1997), but its role in copper transport in human cells has not been defined. Metallothionein is not known to be involved in copper uptake, but it modulates the basolateral transport of copper ions in the intestine (Linder and Hazegh-Azam, 1996). Neither the Menkes MNK gene nor the hCtr1 gene is up-regulated in animals on Cu-deficient diets. Other mammalian copper homeostasis genes including the Wilson (ATP7b) P-type ATPases, ceruloplasmin, and Cu metallochaperones HAH1, CCS, and COX17 are also not transcriptionally regulated by copper ion status (Gittlin *et al.*, 1992; Hishihara *et al.*, 1998).

Cu-deficient animal cells up-regulate the expression of a few genes such as dopamine B-monooxygenase and fatty acid synthase (Prohaska and Brokate, 1999; Wilson *et al.*, 1997). Cu-deficient animals have modest elevations in neuropeptide Y mRNA levels, but no increase was observed in immunoreactive neuropeptide Y (Rutkoski *et al.*, 1999). Differential display was used to identify cytochrome b as one gene up-regulated in Cu-treated cells (Levenson *et al.*, 1999). This observation was intriguing as copper overload and deficiency are known to cause morphological and functional mitochondrial abnormalities. It is unclear whether Cu modulation of gene expression occurs in animal cells as in yeast. It is conceivable that Cu modulation of transcription does occur in animal cells but that target genes have not been identified to date. No sequence homologues of Mac1 have appeared in animal EST databases to date; however, human Cu sensors may be quite distinct from yeast regulatory molecules as is the case for Zn sensors (Zhao and Eide, 1997).

### *C. Response of Cells to Stressful Copper Levels*

Exposure of yeast cells to elevated copper triggers a series of events to limit Cu ion uptake. First, Mac1 function is inhibited as a transcriptional activator of *CTR1* and *FRE1* by the Cu-induced conformer. Second, Cu-induced proteolysis of Mac1 occurs. The Cu-induced turnover of Mac1 is likely important in preventing Cu-induced toxicity in cells with residual Ctr1 transporters. Third, elevated Cu uptake leads to the diminution of Ctr1 levels in the plasma membrane through two processes, internalization and degradation (Ooi *et al.*, 1996). Cells exposed to 10 μM Cu(II) salts internalize preexisting Ctr1 molecules in the plasma membrane as well as induce the degradation of Ctr1. The degradation of Ctr1 proceeds independent of the Cu-induced internalization (Ooi *et al.*, 1996). The C-terminal tail of Ctr1 is important for Cu-induced internalization, but not proteolysis. Cu treatments alters the solubility of Ctr1 within the membrane (Ooi *et al.*, 1996). The protease responsible for the Cu-induced degradation of Ctr1 is unknown, but the process does not involve vacuolar proteases (Ooi *et al.*, 1996).

The combined effects of Cu inhibition of Mac1 function, Cu-induced proteolysis of Mac1, and Cu-triggered internalization and degradation of Ctr1 restrict Cu ion uptake through the high-affinity uptake system. Divalent Cu ions can still be internalized by low-affinity permeases including

Fet4, Ctr2, and Smf1 (Dix *et al.*, 1997; Kampfenkel *et al.*, 1995; Liu *et al.*, 1997), so an additional mechanism of copper detoxification exists. This additional response of *Sa. cerevisiae* cells to stressful copper ion levels is the transcriptional activation of a set of protective genes. These genes are transcriptionally activated when the extracellular copper concentration exceeds 1 μM. This is a protective response to counteract the potential cytotoxicity of Cu ions. The Cu-activated genes include *CUP1*, *CRS5*, and *SOD1* (Culotta *et al.*, 1994; Furst *et al.*, 1988; Gralla *et al.*, 1991). *CUP1* and *CRS5* encode copper-binding, cysteinyl-rich polypeptides in the metallothionein family (Culotta *et al.*, 1994; Jensen *et al.*, 1996). In addition to its role in dismuting superoxide anions, Sod1 has a secondary role of contributing to copper buffering in *Sa. cerevisiae* (Culotta *et al.*, 1995).

*CUP1* is the dominant locus that confers the ability of yeast cells to propagate in medium containing copper salts (Fogel and Welch, 1982; Hamer *et al.*, 1985; Jensen *et al.*, 1996). Cells highly resistant to copper salts contain a *CUP1* locus with tandem repeats of genes encoding the Cup1 metallothionein (Fogel and Welch, 1982). In contrast, cells lacking the *CUP1* locus are hypersensitive to copper salts (Ecker *et al.*, 1986). The Cup1 metallothionein binds 7 Cu(I) ions within a buried polycopper-thiolate cluster, thereby buffering the cytosolic Cu levels to maintain a low reactive pool of Cu(I) (Narula *et al.*, 1991; Peterson *et al.*, 1996; Rae *et al.*, 1999). The second type of metallothionein in yeast, Crs5, is also copper metalloregulated in its expression (Culotta *et al.*, 1994). *CRS5* is present as a single-copy gene, unlike the tandem array of *CUP1* metallothionein genes, so its effectiveness in copper ion buffering is limited (Jensen *et al.*, 1996). Crs5 binds 12 Cu(I) ions presumably within two separate polycopper clusters, as is the case with mammalian metallothioneins (Jensen *et al.*, 1996).

Differential expression of other genes occurs in Cu-stressed cells. High exogenous Cu levels result in the induced expression of *FET3* and *FTR1* in *Sa. cerevisiae* (Gross *et al.*, 2000). Fet3 and Ftr1 function in high-affinity iron uptake and are induced in iron-deficient cells through the Aft1 transcriptional activator (Askwith *et al.*, 1996; Yamaguchi-Iwai *et al.*, 1995). The rapid Cu-induced expression of *FET3* and *FTR1* is an indirect effect arising from a transient Cu-induced diminution in a cellular Fe pool resulting in Aft1 activation (Gross *et al.*, 2000). Cu stress was reported to result in the activation of $H^+$-ATPase activity in yeast plasma membranes (Fernandes *et al.*, 1998). One of the two yeast proton ATPase genes, *PMA2*, was reported to be Cu induced (Fernandes *et al.*, 1998). However, *PMA2* was not observed as a Cu-inducible gene by DNA microarray experiments (Gross *et al.*, 2000) *PMA2* may not have been seen in the DNA microarray experiment as a Cu-induced gene if the differential expression was less than two-fold. This observation raises the possibility that CuAce1 may weakly activate many more yeast genes and microarray technology may not be sensitive enough to detect the weak activation. The microarray experiments did not pick up *SOD1* as a Cu-activated gene, although other studies demonstrated that CuAce1 weakly stimulates *SOD1* expression (Carri *et al.*, 1991; Gralla *et al.*, 1991).

Although CuAce1-induced expression of *CUP1* is the dominant mechanism of copper tolerance in *Sa. cerevisiae*, other factors contribute to copper tolerance in yeast. For example, cellular histidine levels are important for copper tolerance in yeast cells (Pearce and Sherman, 1999). Histidine levels in vacuoles appear to be important for cell propagation in medium containing elevated Cu(II) levels. The yeast vacuole has been implicated in several studies as an important component of copper tolerance (Szczypka *et al.*, 1997).

*1. Copper Activation of Ace1*

Cu(I) activation of *CUP1* expression is mediated by the Ace1 transcription factor in *Sa. cerevisiae* (Buchman *et al.*, 1989; Thiele, 1988) (Fig. 9).

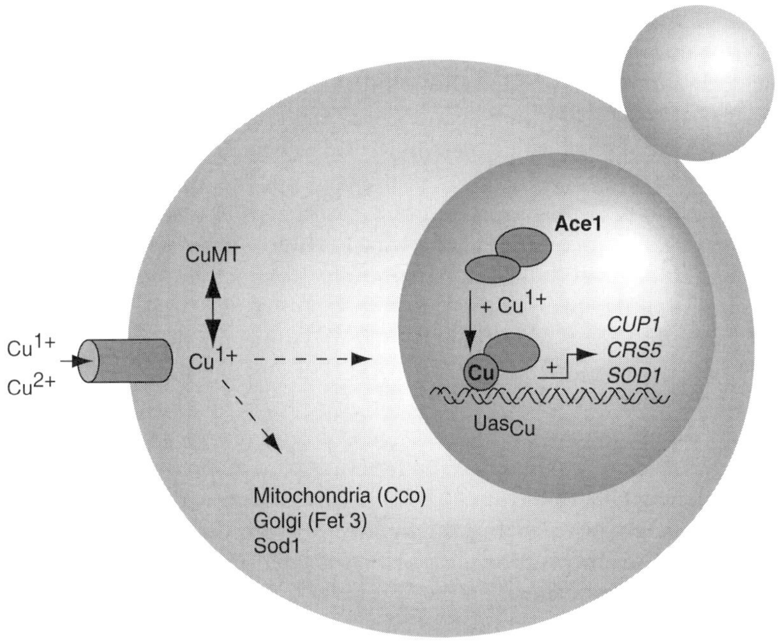


FIG. 9. Copper activation of Ace1 induces expression of three genes in *Saccharomyces cerevisiae*. Crs5 and Cup1 encode metallothionein-like molecules that buffer the cytoplasmic Cu concentration. Sod1 is a Cu-buffering molecule in addition to its role as a superoxide dismutase.

Cells lacking a functional Ace1 are hypersensitive to copper salts (Buchman *et al.*, 1989; Thiele, 1988). Ace1 is a typical fungal transcriptional activator with both DNA-binding and transactivation functions (Furst *et al.*, 1988). These two activities are typically separable and functionally independent (Mitchell and Tjian, 1989). The C-terminal half of Ace1 possesses transactivation activity (Chaudhuri *et al.*, 1995). The N-terminal half of Ace1 contains the DNA-binding moiety (Furst *et al.*, 1988). Eleven cysteinyl residues in the N-terminal DNA-binding moiety are essential for function (Hu *et al.*, 1990). These 11 cysteines are important ligands to one Zn and four Cu ions. Cu(I) ion binding within the N-terminal half of Ace1 stabilizes a specific tertiary fold capable of high-affinity interaction with specific DNA promoter sequences (Furst *et al.*, 1988). Cu(I) triggering involves the formation of a tetracopper-thiolate cluster within the regulatory domain (Winge, 1998). The DNA-binding domain is interdigitated within the Cu-regulatory domain such that DNA binding requires formation of the tetracopper thiolate cluster (Dameron *et al.*, 1991; Hu *et al.*, 1990). It is unlikely that Ace1 regulation involves changes in the population of the single Zn site.

A homologous Cu-activated factor has been identified in the yeast, *Ca. glabrata* (Geierstanger *et al.*, 1994). The factor Amt1 (activator of MT transcription) mediates the Cu-induced expression of three distinct metallothionein genes (Georgatsou and Alexandrati, 1994; Georgatsou *et al.*, 1997). The N-terminal segment of Amt1 is homologous to that of Ace1 and the 11 essential cysteinyl residues in Ace1 are conserved in Amt1 (Geierstanger *et al.*, 1994). The paucity of hydrophobic residues in the Cys-rich N-terminal segments of Ace1 and Amt1 is consistent with the model of the conformation of Ace1 and Amt1 being responsive to Cu(I) binding rather than the usual hydrophobic forces that dominate folding of typical globular proteins.

*CUP1*, *CRS5*, and *SOD1* contain CuAce1-binding sites within 5′ promoter sequences (Buchman *et al.*, 1990; Furst *et al.*, 1988). Footprinting analyses of Ace1 binding to the promoter elements, designated $UAS_{Cu}$, revealed major groove base contacts at the two ends of $UAS_{Cu}$ and minor groove contacts in the middle A/T-rich region (Buchman *et al.*, 1990; Dixon *et al.*, 1996; Dobi *et al.*, 1995). A prediction is that Ace1 lies atop the minor groove, contacting the major groove on both sides (Buchman *et al.*, 1990). Cu-induced expression levels correlate with the number of copper-regulatory elements. The highly Cu-induced *CUP1* MT genes contain multiple Cu-responsive DNA elements in the 5′ sequences, in contrast to the weakly Cu-inducible *CRS5*, which contains only a single element (Jensen *et al.*, 1996).

The N-terminal DNA-binding segment of Ace1 and Amt1 consists of two domains, a 40-residue Zn module seen also in the Mac1 transcription factor and a 60 to 70-residue Cu-regulatory domain that binds the tetrathiolate copper cluster (Farrell *et al.*, 1996; Graden *et al.*, 1996). The conserved Zn module in Ace1 and Amt1 is an independently folded domain consisting of a single Zn(II) ion bound by three cysteines and a single histidine within a C-$x_2$-C-$x_8$-C-x-H motif, as described previously for Mac1 (Fig. 6). The Zn(II) module is essential for *in vivo* function of both Ace1 and Amt1 (Buchman *et al.*, 1990; Posewitz *et al.*, 1996). The likely function involves minor groove DNA binding in the A/T-rich region of $UAS_{Cu}$ by the RGRP motif as in Mac1, as well as major groove DNA binding by Ace1 in the GCG subsite of $UAS_{Cu}$. Amt1 does not appear to make the second major groove DNA interaction that is characteristic of the Ace1:DNA complex (Koch and Thiele, 1996). Three basic residues (RxHR) lying between the second and the third cysteines in the Zn module of Ace1 are likely to be contact residues for GCG DNA contacts. In Amt1 the residues corresponding to the two Ace1 arginines are lysines, which may preclude guanine contacts.

The Cu-regulatory domain consisting of 60 residues in Ace1 makes base contacts within a major groove of the DNA helix in a Cu-dependent manner. The tetracopper domain contains eight essential cysteines present as Cys-$X_{1,2}$-Cys sequence motifs (Graden *et al.*, 1996; Hu *et al.*, 1990) (Fig. 10). From homologous sequences in Ace1, Amt1, Lpz8 in *Sa. cerevisiae*, and Crf1 in *Yarrowia lipolyitica*, a consensus sequence for the Cu-regulatory domain can be derived: C-$X_2$-C-$(X)_{12,14}$-C-X-C-$(X)_{10-27}$ -C-X-C-$X_5$-C-X-C. The presence of related Cys-$X_{1,2}$-Cys motifs in Cu-binding metallothioneins led Furst *et al.* to postulate in 1988 that Cu(I) binding to Ace1 triggered a conformational change to a fold that was poised for DNA binding (Furst *et al.*, 1988). According to the model, binding of Ace1 to $UAS_{Cu}$ elements upstream of *CUP1* allows the transactivation domain of Ace1 to function in the assembly of the preinitiation transcription complex.

Cu activation of Ace1 involves formation of a tetracopper cluster within the Cu-regulatory domain (Fig. 10). Expression of a truncated Ace1 or Amt1 consisting of the tetracopper domain and the GRP motif (residues 37–110 in Amt1) in bacteria yielded a Cu complex with 4 mol eq of Cu(I) bound (Graden *et al.*, 1996). The truncated CuAmt1 complex exhibited high-affinity and high-specificity DNA binding. The truncation reduces the binding affinity by only a factor of 10 (Graden *et al.*, 1996). Luminescence and X-ray absorption spectroscopy of the CuAmt1 and CuAce1 complexes revealed that the bound Cu ions are present as Cu(I) ions (Graden *et al.*, 1996). The Cu(I) ions are coordinated by thiolate ligands as predicted by the Furst *et al.* model. X-ray absorption spectroscopy of model Cu-thiolate complexes CuAmt1 and CuAce1 revealed that the Cu(I) ions are bound in trigonal geometry. Three-coordinate Cu(I)

**41**

**TTCGHCKELR RTKNFNP SGG CMCASARRPA**

**71**

**VGSKEDETRC RCDEGEPCKC HTKRKSSRKS**

Consensus Sequence

**C**-X$_2$-**C**-(X)$_{12,14}$-**C**-X-**C**-(X)$_{10\text{-}27}$-**C**-X-**C**-(X)$_5$-**C**-X-**C**

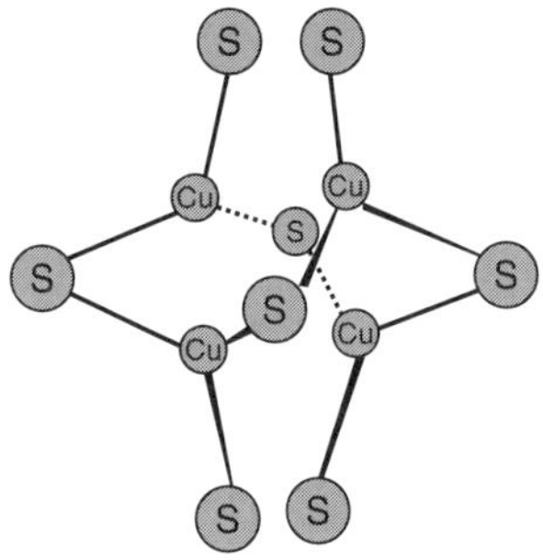


FIG. 10. Sequence of the Cu regulatory domain (CuRD) of Ace 1 from *Saccharomyces cerevisiae*. Eight cysteinyl residues form the CuRD. A consensus sequence is derived from orthologues from *Candida glabrata* (Amt 1) and *Yarrowia lipolytica* (Crf1). The presence of a large variable segment in the middle of the CuRD suggests that the CuRD consists of two lobes with four Cys residues each. The tetracopper cluster is likely buried between these two-lobe polypeptides.

binding is suggested by the X-ray absorption edge spectrum and the mean Cu–S distance of 2.26 Å. Three-coordinate Cu(I) centers with sulfur ligands typically show Cu–S bond distances of 2.26–2.28 Å. This is in contrast to two-coordinate Cu(I) centers in which the mean Cu–S bond distance is 2.16–2.18 Å (Dance, 1986). Three-coordinate Cu(I) binding is seen in a crystallographically defined tetracopper-thiolate cluster $[Cu_4(SPh)_6]^{2-}$ (Fig. 11). An outer shell X-ray scatter peak at 2.7 Å in CuAce1 was fitted only by the inclusion of a heavy scatterer atom indicating a polycopper cluster. Structures of a series of polyhedral Cu-thiolate clusters formed by simple ligands have a common feature of a short 2.7-Å Cu–Cu distance in the clusters (Dance, 1986). The copper nuclearities of the polycopper clusters in synthetic models vary from 2 to 8. Tetracopper cage clusters can form with six thiolate ligands and such clusters are stabilized by bridging thiolate ligands (Dance, 1986). Each thiolate in the $[Cu_4(SPh)_6]^{2-}$ cluster bridges two Cu(I) ions (Fig. 11). Extended X-ray

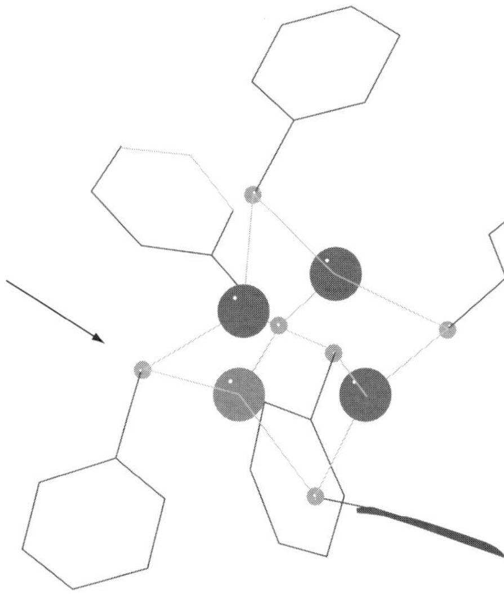

FIG. 11. Structure of a tetracopper-thiolate cluster of a small model compound $[Cu_4(SPh)_6]^{2-}$ cluster (Dance *et al.*, 1983). In this cluster all thiolates shown as small balls (one highlighted by an arrow) coordinate two Cu(I) ions. These are the μ-bridging thiolates that hold the cluster together. Each Cu(I) ion has trigonal planar coordination.

absorption fine structure (EXAFS) analyses of the crystallographically defined tetracopper-thiolate cluster $[Cu_4(SPh)_6]^{2-}$ revealed a pattern of photoelectron scattering similar to that of CuAce1 (Pickering *et al.*, 1993). The transformed EXAFS of the synthetic cluster revealed a first shell peak at 2.28 Å and an outer shell interaction at 2.74 Å (Pickering *et al.*, 1993). The mean Cu–S bond and Cu–Cu distances observed by crystallography of the $[Cu_4(SPh)_6]^{2-}$ complex were 2.287 and 2.74 Å, respectively (Dance *et al.*, 1983). The observation of a short 2.7-Å Cu–Cu distance in Cu,ZnAmt1 and CuAce1 is, therefore, compelling evidence for the existence of a polycopper-thiolate cluster in these transcription factors (Pickering *et al.*, 1993; Thorvaldsen *et al.*, 1994). EXAFS provides no clear information about the nuclearity of the Cu(I)-thiolate cluster in Cu,ZnAmt1. There is solid evidence suggesting that the cluster nuclearity is 4. The all-or-nothing formation of a four Cu(I) ion-bound species and the linear rise in Cu(I) luminescence in titrations of apo-Amt1 with Cu(I) peaking at 4 mol eq are both consistent with a nuclearity of 4 in the polycopper cluster (Thorvaldsen *et al.*, 1994).

Both CuAce1 and CuMac1 appear to form tetracopper-thiolate clusters. Structures are not known for either complex, but the Cu(I)-thiolate cluster in the Cup1 metallothionein is known (Peterson *et al.*, 1996) (Fig. 12). Seven Cu(I) ions are coordinated by 10 thiolates in the Cup1 cluster. As can be seen in Fig. 12, most Cu(I) ions are trigonally bound and most thiolates are μ-bridging. Bridging thiolates stabilize polycopper clusters.

Only six thiolates are needed to form a tetracopper cluster based on the known $Cu_4S_6$ synthetic cage clusters. In the synthetic $Cu_4S_6$ clusters, all thiolates are μ-bridging sulfurs in which the sulfur atom bridges two different Cu(I) ions. The tetracopper clusters in Ace1 and Amt1 may not contain all bridging sulfurs as is observed in the $Cu_4S_6$ cage clusters, since the Cu-regulatory domains contains eight, rather than six, thiolates. One model is that the Cu-thiolate cage cluster in Amt1 and Ace1 is stabilized by four terminal and four μ-bridging thiolates. Bridging thiolates may be the predominant stabilizing force in the integrity of the tetracopper clusters in Ace1 and Amt1. Cu–Cu bonding does not appear to be a significant energetic factor in the stability of polycopper thiolate clusters with Cu–Cu distances near 2.7 Å (Dance, 1986).

The significance of a tetracopper cluster as the structural unit within the activated transcription factors is threefold. First, a polycopper cluster formed by eight cysteinyl residues organizes and stabilizes a larger structural unit than a single bound metal ion. A single Cu(I) site is expected to

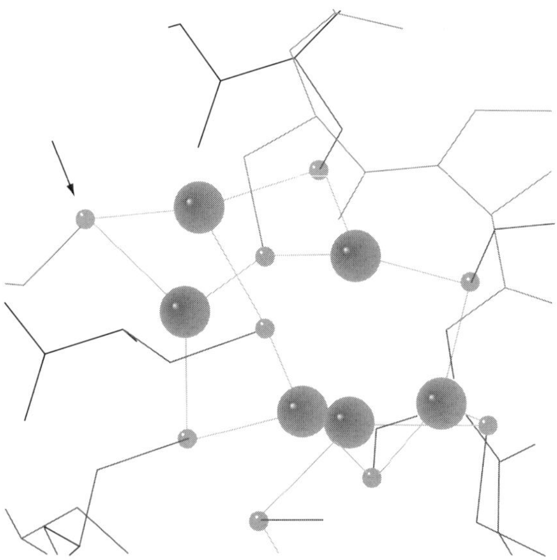

FIG. 12. Structure of the Cup 1 heptacopper-thiolate cluster (Peterson *et al.*, 1996). Most Cu(I) ions are bound by three cysteinyl thiolates, and most thiolates are μ-bridging such as the one indicated by the arrow.

be only three- or four-coordinate and, therefore, would anchor the polypeptide in only three or four places rather than the eight anchor sites in the candidate CuRD of Ace 1 and Amt 1.

A second significant aspect of a tetracopper cluster in the CuRD is that a polycopper cluster provides metal ion specificity. Ace1 is activated by either Cu(I) or Ag(I), but not by other metal ions (Dameron *et al.*, 1991; Furst *et al.*, 1988). Polymetal clusters are also known for Zn(II) and Cd(II) ions, but these clusters are structurally distinct from the polycopper clusters (Dance, 1986). Polycopper-thiolate clusters coordinate Cu(I) ions in either digonal or trigonal geometry. Zn(II)-thiolate clusters are characterized by tetrahedral Zn(II) coordination (Dance, 1986). In both cases, bridging thiolates are key features of cluster stability. Mammalian metallothionein isoforms 1 and 2 consist of two polymetal-thiolate clusters that are distinct depending on whether Zn(II) or Cu(I) ions are bound (Winge *et al.*, 1994). The distinct clusters translate into metal-dependent structures. The observed activation of Ace1 by Cu(I) and Ag(I) is expected, since structurally similar $[Cu_5(SPh)_7]^{2-}$ and $[Ag_5(SPh)_7]^{2-}$ metal-thiolate cage clusters exist (Dance, 1978). Subtle structural differences observed between AgAce1 and CuAce1 (Peterson *et al.*, 1996) may relate to volume differences of the metal-thiolate cages for the two monovalent ions. The mean Cu–S bond distance for a trigonally bound Cu(I) ion is 2.27 Å, whereas the mean Ag–S bond distance is 2.50 Å (Dance, 1978). Cluster volume was implicated as a critical factor in metal ion binding within clusters in metallothionein (Good *et al.*, 1991). Thus, volume constraints as well as cluster geometry may be important factors in dictating metal ion specificity in Ace1 and Amt1.

The third important feature is the observed cooperativity in cluster formation. The tetracopper center in Amt1 was shown to form in an all-or-nothing manner (Thorvaldsen *et al.*, 1994). Cooperativity in Cu(I) binding was also reported for Ace1 in Cu(I) titration studies (Casas-Finet *et al.*, 1992; Dameron *et al.*, 1991). The Cu(I) titration studies monitored by the luminescence of the Cu(I)-thiolate center was biphasic with an inflection point at 4 mol eq Cu(I). A Hill coefficient of 6 was calculated for the overall process of cluster formation (Casas-Finet *et al.*, 1992). Addition of Cu(I) to apo-Ace1 followed by DNA binding with a $UAS_{Cu}$ assay revealed that DNA binding was a sigmoidal function of copper concentration (Furst and Hamer, 1989). A Hill coefficient of 4 was derived from the binding data. Cooperative formation of polymetal clusters in metallothioneins is also known (Good *et al.*, 1988). Cooperativity in cluster formation may be significant in that it permits a direct coupling of the intracellular, exchangeable Cu ion concentration to transcriptional activation of *CUP1* and to a lesser extent *CRS5* and *SOD1*. Cells can respond to small increases in copper ion concentration to activate Ace1 and, therefore, enhance MT biosynthesis.

*2. Processes Involved in Transcriptional Activation*

Cu activation of Ace1 is one component in Cu-induced expression of yeast genes. However, several complex processes, other than activation of a transcriptional activator, control gene expression in eukaryotes. Gene expression is modulated by the chromatin structure and position of nucleosomes on the gene, the extent of posttranslational acetylation of core histones within nucleosomes, the function of chromatin remodeling complexes (RSC and SWI/SNF), and the recruitment of the preinitiation transcription complex. Evidence that the nucleosome structure is important for Ace1 activation of *CUP1* expression comes from the observation that nucleosome loss by repression of histone H4 biosynthesis activates the expression of *CUP1* to the maximal extent independent of Cu induction (Durrin *et al.*, 1992). The activation induced by nucleosome loss is independent of the Cu-responsive promoter elements. One implication of this observation is that CuAce1 activation of *CUP1* may involve nucleosome remodeling (Durrin *et al.*, 1992).

Nucleosome structure is a major factor in the Cu activation of Amt1 gene expression in *Ca. glabrata*. As mentioned, Amt1 is the Ace1 orthologue in *Ca. glabrata* and it exhibits autoregulation (Zhou and Thiele, 1993). Amt1 is present at low levels in Cu-naive *Ca. glabrata* cells, but exposure of cells to elevated Cu(II) concentrations results in greatly increased Amt1 levels. Cu activation of *AMT1* expression is achieved through a single Cu-regulatory promoter element ($UAS_{Cu}$) upstream of the transcription start site (Zhou and Thiele, 1993). Disruption of the Cu-regulatory element in *AMT1* attenuates the copper resistance and Cu induction of metallothionein genes in *Ca. glabrata* (Zhou and Thiele, 1993). The autoregulation of *AMT1* is distinctive in the rapid induction response relative to the kinetics of Cu induction of metallothionein gene expression (Zhou and Thiele, 1993). It is believed that the rapid Cu induction of Amt1 enhances the efficiency of CuAmt1 induction of metallothionein gene expression under conditions of copper stress. The Cu-regulatory element in the 5′ promoter sequence of *AMT1* resembles the elements found in the various metallothionein genes in *Ca. glabrata*. This element is similar to the Ace1-responsive $UAS_{Cu}$ in containing a core major groove GCTG sequence and an adjacent AT-rich minor groove site (Koch and Thiele, 1996). Amt1 does not appear to contact a second major groove site, analogous to Ace1. The N-terminal AT-hook motif and Cu-regulatory domain of Amt1 contact the AT-rich sequence and core GCTG sequence, respectively. Both contact sites are critical for Amt1 function. The rapid kinetics of autoactivation of *AMT1* was found to arise from a specialized nucleosome structure on *AMT1* that does not require remodeling prior to *AMT1* expression (Zhu and Thiele, 1966). A nucleosome is stably positioned on a 5′ segment of *AMT1* containing the Amt1-binding site, and the nucleosome does not move upon *AMT1* induction (Zhu and Thiele, 1966). The Cu-responsive $UAS_{Cu}$ contains an upstream 16-bp dA–dT sequence that is critical to the rapid autoactivation (Zhu and Thiele, 1966). The homopolymeric element distorts the nucleosome structure, thereby enhancing accessibility of the $UAS_{Cu}$ element to Amt1 (Zhu and Thiele, 1966). The position of the dA–dT sequence relative to the $UAS_{Cu}$ element is important. The presence of the dA–dT sequence at a location 3′ to the $UAS_{Cu}$ element rather than at its usual 5′ position impairs the responsiveness of *AMT1* to Cu activation (Koch and Thiele, 1999). Thus, the chromatin structure on *AMT1* prepares it for rapid induction by CuAmt1. Thus, Cu-treated *Ca. glabrata* cells initially induce *AMT1* expression. The increased Amt1 concentration subsequently facilitates expression of metallothionein genes.

Additional levels of regulation may exist, since RNA pol II transcription is regulated by the assembly of an oligomeric protein complex at the site of transcription initiation involving a number of general initiation factors such as TFIIA, TFIIB, TFIID, TFIIE, RNA polymerase, and coactivators in the mediator complex (Malik and Roeder, 2000; Zawel and Reinberg, 1995). An initial step is the binding of the TATA-binding subunit TBP of the TFIID complex to the TATA box of the promoter. Recruitment of TBP to the promoter of *CUP1* is essential to the activation by CuAce1 (Stargell and Struhl, 1995). TAF accessory factors in TFIID interact with activation domains in transcriptional activators (Stargell and Struhl, 1995) and in addition regulate histone acetylation (Brown *et al.*, 2000). Histone acetyltransferases are targeted to promoters through protein:protein interactions with activators (Brown *et al.*, 2000). Hyperacetylation of core histones correlates with enhanced gene transcription. Depletion of a specific TAF protein can modulate gene expression. The TAF molecule(s) responsible for Ace1-activated transcription is not known. However, cells depleted of Taf17 or containing a nonfunctional TFIIA broadly affect pol II transcription but do not affect *CUP1* induction by Cu (Liu *et al.*, 1999; Moqtaderi *et al.*, 1998).

### D. *Copper-Induced Transcription in Other Fungi*

*Candida albicans* is highly resistant to copper salts in the growth medium and is capable of normal growth in medium up to 20 mM $CuSO_4$ (Weissman *et al.*, 2000). Two genes were identified as major determinants of its Cu tolerance. One determinant was a small metallothionein-like molecule with three Cys-x-Cys sequence motifs (Weissman *et al.*, 2000). The second determinant was a P-type ATPase homologue. Disruption of the two metallothionein genes in the diploid *Ca. albicans* resulted in only minimal copper sensitivity, whereas disruption of the two *CRP1* P-type ATPase genes conferred marked copper sensitivity (Weissman *et al.*, 2000). *crp1* mutant cells accumulated excess Cu ions. Expression of *CRP1* and the metallothionein is Cu-inducible, suggesting the presence of a Cu-activated transcription factor. The importance of *CRP1* in copper tolerance is consistent with its function as a Cu(I) effluxer analogous to the CopB of *En. hirae* to limit cellular Cu levels. Consistent with its role, Crp1 was found to be localized to the plasma membrane (Weissman *et al.*, 2000).

Cu-induced transcriptional activation of metallothionein is known in *Neurospora crassa*, but the factor that mediates Cu activation is unknown (Munger *et al.*, 1987). The MT gene does not contain an Ace1-binding site in its 5′ promoter sequences.

Cu-induced transcription of metallothionein genes is observed in many species from fungi, plants, and animals. One species that does not contain metallothioneins is the fission yeast, *Sc. pombe*. Cu ion buffering in *Sc. pombe* as well as plants is achieved by Cu(I) coordination to glutathione-related isopeptides, designated phytochelatins (Rauser, 1990). The isopeptides are of general structure $(\gamma\text{GluCys})_n$ Gly, in which the number of dipeptide repeats varies from 2 to 5 (Reese *et al.*, 1988). Phytochelatin (PC) peptides accumulate in Cu-stressed *Sc. pombe* cells, but the mechanism does not involve transcriptional metalloregulation. Genes that encode enzymes capable of phytochelatin biosynthesis were recently identified in *Sc. pombe* and *Arabidopsis thaliana* (Clemens *et al.*, 1999; Ha *et al.*, 1999; Vatamaniuk *et al.*, 1999). Expression of the two genes in *Sa. cerevisiae* conferred cadmium tolerance and led to the accumulation of $n = 2, 3$ phytochelatins. *Saccharomyces cerevisiae* normally synthesizes only small quantities of $n = 2$ phytochelatins. Phytochelatins are also induced in Cu-treated cells (Clemens *et al.*, 1999). The mechanism of metal-induced phytochelatin synthesis is unclear. The purified *Arabidopsis* phytochelatin synthase (PCS1) was found to stimulate phytochelatin synthesis *in vitro* in a metal-dependent manner in one study (Ha *et al.*, 1999). However, in a second study, PC synthesis by the *Arabidopsis* PC synthase was metal independent (Vatamaniuk *et al.*, 1999). PC synthases contains five conserved Cys residues that may participate in metal activation (Ha *et al.*, 1999). The preferred substrates for the enzymes appear to be gGlu-Cys dipeptide units with blocked thiolate groups (Vatamaniuk *et al.*, 2000). Thus, one role of metal ions in the PC synthase reaction may be in blocking the thiolates of the substrates.

## IV. Copper-Induced Transcription in Animal Cells

Two dominant mechanisms of copper resistance in animal cells include Cu(I) efflux by a P-type ATPase and Cu(I) sequestration by metallothionein. As mentioned, Cu metalloregulation of both systems is known in unicellular organisms. In animal cells, distinct types of copper regulation occur for each pathway. The Menkes P-type ATPase (MNK) in mammalian cells is the dominant molecule involved in copper ion detoxification. Copper-resistant CHO cells were found to contain an amplified MNK locus (Camakaris *et al.*, 1995). MNK is normally localized in post-Golgi vesicles, but is translocated to the plasma membrane in copper-stressed cells (Goodyer *et al.*, 1999; Petris and Mercer, 1999; Petris *et al.*, 1996). Cu-induced relocalization of MNK is the predominant mechanism of Cu regulation of MNK. Wild-type CHO cells transfected with wild-type human MNK showed a diminished cellular Cu content. In contrast, CHO cells transfected with mutant MNK variants lacking a subset of the N-terminal Cu-binding Cys-x-x-Cys modules showed hyperaccumulation of Cu(I) (Voskoboinik *et al.*, 1999). The MNK P-type ATPase, unlike the Cu ATPase genes in *En. hirae* and *Ca. albicans*, is not transcriptionally regulated.

Cu induction of metallothionein (MT) genes has been observed in mammalian cells (Durnam and Palmiter, 1981). Mammalian cells contain multiple metallothionein genes in four distinct protein families, but only the MTs in families I and II are metal inducible (Palmiter *et al.*, 1992). Human cells contain six or seven MT genes in the isoform I family and one gene in isoform family II (Palmiter, 1987). A low steady-state expression of these MT genes occurs in most cells, but exposure to metal ions leads to a rapid and transient induction of MT gene expression. Induction occurs at the level of transcription via several *cis*-acting elements, designated metal regulatory elements (MREs), located within the proximal promoter of each MT gene (Westin and Schaffner, 1988). The conserved MRE enhancer elements consist of approximately 12–15 bp and confer metal responsiveness when fused to foreign genes. Although Cu induction of MT biosynthesis occurs in mammalian cells, MT synthesis is not a primary mechanism of copper tolerance in animal cells (Palmiter, 1998). Mice with targeted disruptions in both MTI and MTII genes are not copper hypersensitive, but are cadmium hypersensitive. Induction of MT is a secondary line of defense against elevated Cu ions, but becomes a significant mechanism of resistance in animals with an impairment in the function of the MNK P-type ATPase (Kelly and Palmiter, 1996).

MT gene expression is activated by binding the MTF-1 transcription factor to the 5′ MRE sequences (Westin and Schaffner, 1988). MTF-1 contains six classical Zn finger motifs and multiple potential C-terminal transactivation domains (Radtke *et al.*, 1993). MTF-1 binds to DNA in a metal-dependent manner with selectivity for Zn(II) ions *in vitro* (Westin and Schaffner, 1988). However, MTF-1 activates MT gene expression *in vivo* by a variety of metals including Cu ions (Heuchel *et al.*, 1994). Introduction of an MTF-1 antisense construct blocked all metal induction of MT expression (Palmiter, 1994). Likewise, disruption of both MTF-1 alleles in mouse embryonic stem cells by homologous recombination eliminates both Zn and Cu induction and basal MT expression (Heuchel *et al.*, 1994). This observation clearly demonstrates the importance of MTF-1 for basal and metal-induced MT expression. Mice lacking MTF-1 die *in utero* at approximately 14 days of gestation (Guncs *et al.*, 1998).

The mechanism of metal activation of MTF-1 remains unclear. Domain mapping studies have been conducted. The activation domains confer constitutive activity when fused to a heterologous DNA-binding domain (Radtke *et al.*, 1995). The Zn finger domain conferred limited Zn-induced transcription when fused to a heterologous transactivator (Radtke *et al.*, 1995). Although this result is consistent with the Zn finger domain being a component of the metal sensor, experiments with human/mouse MTF-1 chimeras suggest that the Zn activation of MTF-1 is more complex. Transfection of MTF-1 null cells with human versus mouse MTF-1 showed that the human factor exhibited a greater Zn responsiveness (Radtke *et al.*, 1995). Using the human/mouse MTF-1 chimera, the segment of human MTF-1 responsible for the greater Zn responsiveness was found to reside in a portion of the transactivation domain (Radtke *et al.*, 1995). Zn activation of MTF-1 is expected to involve two or more domains of the factor. MTF-1 truncates synthesized by a coupled *in vitro* transcription/translation system showed Zn-induced DNA binding in truncates lacking the C-terminal transactivation domains or lacking the N-terminal segment upstream of the Zn finger domain (Dalton *et al.*, 1997), suggesting the importance of the Zn finger domain in Zn metalloregulation. The activation of MTF-1 was found to be Zn specific; no activation was observed with Cu ions (Bittel *et al.*, 1998). Thus, Cu-induced expression of MT genes in mammalian cells is likely to arise from secondary effects, such as the Cu-induced changes in Zn pools.

Cu induction of MT genes occurs efficiently in *Drosophila* and may be a direct effect. *Drosophila melanogaster* contains two distinct metallothioneins, designated Mto and Mtn (Mokdad *et al.*, 1987). The Mto gene is more efficiently induced by Cd salts than by Cu salts, whereas Mtn is more efficiently induced by Cu salts (Silar *et al.*, 1990). The mechanism of Cu induction of Mtn remains unresolved, so a significant unresolved question is whether Cu induction of MT biosynthesis occurs through a direct transcriptional process as in yeast.

## V. Summary of Mechanism of Copper-Modulated Transcription

Cu modulation of transcription occurs in prokaryotes and eukaryotes. Transcription factors binding to *cis*-acting promoter elements may have either a positive or a negative influence on transcription. A number of transcription factors in both bacteria and yeast have been identified that function as metalloregulatory transcription factors in that they sense metal levels and transduce a biological response depending on the metal status of cells. The Cu-regulatory factors function as either transcriptional repressors or transcriptional activators. In general, conditions of Cu deprivation result in enhanced expression of gene products that function in Cu ion uptake.

In contrast, Cu-stressed cells show inhibited expression of Cu uptake gene products, but activated expression of genes whose products are protective against Cu-induced toxicity. Cu tolerance is largely imparted by induced expression of Cu effluxing P-type ATPases or Cu(I) sequestering metallothionein-like molecules. Cu metalloregulation of transcription occurs for ATPase genes in bacteria and certain fungi. Cu metalloregulation of transcription of metallothionein genes occurs in organisms ranging from bacteria to animal cells.

The *En. hirae* CopY is a transcriptional repressor that limits expression of the CopA and CopB P-type ATPases in cells cultured in medium containing minimal Cu(II) levels (Odermatt and Solioz, 1995). An increase in medium Cu(II) levels results in Cu ion uptake, routing of the Cu(I) ions to CopY by the CopZ metallochaperone, and the subsequent dissociation of CuCopY from the *copA* and *copB* genes (Strausak and Solioz, 1997).

In contrast to the *En. hirae* CopY metalloregulation, *Sa. cerevisiae* copper metalloregulation involves two transcriptional activators, Ace1 and Mac1. Both factors reside within the yeast nucleus and are regulated in their function in a Cu-dependent manner. Addition of Cu salts to the growth medium (at greater than micromolar (μm) concentrations of Cu) induces Ace1-dependent transcription within 10 min. Cu binding to Ace1 stabilizes a distinct conformation that enables CuAce1 to bind to DNA in a specific manner. Cu(I) ions activate Ace1 by formation of a tetracopper-thiolate cluster within the DNA-binding domain. The Cu-regulatory domain (CuRD) consists of 60–70 residues and has eight essential cysteinyl sulfur ligands. Polycopper cluster formation is cooperative, permitting a graded response of transcriptional activation to the extent of Cu stress. Mac1 is a transcriptional activator in Cu-deficient cells. Cu(I) binding to Mac1 inhibits both DNA-binding and transactivation activities in Mac1. The Cu-regulatory domain of Mac1 is a short motif consisting of fewer than 20 residues and contains 5 cysteinyl residues and a conserved histidine. Four Cu(I) ions bind to this motif in a polycopper cluster. Thus, both the Cu-activated Ace1 and the Cu-inhibited Mac1 appear to contain regulatory tetracopper clusters. Cluster formation in Mac1 is also likely to be cooperative, permitting a graded transcriptional response to the copper status of cells.

The fact that the two characterized eukaryotic Cu-regulated transcription factors are regulated by apparent formation of tetracopper clusters suggests that this may repeat as a structural motif in other systems. Both Ace1 and Mac1 contain a common structural motif, the small Zn module; thus, it is also conceivable that this structural motif will be duplicated in other species.

Since both Ace1 and Mac1 reside within the nucleus, Cu(I) translocation to the nucleus must occur for metalloregulation to proceed. Yeast cells contain metallochaperones that shuttle Cu(I) ions to distinct destinations, so it is predicted that a shuttle mechanism exists for Cu(I) ion delivery to the nucleus. However, no candidate nuclear copper metallochaperones have been identified.

Mac1 is Cu inhibited in cells cultured in medium containing nanomolar concentrations of Cu(II), whereas Ace1 is only fully activated in cells cultured in medium with μm concentrations of Cu(II). If the Cu-regulatory domains of each form tetracopper clusters, it is curious what factors contribute to the sensitivity of Mac1 to regulation at the nanomolar Cu level in contrast to the Ace1 activation at μm concentrations of Cu. Each transcription factor may have distinct and specific metallochaperone delivery systems that control metalloregulation. Alternatively, Mac1 may be triggered at a lower medium Cu ion concentration because Cu(I) binding to the short CuRD has a lower loss of entropy and less desolvation upon folding than Cu(I) binding to the larger CuRD in Ace1.

Cu metalloregulation of Ace1 and Mac1 is transient (Pena *et al.*, 1998). Cu activation of Ace1 is rapid and transient. The rapid diminution in Cu-induced transcription may arise from changes in chromatin structure, Ace1 modification, or Cu(I) dissociation. One mechanism for CuAce1 inactivation involves Cu(I) dissociation by ligand exchange with the Cup1 metallothionein. CuAce1 induces the transcription of the *CUP1* locus. The newly synthesized Cup1 metallothionein may inactivate Ace1 by competing for the Ace1-bound Cu(I) ions (Wright *et al.*, 1988). Diffusion of Cup1 metallothionein into the nucleus may be sufficient to return CuAce1 to the transcriptionally silent state. Fusion of Cup1 to a large macromolecule that fails to diffuse into the nucleus may be a reasonable test of the autoregulation model of Cup1 and Ace1. Alternatively, CuAce1 inactivation may be facilitated by a nuclear Cu efflux pump that functions to limit Cu levels within the nucleus.

Cu inhibition of Mac1 occurs in a rapid manner (Pena *et al.*, 1998) as does the reactivation of Mac1. *CTR1* transcription is induced in cells treated with the Cu(I) chelator for 15 min. Reactivation of CuMac1 presumably occurs by Cu(I) dissociation. It is unclear whether the Cu(I) conformer of Mac1 is inherently unstable *in vivo* or whether Cu(I) dissociation is facilitated by a ligand exchange reaction as suggested for the Cup1-dependent inactivation of Ace 1.

All cells appear to have a complex thermostat-like control system for sensing copper status. Various setpoints must exist in the thermostat system to achieve copper homeostasis. Distinct Cu sensory/transduction systems respond to the various setpoints. Future research will be needed to establish the signal sensed and what determines the setpoint of the cellular copper thermostat.

## Acknowledgments

Support by National Institutes of Health Grants ES 03817 and CA 62186 is acknowledged. I acknowledge many individuals in my research group for editorial assistance. Special appreciation goes to Drs. Keith McCall, Julian Rutherford, and Thalia Nittis.

## References

Aravind, L., and Landsman, D. (1998). *Nucleic Acid Res.* **26**, 4413–4421.

Arredondo, M., Uauy, R., and Gonzalez, M. (2000). *Biochim. Biophys. Acta* **1474**, 169–176.

Askwith, C., and Kaplan, J. (1997). *J. Biol. Chem.* **272**, 401–405.

Askwith, C. C., de Silva, D., and Kaplan, J. (1996). *Mol. Microbiol.* **20**, 27–34.

Bittel, D., Dalton, T., Samson, S. L. A., Gedamu, L., and Andrews, G. K. (1998). *J. Biol. Chem.* **273**, 7127–7133.

Borghouts, C., Kerschner, S., and Osiewacz, H. D. (2000). *Curr. Genet.* **37**, 268–275.

Borghouts, C., and Osiewacz, H. D. (1998). *Mol. Gen. Genet.* **260**, 492–502.

Brewer, G. J., Yuzbasiyan-Gurkan, F., and Young, A. B. (1987). *Semin. Neurol.* **7**, 209–220.

Brown, C. E., Lechner, T., Howe, L., and Workman, J. L. (2000). *Trends Biochem. Sci.* **25**, 15–19.

Brown, N. L., Barrett, S. R., Camakaris, J., Lee, B. T., and Rouch, D. A. (1995). *Mol. Microbiol.* **17**, 1153–1166.

Buchman, C., Skroch, P., Dixon, W., Tullius, T. D., and Karin, M. (1990). *Mol. Cell. Biol.* **10**, 4778–4787.

Buchman, C., Skroch, P., Welch, J., Fogel, S., and Karin, M. (1989). *Mol. Cell. Biol.* **9**, 4091–4095.

Bull, P. C., and Cox, D. W. (1994). *Trends Genet.* **10**, 246–252.

Bustin, M., and Reeves, R. (1996). *Prog. Nucleic Acid Res. Mol. Biol.* **54**, 35–95.

Camakaris, J., Petris, M. J., Bailey, L., Shen, P., Lockhart, P., Glover, T. W., Barcroft, C., Patton, J., and Mercer, J. F. (1995). *Hum. Mol. Genet.* **4**, 2117–2123.

Carri, M. T., Galiazzo, F., Ciriolo, M. R., and Rotilio, G. (1991). *FEBS Lett.* **278**, 263–266.

Casas-Finet, J. R., Hu, S., Hamer, D., and Karpel, R. L. (1992). *Biochemistry* **31**, 6617–6626.

Chaudhuri, B., Huijbregts, R. P., Coen, J. J., and Furst, P. (1995). *Biochem. Biophys. Res. Commun.* **216**, 1–10.
Clemens, S., Kim, E. J., Neumann, D., and Schroeder, J. I. (1999). *EMBO J.* **18**, 3325–3333.
Cobine, P., Wickramasinghe, W. A., Harrison, M. D., Weber, T., Solioz, M., and Dameron, C. T. (1999). *FEBS Lett.* **445**, 27–30.
Culotta, V. C., Howard, W. R., and Liu, X. F. (1994). *J. Biol. Chem.* **269**, 1–8.
Culotta, V. C:, Joh, H. D., Lin, S. J., Slekar, K. H., and Strain, J. (1995). *J. Biol. Chem.* **270**, 29991–29997.
Dalton, T. P., Bittel, D., and Andrews, G. K. (1997). *Mol. Cell. Biol.* **17**, 27781–2789.
Dameron, C. T., Winge, D. R., George, G. N., Sansone, M., Hu, S., and Hamer, D. (1991). *Proc. Natl. Acad. Sci. USA* **88**, 6127–6131.
Dance, I. G. (1978). *Aust. J. Chem.* **31**, 2195–2206.
Dance, I. G. (1986). *Polyhedron* **5**, 1037–1104.
Dance, I. G., Bowmaker, G. A., Clark, G. R., and Seadon, J. K. (1983). *Polyhedron* **2**, 1031–1043.
Dancis, A., Haile, D., Yuan, D. S., and Klausner, R. D. (1994). *J. Biol. Chem.* **269**, 25660–25667.
Dancis, A., Roman, D. G., Anderson, G. J., Hinnebush, A. G., and Klausner, R. D. (1990). *Proc. Natl. Acad. Sci. USA* **89**, 3869–3873.
Danks, D. M. (1988). *Annu. Rev. Nutr.* **8**, 235–257.
Dix, D., Bridgham, J., Broderius, M., and Eide, D. (1997). *J. Biol. Chem.* **272**, 11770–11777.
Dixon, W. J., Inouye, C., Karin, M., and Tullius, T. D. (1996). *J. Biol. Inorg. Chem.* **1**, 451–459.
Dobi, A., Dameron, C. T., Hu, S., Hamer, D., and Winge, D. R. (1995). *J. Biol. Chem.* **270**, 10171–10178.
Durnam, D. M., and Palmiter, R. D. (1981). *J. Biol. Chem.* **256**, 5712–5716.
Durrin, L. K., Mann, R. K., and Grunstein, M. (1992). *Mol. Cell. Biol.* **12**, 1621–1629.
Ecker, D. J., Butt, T. R., Sternberg, E. J., Neeper, M. P., Debouck, C., Gorman, J. A., and Crooke, S. T. (1986). *J. Biol. Chem.* **261**, 16895–16900.
Farrell, R. A., Thorvaldsen, J. L., and Winge, D. R. (1996). *Biochemistry* **35**, 1571–1580.
Fernandes, A. R., Peixoto, F. P., and Sa-Correia, I. (1998). *Arch. Microbiol.* **171**, 6–12.
Finegold, A. A., Shatwell, K. P., Segal, A. W., Klausner, R. D., and Dancis, A. (1996). *J. Biol. Chem.* **271**, 31021–31024.
Fogel, S., and Welch, J. W. (1982). *Proc. Natl. Acad. Sci. USA* **79**, 5342–5346.
Francis, M. S., and Thomas, C. J. (1997). *Mol. Gen. Genet.* **253**, 484–491.
Furst, P., and Hamer, D. (1989). *Proc. Natl. Acad. Sci. USA* **86**, 5267–5271.
Furst, P., Hu, S., Hackett, R., and Hamer, D. (1988). *Cell* **55**, 705–717.
Geierstanger, B. H., Volkman, B. F., Kremer, W., and Wemmer, D. E. (1994). *Biochemistry* **33**, 5347–5355.
Georgatsou, E., and Alexandraki, D. (1994). *Mol. Cell. Biol.* **14**, 3065–3073.
Georgatsou, E., Mavrogiananis, L. A., Fragiadakis, G. S., and Alexandraki, D. (1997). *J. Biol. Chem.* **272**, 13786–13792.
Gittlin, J. D., Schroeder, J. J., Lee-Ambrose, L. M., and Cousins, R. J. (1992). *Biochem. J.* **282**, 835–839.
Good, M., Hollenstein, R., Sadler, P. J., and Vasak, M. (1988). *Biochemistry* **27**, 7163–7166.
Good, M., Hollenstein, R., and Vasak, M. (1991). *Eur. J. Biochem.* **197**, 655–659.
Goodyer, I. D., Jones, E. E., Monaco, A. P., and Francis, M. J. (1999). *Hum. Mol. Genet.* **8**, 1473–1478.
Graden, J. A., Posewitz, M. C., Simon, J. R., George, G. N., Pickering, I. J., and Winge, D. R. (1996). *Biochemistry* **35**, 14583–14589.
Graden, J. A., and Winge, D. R. (1997). *Proc. Natl. Acad. Sci. USA* **94**, 5550–5555.

Gralla, E. B., Thiele, D. J., Silar, P., and Valentine, J. S. (1991). *Proc. Natl. Acad. Sci. USA* **88**, 8558–8562.
Gross, C., Kelleher, M., Iyer, V. R., Brown, P. O., and Winge, D. R. (2000). *J. Biol. Chem.* **275**, 32310–32316.
Gunes, C., Heuchel, R., Georgiev, O., Muller, K. -H., Lichtlen, P., Bluthmann, H., Marino, S., Aguzzi, A., and Schaffner, W. (1998). *EMBO J.* **17**, 2846–2854.
Ha, S. -B., Smith, A. P., Howden, R., Dietrich, W. M., Bugg, S., O'Connell, M. J., Goldsborough, P. B., and Cobbett, C. S. (1999). *Plant Cell* **11**, 1153–1163.
Hamer, D. H., Thiele, D. J., and Lemontt, J. E. (1985). *Science* **228**, 685–690.
Harrison, M. D., Jones, C. E., Solioz, M., and Dameron, C. T. (2000). *Trends Biochem. Sci.* **25**, 29–32.
Hassett, R., and Kosman, D. J. (1995). *J. Biol. Chem.* **270**, 128–134.
Heuchel, R., Radtke, F., Georgiev, O., Stark, G., Aguet, M., and Schaffner, W. (1994). *EMBO J.* **13**, 2870–2875.
Hill, K. L., Hassett, R., Kosman, D., and Merchant, S. (1996). *Plant Physiol.* **112**, 697–704.
Hishihara, E., Furuyama, T., Yamashista, S., and Mori, N. (1998). *NeuroReport* **9**, 3259–3263.
Hu, S., Furst, P., and Hamer, D. (1990). *New Biologist* **2**, 544–555.
Huffman, D. L., and O'Halloran, T. V. (2000). *J. Biol. Chem.* **275**, 18611–18614.
Huth, J. R., Bewley, C. A., Nissen, M. S., Evans, J. N. S., Reeves, R., Gronenborn, A. M., and Clore, G. M. (1997). *Nat. Struct. Biol.* **4**, 657–665.
Jamison McDaniels, C. P., Jensen, L. T., Srinivasan, C., Winge, D. R., and Tullius, T. D. (1999). *J. Biol. Chem.* **274**, 26962–26967.
Jensen, L. T., Howard, W. R., Strain, J. J., Winge, D. R., and Culotta, V. C. (1996). *J. Biol. Chem.* **271**, 18514–18519.
Jensen, L. T., Posewitz, M. C., Srinivasan, C., and Winge, D. R. (1998). *J. Biol. Chem.* **273**, 23805–23811.
Jensen, L. T., and Winge, D. R. (1998). *EMBO J.* **17**, 5400–5408.
Johnson, P. E. (1989). *Adv. Exp. I Med. Biol.* **258**, 71–79.
Joshi, A., Serpe, M., and Kosman, D. J. (1999). *J. Biol. Chem.* **274**, 218–226.
Jungmann, J., Reins, H. A., Lee, J., Romeo, A., Hassett, R., Kosman, D., and Jentsch, S. (1993). *EMBO J.* **12**, 5051–5056.
Kampfenkel, K., Kushnir, S., Babiychuk, E., Inze, D., and Van Montagu, M. (1995). *J. Biol. Chem.* **270**, 28479–28486.
Kaplan, J., and O'Halloran, T. V. (1996). *Science* **271**, 1510–1512.
Keller, G., Gross, C., Kelleher, M., and Winge, D. R. (2000). *J. Biol. Chem.* **275**, 29193–29199.
Kelly, E. J., and Palmiter, R. D. (1996). *Nat. Genet.* **13**, 219–222.
Knight, S. A. B., Labbe, S., Kwon, L. F., Kosman, D. J., and Thiele, D. J. (1996). *Genes Dev.* **10**, 1917–1929.
Koch, K. A., and Thiele, D. J. (1996). *Mol. Cell. Biol.* **16**, 724–734.
Koch, K. A., and Thiele, D. J. (1999). *J. Biol. Chem.* **274**, 23752–23760.
Labbe, S., Pena, M. M. O., Fernandes, A. R., and Thiele, D. J. (1999). *J. Biol. Chem.* **274**, 36252–36260.
Labbe, S., Zhu, Z., and Thiele, D. J. (1997). *J. Biol. Chem.* **272**, 15951–15958.
Lesuisse, E., Casteras-Simon, M., and Labbe, P. (1996). *J. Biol. Chem.* **271**, 13578–13583.
Levenson, C. W., Song, Y., Narayanan, V. S., Fitch, C. A., and Yeiser, E. C. (1999). *Biol. Trace Elem. Res.* **70**, 149–164.
Lin, C. M., and Kosman, D. J. (1990). *J. Biol. Chem.* **265**, 9194–9200.
Lin, S. -J., Pufahl, R. A., Dancis, A., O'Halloran, T. V. O., and Culotta, V. C. (1997). *J. Biol. Chem.* **272**, 9215–9220.
Linder, M. C. (1991). "Biochemistry of Copper" Plenum, New York.
Linder, M. C., and Hazegh-Azam, M. (1996). *Am. J. Clin. Nutr.* **63**, 797S–818S.

Liu, Q., Gabriel, S. E., Roinick, K. L., Ward, R. D., and Arndt, K. M. (1999). *Mol. Cell. Biol.* **19**, 8673–8685.

Liu, X. F., Supek, F., Nelson, N., and Culotta, V. C. (1997). *J. Biol. Chem.* **272**, 11763–11769.

Lonnerdal, B. (1996). *Am. J. Clin. Nutr.* **63**, 821S–829S.

Lonnerdal, B., Bell, J. G., and Keen, C. L. (1985). *Am. J. Clin. Nutr.* **42**, 836–844.

Lutsenko, S., and Kaplan, J. H. (1995). *Biochemistry* **34**, 15607–15613.

MacLennan, D. H., Rice, W. J., and Green, N. M. (1997). *J. Biol. Chem.* **272**, 28815–28818.

Malik, S., and Roeder, R. G. (2000). *Trends Biol. Sci.* **25**, 277–283.

Marbach, K., Fernandez-Larrea, J., and Stahl, U. (1994). *Curr. Genet.* **26**, 184–186.

Martins, L., Jensen, L. T., Simon, J. R., Keller, G., and Winge, D. R. (1997). *J. Biol. Chem.* **273**, 23716–23721.

Merchant, S., Hill, K., and Howe, G. (1991). *EMBO J.* **10**, 1383–1389.

Mitchell, P. J., and Tijan, R. (1989). *Science* **245**, 371–378.

Mokdad, R., Debec, A., and Wegnez, M. (1987). *Proc. Natl. Acad. Sci. USA* **84**, 2658–2662.

Moqtaderi, Z., Keaveney, M., and Struhl, K. (1998). *Mol. Cell* **2**, 675–682.

Munger, K., Germann, U. A., and Lerch, K. (1987). *J. Biol. Chem.* **262**, 7363–7367.

Narula, S. S., Mehra, R. K., Winge, D. R., and Armitage, I. M. (1991). *J. Am. Chem. Soc.* **113**, 9354–9358.

Odermatt, A., and Solioz, M. (1995). *J. Biol. Chem.* **270**, 4349–4354.

O'Halloran, R. V., and Culotta, V. C. (2000). *J. Biol. Chem.* **275**, 25057–25060.

Ono, B. I., Hazu, T., Yoshida, S., Kawato, T., Shinoda, S., Brzvwczy, J. and Yaszewski, A. (1999). *Yeast* **15**, 1365–1375.

Ooi, C. E., Rabinovich, E., Dancis, A., Bonifacino, J. S., and Klausner, R. D. (1996). *EMBO J.* **15**, 3515–3523.

Osiewacz, H. D., and Nuber, U. (1996). *Mol. Gen. Genet.* **252**, 115–124.

Outten, F. W., Outten, C. E., Hale, J., and O'Halloran, T. V. (2000). *J. Biol. Chem.* **275**, 31024–31029.

Palmiter, R. D. (1987). *Experentia Suppl.* **52**, 63–80.

Palmiter, R. D. (1994). *Proc. Natl. Acad. Sci. USA* **91**, 1219–1223.

Palmiter, R. D. (1998). *Proc. Natl. Acad. Sci. USA* **95**, 8428–8430.

Palmiter, R. D., Findley, S. D., Whitmore, T. E., and Durnam, D. M. (1992). *Proc. Natl. Acad. Sci. USA* **89**, 6333–6337.

Pearce, D. A., and Sherman, F. (1999). *J. Bacteriol.* **181**, 4774–4779.

Pena, M. M. O., Koch, K., and Thiele, D. J. (1998). *Mol. Cell. Biol.* **18**, 2514–2523.

Pena, M. M. O., Lee, J., and Thiele, D. J. (1999). *J. Nutr.* **129**, 1251–1260.

Pena, M. M. O., Puig, S., and Thiele, D. J. (2000). *J. Biol. Chem.* **275**, 33244–33251.

Peterson, C. W., Narula, S. S., and Armitage, I. M. (1996). *FEBS Lett.* **379**, 85–93.

Petris, M. J., and Mercer, J. F. B. (1999). *Hum. Mol. Genet.* **8**, 2107–2115.

Petris, M. J., Mercer, J. F. B., Culvenor, J. G., Lockhart, P., Gleeson, P. A., and Camakaris, J. (1996). *EMBO J.* **15**, 6084–6095.

Pickering, I. J., George, G. N., Dameron, C. T., Kurz, B., Winge, D. R., and Dance, I. G. (1993). *J. Am. Chem. Soc.* **115**, 9498–9505.

Posewitz, M. C., Simon, J. R., Farrell, R. A., and Winge, D. R. (1996). *J. Biol. Inorg. Chem.* **1**, 560–566.

Posey, J. E., and Gherardini, F. C. (2000). *Science* **288**, 1651–1653.

Prohaska, J. R., and Brokate, B. (1999). *J. Nutr.* **129**, 147–153.

Pufahl, R. A., Singer, C. P., Peariso, K. L., Lin, S.-J., Schmidt, P., Fahrni, C., Culotta, V. C., Penner-Hahn, J. E., and O'Halloran, T. V. O. (1997). *Science* **278**, 853–856.

Quinn, J. M., Barraco, P., Eriksson, M., and Merchant, S. (2000). *J. Biol. Chem.* **275**, 6080–6089.

Quinn, J. M., and Merchant, S. (1995). *Plant Cell* **7**, 623–628.
Radtke, F., Georgiev, O., Muller, H. P., Brugnera, E., and Schaffner, W. (1995). *Nucleic Acids Res.* **23**, 2277–2286.
Radtke, F., Heuchel, R., Georgiev, O., Hergersberg, M., Gariglio, M., Dembic, Z., and Schaffner, W. (1993). *EMBO J.* **12**, 1355–1362.
Rae, R. D., Schmidt, P. J., Pufahl, R. A., Culotta, V. C., and O'Halloran, T. V. (1999). *Science* **284**, 805–807.
Rauser, W. E. (1990). *Annu. Rev. Biochem.* **59**, 61–86.
Reese, R. N., Mehra, R. K., Tarbet, E. B., and Winge, D. R. (1988). *J. Biol. Chem.* **263**, 4186–4192.
Rensing, C., Fan, B., Sharma, R., Mitra, B., and Rosen, B. P. (2000). *Proc. Natl. Acad. Sci. USA* **97**, 652–656.
Robinson, N. J., Procter, C. M., Connolly, E. L., and Guerinot, M. L. (1999). *Nature* **397**, 694–697.
Rosenzweig, A. C., Huffman, D. L., Hou, M. Y., Wernimont, A. K., Pufahl, R. A., and O'Halloran, T. V. (1999). *Structure* **7**, 605–617.
Rosenzweig, A. C., and O'Halloran, T. V. (2000). *Curr. Opin. Chem. Biol.* **4**, 140–147.
Rutkoski, N. J., Fitch, C. A., Yeiser, E. C., Dodge, J., Trombley, P. Q., and Levenson, C. W. (1999). *Brain Res. Mol. Brain Res.* **656**, 80–86.
Scheinberg, I. H., and Sternlieb, I. (1968). *N. Engl. J. Med.* **278**, 352–359.
Scheinberg, I. H., and Sternlieb, I. (1976). "Trace Elements in Human Health and Disease," (A. S. Prasad and O. Oberleas, eds.), Vol. 1, pp. 415–436. Academic Press, New York.
Serpe, M., Joshi, A., and Kosman, D. J. (1999). *J. Biol. Chem.* **274**, 29211–29219.
Shatwell, K. P., Dancis, A., Cross, A. R., Klausner, R. D., and Segal, A. W. (1996). *J. Biol. Chem.* **271**, 14240–14244.
Shulman, R. J. (1989). *Am. J. Clin. Nutr.* **49**, 879–883.
Silar, P., Theodore, L., Mokdad, R., Erraiss, N.-E., Cadic, A., and Wegnez, M. (1990). *J. Mol. Biol.* **215**, 217–224.
Solioz, M., and Odermatt, A. (1995). *J. Biol. Chem.* **270**, 217–221.
Stargell, L. A., and Struhl, K. (1995). *Science* **269**, 75–78.
Strausak, D., and Solioz, M. (1997). *J. Biol. Chem.* **272**, 8932–8936.
Szczypka, M. S., Zhu, Z., Silar, P., and Thiele, D. J. (1997). *Yeast* **13**, 1423–1435.
Thiele, D. J. (1988). *Mol. Cell. Biol.* **8**, 2745–2752.
Thorvaldsen, J. L., Sewell, A. K., Tanner, A. M., Peltier, J. M., Pickering, I. J., George, G. N., and Winge, D. R. (1994). *Biochemistry* **33**, 9566–9577.
Toyoshima, C., Nakasako, M., Nomura, H., and Ogawa, H. (2000). *Nature* **405**, 647–655.
Turner, R. B., Smith, D. L., Zawrotny, M. E., Summers, M. F., Posewitz, M. C., and Winge, D. R. (1998). *Nat. Struct. Biol.* **5**, 551–555.
Turnlund, J. R., Keyes, W. R., Anderson, H. L., and Acord, L. L. (1989). *Am. J. Clin. Nutr.* **49**, 870–878.
Vatamaniuk, O. K., Mari, S., Lu, Y.-P., and Rea, P. A. (1999). *Proc. Natl. Acad. Sci. USA* **96**, 7110–7115.
Vatamaniuk, O. K., Mari, S., Lu, Y.-P., and Rea, P. A. (2000). *J. Biol. Chem.* **275**, 31451–31459.
Voskoboinik, I., Strausak, D., Greenough, M., Brooks, H., Petris, M., Smith, S., Mercer, J. F., and Camakaris, J. (1999). *J. Biol. Chem.* **274**, 22008–22012.
Vulpe, C. D., and Packman, S. (1995). *Annu. Rev. Nutr.* **15**, 293–322.
Wapnir, R. A., and Stiel, L. (1987). *Proc. Soc. Exp. Biol. Med.* **185**, 277–282.
Weissman, Z., Berdicevsky, I., Cavari, B.-Z., and Kornitzer, D. (2000). *Proc. Natl. Acad. Sci. USA* **97**, 3520–3525.
Welch, J., Fogel, S., Buchman, C., and Karin, M. (1989). *EMBO J.* **8**, 255–260.

Westin, G., and Schaffner, W. (1988). *EMBO J.* **7**, 3763–3770.
Wilson, J., Kim, S., Allen, K. G. D., Baillie, R., and Clarke, S. D. (1997). *Am. J. Physiol.* **272**, E1124–E1129.
Wimmer, R., Herrmann, T., Solioz, M., and Wuthrich, K. (1999). *J. Biol. Chem.* **274**, 2597–22603.
Winge, D. R. (1998). *Prog. Nucleic Acid Res.* **58**, 165–195.
Winge, D. R., Dameron, C. T., and George, G. N. (1994). *Adv. Inorg. Biochem.* **10**, 1–48.
Winge, D. R., and Mehra, R. K. (1990). *Int. Rev. Exp. Pathol.* **31**, 47–83.
Wright, C. F., Hamer, D. H., and McKenney, K. (1988). *J. Biol. Chem.* **263**, 1570–1574.
Wunderli-Ye, H., and Solioz, M. (1999). *Biochem. Biophys. Res. Commun.* **259**, 443–449.
Yamaguchi-Iwai, Y., Dancis, A., and Klausner, R. D. (1995). *EMBO J.* **14**, 1231–1239.
Yamaguchi Iwai, Y., Serpe, M., Haile, D., Yang, W., Kosman, D. J., Klausner, R. D., and Dancis, A. (1997). *J. Biol. Chem.* **272**, 17711–17718.
Zawel, L., and Reinberg, D. (1995). *Annu. Rev. Biochem.* **64**, 533–561.
Zhao, H., and Eide, D. J. (1997). *Mol. Cell. Biol.* **17**, 5044–5052.
Zhou, B., and Gitschier, J. (1997). *Proc. Natl. Acad. Sci. USA* **94**, 7481–7486.
Zhou, P., and Thiele, D. J. (1991). *Proc. Natl. Acad. Sci. USA* **88**, 6112–6116.
Zhou, P., and Thiele, D. J. (1993). *Genes Dev.* **7**, 1824–1835.
Zhu, Z., Labbe, S., Pena, M. M. O., and Thiele, D. J. (1998). *J. Biol. Chem.* **273**, 1277–1280.
Zhu, Z., and Thiele, D. J. (1966). *Cell* **87**, 459–470.

# BACTERIAL COPPER TRANSPORT

BY ZEN HUAT LU AND MARC SOLIOZ

Department of Clinical Pharmacology, University of Berne, 3010 Berne, Switzerland



## I. INTRODUCTION

The discovery of the oldest known microfossils in deep-sea volcanic rock suggests that hydrothermal vents hosted the first living systems on earth (Rasmussen, 2000). The hot, acidic seawater encountered at these vents releases metals such as iron, manganese, zinc, and copper from the volcanic rock (Zierenberg *et al.*, 2000), and resistance to these metal ions might have been an evolutionary priority for the first life forms. However, copper was probably not an essential trace element in early cells. Rather, it might have become a cellular constituent with the advent of a more oxidized biosphere and could thus be considered a "modern" bioelement (Kaim and Rall, 1996). Cuproenzymes function almost exclusively in the metabolism of $O_2$, $N_2O$, or $NO_2^-$, which became necessary only with the advent of an oxidizing environment $3 \times 10^9$ years ago. The corresponding need for a redox-active metal with potentials between 0 and 0.8 V is ideally fulfilled by the Cu(I)/Cu(II) redox pair. Interestingly, cuproenzymes are associated with every reduction state of oxygen (Fig. 1).



*ADVANCES IN PROTEIN CHEMISTRY, Vol. 60*


0065-3233/02 $35.00

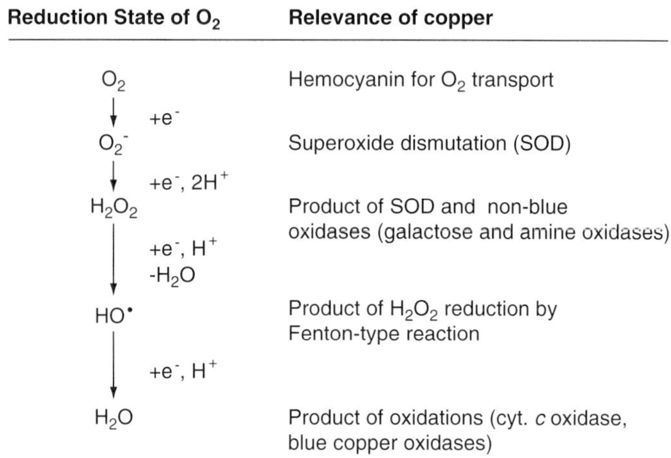


FIG. 1. Cuproenzymes involved at the different reduction states of oxygen.

Excess copper is toxic to cells. On one hand, copper ions can avidly bind to biomolecules by ligand interaction with cysteines or by binding to histidine-rich regions. Copper ions could also be incorporated into proteins instead of zinc or other metal ions during biosynthesis. On the other hand, copper ions can form radicals by a Fenton-type reaction as shown in Eq. (1):

$$Cu^+ + H_2O_2 \rightarrow Cu^{2+} + OH^- + HO^{\cdot}. \quad (1)$$

This reaction generates reactive hydroxyl radicals that can damage biomolecules. However, cellular hydrogen peroxide is rapidly removed by catalase and concentrations are very low, usually in the submicromolar concentration range. A Fenton-type reaction may therefore not be the primary cause of copper toxicity (Kaim and Rall, 1996). An alternative route of copper-induced cell damage is the depletion of sulfhydryls by redox cycling as described in reactions (2) and (3):

$$2Cu^{2+} + 2RSH \rightarrow 2Cu^+ + RSSR + 2H^+ \quad (2)$$

$$2Cu^+ + 2H^+ + O_2 \rightarrow 2Cu^{2+} + H_2O_2. \quad (3)$$

Reaction (3) will in fact generate hydrogen peroxide, which could fuel a Fenton-type reaction and thus enhance copper-induced damage by reactions (2) and (3).

Whatever the underlying mechanism(s) of copper toxicity is it makes tight control of copper levels a cellular necessity. While cuproenzymes that catalyze oxygen transport or redox reactions have been under investigation for several decades, the questions of how cells take up copper, route it intracellularly to sites of utilization, and secrete it when in excess remained unanswered until a few years ago. With the advent of the discovery of copper-pumping ATPases in *Enterococcus hirae* in 1992 (Odermatt *et al.*, 1992), the field of copper homeostasis has virtually exploded. Today, we have a fairly detailed, although not yet complete, picture of copper homeostasis in eukaryotic and prokaryotic cells. Some of the key elements of copper homeostasis and transport will be summarized in the remainder of this chapter.

## II. The New Subclass of Heavy Metal CPx-type ATPases

P-type ATPases (previously called $E_1E_2$-type ATPases) acquired their name from the fact that they form a phosphorylated intermediate in the course of the reaction cycle (Pedersen and Carafoli, 1987b). P-type ATPases are classically represented by the $Na^+$, $K^+$-ATPases of the plasma membrane and the $Ca^{2+}$-ATPases of the sarcoplasmic reticulum. Other metal ions transported by P-type ATPases are $Mg^{2+}$ in bacteria and $H^+$ in plants and fungi (Maguire, 1992; Fagan and Saier, 1994). With the discovery of cadmium- and copper-transporting P-type ATPases, it has become clear that there exists a subclass of these ATPases involved in the transport of heavy metal ions. Members of this subclass differ from the classical P-type ATPases not only in their transport specificities, but also in their membrane topology. Accordingly, they form a distinct evolutionary branch. Figure 2 shows an unrooted phylogenetic tree with representative members of heavy metal and non-heavy metal P-type ATPases. Based on this divergence, it has been proposed to call the heavy metal ATPases $P_1$-type or CPx-type ATPases, respectively (Lutsenko and Kaplan, 1995; Solioz and Vulpe, 1996). The division into heavy metal and non-heavy metal ATPases probably took place before the division into prokaryotes and eukaryotes.

Early life forms thriving near thermal vents in waters enriched in heavy metal ions would have had to have been endowed with mechanisms to deal with toxic metal ions and it is conceivable that efflux mechanisms for these metals evolved before or concomitantly with their use as cofactors. In line with such a hypothesis, the CPx-type ATPases encompass a wider spectrum of ion specificities than the non-heavy metal ATPases, now including $Cu^+$, $Ag^+$, $Zn^{2+}$, $Cd^{2+}$, and $Pb^{2+}$. It is to be expected that other metal ions will be added to this list. ATPases transporting silver, zinc, cadmium, and lead are involved in bacterial resistance to these toxic metal ions, while copper-transporting ATPases have a role both in copper uptake to meet cellular demands and in copper extrusion when ambient

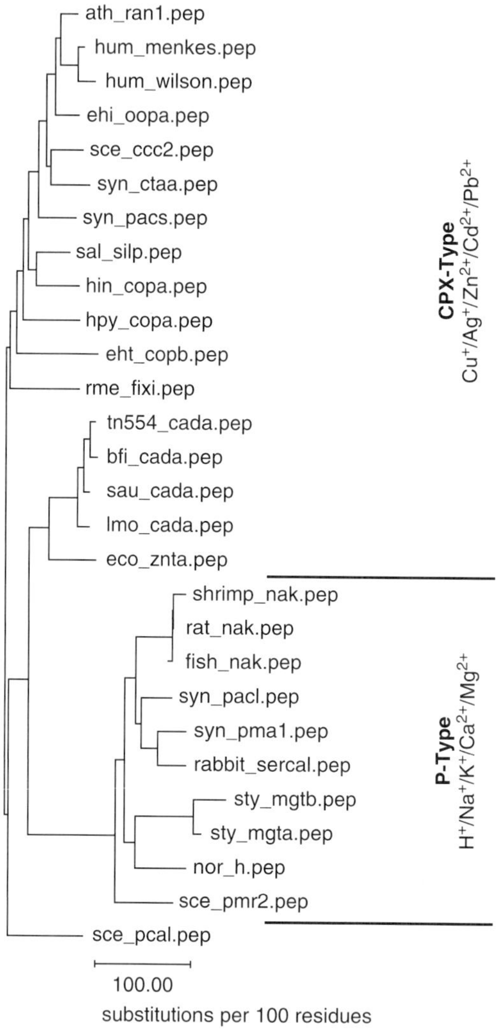


FIG. 2. Phylogram of ATPases. Divergence was calculated for a selected sample of P-type and CPx-type ATPases by the method of Kimura (1980), using 70 amino acids of the most highly conserved ATPase core. The phylogram was constructed by nearest-neighbor joining. The known ion specificities of non-heavy metal transporting P-type ATPases and heavy metal transporting CPx-type ATPases are indicated below the respective groups. The enzymes of the tree from top to bottom are as follows: *Arabidopsis thaliana* RESPONSE-TO-ANTAGONIST1 Cu-ATPase, human Menkes *ATP7A* Cu-ATPase, human Wilson *ATP7B* Cu-ATPase, *Enterococcus hirae* CopA Cu-ATPase, *Saccharamyces cerevisiae* CCC2 Cu-ATPase, *Synechococcus* CtaA Cu-ATPase, *Synechococcus* PacS Cu-ATPase, *Salmonella* SilP Ag-ATPase, *Hemophilus influenza* CopA Cu-ATPase, *Helicobacter pylori* CopA Cu-ATPase, *En. hirae* CopB Cu-ATPase, *Rhizobium*

copper is excessive. Some of the key features of heavy metal ATPases will be discussed in the next section.

### *A. Membrane Topology of CPx-type ATPases*

Figure 3 shows a comparison of the membrane topologies of the *En. hirae* CopB copper ATPase and the $Ca^{2+}$-ATPase of sarcoplasmic reticulum. The three-dimensional structure of the latter has recently been derived to a resolution of 2.6 Å (Toyoshima *et al.*, 2000). The residue that has been demonstrated to be phosphorylated is the aspartic acid in the conserved sequence DKTGT (given in single-letter amino acid code,

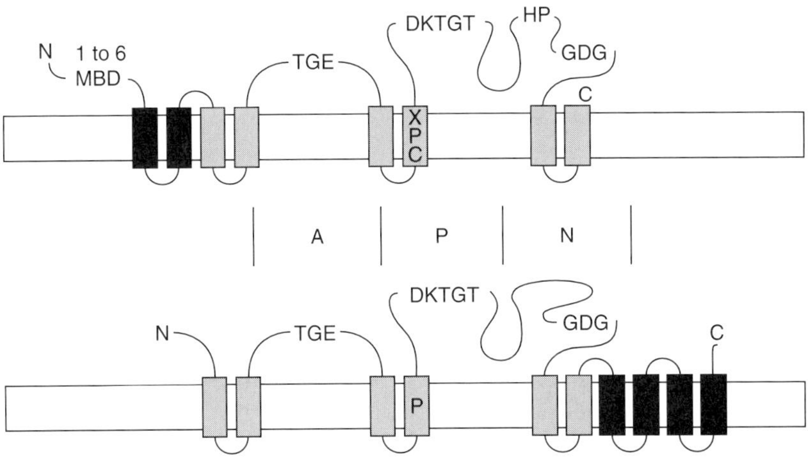


FIG. 3. Comparison of the membrane topology of a CPx-type ATPase and a non-heavy metal ATPase. Shown are CopB of *Enterococcus hirae* and the $Ca^{2+}$-ATPase of sarcoplasmic reticulum. Helices common to both types of ATPases are in gray and helices unique to one type of ATPase are in black. Key sequence motifs are indicated in single-letter amino acid code. In the center of the figure, the approximate locations of the three cytoplasmic domains, A, P, and N, are indicated. MBD, metal-binding domain containing repeat metal-binding sites; TGE, conserved site in transduction domain (A); CPx, putative copper-binding site; DKTGT, phosphorylation site in domain P; HP, motif of unknown function, probably in domain N; GDG, nucleotide-binding site residues in domain N.

---

*meliloti* FixI ATPase, Tn*554* CadA Cd-ATPase, *Boletus firmus* CadA Cd-ATPase, *Staphylococcus aureus* CadA Cd-ATPase, *Listeria monocytogenes* CadA Cd-ATPase, *Escherichia coli* ZntA Zn/Cd/Pb-ATPase, shrimp $Na^+$, $K^+$-ATPase, rat $Na^+$, $K^+$-ATPase, fish $Na^+$, $K^+$-ATPase, *Synechococcus* PacL Ca-ATPase, *Synechocystis* Pmal $H^+$-ATPase, rabbit SERCA1 Ca-ATPase, *Salmonella typhimurium* MgtB Mg-ATPase, *Sal. typhimurium* MgtB Mg-ATPase, *Sac. cerevisiae* Pmr2 $Na^+$-ATPase, and *Sac. cerevisiae* Pca1 Cu-ATPase.

used throughout this chapter). This sequence, also called the aspartyl kinase domain, is present in all of the more than 100 P-type ATPases sequenced todate. Other motifs common to P-type ATPases are the so-called phosphatase domain of consensus sequence TGES and the ATP-binding domains of consensus GDGINDAP, discussed in more detail elsewhere in this book (see also MacLennan *et al.*, 1997). However, the copper ATPases as well as the cadmium ATPases exhibit several striking features not found in any of the other P-type ATPases: (i) they have putative heavy metal-binding sites in the polar N-terminal region; (ii) they have a conserved intramembranous CPC, CPH, or CPS motif; (iii) they have a conserved HP motif 34 to 43 amino acids C-terminal to the CPC motif; (iv) they have two additional predicted transmembrane helices on the N-terminal end of the protein; and (v) they have only two membrane spans at the C-terminal end, thus lacking four membrane spans present at this position in other P-type ATPases. Clearly, the heavy metal ATPases form a subgroup of the P-type ATPases that is quite distinct.

### *B. Role of the CPx Motif*

The CPx motif, which is CPC in most known CPx-type ATPases, including the human Menkes and Wilson copper ATPases, and CPH in CopB and a few others, has been postulated to be part of the ion channel through the membrane, chiefly based on site-directed mutagenesis studies with the $Ca^{2+}$-ATPase (Solioz and Vulpe, 1996; Vilsen *et al.*, 1989). From the three-dimensional structure of the $Ca^{2+}$-ATPase of sarcoplasmic reticulum, the role of this domain in ion binding is now apparent (Toyoshima *et al.*, 2000). In this ATPase, the type II calcium-binding site is made up primarily of residues in transmembrane helix 4 (TM4). The key residues forming the calcium-binding site are VAA**IPE**-309. The main-chain carbonyl oxygen atoms of V304, A305, and I307 and a side-chain oxygen atom of E309 (plus oxygens from N796 and D800) contribute to the site, requiring some unwinding of helix TM4. Table I shows a comparison of the corresponding motifs in relation to the transport specificities of ion motive ATPases. Due to the different membrane topologies of heavy metal and non-heavy metal ATPases, the critical residues are located in helix TM4 in the calcium ATPase and other non-heavy metal ATPases, but in helix TM6 in heavy metal ATPases (cf. Fig. 3).

Since nitrogen and sulfur atoms are much better copper ligands than oxygen, it is fitting that the residues corresponding to I307 and E309 in the $Ca^{2+}$-ATPase are cysteines or histidines in heavy metal ATPases. In agreement with the proposed key role of the CPx motif in ATPase function, the C369S mutation of CopB showed no function *in vivo* and

TABLE I
*Sequence Motifs in the Ion Channels of P-type and CPx-type ATPases*

| Organism | ATPase | Type | Ions | Membrane helix | Motif |
|---|---|---|---|---|---|
| Rat | $Na^+$, $K^+$-ATPase | P | $Na^+/K^+$ | 4 | VANV**PE** |
| Rabbit | SERCA1 | P | $Ca^{2+}$ | 4 | VAAI**PE** |
| *Escherichia coli* | MgtA | P | $Mg^{2+}$ | 4 | VGLT**PE** |
| *Neurospora crassa* | $H^+$-ATPase | P | $H^+$ | 4 | IIGV**PV** |
| *Enterococcus hirae* | CopB | CPx | $Cu^+/Ag^+$ | 6 | IIA**CPH** |
| *Salmonella* | SilP | CPx | $Ag^+$ | 6 | IIA**CPC** |
| *En. hirae* | CopA | CPx | $Cu^+/Ag^+$ | 6 | VIA**CPC** |
| Human | Menkes | CPx | $Cu^+$ | 6 | CIA**CPC** |
| *Es. coli* | ZntA | CPx | $Cd^{2+}/Zn^{2+}/Pb^{2+}$ | 6 | LIG**CPC** |
| *Staphylococcus aureus* | CadA | CPx | $Cd^{2+}$ | 6 | VVG**CPC** |

the purified enzyme had no detectable ATPase activity (Bissig *et al.*, in press). Also, the corresponding Menkes disease mutation C1000R changing the conserved CPC motif to RPC has been described as causing a severe phenotype, although with a long survival time (Horn and Tümer, 1999). Direct binding measurements would be required to demonstrate the involvement of the CPC motif in high-affinity copper binding. However, preliminary studies in our laboratory indicated that the binding affinity of CopB for copper is in the low nanomolar range. Current instrumentation does not allow measurement of such low copper concentrations. A detailed understanding of the CPx motif in copper binding and translocation will thus have to await further technical developments.

### *C. CxxC Heavy Metal-Binding Motifs*

A conspicuous feature of CPx-type ATPases is the occurrence of one to six copies of conserved metal-binding domains in the polar N-terminus preceding the first predicted membrane span. These metal-binding sites are of two types. Usually, they feature a CxxC motif in a conserved domain encompassing 40 to 60 amino acids, but in some instances (e.g., CopB of *En. hirae*) an exceptionally histidine-rich N-terminus is present instead. A role of the CxxC motif in heavy metal binding was first proposed by Silver *et al.*, (1989). They pointed to the presence of this motif in the cadmium ATPase of *Staphylococcus aureus* in the periplasmic mercury-binding protein MerP, and in three different MerA mercuric reductases. In the meantime, new proteins containing this sequence element have been found, notably the copper ATPases and the copper chaperones. The latter are proteins of approximately 70 amino acids that bind copper and route it intracellularly (Harrison *et al.*, 2000). They are members of a homologous family that includes representatives from humans (HAH1/ATOX1), *Caenorhabditis elegans* (CUC-I), *Arabidopsis* (CCH), yeast (Atx1), and bacteria (CopZ) (Klomp *et al.*, 1997; Wakabayashi *et al.*, 1998; Himelblau *et al.*, 1998; Lin and Culotta, 1995; Wimmer *et al.*, 1999). A related bacterial chaperone is MerP, which routes mercury in the intermembrane space to the uptake system of mercury-resistant bacteria (Steele and Opella, 1997). The function of copper chaperones will be discussed in more detail in Section III, C of this chapter.

Figure 4A gives a schematic overview of the occurrence of CxxC motifs, suggesting that the sequence is a feature of proteins that interact with heavy metal ions. Detailed structural studies of Atx1 of yeast, the human homologue HAH1 (more recently called ATOX1), MerP and CopZ of bacteria, and MNKr4, the fourth metal-binding domain of the Menkes copper ATPase, have shown that these proteins/domains have the same fold (Rosenzweig *et al.*, 1999; Wernimont *et al.*, 2000; Steele and Opella, 1997; Wimmer *et al.*, 1999; Gitschier *et al.*, 1998). The fold consists of four β-strands forming an antiparallel β-sheet, situated below two α-helices (Fig. 4B). The αβααβα arrangement of secondary structure elements is characteristic of the ferredoxin-like proteins and is colloquially known as an "open-faced β-sandwich" (Gitschier *et al.*, 1998). The CxxC metal-binding motif occurs on the mobile loop between the first β-strand and the first α-helix. This motif may bind $Hg^{2+}$ as in the periplasmic MerP mercury chaperone and mercuric reductases or $Cu^{+}$, $Cd^{2+}$, or $Ag^{+}$ in other proteins shown in Fig. 4A.

### *D. The HP Locus*

A HP dipeptide motif is universally present in the CPx-type ATPases but absent in other P-type ATPases (Solioz and Vulpe, 1996). It is located 34 to 43 amino acids C-terminal to the phosphorylated aspartic acid residue. In the $Ca^{2+}$-ATPase, this region is divided into two clearly separated domains, the phosphorylation domain (P), extending roughly 8 amino acids beyond the DKTGT phosphorylation site, and the nucleotide-binding domain (N), formed by the remainder of the large cytoplasmic loop (cf. Fig. 3). By analogy, the HP motif of heavy metal ATPases would be located in the N-domain near the ATP-binding site, but there is no recognizable sequence similarity between the $Ca^{2+}$-ATPase and copper ATPases in the region of the HP motif. In Wilson copper ATPase (*ATP7B*), which is expressed mainly in the liver and is required for copper secretion via the bile, more than 100 point mutations have been identified. Mutation of the histidine in the HP motif H1069Q is found in

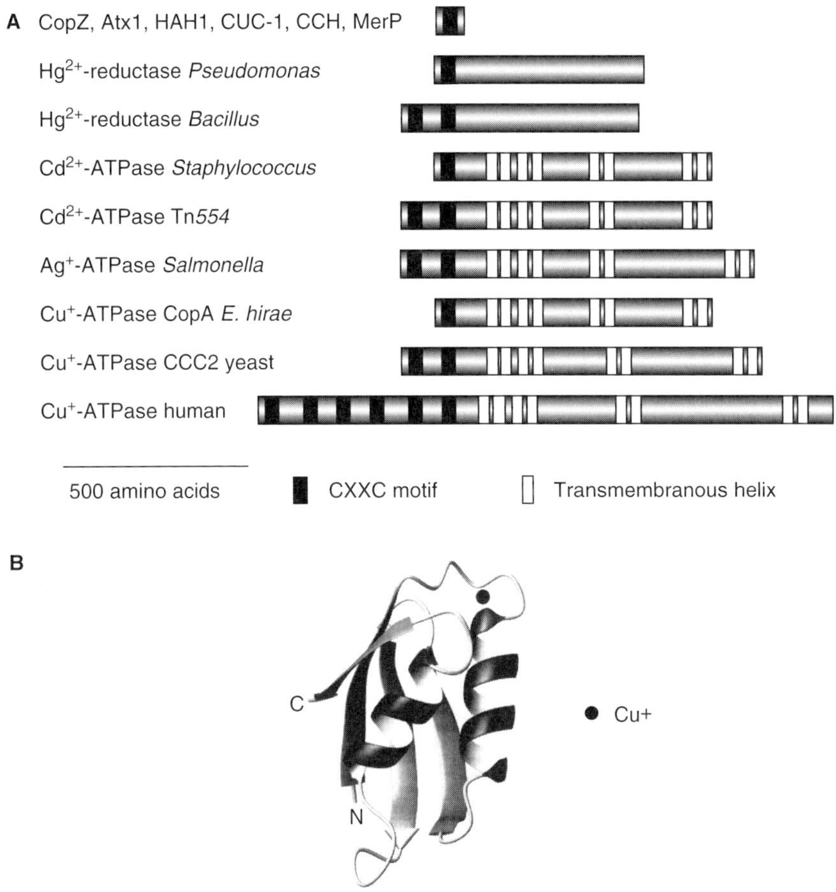


FIG. 4. (A) Schematic representation of the occurrence of CxxC motifs in various proteins. The polypeptide chains are drawn to scale as boxes. Transmembrane helices are indicated by empty rectangles and CxxC motifs by filled rectangles. (B) Ribbon model of the structure of CopZ. The position of the copper ion is inferred. Other metallochaperones and the CopZ-like building blocks shown in (A) probably all have a very similar structure.

30–40% of the patients in North America and northern Europe (Shah *et al.*, 1997; Forbes and Cox, 1998). Presumably, this renders the Wilson copper ATPase nonfunctional, thus causing the accumulation of copper in the liver and subsequent liver damage as well as neurological symptoms.

The impact of the H1069Q mutation on Wilson ATPase function has previously been tested by functional complementation, but the findings remained contradictory. When expressed in fibroblast of the mottled mouse, a model for Menkes disease, the Wilson ATPase gene carrying the H1069Q mutation could not rescue the mottled phenotype, while a wild-type Wilson gene did. The mutant enzyme mislocalized to the endoplasmic reticulum at normal growth temperatures and was degraded more rapidly than wild-type Wilson ATPase (Payne *et al.*, 1998). In contrast to this finding, several groups have shown that the H1069Q-ATPase could rescue the iron uptake-deficient phenotype of a yeast *Ccc2* knockout strain (Iida *et al.*, 1998; Forbes and Cox, 1998). It had been speculated that this was due to overexpression of the mutated, yet slightly active protein (Forbes and Cox, 1998). Since expression was determined with antibodies on Western blots, it is inherently not possible to compare expression of complementing ATPase to normal expression of the endogenous Ccc2 copper ATPase in this system.

In CopB of *En. hirae*, it was shown that H480Q-CopB, corresponding to the H1069Q Wilson ATPase, did not complement a CopB knockout strain. *In vitro*, H480Q-CopB exhibited residual ATPase activity. The mutation did not significantly affect the $K_m$ for ATP, but reduced $V_{max}$ over 40-fold (Bissig *et al.*, in press). This suggests that the HP motif is not involved in ATP binding, but is essential in a later step of the pump cycle, such as in coupling ATP hydrolysis to copper transport. Interestingly, an HP locus is also conserved between a group of related copper proteins, including CopA from *Pseudomonas syringae* and a similar protein from *Xanthomonas campestris* (see below), laccase from four different fungi, and ascorbate oxidase from cucumber (Lee *et al.*, 1994). The function of this motif in these proteins has thus far not been investigated.

## III. Copper Homeostasis in *Enterococcus Hirae*

The *En. hirae cop* operon consists of four closely spaced genes in the order: *cop*Y, *cop*Z, *cop*A and *cop*B. The openn is located on the chromosome, in contrast to many bacterial resistance systems, which are typically plasmid-borne. *Cop*Y and *cop*Z encode regulatory proteins, whose function is described below, while *cop*A and *cop*B encode CPx-type ATPases of 727 and 745 amino acids, respectively (Odermatt *et al.*, 1993). Figure 5 provides a summary of the current understanding of copper homeostasis in *En. hirae*. CopA and CopB were the first copper ATPases to be described (Odermatt *et al.*, 1992) and were cloned fortuitously during attempts to clone a potassium ATPase, using an antibody of low specificity. The sequence similarity of the histidine-rich N-terminus of CopB to the 120-amino-acid periplasmic CopP copper-binding protein of *P. syringae* initially gave the clue to an involvement of CopB in copper homeostasis.

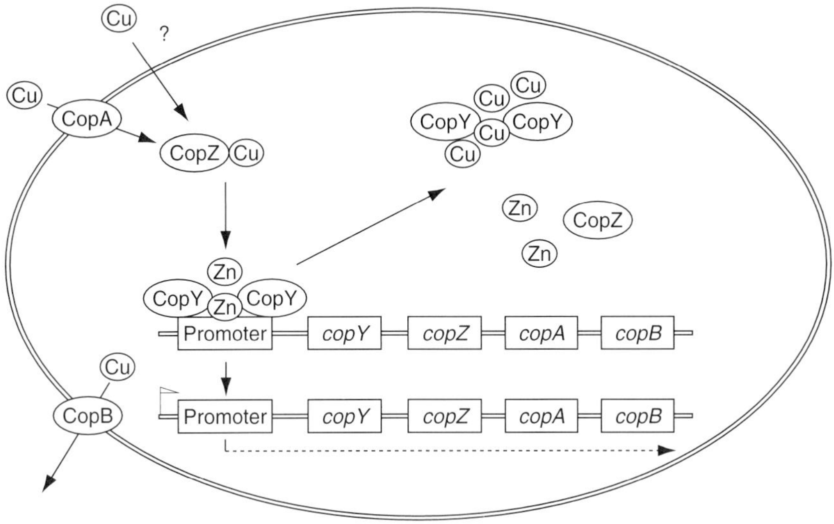


FIG. 5. Copper homeostasis in *Enterococcus hirae*. Under copper-limiting conditions, copper is pumped into the cell by CopA. The CopZ copper chaperone picks up copper at this site of entry. Under physiological copper conditions, Zn(II)CopY binds to the promoter and represses transcription of the *cop* operon. Under conditions of copper excess, Cu-CopZ donates Cu(I) to CopY, which leads to the replacement of the Zn(II), loss of DNA-binding affinity, and ultimately synthesis of the operon products. Excess copper is secreted by the CopB efflux pump. The substrate for this pump may be a copper–glutathione (GSH) complex, rather than Cu-CopZ.

### *A. Function of CopA in Copper Uptake*

CopA of *En. hirae* exhibits 43% sequence identity with the human Menkes and Wilson copper ATPases; in the transduction domain, sequence identity between these enzymes is even 92%. This suggests that CopA is a representative model of a copper ATPase. Based on indirect evidence, CopA appears to function in copper uptake. Cells disrupted in *cop*A cease to grow in medium in which the copper has been complexed with 8-hydroxyquinoline or *o*-phenanthroline. This growth inhibition can be overcome by adding copper to the growth medium. Null mutants in *cop*A could grow in the presence of 5 μM $AgNO_3$, which fully inhibits the growth of wild-type cells. The CopA ATPase thus appears to be a route for the entry of copper as well as silver into the cell (Odermatt *et al.*, 1993). Silver transport by CopA is probably fortuitous as silver has no known biological role. The transport of Ag(I) by CopA is an indication that Cu(I) rather than Cu(II) is transported by CopA.

Interestingly, a screen for virulence genes in *St. aureus* revealed a gene, *ivi44*, that encodes a protein with 50% sequence identity to CopA of *En. hirae* (Lowe *et al.*, 1998). This suggests that copper is a limiting nutrient in pathogenesis and copper import becomes a cellular priority when bacteria infect a host. It would be interesting to test a *copA* knockout strain of *En. hirae* for its ability to infect a host.

CopA of *En. hirae* could be expressed in *Escherichia coli* and purified to homogeneity by Ni-NTA affinity chromatography by means of an added histidine tag (Wunderli-Ye and Solioz, in press). Purified CopA has a pH optimum of 6.3 and a $K_m$ for ATP of 0.2 mM. The enzyme forms an acylphosphate intermediate, which is a hallmark of P- and CPx-type ATPases (Pedersen and Carafoli, 1987b). Purified CopA can now serve in the analysis of mechanistic aspects of copper transport and in the characterization of structure–function relationships.

Using purified CopA and CopZ, it could be shown by surface plasmon resonance analysis that the two proteins interact directly with one another (Multhaup and Solioz, unpublished results). This interaction was significantly reduced, but not abolished, when the CxxC copper-binding motif of CopA was mutated to SxxS. Thus, the interaction of CopA and CopZ not only takes place by virtue of bound copper, but also involves direct protein–protein interaction between CopA and CopZ. This allows refinement of the model of copper circulation in *En. hirae* to include copper transfer from CopA as a site of copper entry into the cell to the CopZ chaperone, which in turn donates copper to the CopY repressor and probably other cuproenzymes (cf. Fig. 5).

### *B. Function of CopB in Copper Excretion*

CopB was shown to catalyze ATP-driven copper(I) and silver(I) transport into native membrane vesicles of *En. hirae*. Since only inside-out oriented ATPase molecules were active in this transport assay, this corresponds to copper extrusion by CopB *in vivo*. Copper transport by vesicles took place only under reducing conditions. Cu(I) rather than Cu(II) was thus the transported species. Use of null mutants in *cop*A, *cop*B, or *cop*A and *cop*B made it possible to attribute the observed transport to the activity of the CopB ATPase. Copper transport exhibited an apparent $K_m$ for $Cu^+$ of 1 μM and a $V_{max}$ of 0.07 nmol/min/mg of membrane protein. $^{110m}Ag^+$ was transported with a similar affinity and at a similar rate (Solioz and Odermatt, 1995). Since $Cu^+$ and $Ag^+$ were complexed to Tris buffer and dithiothreitol present in the assay, the $K_m$ values must be considered as relative only. The results obtained with membrane vesicles were further supported by evidence of $^{110m}Ag^+$ extrusion from whole cells, preloaded with this isotope. Again, transport depended on the presence of functional

CopB, and the corresponding knockout strains exhibited no silver extrusion (Odermatt *et al.*, 1994). These findings suggest that CopB functions as a $Cu^+/Ag^+$-ATPase for the export of $Cu^+$ and $Ag^+$ *in vivo*.

Vanadate, a diagnostic inhibitor of P-type ATPases, showed an interesting biphasic pattern of inhibition of ATP-driven copper and silver transport by CopB: maximal inhibition was observed at 40 μM $VO_4^{3-}$ for $Cu^+$ transport and at 60 μM $VO_4^{3-}$ for $Ag^+$ transport. At higher concentrations of vanadate the inhibition of transport was reversed. This behavior is unexplained at present, but may relate to the complex chemistry of vanadate, involving many oxidation states and polymeric forms of vanadate (Pope and Dale, 1968).

### *C. Regulation of Expression by Copper and Copper Chaperone Function*

The two copper ATPases of *En. hirae* exhibited biphasic regulation: induction of the genes is lowest in standard growth medium (average copper content, 10 μM). If medium copper is increased, expression is increased up to 50-fold at 2 mM extracellular copper. Induction was also observed by 5 μM $Ag^+$ or 5 μM $Cd^{2+}$. The induction by silver and cadmium was in all likelihood fortuitous, since *En. hirae* does not exhibit significant resistance to these highly toxic metal ions. Interestingly, a high level of induction was also observed when copper was depleted from the medium. Since CopA serves in copper uptake and CopB in its extrusion according to the current model, this coinduction of CopA and CopB by high and low concentrations of copper seems puzzling. It could be a safety mechanism: if cells express, under copper-limiting conditions, only the import ATPase, they would become highly vulnerable to copper poisoning in the event of a sudden increase in ambient free copper, such as by acidification of the ambient.

Regulation of the *cop* operon is accomplished by *Cop*Y, a repressor protein of 145 amino acids (Odermatt and Solioz, 1995). The N-terminal half of CopY exhibits approximately 30% sequence identity to the bacterial repressors of β-lactamases, MecI, PenI, and BlaI (Himeno *et al.*, 1986; Suzuki *et al.*, 1993; Hackbarth and Chambers, 1993). In the best studied of these, PenI, the N-terminal portion appears to be the domain that recognizes the operator (Wittman and Wong, 1988). In the C-terminal half of CopY, there are multiple cysteine residues, arranged as $CxCx_4CxC$. A consensus motif, $CxCx_{4-5}CxC$, is also found in the yeast copper-responsive transcriptional activators for metallothionein, ACE1 and AMT1 (Zhou and Thiele, 1991; Dobi *et al.*, 1995), in the MAC1 transcription factor for the Ctrl copper transporter of *Saccharomyces cerevisiae* (Jungmann *et al.*, 1993), in Grisea, the MAC1 orthologue of *Podospora anserina*, and in the N-terminal β-domain of human metallothionein-2 (Fig. 6).

| | |
|---|---|
| CopY | etie**C**n**C**ipgq.**C**e**C**kkq |
| Mac1 | lstq**C**s**C**.edes**C**p**C**vnc |
| AMT1 | nmdm**C**l**C**vrgep**C**r**C**har |
| ACE1 | detr**C**r**C**degep**C**k**C**htk |
| Grisea | plst**C**p**C**ppgkp**C**v**C**ggv |
| MT-2 β-domain | mdpn**C**s**C**atggs**C**t**C**tgs |
| **Consensus** | **CxC** $(X_{4-5})$ **CxC** |

FIG. 6. Occurrence of the $CXCX_{(4-5)}$ CXC consensus motif. CopY, *cop* operon repressor protein from *Enterococcus hirae*; Mac1, transcription factor for the Ctrl copper transporter of *Saccharomyces cerevisiae*; AMT1, transcription factor for metallothionein from *Candida albicans*; ACE1, transcription factor for metallothionein from *Sa. cerevisiae*; Grisea, MAC1 orthologue of *Podospora anserina*; MT-2 β-domain, N-terminal domain of human metallothionein-2.

Disruption of the *En. hirae cop*Y gene resulted in constitutive overexpression of the *cop* operon (Odermatt and Solioz, 1995). Binding of CopY to an inverted repeat sequence upstream of the *cop*Y gene has been demonstrated *in vitro* (Strausak and Solioz, 1997). CopY binds to DNA as a Zn(II)CopY complex. For the release of CopY from the DNA and induction of the operon, Cu(I)CopZ donates copper to CopY, thereby displacing the bound Zn(II) and releasing CopY from the DNA (Cobine *et al.*, 1999). The release of zinc was monitored by the spectral shift of the absorbance maximum from 412 to 500 nm of 2, 4-pyridylazoresorcinol (PAR) upon zinc binding (Cobine and Dameron, unpublished results). Zinc is presumably bound to the CxCxxxxCxC motif of CopY, which has been shown by EXAFS to be the site for copper insertion. PAR by itself was unable to extract Zn from this site, suggesting that the site is poorly accessible or that the zinc ions are very tightly chelated. PAR titration of copper-induced zinc release by CopY showed a stoichiometry of 1 Zn(II) per CopY monomer. The Cu(I)CopY complex exhibited luminescence, indicating that Cu(I) was sequestered in an environment where it was protected from solvent. It is plausible that the Cu(I) ions are being sequestered in a Cu(I)-thiolate cluster as found in the Cu(I)-regulated transcription factor ACE1 and the metallothioneins (Dobi *et al.*, 1995). The inability of the displaced Zn(II) to bind to CopZ indicates that the metal-binding site in CopZ is specific for Cu(I). Hence, the combined findings of X-ray absorption spectroscopy and luminescence studies show that CopY binds one zinc ion to the CxCxxxxCxC motif, located in a protected, solvent-inaccessible core (Cobine *et al.*, submitted for publication). This cysteine consensus motif also occurs in other metal-binding proteins, as summarized in Fig. 6.

The transfer of Cu(I) from the CxxC metal-binding site of CopZ is probably driven by the higher affinity of the more cysteine-rich

CxCxxxxCxC metal-binding site in CopY. The failure of the structural analog MNKr2, the second N-terminal metal-binding domain of the Menkes ATPase, to deliver its copper to CopY suggests that the presence of the conserved metal-binding site and global fold is not sufficient to effect copper transfer. Rather, the mechanism probably involves the specific docking of the chaperone to the recipient protein, with subsequent transfer of the metal ion. A two-step mechanism of this type would help protect the cell from the toxic effects of copper ions by preventing their nonspecific release to inappropriate sites.

Figure 7 shows a scheme for the proposed steps in copper transfer from CopZ to CopY. Initially, the first $Cu^+$-CopZ docks on CopY by electrostatic interaction of the two proteins involving sites away from the metal-binding domains of the two proteins. (1) The flexible loop of $Cu^+$-CopZ is positioned close enough to allow for ligand attack by one of the metal-binding sulfurs of CopY. (2) This interaction leads to a cascade of ligand exchanges from the less favorable diagonal to the more stable triagonal coordination of $Cu^+$ in CopY. (3) The exchange of $Cu^+$ from $Cu^+$-CopZ to CopY then causes *apo*-CopZ to change the positioning of charged residues, resulting in dissociation from CopY. (4–6) The interaction of a second $Cu^+$-CopZ causes a renewed cascade of ligand exchanges, which results in the displacement of $Zn^{2+}$ from CopY and movement of the first $Cu^+$ deeper into CopY, followed by the second $Cu^+$. Clearly, understanding the molecular steps in the $Cu^+$-CopZ to $Zn^{2+}$-CopY copper transfer is a complicated reaction with many intermediates and requires further investigation.

## IV. Copper Resistance in *Escherichia coli*

### *A. The ecCopA Copper ATPase*

ORF *f834* of *Es. coli* encodes an 834-residue P-type ATPase that exhibits 36% identity with CopA from *En. hirae* (Rensing *et al.*, 2000). Since the gene product of *f834* could be shown to catalyze copper export, the gene was renamed *copA* and the gene product CopA (hereinafter called ecCopA to differentiate it from *En. hirae* CopA). EcCopA exhibits all the structural features of CPx-type ATPases. Interestingly, ecCopA possesses two N-terminal CxxC motifs compared to related bacterial copper ATPases, which contain only one such motif (Odermatt *et al.*, 1993; Ge *et al.*, 1995; Phung *et al.*, 1994). The presence of these two copper-binding motifs may serve as a better working model for the human copper ATPases, which have six motifs. However, the exact role of the multiple motifs remains controversial since conflicting results on their

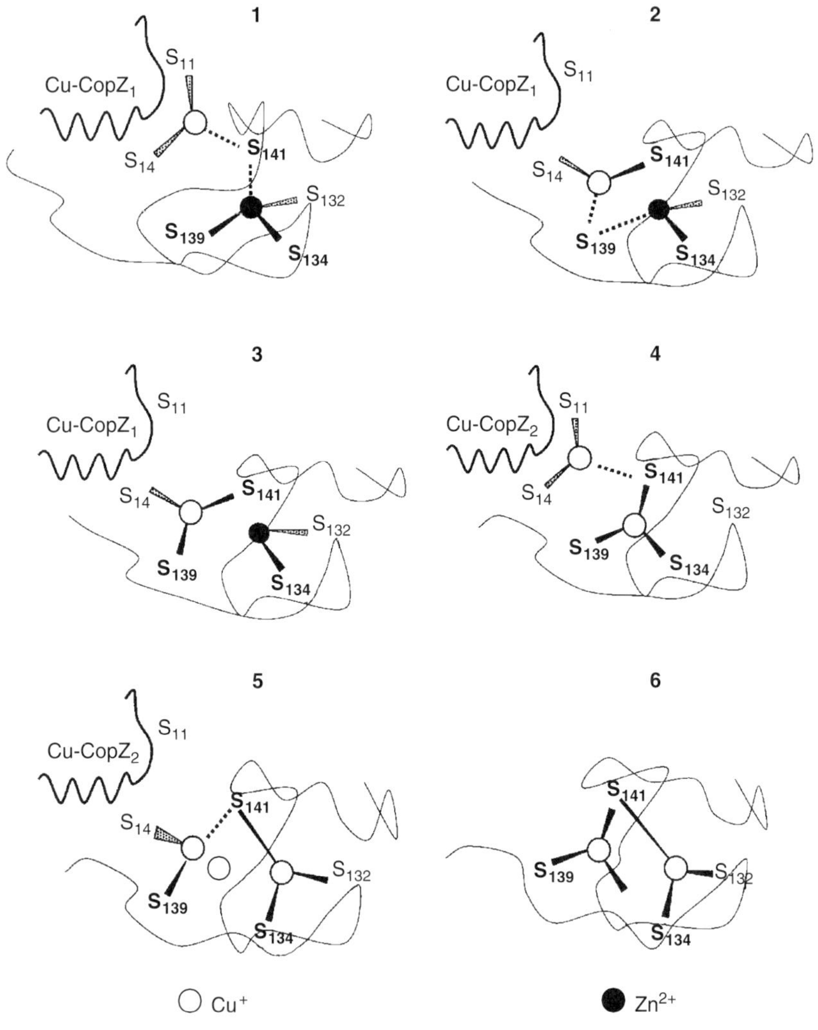


FIG. 7. Schematic representation of copper transfer from CopZ to CopY. $Cu^+$ ions from two subsequently docking Cu-CopZ molecules are transferred to the trigonal binuclear cluster in CopY, which results in the displacement of the zinc ion and reorientation of the sulfur residues. Bold "S" residues are in the plane; "S" residues are out of the plane of the drawing. The boldface curly line represents part of the backbone of CopZ and the fine curly line represents part of the backbone of CopY. See text for further details of the transfer intermediates 1–6 (courtesy of Charles T. Dameron).

copper-translocating function have been reported (Voskoboinik *et al.*, 1999; Payne and Gitlin, 1998; Iida *et al.*, 1998).

*Escheridia coli* cells with a disrupted *copA* gene exhibit decreased resistance to copper. This apparent copper sensitivity could be complemented by introduction of a plasmid expressing ecCopA or CopB of *En. hirae*. *Cop*A-disrupted strains were still relatively resistant to copper salts, which can be attributed to other genes involved in copper tolerance in *Es. coli* (Silver and Phung, 1996). However, these other functions that participate in copper resistance are still largely unclear.

ATP-dependent uptake of copper into everted membrane vesicles from cells expressing ecCopA could be demonstrated. Transport was inhibited by the classical P-type ATPase inhibitor vanadate. Dithiothreitol, a strong reductant, was required for ecCopA-catalyzed $^{64}Cu$ uptake, suggesting that the substrate of ecCopA is Cu(I). Thus the function of ecCopA resembles that of the *En. hirae* CopB ATPase by functioning as a copper efflux pump *in vivo* when excess copper is present in the cytoplasm (Rensing *et al.*, 2000).

A role in copper transport in *En. Coli* has also been attributed to the products of at least six chromosomal genes, *cutA*, *cutB*, *cutC*, *cutD*, *cutE*, and *cutF* (Brown *et al.*, 1991). Mutation of one or more of these genes resulted in an increased copper sensitivity. Only the *cutC* and *cutF* genes were cloned and sequenced. The *cutC* gene encodes a cytoplasmic protein of 146 amino acids with an N-terminus containing an MxxMxxxM motif similar to the copper-binding motif MxxxMxxM of the *En. hirae* CopB ATPase. The *cutF* gene, which is identical to the *nlpE* gene, encodes a putative outer membrane protein with a metal-binding motif that is also similar to putative metal-binding motifs in CopB of *En. hirae* (Gupta *et al.*, 1995). *CutC* and *cutF* mutants accumulated copper and had decreased copper efflux. The two genes were also shown to contribute to copper tolerance by *Es. coli*. CutC was postulated to remove excess cytoplasmic copper by acting as an efflux protein and the CutF was postulated to be involved in copper efflux by acting as a copper metallochaperone; no further information on the function of these proteins is currently available. Also, no copper ATPase uptake system has been identified in *En. coli* thus far and neither has a copper metallochaperone similar to CopZ of *En. hirae* been discovered.

### *B. Regulation of the Escherichia coli Copper ATPase*

The copper-inducible regulator CueR, which activates the transcription of the ecCopA copper efflux system, has recently been identified in

*Es. coli* DH5α and K-12 (Outten *et al.*, 2000; Stoyanov *et al.*, 2001). The protein is encoded by the *ybbI* gene, which was renamed *cueR* for copper efflux regulator. The CueR protein is related to the MerR family of metalloregulatory proteins (Summers, 1992). In fact, the gene was discovered through the inspection of the *copA* promoter, which unveiled signature elements of the promoter regulated by MerR. These same elements are also present upstream of *yacK* (renamed *cueO* for Cu efflux oxidase). Homologues of this putative multicopper oxidase are also found in the *P. syringae* and *Es. coli* plasmid-based copper resistance operons (Cooksey, 1994) described in more detail in Section V.

The promoters of ecCopA and YacK are apparently regulated by CueR. Both copper and silver are inducers, but not zinc or mercury. The loss of copper activation at both promoters in the *cueR* deletion strain can be rescued by complementing it with a plasmid carrying the gene. Interaction between CueR and the ecCopA promoter has been shown with a DNase I protection assay. It showed that CueR binds *in vitro* to a sequence with dyad symmetry within a 19-spacer sequence in the promoter (Stoyanov *et al.*, 2001). CueR is thus the primary copper-responsive activator of the ecCopA copper efflux system of *Es. coli*.

## V. Other Bacterial Copper ATPases

### *A. Synechococcus Copper ATPases*

*Synechococcus* PCC7942 is a Gram-negative bacterium that harbors a photosynthetic apparatus (thylakoid) similar in structure and function to the chloroplasts of phototrophic eukaryotes. Two CPx-type copper-importing and -exporting ATPases, CtaA (Phung *et al.*, 1994) and PacS (Kanamaru *et al.*, 1993, 1994), have recently been cloned from this organism.

The *cta*A gene was found fortuitously during an attempt to clone the biotin-carboxyl carrier protein with a DNA probe. This probe, by coincidence, had a perfect 17-bp match with *cta*A. *Cta*A encodes a CPx-type ATPase of 790 amino acids with a conserved intramembranous CPC sequence and the N-terminal CxxC metal-binding motif (Phung *et al.*, 1994). Disruption of the *cta*A gene by cassette mutagenesis resulted in a strain that still showed some growth in 10 μM $Cu^{2+}$ while wild-type cells were completely inhibited. It also retained better viability in the presence of copper, while other cations tested had no effect on both the wild-type and the mutant strains. However, the change in copper resistance was marginal, with wild-type and mutant showing nearly the same growth behavior in 3 μM $Cu^{2+}$. Nevertheless, the fact that the mutant cells are more tolerant of $Cu^{2+}$ and the fact that CtaA has the most extensive sequence similarity to the CCC2 copper ATPase of yeast suggest that CtaA is a copper uptake ATPase.

The *pac*S gene, on the other hand, was discovered in a systematic search for P-type ATPases in *Synechococcus* PCC 7942 (Kanamaru *et al.*, 1993). PCR fragments were initially generated from chromosomal DNA, using degenerated primers for the phosphorylation and the ATP-binding domains of P-type ATPases. These fragments were then used to screen a library and one of the genes cloned was *pac*S. It encodes a protein of 747 amino acids with all the common conserved features of copper-transporting CPx-type ATPase (Kanamaru *et al.*, 1994). The level of *pac*S mRNA increased specifically in response to copper and silver: it was induced 20- to 30-fold by 5 μM $Cu^{2+}$ or 40 μM $Ag^{+}$, but not by metal-depleted growth medium. Growth of a *pac*S-deleted strain was inhibited by 5 μM $Cu^{2+}$ or 25 μM $Ag^{+}$ while the wild-type was not adversely affected. The significance of PacS, however, is its localization in the thylakoid membranes. Although PacS has been speculated to function as a copper uptake system in the thylakoid lumen, the copper hypersensitivity of the mutant suggested the opposite role. In addition, since the expression level of plastocyanin, the electron carrier of the photosystem in *Synechococcus*, is unaffected by variable copper concentration (Clarke and Campbell, 1996), the ability to avoid excess accumulation of copper in the thylakoid becomes more important. Therefore, unlike most of the reported bacterial P-type ATPases, PacS is involved in copper equilibrium within intracellular subcompartments instead of between extra- and intracellular compartments. Although two copper ATPases are now known in *Synechococcus* PCC 7942, the picture of copper homeostasis in this organism is far from complete.

### *B. The Helicobacter pylori Copper ATPases*

*Helicobacter pylori* is a curved, microaerophilic Gram-negative bacterium that currently receives attention because of its association with chronic active type B gastritis in humans (Dick, 1990). Efficient treatment of *H. pylori* infections can be accomplished with a combination of the antibiotics roxithromycin and omeprazole (Cellini *et al.*, 1991). Omepra zole is a pro-drug that is converted to the active form in the acid environment of the stomach. When activated, it reacts with sulfhydryls and strongly inhibits the gastric P-type $K^{+}$, $H^{+}$-ATPase and also inhibits the growth of *H. pylori*.Because of this dual effect, it seemed reasonable to search for an essential, omeprazole-sensitive ATPase in *H. pylori*. Furthermore, a deficiency of copper in local gastric epithelial and endothelial cells has been reported in *H. pylori*-associated gastritis. It is probable that copper may play a role in the pathogenesis of this organism (Taha *et al.*, 1995). Therefore, many laboratories have set out to clone P-type ATPases from this bacterium.

A *copAP* operon was recently cloned from *H. pylori* by two independent groups (Ge *et al.*, 1995; Bayle *et al.*, 1998). The first gene on this operon, *hpcopA*, encodes a protein of 741 amino acids that exhibits signature features common to all CPx-type ATPases, namely, the phosphorylation sequence, the ATP-binding domain, the conserved CPC intramembrane motif, and the N-terminal metal-binding CxxC motifs. The membrane topology of hpCopA was experimentally demonstrated by an *in vitro* transcription/translation/glycosylation system (Bamberg and Sachs, 1994). The ATPase was found to have eight transmembrane segments (H1 to H8), with the phosphorylation and ATP-binding domains localizing on the loop between H6 and H7. This topology is in line with previous models and it is so far the only data available on the membrane topology of CPx-type ATPases. The function of hpCopA was first determined with a *hpcopA* gene knockout mutant. It was found to be more sensitive to $Cu^{2+}$: growth of wild-type was inhibited by 50 μM $Cu^{2+}$ while the mutant could withstand only 7.5 μM $Cu^{2+}$, suggesting that hpCopA is a copper efflux pump (Bayle *et al.*, 1998). Interestingly, when only the N-terminal metal-binding motif was deleted, the mutant was able to tolerate almost 10 times more $Cu^{2+}$ (Ge and Taylor, 1996). Studies using ion affinity chromatography and electrospray ionization mass spectrometry have shown that the N-terminal motif exhibits affinity for $Cu^{2+}$, further supporting the role of this enzyme in copper transport.

The second gene downstream of *hpcop*A, termed *hpcop*P, encodes a protein of 66 amino acids. HpCopP has the most extensive amino acid similarity to the *En. hirae* CopZ copper chaperone, but expression of hpCopP or a function similar to that of CopZ in intracellular copper routing has not been demonstrated (Ge and Taylor, 1996). It may be worthy to note that the *hpcop* operon, in contrast to the *En. hirae cop* operon, lacks a CopB-like copper ATPase and a repressor. In addition, the CopP chaperone is encoded downstream of the ATPase in *H. pylori*, while in *En. hirae* it is located upstream of the ATPase genes. Given these and other differences, the two organisms may differ significantly in their copper homeostatic mechanisms.

### *C. The Copper ATPase of Listeria monocytogenes*

*Lesteria monocytogenes* is a gram-positive bacterium responsible for serious foodborne diseases, such as the commonly recognized meningitis, in humans and animals. Among the various virulence factors investigated to play a role in the pathogenicity of the bacterium is the regulation of virulence gene expression according to the variations in the environmental metal ion concentrations. In fact, a significant number of *L. monocytogenes* strains are resistant to heavy metals, such as cadmium (Lebrun *et al.*, 1992).

CtpA of *L. monocytogenes* appears to be a CPx-type ATPase involved in copper homeostasis (Francis and Thomas, 1997b). However, it is different from the other CPx-type ATPases in that CtpA lacks the common N-terminal metal-binding motif. This may have been lost during cloning. Nevertheless, it possesses all the other typical features, such as the CPC motif in membrane helix 6 and the HP motif in the second cytoplasmic loop. Second, Francis and Thomas (1997b) proposed a membrane topology for CtpA with six transmembrane helices only, but a membrane helix assignment in accord with the eight membrane spans exhibited by other CPx-type ATPases is possible.

Growth of *ctpA* insertion mutants on agar medium was significantly slower than that of wild-type strains and was inhibited by a 10 μM concentration of the copper-chelating agent 8-hydroxyquinoline. On the other hand, growth of the knockout mutants at 4 mM$CuSO_4$ was comparable to that of wild-type. Another interesting feature of CtpA is that the expression levels of its mRNA were increased by growth in medium containing both low (5 μM 8-hydroxyquinoline) and high copper (4 mM$CuSO_4$) concentrations. This biphasic regulation resembles that observed for the *cop* operon of *En. hirae* (see above).

Francis and Thomas (1997a) then proceeded to investigate the virulence nature of CtpA. A mutant strain with the *ctpA* gene disrupted by an antibiotic-resistance cartridge was compared to the wild-type in tissue culture invasion assays and mouse infection studies. Growth of mutant and wild-type strains in J774 and HeLa cell was not significantly different. However, recovery of mutants from tissue, specifically liver, of infected mice was dramatically reduced compared to wild-type. Also, in the *in vivo* mixed-infection competition experiments, persistence of the mutants in livers and spleens of the infected mice was dramatically impaired. These results demonstrate a role of CtpA in establishing an *in vivo* infection by *L. monocytogenes*. Possibly, CtpA has a role in copper accretion by *L. monocytogenes*, which is expected to be more difficult during infection of a host than under culture conditions because in infected human and laboratory animals, concentrations of trace metals can vary significantly in response to systemic inflammation (Beisel, 1977). CtpA would thus be the first P-type ATPase described to be associated with pathogenicity. In support of such an interpretation are the interesting findings by Lowe *et al.*, (1998). By searching for virulence genes in *St. aureus* by *in vivo* expression technology, one of the virulence genes they identified was homologous to *copA* of *En. hirae* and thus possibly a copper uptake ATPase.

## VI. Mechanism of Copper ATPases

Studies on the mechanism of copper ATPases require highly enriched membrane preparations or purified enzyme. Naturally enriched membranes can be obtained only from specialized membrane compartments such as the sarcoplasmic reticulum or the electric organ of eels. Since copper is a toxic trace element, it is never encountered in large quantities in cells and copper ATPases are expressed at only low levels. However, the purification of the CopA and CopB copper ATPases of *En. hirae* has recently been reported (Wunderli-Ye and Solioz, in press; Bissig *et al.*, in press). Both enzymes have been shown to form acylphosphate intermediates in purified form, reconstituted into proteoliposomes (Wunderli-Ye and Solioz, in press; Wyler-Duda and Solioz, 1996). Acylphosphate formation has also been demonstrated for the human Menkes ATPase in native membrane vesicles (Solioz and Camakaris, 1997). Further mechanistic details are, however, not available.

Acylphosphate formation is characteristic for P-type ATPases and involves the transfer of the $\gamma$-phosphate of ATP to an aspartic acid residue to form a high-energy enzyme intermediate. The phosphorylated aspartic acid residue is located in the sequence DKTGT, which is universally conserved in all members of the P-type superfamily. By this criterion, CopA and CopB of *En. hirae* are clearly members of the P-type superfamily of ATPases and probably function by the same underlying mechanism. Vanadate sensitivity is another hallmark of P-type ATPases. CopA and CopB were inhibited by vanadate with $I_{50}$ values of around 0.1 mM. This is a low vanadate sensitivity compared to $I_{50}$ values in the micromolar to submicromolar range observed for non-heavy metal P-type ATPases.

Figure 8 shows a scheme of the reaction cycle of copper ATPases, assuming that they work by a mechanism analogous to that of $Ca^{2+}$- or $Na^{+}$, $K^{+}$-ATPases. To pump ions, the enzyme must cycle between a state with a high-affinity copper-binding site accessible from only one side of the membrane and a low-affinity state in which the copper cavity is accessible from the other side of the membrane. The high- and low-affinity forms of P-type ATPases were initially named $E_1$ and $E_2$ by Racker (1980) and for many years these ATPases were called $E_1E_2$-ATPases, until they were renamed P-type by Pedersen and Carafoli (1987a).

## VII. Other Copper-Resistance Systems

Many microbes that thrive in environments contaminated with copper contain plasmid-borne copper-resistance systems. Such plasmid-borne

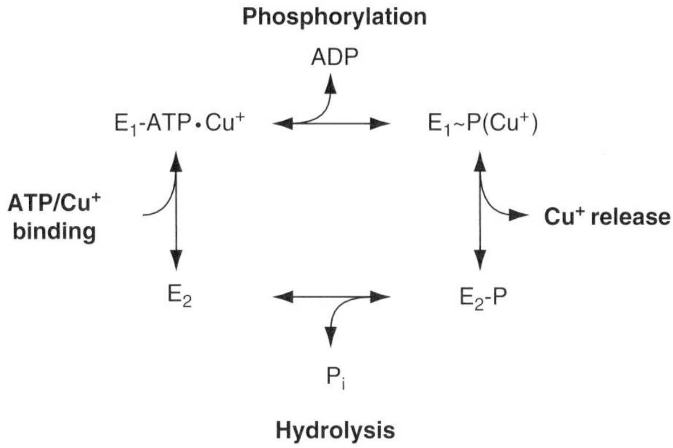


FIG. 8. Putative reaction cycle of a $Cu^+$ export ATPase. In analogy to non-heavy metal P-type ATPases, at least four interconvertible conformations probably occur. Cytoplasmic copper binds to a high-affinity $Cu^+$-binding site. This triggers the formation of a high-energy acylphosphate intermediate ($\sim$ P) by transferring the $\gamma$-phosphate of ATP to a designated aspartic acid residue. This occludes the $Cu^+$ within the ATPase and makes it inaccessible from either side of the membrane. The high-energy acylphosphate intermediate then fuels a conformational change that converts the high-affinity copper-binding site to a low-affinity site. At the same time, a gating mechanism opens this site to the exterior space and copper can be released. Water enters the catalytic site and hydrolyzes the aspartyl phosphate, followed by the release of inorganic phosphate ($P_i$).

systems have been studied in some detail in *Es. coli, P. syringae*, and *X. campestris* (Silver and Phung, 1996).

A plasmid carrying the *pco* operon has been isolated from a strain of *Es. coli* from an Australian pig farm where the diet of piglets had been supplemented with copper to increase growth (Williams *et al.*, 1993). Another similar operon, called the *cop* operon, has been identified on a plasmid of copper-resistant *P. syringae*, isolated from tomato cultures in California that had been treated with copper sprays to combat fungal infections (Cooksey, 1993). A plasmid carrying a related resistance system was also identified in copper-resistant strains of *X. campestris* from northern California (Lee *et al.*, 1994). The *pco* and *cop* operons encode four related structural genes, *pcoABCD* and *copABCD*, which are expressed from the upstream copper-inducible promoters P*pcoA* and P*copA*, respectively. The structural genes encode periplasmic and membrane proteins. They do not belong to any known family of cation transporters and their function in copper resistance and transport has not been firmly established. Despite the similarity of the predicted *pco* and *cop* gene products, the *pco* operon enhances copper efflux (Brown *et al.*, 1995), while the *cop* operon appears to primarily lead to copper sequestration (Cha and Cooksey, 1991). These differences are thus far unexplained and the mechanisms are not understood. It has been proposed that PcoA is a multicopper oxidase, but again firm proof is missing (Kaplan and O'Halloran, 1996). An additional open reading frame, *pcoE*, has been identified in the *Es. coli pco* operon, but not in the *P. syringae cop* operon (Brown *et al.*, 1995). *PcoE* is transcribed from its own promoter and does not appear to be a stringent requirement for copper resistance.

The *pco* and *cop* systems carry two-component regulatory systems required for copper induction of the resistance. The genes encoding PcoRS and CopRS are located immediately downstream of the *cop* operon and are expressed from a separate constitutive promoter. PcoRS and CopRS show homology to the family of two-component sensor/responder phosphokinase regulatory systems (Russo and Silhavy, 1993). PcoS and CopS are homologous to sensor histidine kinases and are predicted to be located in the cytoplasmic membrane with two loops extending into the periplasm. These proteins are envisioned to sense the ambient copper levels and to respond by phosphorylating the response regulators PcoR and CopR, respectively, thereby activating transcription (Mills *et al.*, 1993). CopR has been purified and shown to bind specifically to the promoter from the plasmid-borne *cop* operon to activate transcription (Mills *et al.*, 1994). Interestingly, the same activator also bound to the promoter from the homologous chromosomal *cop* locus. However, mutations in the plasmid-borne *copR* gene could not be complemented by its chromosomal homologue, suggesting that the plasmid-encoded and the chromosomally encoded CopR proteins do not function in the same manner (Mills *et al.*, 1993).

By hybridization and *in vivo* transcription of an RS-regulated promoter, it has been shown that some strains of *P. syringae* carry chromosomal homologues of the *copRS* genes (Lim and Cooksey, 1993). Using a genetic screen, two related chromosomal genes, *cusRS* (*ylcA ybcZ*), were more recently identified as being required for copper-induced expression of *pcoE* (Munson *et al.*, 2000). CusR and CusS are also similar to CopR and CopS of the plasmid-borne *cop* operon of *P. syringae* (see below) and SilR and SilS of the *sil* locus, isolated from a silver-resistant strain of *Salmonella* (Gupta *et al.*, 1999). The cusRS genes are also required for the copper-dependent expression of at least one chromosomal gene, designated *cusC* (*ylcB*), which is allelic to the recently identified virulence gene *ibeB* in *Es. coli* K1. The *cus* locus may comprise a copper ion efflux system, because the expression of *cusC* is induced by high concentrations of copper ions. Furthermore, the translation products of *cusC* and additional downstream genes are homologous to known metal ion antiporters (Munson *et al.*, 2000).

Copper resistance due to the systems under discussion here has been best characterized at the biochemical level for the *cop* system of *P. syringae*. The copper-resistant strains of this organism can grow in minimal medium with up to 2 mM added $Cu^{2+}$, while sensitive strains can survive only in 0.4 to 0.6 mM $Cu^{2+}$ (Bender and Cooksey, 1986). Resistant colonies grown on solid medium containing high copper turn bright blue and accumulate copper up to 1.8 mg/g dry weight of cells (Cha and Cooksey, 1991). The *cop* operon has been shown to be specifically induced by copper and all four genes must be expressed for full copper resistance (Mellano and Cooksey, 1988). CopA, CopB, and CopC of *P. syringae* have also been purified and characterized (Cha and Cooksey, 1991). CopA is a periplasmic protein containing the copper-binding motif DHxxMxxM and those found in type 1, type 2, and type 3 copper sites of eukaryotic multicopper oxidases. It was shown to bind 11 copper ions per monomer. CopB is an outer membrane protein and contains five of the copper-binding motifs that occur in CopA. However, the copper-binding stoichiometry of this protein remains to be determined. CopC is also a periplasmic protein and has been shown to bind one copper ion per monomer. CopD finally is localized at the inner membrane and is of unknown function. Interestingly, mutant strains of *P. syringae* that expressed only *copC* and *copD* were hypersensitive to copper, suggesting that these genes have a role in the uptake of copper into the cell (Cooksey, 1994). A model of the plasmid-borne *cop* system of *P. syringae* is illustrated in Fig. 9.

Copper-resistant as well as copper-sensitive strains of *P. syringae* and other pseudomonads contain chromosomal homologues of the plasmid-borne *cop* operon. Two chromosomal *cop* homologous regions were cloned. They hybridized either with copA and copB or with copA, copB, copC, and the copper-responsive regulatory gene copRS. Only the last gene conferred low levels of expressed copper-resistant proteins related to CopA and CopC. Interestingly, this operon displayed a high frequency of mutation to full copper resistance by mutation of the *copRS* homologous genes, resulting in increased CopA expression. The chromosomal *cop* homologue did not, however, complement site-specific mutations in the plasmid-borne *cop* genes (Lim and Cooksey, 1993).

## VIII. Conclusions

Today, copper homeostasis is a research area of intense interest and work in this field has recently uncovered several surprising new concepts of trace metal homeostasis, and more are likely to emerge. The molecular defects in the inherited disorders of copper metabolism, Menkes disease

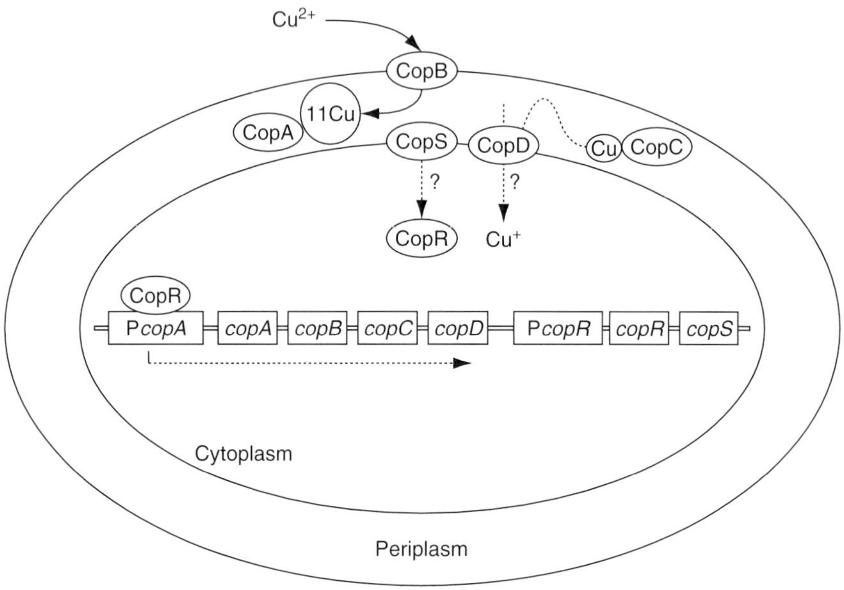


FIG. 9. Model of plasmid-encoded copper resistance in *Pseudomonas syringue*. The plasmid-encoded *P. syringae cop* system encompasses two transcriptional units. The *copRS* genes, transcribed from the constitutive P*copR* promoter, encode a two-component regulatory system that regulates the transcription of the *copABCD* operon from the P*copA* promoter. CopA may contribute to copper resistance by sequestering up to 11 copper ions in the periplasmic space. CopB is an outer membrane protein of unknown function that contains five copper-binding motifs. CopC is a periplasmic protein that binds one copper ion per monomer and may, together with the inner membrane protein CopD, have a role in copper uptake by the cell. See text for more details.

and Wilson disease, have been elucidated and clinical treatment can now be approached or improved. Study of the *En. hirae* model system has significantly contributed to the current understanding. It has shown modes of copper entry into and out of the cell by the action of copper ATPases, transcriptional control of copper homeostatic genes by a copper-responsive repressor, and intracellular copper routing by a copper chaperone. *Entencoccus hirae* CopA and CopB are the first copper ATPases to be purified. However, there is still a lack of information on copper ATPase structure and function. For some copper-resistance systems, like those encoded by the *Pseudomonas cop* genes or the *Es. coli cut* and *pco* genes, the mechanisms are still unclear and further work is needed.

## Acknowledgments

Part of the work described here was supported by Grant 32-56716.99 from the Swiss National Foundation and by a grant from the International Copper Association.

## References

Bamberg, K., and Sachs, G. (1994). *J. Biol. Chem.* **269**, 16909–16919.
Bayle, D., Wangler, S., Weitzenegger, T., Steinhilber, W., Volz, J., Przybylski, M., Schafer, K. P., Sachs, G., and Melchers, K. (1998). *J. Bacteriol.* **180**, 317–329.
Beisel, W. R. (1977). *Am. J. Clin. Nutr.* **30**, 1236–1247.
Bender, C. L., and Cooksey, D. A. (1986). *J. Bacteriol.* **165**, 534–541.
Bissig, K. -D., Wunderli-Ye, H., Duda, P., and Solioz, M. (2001). *Biochem. J.*, in press.
Brown, N. L., Barrett, S. R., Camakaris, J., Lee, B. T. O., and Rouch, D. A. (1995). *Mol. Microbiol.* **17**, 1153–1166.
Brown, N. L., Camakaris, J., Lee, B. T., Williams, T., Morby, A. P., Parkhill, J., and Rouch, D. A. (1991). *J. Cell Biochem.* **46**, 106–114.
Cellini, L., Marzio, L., Di Girolamo, A., Allocati, N., Grossi, L., and Dainelli, B. (1991). *FEMS Microbiol. Lett.* **68**, 255–257.
Cha, J. S., and Cooksey, D. A. (1991). *Proc. Natl. Acad. Sci. USA* **88**, 8915–8919.
Clarke, A. K., and Campbell, D. (1996). *Plant Physiol.* **112**, 1551–1561.
Cobine, P., Wickramasinghe, W. A., Harrison, M. D., Weber, T., Solioz, M., and Dameron, C. T. (1999). *FEBS Lett.* **445**, 27–30.
Cooksey, D. A. (1993). *Mol. Microbiol.* **7**, 1–5.
Cooksey, D. A. (1994). *FEMS Microbiol. Rev.* **14**, 381–386.
Dick, J. D. (1990). *Annu. Rev. Microbiol.* **44**, 249–269.
Dobi, A., Dameron, C. T., Hu, S., Hamer, D., and Winge, D. R. (1995). *J. Biol. Chem.* **270**, 10171–10178.
Fagan, M. J., and Saier, M. H., Jr. (1994). *J. Mol. Evol.* **38**, 57–99.
Forbes, J. R., and Cox, D. W. (1998). *Am. J. Hum. Genet.* **63**, 1663–1674.
Francis, M. S., and Thomas, C. J. (1997a). *Microb. Pathog.* **22**, 67–78.
Francis, M. S., and Thomas, C. J. (1997b). *Mol. Gen. Genet.* **253**, 484–491.
Ge, Z., Hiratsuka, K., and Taylor, D. E. (1995). *Mol. Microbiol.* **15**, 97–106.
Ge, Z., and Taylor, D. E. (1996). *FEMS Microbiol. Lett.* **145**, 181–188.
Gitschier, J., Moffat, B., Reilly, D., Wood, W. I., and Fairbrother, W. J. (1998). *Nat. Struct. Biol.* **5**, 47–54.
Gupta, A., Matsui, K., Lo, J. F., and Silver, S. (1999). *Nat. Med.* **5**, 183–188.
Gupta, S. D., Lee, B. T., Camakaris, J., and Wu, H. C. (1995). *J. Bacteriol.* **177**, 4207–4215.
Hackbarth, C. J., and Chambers, H. F. (1993). *Antimicrob. Agents Chemother.* **37**, 1144–1149.
Harrison, M. D., Jones, C. E., Solioz, M., and Dameron, C. T. (2000). *Trends Biochem. Sci.* **25**, 29–32.
Himelblau, E., Mira, H., Lin, S. J., Cizewski Culotta, V., Penarrubia, L., and Amasino, R. M. (1998). *Plant Physiol.* **117**, 1227–1234.
Himeno, T., Imanaka, T., and Aiba, S. (1986). *J. Bacteriol.* **168**, 1128–1132.
Horn, N., and Tümer, Z. (1999). *J. Trace Elem. Exp. Med.* **12**, 297–313.
Iida, M., Terada, K., Sambongi, Y., Wakabayashi, T., Miura, N., Koyama, K., Futai, M., and Sugiyama, T. (1998). *FEBS Lett.* **428**, 281–285.
Jungmann, J., Reins, H. A., Lee, J. W., Romeo, A., Hassett, R., Kosman, D., and Jentsch, S. (1993). *EMBO J.* **12**, 5051–5056.

Kaim, W., and Rall, J. (1996). *Angew. Chem. Int. Ed. Engl.* **35**, 43–60.
Kanamaru, K., Kashiwagi, S., and Mizuno, T. (1993). *FEBS Lett.* **330**, 99–104.
Kanamaru, K., Kashiwagi, S., and Mizuno, T. (1994). *Mol. Microbiol.* **13**, 369–377.
Kaplan, J., and O'Halloran, T. V. (1996). *Science* **271**, 1510–1512.
Kimura, M. (1980). *J.Mol. Evol.* **16**, 111–120.
Klomp, L. W., Lin, S. J., Yuan, D. S., Klausner, R. D., Culotta, V. C., and Gitlin, J. D. (1997). *J. Biol. Chem.* **272**, 9221–9226.
Lebrun, M., Loulergue, J., Chaslus Dancla, E., and Audurier, A. (1992). *Appl. Microbiol. Biotechnol.* **58**, 3183–3186.
Lee, Y. A., Hendson, M., Panopoulos, N. J., and Schroth, M. N. (1994). *J. Bacteriol.* **176**, 173–188.
Lim, C. K., and Cooksey, D. A. (1993). *J. Bacteriol.* **175**, 4492–4498.
Lin, S. J., and Culotta, V. C. (1995). *Proc. Natl. Acad. Sci. USA* **92**, 3784–3788.
Lowe, A. M., Beattie, D. T., and Deresiewicz, R. L. (1998). *Mol. Microbiol.* **27**, 967–976.
Lutsenko, S., and Kaplan, J. H. (1995). *Biochemistry* **34**, 15607–15613.
MacLennan, D. H., Rice, W. J., and Green, N. M. (1997). *J. Biol. Chem.* **272**, 28815–28818.
Maguire, M. E. (1992). *J. Bioenerg. Biomembr.* **24**, 319–328.
Mellano, M. A., and Cooksey, D. A. (1988). *J. Bacteriol.* **170**, 4399–4401.
Mills, S. D., Jasalavich, C. A., and Cooksey, D. A. (1993). *J. Bacteriol.* **175**, 1656–1664.
Mills, S. D., Lim, C. -K., and Cooksey, D. A. (1994). *Mol. Gen. Genet.* **244**, 341–351.
Munson, G. P., Lam, D. L., Outten, F. W., and O'Halloran, T. V. (2000). *J. Bacteriol.* **182**, 5864–5871.
Odermatt, A., Krapf, R., and Solioz, M. (1994). *Biochem. Biophys. Res. Commun.* **202**, 44–48.
Odermatt, A., and Solioz, M. (1995). *J. Biol. Chem.* **270**, 4349–4354.
Odermatt, A., Suter, H., Krapf, R., and Solioz, M. (1992). *Ann. N. Y. Acad. Sci.* **671**, 484–486.
Odermatt, A., Suter, H., Krapf, R., and Solioz, M. (1993). *J. Biol. Chem.* **268**, 12775–12779.
Outten, F. W., Outten, C. E., Hale, J., and O'Halloran, T. V. (2000). *J. Biol. Chem.* **275**, 31024–31029.
Payne, A. S., and Gitlin, J. D. (1998). *J. Biol. Chem.* **273**, 3765–3770.
Payne, A. S., Kelly, E. J., and Gitlin, J. D. (1998). *Proc. Natl. Acad. Sci. USA* **95**, 10854–10859.
Pedersen, P. L., and Carafoli, E. (1987a). *Trends Biochem. Sci.* **12**, 146–150.
Pedersen, P. L., and Carafoli, E. (1987b). *Trends Biochem. Sci.* **12**, 186–189.
Phung, L. T., Ajlani, G., and Haselkorn, R. (1994). *Proc. Natl. Acad. Sci. USA* **91**, 9651–9654.
Pope, M. T., and Dale, B. W. (1968). *Rev. Chem. Soc.* **22**, 527–545.
Racker, E. (1980). *Fed. Proc.* **39**, 2422–2426.
Rasmussen, B. (2000). *Nature* **405**, 676–679.
Rensing, C., Fan, B., Sharma, R., Mitra, B., and Rosen, B. P. (2000). *Proc. Natl. Acad. Sci. USA* **97**, 652–656.
Rosenzweig, A. C., Huffman, D. L., Hou, M. Y., Wernimont, A. K., Pufahl, R. A., and O'Halloran, T. V. (1999). *Structure* **7**, 605–617.
Russo, F. D., and Silhavy, T. J. (1993). *Trends Microbiol.* **1**, 306–310.
Shah, A. B., Chernov, I., Zhang, H. T., Ross, B. M., Das, K., Lutsenko, S., Parano, E., Pavone, L., Evgrafov, O., Ivanova-Smolenskaya, I. A., Anneren, G., Westermark, K., Urrutia, F. H., Penchaszadeh, G. K., Sternlieb, I., Scheinberg, I. H., Gilliam, T. C., and Petrukhin, K. (1997). *Am. J. Hum. Genet.* **61**, 317–328.
Silver, S., Nucifora, G., Chu, L., and Misra, T. K. (1989). *Trends Biochem. Sci.* **14**, 76–80.
Silver, S., and Phung, L. T. (1996). *Annu. Rev. Microbiol.* **50**, 753–789.
Solioz, M., and Camakaris, J. (1997). *FEBS Lett.* **412**, 165–168.
Solioz, M., and Odermatt, A. (1995). *J. Biol. Chem.* **270**, 9217–9221.
Solioz, M., and Vulpe, C. (1996). *Trends Biochem. Sci.* **21**, 237–241.
Steele, R. A., and Opella, S. J. (1997). *Biochemistry* **36**, 6885–6895.

Stoyanov, J. V., Hobman, J. L., and Brown, N. L. (2001). *Mol. Microbiol.* **39**, 502–512.
Strausak, D., and Solioz, M. (1997). *J. Biol. Chem.* **272**, 8932–8936.
Summers, A. O. (1992). *J. Bacteriol.* **174**, 3097–3101.
Suzuki, E., Kuwahara Arai, K., Richardson, J. F., and Hiramatsu, K. (1993). *Antimicrob. Agents Chemother.* **37**, 1219–1226.
Taha, A. S., Huxham, I. M., Park, R. H., and Beattie, A. D. (1995). *Gut* **36** (Suppl. 1), A10.
Toyoshima, C., Nakasako, M., Nomura, H., and Ogawa, H. (2000). *Nature* **405**, 647–655.
Vilsen, B., Andersen, J. P., Clarke, D. M., and MacLennan, D. H. (1989). *J. Biol. Chem.* **264**, 21024–21030.
Voskoboinik, I., Strausak, D., Greenough, M., Brooks, H., Petris, M. J., Smith, S., Mercer, J. F., and Camakaris, J. (1999). *J. Biol. Chem.* **274**, 22008–22012.
Wakabayashi, T., Nakamura, N., Sambongi, Y., Wada, Y., Oka, T., and Futai, M. (1998). *FEBS Lett.* **440**, 141–146.
Wernimont, A. K., Huffman, D. L., Lamb, A. L., O'Halloran, T. V., and Rosenzweig, A. C. (2000). *Nat. Struct. Biol.* **7**, 766–771.
Williams, J. R., Morgan, A. G., Rouch, D. A., Brown, N. L., and Lee, B. T. (1993). *Appl. Environ. Microbiol.* **59**, 2531–2537.
Wimmer, R., Herrmann, T., Solioz, M., and Wüthrich, K. (1999). *J. Biol. Chem.* **274**, 22597–22603.
Wittman, V., and Wong, H. C. (1988). *J. Bacteriol.* **170**, 3206–3212.
Wunderli-Ye, H., and Solioz, M. (2001). *Biochem. Biophys. Res. Commun.* in press.
Wyler-Duda, P., and Solioz, M. (1996). *FEBS Lett.* **399**, 143–146.
Zhou, P. B., and Thiele, D. J. (1991). *Proc. Natl. Acad. Sci. USA* **88**, 6112–6116.
Zierenberg, R. A., Adams, M. W., and Arp, A. J. (2000). *Proc. Natl. Acad. Sci. USA.*

# UNDERSTANDING THE MECHANISM AND FUNCTION OF COPPER P-TYPE ATPases

**BY ILIA VOSKOBOINIK,* JAMES CAMAKARIS,* AND JULIAN F. B. MERCER**[†]

***Department of Genetics, University of Melbourne, Parkville, Victoria 3010, Australia and** [†]**Centre for Cellular and Molecular Biology, School of Biological and Chemical Sciences, Deakin University, Burwood, Victoria 3125, Australia**



## I. INTRODUCTION

Mechanisms by which cells regulate the uptake, distribution, and detoxification of heavy metals are present in all life forms and are probably of ancient evolutionary origin as the earliest life forms were presumably exposed to these metals (Rensing *et al.*, 1999; Rosen, 1999b). Heavy metals are classified chemically as soft Lewis acids. Similar physicochemical properties of certain heavy metal may have resulted in the cross-specificity of heavy metal transporters in relation to these closely related ions, e.g., $Zn^{2+}/Cd^{2+}/Pb^{2+}$, $Ag^{+}/Cu^{+}$, and $As^{3+}/Sb^{3+}$. Thus, some of the recently characterized transporters found provide biological systems with resistance to metals that are not present in bioavailable form in the current environment, but may have been present in the anaerobic period of the evolution of life (Rensing, *et al.*, 1999; Rosen, 1999a, 1999b). There are toxic heavy metals that do not play any known physiological role (e.g., Hg, Cd), but there are some that are essential for life and are used in a great range of biochemical roles, e.g., Cu, Zn, Ni, and Co. However, like all heavy metals, even the essential elements are potentially toxic in excess. Thus, appropriate transport mechanisms have evolved to provide essential amounts of these metals in the cell and to respond to and to detoxify their excess. These mechanisms involve intracellular chelation, intracellular compartmentalization, or efflux from the cell.



*ADVANCES IN PROTEIN CHEMISTRY, Vol. 60*


0065-3233/02 $35.00

## II. Heavy Metal Toxicity and Essentiality

Soluble salts of heavy metals are generally toxic to biological systems at submicromolar (e.g., Hg) to high micromolar (e.g., Zn) concentrations. Substantial variability in the susceptibility of different organisms to heavy metals is likely to be due to the difference in mechanisms of uptake and detoxification. The fact that heavy metals are soft Lewis acids predisposes their strong association with weak Lewis bases, such as the amino acids cysteine and histidine. Nonspecific high-affinity binding of heavy metals to proteins can alter their structure, inactivate them, and have an adverse effect on their physiological function.

High concentrations of heavy metals that affect, directly or indirectly, the redox potential in the reducing intracellular milieu, particularly copper, induce the formation of reactive oxygen species. These, in turn, can trigger the chain reaction causing lipid peroxidation, which affects membrane integrity and induces structural changes to proteins and nucleic acids that can result in cell death.

While the redox cycling of copper makes it toxic at high concentrations, in small amounts this same property is essential to most organisms. The high redox potential of copper and its ability to be coordinated efficiently by proteins result in the utilization of copper by a number of enzymes catalyzing redox reactions, e.g., Cu, Zn-superoxide dismutase, cytochrome *c* oxidase, lysyl oxidase, and tyrosinase (Linder and Hazegh Azam, 1996). To allow a specific delivery of copper to physiological targets, a complex system of regulated high-affinity copper uptake proteins, low-molecular-weight copper chaperones, and active copper transporters has evolved and is found in organisms ranging from bacteria to humans (Camakaris *et al.*, 1999). As will be discussed below, copper P-type ATPases have evolved from the largely detoxifying role in unicellular organisms to satisfying the physiological requirements of multicellular differentiated systems. While unicellular organisms function as independent units that communicate with the environment directly and need to respond to environmental changes very rapidly, the function and viability of a multicellular differentiated organism rely on intercellular interactions, where the majority of cells are not directly exposed to the nutrients from the environment. Instead, they are received from the circulation and other cells and tissues, which therefore may be regarded as natural barriers for the nutrients. As a result, copper transport mechanisms viewed as a part of a detoxification system for a single-cell organism may be important for intercellular copper transport in differentiated biological systems (Camakaris *et al.*, 1999). This difference could drive the evolution of copper transporters in higher organisms including humans.

This chapter reviews current knowledge of copper homeostasis, with particular emphasis on the role of mammalian copper P-type ATPases in that process.

## III. Vectorial Copper Transport

Intracellular abundance of Lewis soft bases capable of high-affinity copper binding, e.g., the amino acids cysteine and histidine, the tripeptide glutathione, and the protein metallothionein, is responsible for maintaining negligible intracellular concentrations of ionic copper under basal conditions (one of the estimates suggests that there is $<1$ molecule of ionic copper per cell) (Rae *et al.*, 1999). As copper is an essential constituent of active sites in cuproenzymes localized in various subcellular compartments, the cytosol, the Golgi network, and the mitochondria, specific delivery systems are required to avoid nonspecific binding of copper. This delivery is believed to be mediated by cytoplasmic copper chaperones, high-affinity copper-binding proteins whose structure allows docking and copper transfer to target proteins (Rosenzweig and O'Halloran, 2000; see Chapter by Elam *et al.*, this volume). Crystal structures of copper chaperones for copper P-type ATPases, Atx1 in yeast and ATOX1 (HAH1) in humans, were resolved recently (Rosenzweig *et al.*, 1999). Structure–function analysis provided a theoretical basis for high-affinity copper binding, docking to appropriate copper-binding sites in target proteins, and the transfer of copper to the putative metal-binding sites (MBSs), with the common sequence GMxCxxC, at the N-terminal domain of copper P-type ATPases (Banci *et al.*, 2001; Huffman and O'Halloran, 2000; Portnoy *et al.*, 1999; Pufahl *et al.*, 1997; Wernimont *et al.*, 2000). While delivery of copper to cuproenzymes is essential for their catalytic function, the importance of copper chaperones for copper P-type ATPases is, probably, in the regulation of active copper transport and copper homeostasis at physiological, very low, concentrations (Huffman and O'Halloran, 2000). Copper P-type ATPases are involved in the delivery of copper to cuproenzymes of the secretory pathway, e.g., lysyl oxidase and tyrosinase in mammals, in the intercellular transport of copper, and in its detoxification by efflux from the cell (Camakaris *et al.*, 1995, 1999; Danks, 1995; Petris *et al.*, 2000; Royce *et al.*, 1980). Despite their structural similarity, copper P-type ATPases have acquired different functions in various organisms or cell types. This is of particular relevance to human copper P-type ATPases, ATP7A (MNK) and ATP7B (ATP7B), affected in the inherited copper disorders Menkes and Wilson's diseases, respectively (Danks, 1995). These proteins are expressed in different tissues and have distinct physiological roles. The

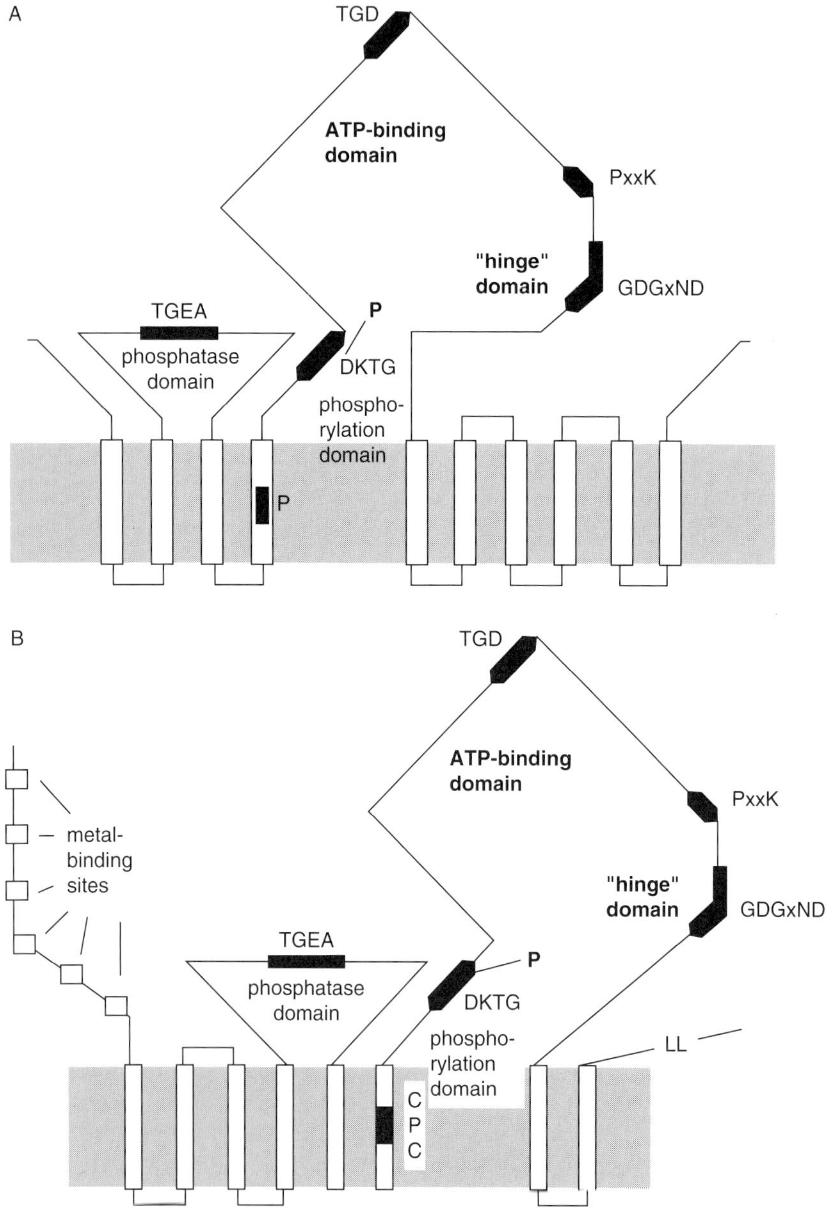


FIG. 1. General structure of (A) non-heavy metal P-type ATPase and (B) heavy metal P-type ATPase. "Metal-binding sites" are putative copper-binding motifs GMxCxxC; "phosphatase domain," a cytosolic loop with a conserved region(s), e.g., TGEA, essential for dephosphorylation of the transient acyl phosphate; P-DKTG is the

rest of this chapter will examine the mechanisms of regulation of copper homeostasis by mammalian copper P-type ATPases.

## IV. P-TYPE ATPASES

P-type ATPases constitute a ubiquitous family of integral membrane proteins that facilitate the translocation of cations against concentration and electrochemical gradients across otherwise impermeable biological membranes (Carafoli, 1991; MacLennan *et al.*, 1997). These proteins bind ATP, and its terminal γ-phosphate binds covalently to the invariant aspartate residue in their conserved DKTG motif, causing a conformational transition that drives the unidirectional transport of a cation across the lipid bilayer (Figs. 1 and 2). There are more than 200 P-type ATPase-like sequences identified in the various sequence databases. Although these enzymes have a common reaction mechanism, there are often considerable differences in their amino acid sequences, structure, and regulation. P-type ATPases have 8 or 10 interconnected transmembrane domains and two cytosolic loops, one forming the "phosphatase" domain and the other large cytosolic loop containing the ATP-binding/phosphorylation domain (Fig. 1). Eight conserved regions have been identified in all P-type ATPases and are thought to constitute essential catalytic elements in relation to ATP binding, phosphorylation, and dephosphorylation of these enzymes (Axelsen and Palmgren, 1998; Møller *et al.*, 1996). With the exception of some heavy metal transporters, P-type ATPases are cation-specific. This explains why there is little conservation in the amino acid composition of cation-binding sites in transmembrane channels of different P-type ATPases.

Hundreds of mutant P-type ATPases have been generated in an effort to understand the catalytic mechanism and regulation of these enzymes. Functional analysis of these mutants has established the roles of particular domains and amino acid residues in the overall catalysis of cation translocation and allowed structure prediction (MacLennan *et al.*, 1997) (Figs. 1 and 2). Recently the 3-D or crystal structures of $Ca^{2+}$, $H^+$, and $Na^+/K^+$ P-type ATPases have been resolved, and they confirmed findings

phosphorylation domain, where P indicates transient aspartyl phosphate; "ATP-binding domain" and "hinge domain," a large cytosolic loop with several conserved regions (e.g., TGD, PxxK, GDGxND) essential for ATP binding and the transfer of the terminal γ-phosphate of ATP to the invariant aspartate residue of the phosphorylation domain; the CPC motif is believed to constitute part of the cation channel in heavy metal P-type ATPases; the LL motif is the internalization signal in ATP7A (position 1487/1488) and ATP7B (position 1454/1455).

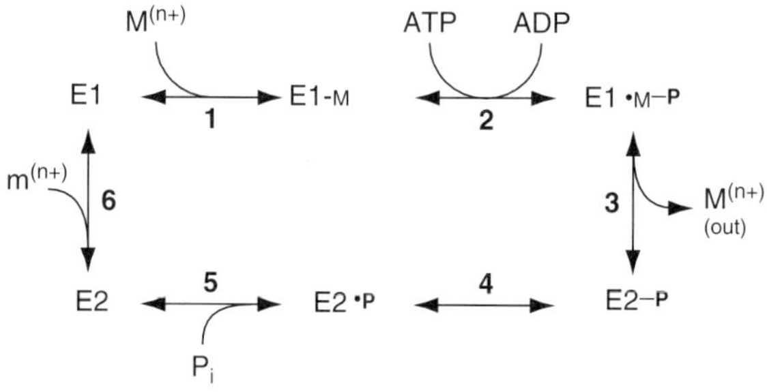


FIG. 2. General concept of the catalytic cycle of P-type ATPases. $M^{(n+)}$, cation to be translocated; $m^{(n+)}$, counterion; P, terminal phosphate group of ATP forming an acyl phosphate bond with an invariant aspartate residue in the DKTG motif. (1) High-affinity cation binding; (2) high-affinity ATP binding and the transfer of γ-P from ATP to the invariant aspartate residue, the formation of the acyl phosphate bond, (3) conformational change from the E1 to the E2 state, translocation of the cation to the lumen; (4) transition to the low-energy acyl phosphate bond; (5) dissociation of inorganic phosphate from the invariant aspartate residue; (6) potential counterion translocation, transition from the E1 to the E2 conformation.

reported in the earlier structure–function studies (Auer *et al.*, 1998; Sweadner and Donnet, 2001; Toyoshima *et al.*, 2000).

The catalytic cycle of P-type ATPases is represented generally by the coupled reaction of ATP hydrolysis, transient phosphorylation of an invariant aspartate residue, and cation translocation across the lipid bilayer (Fig. 2). The catalytic cycle of calcium P-type ATPases is regarded as a paradigm for heavy metal P-type ATPases (HMPAs) (Camakaris *et al.*, 1999; Silver and Phung, 1996). The cation to be translocated binds to high-affinity binding sites, commonly formed by four transmembrane domains adjacent to large cytosolic loops (Fig. 2, step 1). Subsequently, the conformation acquired by the enzyme allows the high-affinity binding of ATP to the second large cytosolic loop (ATP-binding domain) and the transfer of the terminal γ-phosphate of ATP to the invariant aspartate residue within the DKTG motif (Fig. 1 and Fig. 2, step 2). On the transfer, the affinity of ADP for the ATP-binding domain decreases and it dissociates from the protein (Fig. 2, step 2). The high-affinity cation- and ATP-binding state of the enzyme is called the E1 or ADP-sensitive conformation as the phosphorylated ATPase can be dephosphorylated by ADP (Fig. 2, step 2). Following the dissociation of ADP, the conformation of the phosphorylated enzyme changes to the E2 conformation (Fig. 2, step 3). That transition is associated with the decreasing affinity of the cation for the high-affinity binding sites in the transmembrane domains and the transport of the cation across the channel to the lumenal side of the membrane (Fig. 2, step 3). That transition is the only unidirectional step in the catalytic cycle and it ensures a one-way translocation of the cation across the membrane (Fig. 2). The E2 conformation is also characterized by insensitivity to ADP. Following cation transport, the aspartyl–phosphate bond is hydrolyzed (Fig. 2, steps 4 and 5) and after the potential counterion transport from the lumenal to aqueous phase, the enzyme returns to its original E1 conformation (Carafoli, 1991; MacLennan *et al.*, 1997) (Fig. 2, step 6).

An important characteristic of P-type ATPases is their ability to be transiently phosphorylated at the invariant aspartate residue by inorganic orthophosphate in the absence of the cation (Fig. 2, step 5). A diagnostic inhibitor of P-type ATPases, orthovanadate, can also bind to the invariant aspartate but, unlike orthophosphate, cannot participate in the reaction cycle (Carafoli, 1991; MacLennan *et al.*, 1997).

In humans, when candidate genes whose mutations were associated with copper deficiency and copper toxicity disorders, Menkes and Wilson's disease, respectively, were cloned, their predicted amino acid sequence analysis revealed features shared with bacterial heavy metal P-type ATPases (Silver and Phung, 1996) (Fig. 1). Clinical features of Menkes and Wilson's disease, together with the predicted amino acid sequence, suggested that the Menkes (ATP7A) and Wilson (ATP7B) proteins were copper-translocating P-type ATPases (Bull *et al.*, 1993; Chelly *et al.*, 1993; Mercer *et al.*, 1993; Vulpe *et al.*, 1993; Yamaguchi *et al.*, 1993).

## V. Heavy Metal P-type ATPases

While HMPAs, including copper P-type ATPases (CuPAs), share common features of all P-type ATPases, they are sufficiently unique to be classified in a distinct group of enzymes, Type IB P-type ATPases (Axelsen and Palmgren, 1998; Møller *et al.*, 1996). Unlike most of the other P-type ATPases, HMPAs have 8 rather than 10 transmembrane helices. Their overall structure is **m-m-m-m-C-m-m-C-m-m**, while the structure of non-heavy metal transporters is **m-m-C-m-m-C-m-m-m-m-m-m**, where **m**s are transmembrane domains and **C**s are cytosolic loops (Axelsen and Palmgren, 1998; Møller *et al.*, 1996) (Fig. 1). Moreover, distinct mechanisms of regulation of catalytic activity and expression emerged within each of these groups of P-type ATPases due to substantial differences between their biological roles.

The expression of HMPAs is commonly controlled by their ligands, heavy metals, which may interact with a regulatory element(s) up- or down-regulating their gene expression (Rensing *et al.*, 2000). Also, in bacteria, some of the genes encoding heavy metal transporters form a part of operons that include genes, other than CuPAs, responsible for the detoxification of very high concentrations of copper. Thus the *cusCFBA* copper- and silver-responsive operon was characterized recently in *Escherichia coli*. The resultant CusCBA transmembrane complex, which spans the inner and outer membrane of the cell, is responsible for the efflux of copper outside the cell (Outten *et al.*, 2001). A similar system, which includes an apparent silver P-type ATPase, a periplasmic silver-specific binding protein, and a cation/proton antiporter, was reported to constitute a silver-resistance determinant in *Salmonella* (Gupta *et al.*, 1999).

In eukaryotes, HMPAs were identified in wide variety of organisms, from yeast to plants and humans. Interestingly, in eukaryotes, only copper P-type ATPases were identified in yeast and mammals, while a variety of open reading frames for heavy metal transporters were found in plants. Clearly, this difference is determined by environmental factors, such as the presence of relatively high concentrations of heavy metals in soil and the ability of plants to solubilize heavy metal-containing minerals, often without any detrimental effect. There is a possibility that some of the heavy metal transporters in plants may be derived from symbiotic bacteria adapted to particular environmental conditions. Interestingly, a recently discovered copper P-type ATPase in *Arabidopsis*, RAN1, has an essential role in the ethylene signaling pathway in plants. It is predicted that ethylene signaling requires a functional copper transporter, i.e., RAN1, in order to produce functional hormone receptor (Hirayama *et al.*, 1999).

One of the characteristic motifs of heavy metal P-type ATPases is the CPX (commonly CPC) motif within transmembrane domain 6, widely regarded as a part of the cation channel (Silver and Phung, 1996) (Fig. 1). The possible role for these cysteines in the transmembrane domain may be the coordination of a heavy metal during its translocation through the cation channel. While these cysteines have proved to be essential for the function of some HMPAs, their exact role in catalysis is yet to be fully understood (Bissig *et al.*, 2001; Forbes and Cox, 2000).

Another common feature of heavy metal P-type ATPases, which has attracted a great deal of interest, is the presence of at least one putative MBS, commonly GMxCxxC, at the N-terminus of these proteins. Interestingly, the number of MBSs increases from 1 or 2 in bacteria and yeast to 3 in *Caenorhabditis elegans* and 5 or 6 in mammalian CuPAs (Fig. 1). This increase was probably due to the amplification of a portion of coding sequence, as MBSs are highly conserved between various heavy metal transporters, including the conserved leucine residue in the position +8 or +11, isoleucine in the position +8, and phenylalanine in the position

+57 relative to the second Cys residue in the CxxC motif. The possible evolutionary significance of increasing numbers of MBSs will be discussed below.

### *A. Copper P-type ATPases and Their Role in Human Diseases*

Clinical studies on Menkes and Wilson's diseases, copper deficiency and copper toxicity disorders, respectively, provide invaluable information on the role of copper in human physiology. Menkes disease is a rare X-linked disorder with an estimated frequency of 1 in 100,000–300,000 live births (Danks, 1995). Pleiotropic clinical features of Menkes disease are associated with the reduced activity of several copper-dependent enzymes in various tissues, which is thought to be due to systemic copper deficiency as a result of malabsorption of dietary copper (Danks, 1995). This is associated with a build-up of copper in gut epithelial cells. In support of this hypothesis, cultured cells from Menkes disease patients were shown to accumulate high concentrations of copper, apparently due to the impaired efflux mechanism (Camakaris *et al.*, 1980; Horn, 1976; Royce *et al.*, 1980). Low levels of activity of important cuproenzymes cause severe symptoms of copper deficiency in affected boys. Pathological implications of cuproenzyme deficiencies are arterial rupture (for lysyl oxidase), hypothermia (for cytochrome *c*-oxidase), hypopigmentation (for tyrosinase), severe neurological abnormalities (for Cu, Zn-superoxide dismutase, dopamine β-hydroxylase, and peptidylglycine-α-amidating monooxygenase) (Danks, 1995). There is a variable response to parenteral copper therapy and the reason for this is not clearly understood.

Three groups independently cloned the candidate gene for Menkes disease using positional cloning. The nucleotide sequence-derived amino acid sequence revealed features common for P-type ATPases, particularly heavy metal P-type ATPases in bacteria (Chelly *et al.*, 1993; Mercer *et al.*, 1993; Vulpe *et al.*, 1993) (Fig. 1). Clinical features of the disease allowed the prediction that the Menkes disease protein (ATP7A) is a copper P-type ATPase. Mutations in the *ATP7A* gene often result in a truncated protein that lacks transmembrane and catalytic domains (Kaler, 1998) (Fig. 1). Specific effects of missense mutations on protein function are often difficult to predict, particularly when the mutations are located outside domains with a known predicted function. These mutations may cause protein misfolding, which affects the catalytic activity, stability, intracellular localization, or trafficking of the protein. There is a possibility that certain missense or frameshift mutations have a moderate effect on catalytic activity or affect compartmentalization of the protein. In these cases, patients may have a mild Menkes disease phenotype (or occipital horn syndrome) (Dagenais *et al.*, 2001), and copper therapy is predicted to have some beneficial effect. Recent studies on an ATOX1 knockout mouse model have shown that the ATOX1 copper chaperone is essential for absorption of dietary copper as in the absence of copper supplementation the transgenic mice suffer from Menkes disease-like abnormalities (Hamza *et al.*, 2001). However, there is no documented evidence for ATOX1 disease-causing mutations in humans.

Wilson's disease is a more common recessive genetic disorder of copper metabolism with a frequency of approximately 1 in 30,000 live births. It is believed to be caused by a mutation in the *ATP7B* gene, which encodes a copper P-type ATPase (ATP7B; Wilson's protein) expressed, predominantly, in the liver (Danks, 1995). This protein plays an essential role in the biliary excretion of copper and its incorporation into ceruloplasmin (Schaefer and Gitlin, 1999). The disease is characterized, primarily, by intrahepatic copper accumulation leading to cirrhosis and liver failure and neurological disorders as a result of copper accumulation in the brain due to its unregulated uptake (Danks, 1995). Wilson's disease patients can be treated with copper chelators, e.g., penicillamine. Missense mutations appear to be an important factor in the loss of physiological function of ATP7B. One of the most common mutations is His1069Gln, which is found in at least 30% of the patients in northern Europe and North America. An interesting mutation associated with Wilson's disease, where the C-terminal part of the protein was truncated, has been found in Saudi Arabia (Majumdar *et al.*, 2000). In addition an occipital horn syndrome patient appeared to have a frameshift mutation in the C-terminus of *ATP7A* (Dagenais *et al.*, 2001). While there are no known catalytically important elements in that region of ATP7A and ATP7B, a signaling dileucine motif, $^{1454}$LL, is associated with the internalization of ATP7A (Francis *et al.*, 1999; Petris *et al.*, 1998) (Fig. 1). The role of a similar motif in ATP7B has yet to be clarified.

### B. *Catalytic Mechanism of Copper P-type ATPases*

Very little is known about the catalytic mechanism of copper P-type ATPases. A large amount of experimental and clinical information is available on acquired or lost copper (and other heavy metal) resistance and transport as a result of the overexpression or disruption of genes encoding HMPAs (Mercer and Camakaris, 1997; Rosen, 1999b; Silver, 1998). In addition, the expression of certain HMPAs in prokaryotes can be up-regulated by the heavy metal transported by this enzyme. The overexpression resulting from gene amplification has been reported in cultured mammalian cells selected in elevated concentrations of copper. However, copper does not appear to regulate the expression of the

*ATP7A* gene (Camakaris *et al.*, 1995). While these data are indicative of the role of a particular heavy metal ATPase in the detoxification of the toxic cation, it provides little information on the catalytic function of the protein.

The catalytic activity of copper P-type ATPase was first reported by Solioz and colleagues, who estimated the ATP-stimulated copper transport by CopB from *Enterococcus hirae* using everted membrane vesicles (Solioz and Odermatt, 1995). Interestingly, CopB could also actively translocate silver Ag(I), which agrees with its ability to induce the expression of CuPA (Rensing *et al.*, 2000) and induce the trafficking of ATP7A (Petris *et al.*, 1996). Although the rate of the reactions was lower than could be expected for a P-type ATPase, the weight of evidence suggested that the enzyme was an active copper pump. Recently, the same group has reported copper-stimulated transient phosphorylation of purified CopA and CopB reconstituted in liposomes (Bissig *et al.*, 2001; Wunderli-Ye and Solioz, 2001). While the rate of turnover of CopB appeared to be as fast as for other P-type ATPases, the kinetics of CopA was considerably slower (≈3 min, while non-heavy metal P-type ATPases turn over within 30 s). In addition, the apparent $K_m$ values for CopA and CopB were 0.2 and 0.5 mM, respectively, considerably higher than in the case of other P-type ATPases, including ATP7A (Voskoboinik *et al.*, 2001b). These results suggest that CopA and CopB may require a cofactor, which can be available intracellularly since these proteins are essential for copper resistance, or their activity was altered as a result of purification and reconstitution in liposomes. The acyl phosphorylation of both CopA and CopB from *E. hirae* was stimulated by submicromolar concentrations of copper, and the enzyme appeared to be inhibited by 20 μM copper.

Recent studies in our laboratory provided the first insight into the catalytic activity of a mammalian copper P-type ATPase, the Menkes (ATP7A) and Wilson's (ATP7B) protein (Voskoboinik *et al.*, 1998, 1999, 2001a, 2001b). *In vitro* kinetics studies have been conducted using everted membrane vesicles from mammalian cells overexpressing the endogenous and recombinant ATP7A or ATP7B protein or from yeast transformed with the *ATP7A* cDNA. Similar to its bacterial counterparts, ATP7A followed Michaelis–Menten kinetics with respect to $^{64}$Cu translocation and ATP hydrolysis. Apparent kinetic constants have been identified as $K_m^{Cu} = 2–7\ \mu M$ Cu, $V_{max} = 0.6–1.2$ nmol/min/mg total protein, and $K_m^{ATP} = 10–20\ \mu M$ ATP, which is close to the values established for classical P-type ATPases (Daly *et al.*, 1996; Mollman and Pleasure, 1980). The values for apparent catalytic constants in relation to copper are extrapolated as copper in the *in vitro* assay inhibited the catalytic activity of the enzyme at >5–6 μM Cu. The apparent $V_{max}$ values differed depending on the level of overexpression of ATP7A. There appeared to be no significant amounts of any other copper ATPase in membrane vesicles prepared from mammalian cells or yeast that would interfere with the biochemical assay (Voskoboinik *et al.*, 1998, 1999, 2001a; 2001b). Assuming that all molecules of ATP7A in the vesicles are fully active and that the relative amount of ATP7A is $<1\%$ of the total protein, the copper-translocating activity of ATP7A would be $>0.1$ μmol/mg/min, slower than the rate of cation translocation by non-heavy metal P-type ATPases. This is consistent with a relatively lower rate of ATP7A turnover compared to other P-type ATPases. Importantly, ATP7A and ATP7B translocate copper only under reducing conditions, suggesting that Cu(I), rather than Cu(II), is the substrate for these enzymes (Voskoboinik *et al.*, 1998, 2001a, 2001b). Whether Cu(II) is reduced during and/or after the translocation is yet to be determined. The *in vitro* studies indicated that $^{64}$Cu accumulated inside the vesicles in a non-protein-bound form. Thus the lysis of vesicles with detergent during the course of catalytic reaction resulted in leakage of $^{64}$Cu from the vesicles with very little above background level $^{64}$Cu binding to the nitrocellulose membrane (Fig. 3). Extrapolating these findings to intracellular conditions, one can propose that on the translocation to the lumenal side of the membrane, copper dissociates from the protein in bioavailable form. Moreover, based on the *in vitro* data, the copper concentration in vesicles (or lumen) may be considerably higher than that in the cytoplasm. Consequently, copper-dependent proteins of the secretory pathway may acquire copper directly from the lumen rather than through other, intermediate, copper

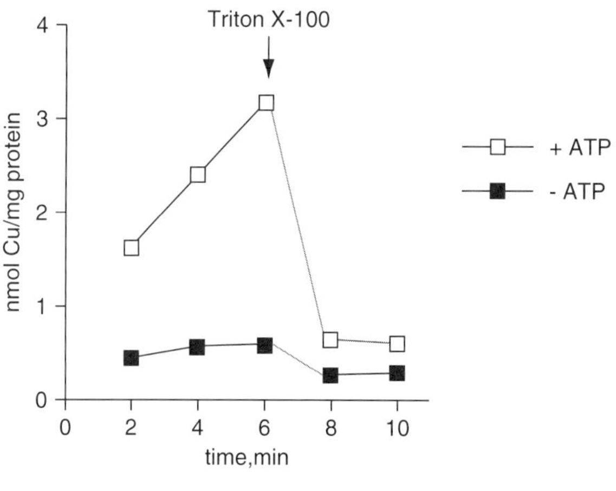


FIG. 3. Lysis of the ATP7A-enriched membrane vesicles with Triton X-100 results in the leakage of non-protein-bound $^{64}$Cu.

transporters or chaperones (Petris *et al.*, 2000). Therefore one of the physiological roles of eukaryotic CuPA may be to compartmentalize copper in the lumen under limited physiological copper concentrations. In support of that, recent findings suggested that the activation of tyrosinase, a cuproenzyme that matures at the Golgi compartment, depends on the presence of catalytically active ATP7A at the *trans*-Golgi network of the cell or increased copper supplementation (Petris *et al.*, 2000). The efflux of bioavailable copper from the cell, which depends predominantly on functional ATP7A protein at the plasma membrane (PM), is also likely to rely on the release of non-protein-bound copper at the lumenal side of ATP7A (Camakaris *et al.*, 1995).

The catalytic activity of ATP7A was sensitive to inhibition by orthovanadate, a classical inhibitor of P-type ATPases (O'Neal *et al.*, 1979). Orthovanadate is a structural homologue of inorganic phosphate and can bind to the invariant aspartate residue within the DKTG motif following the identical pathway of phosphorylation to inorganic phosphate. As a result, transient acyl phosphate cannot be formed, and the reaction cycle, as described above, is blocked (Carafoli, 1991; Møller *et al.*, 1996). Although non-heavy metal P-type ATPases are inhibited by low micromolar concentrations of orthovanadate, the $IC_{50}$ value for CuPAs is generally higher, typically $\approx 50\,\mu M$ (Rensing *et al.*, 2000; Solioz and Odermatt, 1995; Voskoboinik *et al.*, 1998, 2001a). The reasons for this finding may be residual intracellular or environmental copper that binds to high-affinity copper-binding sites in CuPAs and shifts the E1 $\leftrightarrow$ E2 equilibrium toward the E1 conformation or CuPAs may generally have a low affinity for inorganic phosphate and, consequently, for orthovanadate.

A high degree of similarity between conserved domains in the ATP-binding cytosolic loops of all P-type ATPases suggests that they would play the same role in the catalysis of CuPA as in well-characterized $Ca^{2+}$, $Na^+/K^+$, and $H^+$ P-type ATPases. However, certain motifs, particularly in the phosphatase domain, and some conserved, among HMPAs, elements in the ATP-binding domain are not found in non-heavy metal P-type ATPases, and vice versa (Axelsen and Palmgren, 1998; Møller *et al.*, 1996). Nevertheless the most puzzling element in the structure of CuPAs is the composition of the cation channel and cation recognition site(s), since transmembrane domains of CuPAs bear little resemblance to similar regions of non-heavy metal P-type ATPases. A recent study on a yeast plasma membrane CuPA, PcaI, found that the mutation of non-conserved, among other CuPAs, Arg-970 to Gly has provided yeast with cadmium resistance at the expense of copper resistance (Shiraishi *et al.*, 2000). This suggested that Arg-970 may form a part of an unknown functional domain that determines the cation specificity of heavy metal

P-type ATPases. This is an unexpected finding as generally the cation specificity of P-type ATPases is restricted to the amino acid composition of cation channels in transmembrane domains (Axelsen and Palmgren, 1998; Møller *et al.*, 1996). It is probable that other, as yet to be identified, intramolecular interactions may be important for the specific cation transport by heavy metal P-type ATPases. Cysteine residues in transmembrane domain 6, which form a characteristic CPx (commonly CPC) motif, are commonly regarded as core elements of the cation channel in CuPAs. However, there is no direct evidence to date to support that assumption.

### *C. The Role of Putative Metal-Binding Sites in the Regulation of ATP7A/ATP7B*

While the overall structure of CuPA is poorly investigated, there has been a great deal of interest in the structure and function of the N-terminal cytosolic domain, which contains putative MBSs with the general sequence GMxCxxC. The number of these motifs varies from 1–2 in prokaryotes and lower eukaryotes to 6 in mammals. The increase in the number of MBSs is thought to be due to amplification as there is considerable conservation in the position and structure of the MBSs in various CuPAs.

The secondary structure of all MBSs of ATP7A/ATP7B has been predicted as four ß-strands and two α-sheets in the order β-α-β-β-α-β (Gitschier *et al.*, 1998). Copper(I) is predicted to bind within a surface "pocket" commonly defined by the sequence TCxSC. The conserved Ile (+4) and Phe (+49) have been predicted to stabilize the MBS (Gitschier *et al.*, 1998). The side chain of the second cysteine is disordered in the apo form of the peptide, while the conformation of the first cysteine is not changed on the binding of copper. The conserved methionine preceding the MBS is not involved in the coordination of copper, which binds to the two cysteines in the MBS in a linear bicoordinate manner (Gitschier *et al.*, 1998). A similar type of copper coordination has been identified for copper P-type ATPase chaperones, Atx1 in yeast and Atox1 in humans (Portnoy *et al.*, 1999; Pufahl *et al.*, 1997; Rosenzweig *et al.*, 1999; Wernimont *et al.*, 2000).

The purified N-terminal domain of ATP7A and ATP7B was reported to bind copper stoichiometrically, six atoms of copper per N-terminal domain, with high affinity in the presence of reducing agents, indicating that Cu(I) is the preferred form of copper for the MBSs (Lutsenko *et al.*, 1997). However, recently Cobine *et al.* (2000) demonstrated that, in fact, the N-terminus of ATP7A binds only four atoms of Cu(I). Despite the discrepancy between these *in vitro* studies, the binding of copper *in vivo* depends on multiple factors, including the bioavailability of copper and the accessibility of the MBSs. The fact that human ATP7A and ATP7B have a larger number of MBSs than lower organisms is intriguing, and it inspired several detailed studies that aimed to elucidate the regulatory role of MBSs in the physiological function of ATP7A/ATP7B and their role in copper homeostasis and cell physiology in general.

### *D. The Interaction of Putative Metal-Binding Sites with Copper Chaperones*

O'Halloran and co-workers have found that a yeast multicopy suppressor of oxidative injury in yeast, a copper-binding protein Atx1, can deliver copper to MBSs of a yeast copper P-type ATPase Ccc2 (see Huffman and O'Halloran, 2000; Pufahl *et al.*, 1997). A human homologue of Atx1, ATOX1 (HAH1), has also been identified (Klomp *et al.*, 1997) and, by analogy with yeast, implicated in the delivery of copper to the MBSs of ATP7A and ATP7B (Hamza *et al.*, 1999). Detailed studies, including crystallography, have provided the theoretical basis, which has been confirmed experimentally, for the delivery of copper by Atx1/Atox1 to the MBSs. Thus, copper binds with high affinity to the copper-binding site (the GMxCxxC motif) of the chaperone, which has a flexible structure and allows copper complexing in both two- and three-coordinate geometries. This flexibility is important for the transfer of copper to the acceptor via electrostatic interactions between the positively charged residues of the chaperone and the negatively charged docking region of the CuPA (Huffman and O'Halloran, 2000; Portnoy *et al.*, 1999; Wernimont *et al.*, 2000). The transfer of copper between the two proteins appeared to be reversible, which may be important for the control of intracellular copper concentrations, as will be discussed below.

Interestingly, the affinity constant for copper exchange between the chaperone and the MBS of a CuPA was low, suggesting that the delivery of copper to the target protein was not based on a high affinity of copper for the MBSs or chaperone. Huffman and O'Halloran (2000) have proposed that, at least in the case of the yeast system, complementary electrostatic forces orient the donor (Atx1) and acceptor (the MBSs of Ccc2) proteins so that the activation barrier between the proteins is lowered, which allows a rapid copper transfer. Their study has also indicated that the copper chaperone protects Cu(I) from non specific binding by other copper ligands, such as glutathione, whose intracellular concentrations may significantly exceed those of the copper chaperone (Huffman and O'Halloran, 2000; Portnoy *et al.*, 1999). Altogether, Atx1 was proposed to function as an enzyme that decreases the kinetic barrier for copper transfer between the specific target proteins. As a result, the equilibrium [Cu] → [Cu-ligand] ↔ [Cu-Atx1] ↔ [Cu-MBS] may be established (Camakaris *et al.*, 1999; Huffman and O'Halloran, 2000).

The analysis of the crystal structure of human ATOX1 (HAH1) determined in the presence of Cu(I), Hg(II), and Cd(II) has allowed elucidation of the structural basis for the cation exchange between human homologues of yeast proteins, ATP7A and ATOX1 (Wernimont *et al.*, 2000). While Cd(II) formed four thiolate bonds within the metal-binding site of ATOX1, which appeared to be so stable that they prevented its docking with the target sequence, both Cu(I) and Hg(II) formed a less stable transient thiolate complex that assisted in docking and copper transfer to an acceptor. A similar mechanism may be potentially involved in copper transfer between MBSs of ATP7A or ATP7B and copper exchange reactions between ATOX1 molecules (Wernimont *et al.*, 2000).

Yeast and mammalian cell two-hybrid assays both have shown the interaction between ATOX1 and the N-terminus of ATP7A and ATP7B. However, the data suggested that only MBSs 1 to 4 of ATP7A/ATP7B interacted with the chaperone in the presence of copper, while MBSs5 and 6 either had no effect or inhibited the interaction (Hamza *et al.*, 1999; Larin *et al.*, 1999). These results indicated that various MBSs of the CuPA may have different roles in the regulation of ATP7A/ATP7B. Functional studies have shown that MBS1–4 are not essential for catalytic activity and trafficking (see Section V, E), while MBS5 or MBS6 appears to play an important role in the copper-dependent exocytosis of ATP7A (Strausak *et al.*, 1999).

These reports allow the proposal of the following model for MBS in the regulation of ATP7A/ATP7B activity and physiological function (Fig. 4). Under basal conditions the concentration of free ionic copper inside the cell (at least, in yeast) is predicted to be $<10^{-18}$ M, i.e., less than 1 atom of copper per cell (Rae *et al.*, 1999). Copper is sequestered inside the cell by high-affinity copper-binding ligands, including glutathione, amino acids, metallothioneins, and copper chaperones. In yeast the amount of Atx1 has been estimated as $1.5 \times 10^4$ molecules per cell (Portnoy *et al.*, 1999). Given the high affinity of Atx1/ATOX1 for copper and its ability to protect copper from interaction with other ligands, there is a possibility that under basal conditions at least some of the chaperones are present in the copper-bound form (the total intracellular concentration of copper is 5–20 $\mu$M). This would allow constitutive docking of the (ATOX1–Cu) complex with MBS1–4 of ATP7A/ATP7B (Hamza *et al.*, 1999; Larin *et al.*, 1999), although recent studies suggest that these MBSs are not involved in ATP7A trafficking (see Section V, E; Goodyer *et al.*, 1999; Strausak *et al.*, 1999). Importantly, the copper-stimulated trafficking of ATP7A is observed only at increased concentrations of extracellular copper ($>20\,\mu$M compared to $2\,\mu$M under basal conditions). Under these conditions copper may saturate ATOX1, and copper will start binding to other intracellular ligands. Some of these complexes may

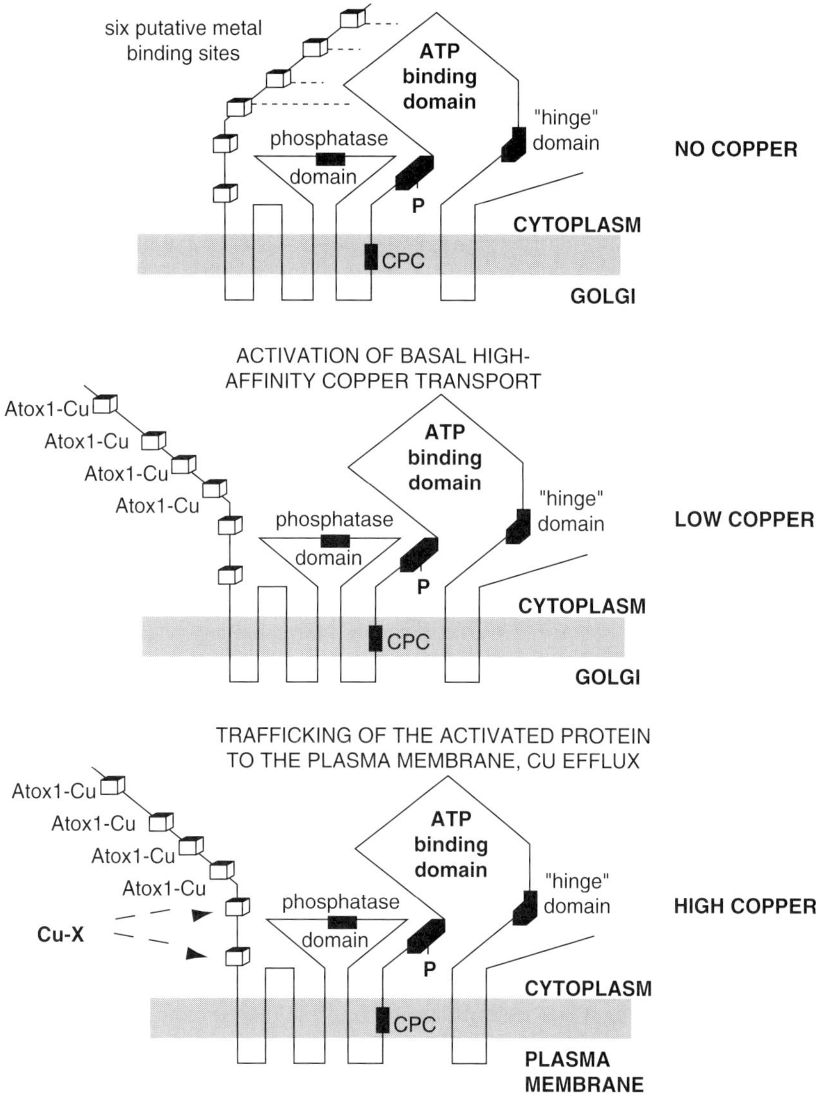


FIG. 4. Proposed regulation of catalytic activity and trafficking of ATP7A by the N-terminal putative copper-binding sites.

deliver copper, possibly randomly, to MBS5 and/or MBS6 or directly to the cation channel of ATP7A, thus signaling that the cell has accumulated excess copper. The binding of copper to MBS5 and/or MBS6 would cause appropriate conformational changes that can trigger the recruitment of

ATP7A in exocytic vesicles and stimulate the trafficking of ATP7A to the PM (Strausak *et al.*, 1999), where the excess of copper is effluxed from the cell (Fig. 4).

Consistent with this hypothesis, the nontrafficking ATP7A mutant with all MBSs mutated was catalytically active but its affinity for copper was decreased (Goodyer *et al.*, 1999; Strausak *et al.*, 1999; Voskoboinik *et al.*, 1999, 2001b). As a result, ATP7A with all the putative metal-binding sites mutated could transport copper only when it was present at higher than physiological concentrations. In recent studies, ATOX1 knock out mice exhibited the Menkes disease-like phenotype associated with lack of copper absorption (Hamza *et al.*, 2001). At least a partial recovery was reported in those mice who received an intraperitoneal injection of copper, suggesting that elevated copper concentrations can result in sufficient copper absorption (presumably via ATP7A) to overcome the lack of ATOX1 (Hamza *et al.*, 2001).

Despite these advances in understanding the mechanism of vectorial copper transport, the role of copper binding to the MBSs of ATP7A in catalytic activity is yet to be fully understood. The majority of studies on the functional role of specific domains/residues in the catalysis of ATP7A or ATP7B are based on the yeast Δ*ccc*2 growth complementation assay and the Fet3 multicopper ferroxidase activation assays. The empirical basis for employing these experimental systems was the observation that the disruption of the gene for yeast copper P-type ATPase, Ccc2, led to the inability of the mutant strain to grow on Cu/Fe-deficient medium due to abolition of the mechanism of high-affinity delivery of copper by Ccc2 to Fet3. The lack of activity of the latter prevents the high-affinity Fe uptake through Ftr1 and, consequently, leads to the inability of yeast to grow on iron-deficient medium. Human wild-type ATP7A and ATP7B have been shown to complement the Δ*ccc*2 phenotype and the assay has been commonly used as an indicator of catalytic activity of human copper P-type ATPases and their mutants (Yuan *et al.*, 1995). Despite the large number of MBS mutants produced for ATP7A and ATP7B, there was no consensus finding on the essentiality of one or another MBS even based on the Δ*ccc*2 complementation assay (Table I). The only consistent outcome of these studies was that the mutation/deletion of all MBSs resulted in the loss of the Δ*ccc*2 complementation (Forbes *et al.*, 1999; Iida *et al.*, 1998; Payne and Gitlin, 1998). However, the major obstacle in the interpretation of results obtained using the Δ*ccc*2 system is the ambiguity of discrimination between an inactive copper transporter and one with reduced affinity for copper. Conversely, mutations within the ATP-binding region are likely to produce less ambiguous results in terms of catalytic activity of the mutant protein with respect to the yeast Δ*ccc*2 complementation assay (Table I).

TABLE I
*Mutational Analysis of Eukaryotic Copper-Translocating P-type ATPases Using the Δccc2 Yeast Growth Complementation Assay*[a]

| Protein | **Δ***ccc2* complementation in *Saccharomyces cerevisiae* | | | | | | Reference |
|---|---|---|---|---|---|---|---|
| ATP7B | *ΔCu1–6* | *ΔCu3–6* | *ΔCu4–6* | *ΔCu5–6* | *ΔCu6* | **ΔCu1–5** | Iida, *et al.* (1998) |
| ATP7B | *D1027A (DKTG)* | *T1029A (DKTG)* | **H1069Q (SEHPL)** patient | | *N1270S (GDGVND)* | | Iida, *et al.* (1998) |
| ATP7B | **D765N** patient | **M769V** patient | **L776V** patient | *R778L* patient | R778Q patient | | Forbes and Cox (2000) |
| ATP7B | **G943S** patient | *T977M* patient | *P992L* patient | **V995A** variant | *CPC(983, 985)/S* | | Forbes and Cox (1998) |
| ATP7B | **mCu1** | **mCu1–2** | **mCu1–3** | **mCu1–4** | **mCu1–5** | *mCu1–6* | Forbes, *et al.* (1999) |
| ATP7B | *mCu4–6* | *mCu3–6* | **mCu3–5** | **mCu4** | **mCu6** | | Forbes, *et al.* (1999) |
| ATP7B | **ΔCu1–5** | *ΔCu4–6* | *ΔCu3–6* | *ΔCu3–5* | *ΔCu1–6* | | Forbes, *et al.* (1999) |
| ATP7A | *P1001A* | *H1086Q* | ***A629P*** patient | *G1019D* patient | | | Payne and Gitlin (1998) |
| ATP7A | **mCu1** | **mCu1–2** | *mCu1–3* | *mCu1–4* | *mCu1–5* | *mCu1–6* | Payne and Gitlin (1998) |
| PINA (ATP7B) | ***ΔCu1–6***, | ***Δtrans-membrane domains 1–4*** | | | | | Borjigin *et al.* (1999) |
| CCC2 (*S. cerevisiae*) | **mCu1** | **mCu2** | *mCu1–2* | | | | Vulpe *et al.* (1997) |
| CUA-1 (*C.* | *D786N (DKTG)* | | | | | | Sambon *gi. et al* (1997) |
| PCA1 (*S. cerevisiae*) | *G970R* (changes copper resistance to cadmium resistance) | | | | | | Shiraishi *et al.* (2000) |
| ATP7A | **mCu1–3** | *mCu1–6* | *GDG/A (MVGDGIND)* | **H1086Q** | *D1044E (DKTG)* | *M1393V* | Voskoboinik *et al.* (2001b) |

[a]Mutants in boldface can complement the Δ*ccc*2 phenotype, mutants in italics cannot complement the Δ*ccc*2 phenotype, and mutants in boldface italics have an intermediate effect.

Direct studies on the catalysis of copper translocation, ATP hydrolysis, and transient acyl phosphorylation of the ATP7A protein are expected to elucidate the role of MBSs in the regulation of catalytic activity of ATP7A. The *in vitro* $^{64}$Cu translocation assay using the mutant ATP7A-enriched membrane vesicles has revealed that the mutation of MBS1 to 3 had little effect on the catalytic activity of ATP7A (Voskoboinik *et al.*, 1999, 2001b), in contrast to the studies using yeast (Payne and Gitlin, 1998) (Table I). Most surprisingly, the mutation of all six MBSs has reduced but not abolished the $^{64}$Cu-translocating activity of ATP7A (Voskoboinik *et al.*, 1999). Subsequently, the formation of transient acyl phosphate was investigated, and the detailed analysis revealed that while the mutant ATP7A was transiently phosphorylated in a copper-dependent manner, it required a higher concentration of copper to stimulate phosphorylation than its wild-type counterpart (Voskoboinik *et al.*, 2001b). The inability of the ATP7A mutant with all six MBSs mutated to undergo copper-stimulated trafficking to the plasma membrane (Strausak *et al.*, 1999) (see Section V, E) was expected to prevent it from effluxing copper from the cell. Indeed, the analysis of $^{64}$Cu accumulation by cultured mammalian cells overexpressing the mutant protein has shown that in the presence of low to medium concentrations of extracellular copper, ATP7A with six MBSs mutated accumulated as much copper as the parental cells, suggesting lack of activity (overexpression of wild-type ATP7A results in increased copper efflux and reduced intracellular copper). However, the exposure of cells to higher, subtoxic concentrations of copper resulted in a significantly higher accumulation of copper compared to parental cells. In addition, the mutant-expressing cells appeared to be more copper-sensitive than controls (our unpublished observations) (Voskoboinik *et al.*, 1999). These findings suggested that, in agreement with the *in vitro* assays, the mutant ATP7A without MBSs remained catalytically active, but its affinity for copper decreased and its abnormal trafficking behavior resulted in unregulated intracellular copper accumulation.

Tsivkovskii *et al.* (2001) reported recently that the interaction between the MBSs and the ATP-binding domain of ATP7B was weaker in the presence of copper than in the copper-free environment. Consistent with this finding, the affinity of the ATP-binding domain for ATP was increased in the presence of copper-bound MBSs. This finding supported the notion that the MBSs can modulate the catalytic activity of the copper P-type ATPase through intramolecular interactions (Fig. 4).

In summary, the role of the ATOX1–ATP7A/ATP7B interaction may be to facilitate the vectorial transport of copper to ATP7A/ATP7B, which may activate the protein and allow the delivery of very low physiological concentrations of copper to target proteins. In addition, in higher organisms, some MBSs appear to have a signaling role: the elevation of intracellular copper to potentially toxic concentrations results in apparently ATOX1-independent copper binding to these MBSs, which subsequently stimulates trafficking to the plasma membrane where copper is effluxed from the cell. However, the relationship between copper-stimulated exocytosis of ATP7A and the catalytic activity is still unknown.

### *E. Trafficking of the Menkes P-type ATPase (ATP7A)*

The discovery of copper-mediated trafficking of ATP7A has been one of the major breakthroughs in unraveling the mechanisms of copper homeostasis and the regulation of ATP7A (Petris *et al.*, 1996). ATP7A was found to localize at the *trans*-Golgi network (TGN) under basal conditions, where it is believed to be responsible for the supply of copper to cuproenzymes of the secretory pathway, such as lysyl oxidase and tyrosinase (Petris *et al.*, 1996; Yamaguchi *et al.*, 1996). However, on the addition of extracellular copper (at concentrations as low as 20 μM) the protein relocalizes away from the TGN and within 30 min can be clearly visualized at the PM of the cell (Petris *et al.*, 1996). This was the first report of a metal ligand inducing the trafficking of its own transporter, as copper-stimulated trafficking appears to result directly from copper interaction with ATP7A. More detailed studies, including electron microscopy, have established that ATP7A was undergoing vesicular trafficking and fusion with the PM where it, presumably, effluxes excess copper from the cell (Petris *et al.*, 1996, 1998; Petris and Mercer, 1999). The process has been subsequently observed in a variety of cell types from different organisms and has also been reported for ATP7B (Roelofsen *et al.*, 2000; Schaefer *et al.*, 1999a,b).

Following elevation of copper concentrations, a relatively moderate increase in the steady-state level of ATP7A is observed at the PM (Petris *et al.*, 1996), indicating continuous recycling of ATP7A between the TGN and the PM (Petris and Mercer, 1999). Upon removal of extracellular copper, ATP7A returned to its original TGN location within 30 min (Petris *et al.*, 1996). The fusion of ATP7A with the PM has been confirmed by observing the uptake of anti-myc antibodies by cells transfected with an *ATP7A* that had a myc tag at the exofacial loop. These studies have also indicated that under basal conditions, ATP7A undergoes constitutive recycling (Petris and Mercer, 1999).

Analysis of the C-terminus of ATP7A has revealed three dileucine motifs. Through site-directed mutagenesis studies, the most C-terminal dileucine motif at position 1487/1488 appeared to be an internalization motif, as the mutant protein (LL to AA) was retained at the PM (Francis *et al.*, 1999; Petris *et al.*, 1998). The addition of extracellular copper had no effect on the trafficking of the mutant ATP7A to or from the TGN. The dileucine signaling motif indicates the involvement of clathrin-mediated endocytosis in ATP7A trafficking, as numerous membrane proteins have been shown to be internalized from the PM via dileucine motifs commonly associated with the binding to the adaptor protein complex AP-2 (Francis *et al.*, 1999; Petris *et al.*, 1998).

The mechanism of copper-stimulated exocytosis of ATP7A is poorly understood. While the importance of MBSs in the copper-regulated exocytosis of ATP7A has been reported, there is little understanding of how the binding of copper to certain MBSs stimulates the trafficking of ATP7A to the PM. One will be tempted to propose that the catalytic activity of

ATP7A is intrinsically linked with its ability to traffic, as certain catalytically inactive mutants of ATP7A could not undergo copper-stimulated exocytosis (our unpublished observations). However, some ATP7A mutants associated with the mild or treated forms of Menkes disease (suggesting at least some catalytic activity) are either mislocalized to the plasma membrane or do not appear to traffic from the TGN *in vitro* (Ambrosini and Mercer, 1999). Moreover, the mutation of all six MBSs abolishes copper-stimulated trafficking, while allowing the transport of copper (Strausak *et al.*, 1999; Voskoboinik *et al.*, 1999). Overall, it is likely that ATP7A trafficking is regulated by intramolecular conformational changes associated with the catalytic cycle and/or copper binding.

Unlike other P-type ATPases whose functions are restricted to a particular compartment, the mammalian copper-transporting system evolved the regulatory mechanism that allowed the cell to utilize a single transporter to donate copper to cuproenzymes at physiological levels of copper and to protect the cell from detrimental effects when copper concentrations reach potentially toxic levels. Such a level of sophistication is important for multicellular organisms, where alimentary copper must be transported between various cells and tissues. It is plausible that the ability of ATP7A to traffic allows the delivery of copper to the enzymes at the TGN and, at the same time, facilitates the efflux of copper from the cell either as a means of detoxification or as a means of intercellular transfer of copper, e.g., at the basolateral membrane of gut epithelial cells. The CuPAs characterized in unicellular organisms normally have a distinct function of intracellular copper transport or efflux out of the cell, e.g., the Golgi membrane Ccc2 and the plasma membrane Pca1 protein in yeast.

### *F. Trafficking of the Wilson P-type ATPase (ATP7B)*

Along with ATP7A, the trafficking of the Wilson protein has recently become the subject of intensive studies. The trafficking of ATP7B has been analyzed using animal tissues and human primary and established hepatocyte cell cultures (Roelofsen *et al.*, 2000; Schaefer *et al.*, 1999a,b). Similar to ATP7A, ATP7B has been localized to the TGN compartment in hepatocytes in whole animals and cultured cells. Copper administration to animals caused redistribution of ATP7B from the TGN to a vesicular compartment localized in proximity to the canalicular membrane of hepatocytes, which constitute bile ducts (Schaefer *et al.*, 1999a). Similar results were observed in sections of human liver, where ATP7B was predominantly associated with the *trans*-Golgi vesicles close to the pericanalicular membrane and small amounts of ATP7B were associated with the membrane (Schaefer *et al.*, 1999b). This finding has led to a proposal that ATP7B is essential for both the delivery of copper to ceruloplasmin and the biliary excretion of copper, consistent with the reduction of these functions in patients with Wilson's disease (Schaefer *et al.*, 1999a,b). In agreement with that report was the finding that ATP7B in a polarized hepatoma cell line, HepG2, localizes to the *trans*-Golgi compartment and undergoes copper-induced trafficking to the apical membrane in the presence of elevated copper. The apical membrane in these cells is related to the canalicular membrane in the liver, where ATP7B may facilitate biliary copper excretion (Roelofsen *et al.*, 2000). Recent studies on mammary gland cells of wild-type and mutant mice that had a point mutation in ATP7B in transmembrane domain 8 (Met1359Val) were, generally, in agreement with earlier reports. Thus, both the wild-type and the mutant ATP7B were localized to the TGN under normal conditions. However, lactating mice had ATP7B trafficking toward the plasma membrane, while the mutant form of the protein remained at the TGN (Michalczyk *et al.*, 2000). Interestingly, copper supplementation resulted in the relocalization of both the wild-type and the mutant proteins from the TGN toward the plasma membrane. In contrast, a recent report on the trafficking of the wild-type and mutant ATP7B overexpressed in CHO cells indicated that the mutant protein was unable to undergo exocytic trafficking even in the presence of elevated copper (La Fontaine *et al.*, 2001). Together, these results suggest the possibility of hormonal regulation of ATP7B trafficking, which has not been considered before, and indicate that the mechanisms regulating this process may be cell-type specific.

In contrast to these reports, green fluorescent protein-tagged ATP7B and endogenous ATP7B have been localized to the late endosome compartment in a human hepatoma cell line and isolated rat hepatocytes (Harada *et al.*, 2000), while no ATP7B was detected at the *trans*-Golgi network or plasma membrane. That finding has led the authors to propose the following pathway for ATP7B-facilitated copper efflux from the cell: late endosomes, lysosomes, and, finally, excretion into the bile by biliary lysosomal excretion (Harada *et al.*, 2000).

The $^{1454}$LLL motif at the C-terminus of ATP7B is located in the position similar to the $^{1487}$LL TGN internalization motif of ATP7A. While no studies on the role of that motif in ATP7B have been reported, it is likely that this motif is involved in the regulation of trafficking of ATP7B. More detailed studies are required to identify the exact pathway of ATP7B in the cell.

## VI. Conclusion

The studies on physiological and biochemical properties of mammalian ATP7A and ATP7B suggest that the combination of copper-translocating

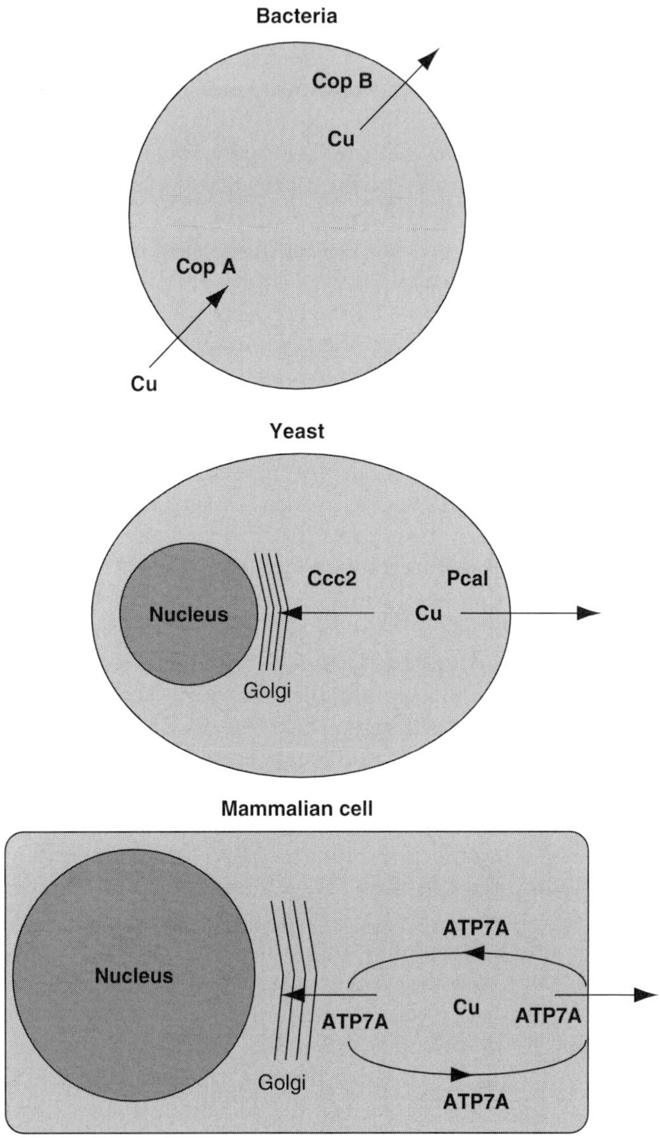


FIG. 5. Evolution of copper P-type ATPases.

activity and copper-stimulated trafficking is a key regulator of intracellular copper homeostasis by these transporters. While the structure of prokaryotic and early eukaryotic CuPA resembles that of their mammalian counterparts, one can hypothesize about the evolution of this ubiquitous family of heavy metal transporters (Fig. 5). First, in bacteria, which have a relatively primitive subcellular compartmentalization, CuPAs are localized at the plasma membrane of the cell and appear to be partly responsible for the intracellular uptake and efflux of copper, e.g., CopA and CopB in *E. hirae*. With the differentiation of compartments in eukaryotes, there was a need not only to take up copper inside the cell but also to deliver it to copper-dependent enzymes in the lumen of the Golgi compartment. Thus while *Saccharomyces cerevisiae* retained a plasma membrane copper efflux pump, PcaI, it also evolved an intracellular copper transporter, Ccc2, which appears to be responsible for the delivery of copper to multicopper ferroxidase, Fet3, at the Golgi compartment. However, copper homeostasis of the whole organism relies on its absorption and intercellular distribution. Mammalian CuPAs have evolved a trafficking function that allows both copper delivery to cuproenzymes of secretory pathway and copper transport to adjacent cells (e.g., systemic copper absorption from gut epithelial cells and reabsorption in kidney epithelial cells) or efflux in response to copper stress. The mechanism of regulation of copper-ATPase trafficking is not fully understood but it appears that the evolutionary amplification of MBSs at the N-terminus of ATP7A and ATP7B plays some role in that process; i.e., MBS1 to MBS4 appear to interact with ATOX1 but are not involved in copper-stimulated trafficking, while MBS5 and MBS6 do not interact with ATOX1 but are critical for exocytosis of ATP7A. The studies on mammalian CuPA are still at an early stage. However, as the role of copper in human physiology and pathology, e.g., neurodegenerative disorders (Alzheimer's and prion diseases), osteoporosis, and cardiovascular diseases, becomes more appreciated and attracts a great deal of attention, there is a need for detailed studies into the mechanisms of catalysis, regulation, and trafficking of these transporters.

## References


Ambrosini, L., and Mercer, J. F. (1999). *Hum. Mol. Genet.* **8**, 1547–1555.

Auer, M., Scarborough, G. A., and Kuhlbrandt, W. (1998). *Nature* **392**, 840–843.

Axelsen, K. B., and Palmgren, M. G. (1998). *J. Mol. Evol.* **46**, 84–101.

Banci, L., Bertini, I., Ciofi-Baffoni, S., Huffman, D. L., and O'Halloran, T. V. (2001). *J. Biol. Chem.* **276**, 8415–8426.

Bissig, K. D., Wunderli-Ye, H., Duda, P. W., and Solioz, M. (2001). *Biochem. J.* **357**, 217–223.

Borjigin, J., Payne, A. S., Deng, J., Li, X., Wang, M. M., Ovodenko, B., Gitlin, J. D., and Snyder, S. H. (1999). *J. Neurosci.* **19**, 1018–1026.

Bull, P. C., Thomas, G. R., Rommens, J. M., Forbes, J. R., and Cox, D. W. (1993). *Nat. Genet.* **5**, 327–337.

Camakaris, J., Danks, D. M., Ackland, L., Cartwright, E., Borger, P., and Cotton, R. G. H. (1980). *Biochem. Genet.* **18**, 117–131.

Camakaris, J., Petris, M. J., Bailey, L., Shen, P., Lockhart, P., Glover, T. W., Barcroft, C., Patton, J., and Mercer, J. F. (1995). *Hum. Mol. Genet.* **4**, 2117–2123.
Camakaris, J., Voskoboinik, I., and Mercer, J. F. (1999). *Biochem. Biophys. Res. Commun.* **261**, 225–232.
Carafoli, E. (1991). *Physiol. Rev.* **71**, 129–153.
Chelly, J., Tumer, Z., Tonnesen, T., Petterson, A., Ishikawa Brush, Y., Tommerup, N., Horn, N., and Monaco, A. P. (1993). *Nat. Genet.* **3**, 14–19.
Cobine, P. A., George, G. N., Winzor, D. J., Harrison, M. D., Mogahaddas, S., and Dameron, C. T. (2000). *Biochemistry* **39**, 6857–6863.
Dagenais, S. L., Adam, A. N., Innis, J. W., and Glover, T. W. *Am. J. Hum. Genet.* **69**, 420–427.
Daly, S. E., Lane, L. K., and Blostein, R. (1996). *J. Biol. Chem.* **271**, 23683–23689.
Danks, D. M. (1995). *In* "The Metabolic Basis of Inherited Disease" (Scriver, C. R., Beaudet, A. L., Sly, W. V., and Valle, D., Eds.), pp. 2211–2235. McGraw-Hill, New York.
Forbes, J. R., and Cox, D. W. (1998). *Am. J. Hum. Genet.* **63**, 1663–1674.
Forbes, J. R., and Cox, D. W. (2000). *Hum. Mol. Genet.* **9**, 1927–1935.
Forbes, J. R., Hsi, G., and Cox, D. W. (1999). *J. Biol. Chem.* **274**, 12408–12413.
Francis, M. J., Jones, E. E., Levy, E. R., Martin, R. L., Ponnambalam, S., and Monaco, A. P. (1999). *J. Cell Sci.* **112**, 1721–1732.
Gitschier, J., Moffat, B., Reilly, D., Wood, W. I., and Fairbrother, W. J. (1998). *Nat. Struct. Biol.* **5**, 47–54.
Goodyer, I. D., Jones, E. E., Monaco, A. P., and Francis, M. J. (1999). *Hum. Mol. Genet.* **8**, 1473–1478.
Gupta, A., Matsui, K., Lo, J. F., and Silver, S. (1999). *Nat. Med.* **5**, 183–188.
Hamza, I., Faisst, A., Prohaska, J., Chen, J., Gruss, P., and Gitlin, J. D. (2001). *Proc. Natl. Acad. Sci. USA* **98**, 6848–6852.
Hamza, I., Schaefer, M., Klomp, L. W., and Gitlin, J. D. (1999). *Proc. Natl. Acad. Sci. USA* **96**, 13363–13368.
Harada, M., Sakisaka, S., Terada, K., Kimura, R., Kawaguchi, T., Koga, H., Taniguchi, E., Sasatomi, K., Miura, N., Suganuma, T., Fujita, H., Furuta, K., Tanikawa, K., Sugiyama, T., and Sata, M. (2000). *Gastroenterology* **118**, 921–928.
Hirayama, T., Kieber, J. J., Hirayama, N., Kogan, M., Guzman, P., Nourizadeh, S., Alonso, J. M., Dailey, W. P., Dancis, A., and Ecker, J. R. (1999). *Cell* **97**, 383–393.
Horn, N. (1976). *Lancet* **1**, 1156–1158.
Huffman, D. L., and O'Halloran, T. V. (2000). *J. Biol. Chem.* **275**, 18611–18614.
Iida, M., Terada, K., Sambongi, Y., Wakabayashi, T., Miura, N., Koyama, K., Futai, M., and Sugiyama, T. (1998). *FEBS Lett.* **428**, 281–285.
Kaler, S. G. (1998). *Pediatr. Dev. Pathol.* **1**, 85–98.
Klomp, L. W., Lin, S. J., Yuan, D. S., Klausner, R. D., Culotta, V. C., and Gitlin, J. D. (1997). *J. Biol. Chem.* **272**, 9221–9226.
La Fontaine, S., Theophilos, M. B., Firth, S. D., Gould, R., Parton, R. G., and Mercer, J. F. (2001). *Hum. Mol. Genet.* **10**, 361–370.
Larin, D., Mekios, C., Das, K., Ross, B., Yang, A. S., and Gilliam, T. C. (1999). *J. Biol. Chem.* **274**, 28497–28504.
Linder, M. C., and Hazegh Azam, M. (1996). *Am. J. Clin. Nutr.* **63**, 797S–811S.
Lutsenko, S., Petrukhin, K., Cooper, M. J., Gilliam, C. T., and Kaplan, J. H. (1997). *J. Biol. Chem.* **272**, 18939–18944.
MacLennan, D. H., Rice, W. J., and Green, N. M. (1997). *J. Biol. Chem.* **272**, 28815–28818.
Majumdar, R., Al Jumah, M., Al Rajeh, S., Fraser, M., Al Zaben, A., Awada, A., Al Traif, I., and Paterson, M. (2000). *J. Neurol. Sci.* **179**, 140–143.
Mercer, J. F., Livingston, J., Hall, B., Paynter, J. A., Begy, C., Chandrasekharappa, S., Lockhart, P., Grimes, A., Bhave, M., Siemieniak, D., and Glover, T. W. (1993). *Nat. Genet.* **3**, 20–25.

Mercer, J. F. B., and Camakaris, J. (1997). *In* "Metal Ions in Gene Regulation" (Silver, S., and Walden, W., Eds.), pp. 250–276. Chapman and Hall, New York.
Michalczyk, A. A., Rieger, J., Allen, K. J., Mercer, J. F., and Ackland, M. L. (2000). *Biochem. J.* **352**, 565–571.
Møller, J. V., Juul, B., and le Maire, M. (1996). *Biochim. Biophys. Acta.* **1286**, 1–51.
Mollman, J. E., and Pleasure, D. E. (1980). *J. Biol. Chem.* **255**, 569–574.
O'Neal, S. G., Rhoads, D. B., and Racker, E. (1979). *Biochem. Biophys. Res. Commun.* **89**, 845–850.
Outten, F. W., Huffman, D. L., Hale, J. A., and O'Halloran, T. V. *J. Biol. Chem.* **276**, 30670–30677.
Payne, A. S., and Gitlin, J. D. (1998). *J. Biol. Chem.* **273**, 3765–3770.
Petris, M. J., Camakaris, J., Greenough, M., LaFontaine, S., and Mercer, J. F. B. (1998). *Hum. Mol. Genet.* **7**, 2063–2071.
Petris, M. J., and Mercer, J. F. (1999). *Hum. Mol. Genet.* **8**, 2107–2115.
Petris, M. J., Mercer, J. F., Culvenor, J. G., Lockhart, P., Gleeson, P. A., and Camakaris, J. (1996). *EMBO J.* **15**, 6084–6095.
Petris, M. J., Strausak, D., and Mercer, J. F. (2000). *Hum. Mol. Genet.* **9**, 2845–2851.
Portnoy, M. E., Rosenzweig, A. C., Rae, T., Huffman, D. L., O'Halloran, T. V., and Culotta, V. C. (1999). *J. Biol. Chem.* **274**, 15041–15045.
Pufahl, R. A., Singer, C. P., Peariso, K. L., Lin, S. -J., Schmidt, P., Cizewski Culotta, V., Penner-Hahn, J. E., and O'Halloran, T. V. (1997). *Science* **278**, 853–856.
Rae, T. D., Schmidt, P. J., Pufahl, R. A., Culotta, V. C., and O'Halloran, T. V. (1999). *Science* **284**, 805–808.
Rensing, C., Fan, B., Sharma, R., Mitra, B., and Rosen, B. P. (2000). *Proc. Natl. Acad. Sci. USA* **97**, 652–656.
Rensing, C., Ghosh, M., and Rosen, B. P. (1999). *J. Bacteriol.* **181**, 5891–5897.
Roelofsen, H., Wolters, H., Van Luyn, M. J., Miura, N., Kuipers, F., and Vonk, R. J. (2000). *Gastroenterology* **119**, 782–793.
Rosen, B. P. (1999a). *Trends Microbiol.* **7**, 207–212.
Rosen, B. P. (1999b). *Essays Biochem.* **34**, 1–15.
Rosenzweig, A. C., Huffman, D. L., Hou, M. Y., Wernimont, A. K., Pufahl, R. A., and O'Halloran, T. V. (1999). *Struct. Fold Des.* **7**, 605–617.
Rosenzweig, A. C., and O'Halloran, T. V. (2000). *Curr. Opin. Chem. Biol.* **4**, 140–147.
Royce, P. M., Camakaris, J., and Danks, D. (1980). *Biochem. J.* **192**, 579–586.
Sambongi, Y., Wakabayashi, T., Yoshimizu, T., Omote, H., Oka, T., and Futai, M. (1997). *J. Biochem.* **121**, 1169–1175.
Schaefer, M., and Gitlin, J. D. (1999). *Am. J. Physiol.* **276**, G311–G314.
Schaefer, M., Hopkins, R. G., Failla, M. L., and Gitlin, J. D. (1999a). *Am. J. Physiol.* **276**, G639–G646.
Schaefer, M., Roelofsen, H., Wolters, H., Hofmann, W. J., Muller, M., Kuipers, F., Stremmel, W., and Vonk, R. J. (1999b). *Gastroenterology* **117**, 1380–1385.
Shiraishi, E., Inouhe, M., Joho, M., and Tohoyama, H. (2000). *Curr. Genet.* **37**, 79–86.
Silver, S. (1998). *J. Ind. Microbiol. Biotechnol.* **20**, 1–12.
Silver, S., and Phung, L. T. (1996). *Annu. Rev. Microbiol.* **50**, 753–789.
Solioz, M., and Odermatt, A. (1995). *J. Biol. Chem.* **270**, 9217–9221.
Strausak, D., La Fontaine, S., Hill, J., Firth, S. D., Lockhart, P. J., and Mercer, J. F. (1999). *J. Biol. Chem.* **274**, 11170–11177.
Sweadner, K. J., and Donnet, C. (2001). *J. Biochem.* **356**, 685–704.
Toyoshima, C., Nakasako, M., Nomura, H., and Ogawa, H. (2000). *Nature* **405**, 647–655.

Tsivkovskii, R., MacArthur, B. C., and Lutsenko, S. (2001). *J. Biol. Chem.* **276**, 2234–2242.

Voskoboinik, I., Brooks, H., Smith, S., Shen, P., and Camakaris, J. (1998). *FEBS Lett.* **435**, 178–182.

Voskoboinik, I., Greenough, M., La Fontaine, S., Mercer, J. F., and Camakaris, J. (2001a). *Biochem. Biophys. Res. Commun.* **281**, 966–970.

Voskoboinik, I., Mar, J., Strausak, D., and Camakaris, J. (2001b). *J. Biol. Chem.* **276**, 28620–28627.

Voskoboinik, I., Strausak, D., Greenough, M., Brooks, H., Petris, M., Smith, S., Mercer, J. F., and Camakaris, J. (1999). *J. Biol. Chem.* **274**, 22008–22012.

Vulpe, C., Levinson, B., Whitney, S., Packman, S., and Gitschier, J. (1993). *Nat. Genet.* **3**, 7–13.

Vulpe, C., Yuan, D., Ibom, V., and Gitschier, J. (1997). "Copper and Zinc Receptors in Signalling, Trafficking and Disease," p. 35.

Wernimont, A. K., Huffman, D. L., Lamb, A. L., O'Halloran, T. V., and Rosenzweig, A. C. (2000). *Nat. Struct. Biol.* **7**, 766–771.

Wunderli-Ye, H., and Solioz, M. (2001). *Biochem. Biophys. Res. Commun.* **280**, 713–719.

Yamaguchi, Y., Heiny, M. E., and Gitlin, J. D. (1993). *Biochem. Biophys. Res. Commun.* **197**, 271–277.

Yamaguchi, Y., Heiny, M. E., Suzuki, M., and Gitlin, J. D. (1996). *Proc. Natl. Acad. Sci. USA* **93**, 14030–14035.

Yuan, D. S., Stearman, R., Dancis, A., Dunn, T., Beeler, T., and Klausner, R. D. (1995). *Proc. Natl. Acad. Sci. USA* **92**, 2632–2636.

# COPPER CHAPERONES

BY JENNIFER STINE ELAM, SUSAN T. THOMAS, STEPHEN P. HOLLOWAY, ALEXANDER B. TAYLOR, AND P. JOHN HART

Center for Biomolecular Structure Analysis, Department of Biochemistry, University of Texas Health Science Center at San Antonio, San Antonio, Texas 78229



## I. Introduction

### *A. Background and Scope of Review*

Copper, the third most abundant trace element in humans after iron and zinc, is required for the activation of dioxygen, a function essential for the survival of all aerobic organisms (Underwood, 1977; Adman, 1991; Linder and Goode, 1991; Solomon and Lowery, 1993). Because it can easily cycle through the oxidized [Cu(II)] and reduced [Cu(I)] states, it is a versatile cofactor for a variety of enzymes, including lysine oxidase (Knowles and Yadav, 1984), cytochrome *c* oxidase (Capaldi, 1990), dopamine β-hydroxylase (Ljones and Skotland, 1984), copper–zinc superoxide dismutase (Cu,ZnSOD) (Valentine and Pantoliano, 1981; Fielden and Rotilio, 1984), tyrosinase (Robb, 1984), ceruloplasmin (Ryden, 1984), and blood coagulation factor V (Ryden, 1988; Adman, 1991; Linder and Goode, 1991). Paradoxically, the electronic structure of copper that permits its direct interaction with oxygen also renders it quite toxic. "Free" copper ions, those corresponding to hydrated Cu(I) or Cu(II) complexes not coordinated by amino acids or other organic



ADVANCES IN PROTEIN CHEMISTRY, Vol. 60


0065-3233/02 $35.00

molecules, can initiate hydroxyl radical formation via Fenton-like reactions in the presence of superoxide anion or hydrogen peroxide, both of which are formed during aerobic cellular metabolism (Fridovich, 1978; Halliwell and Gutteridge, 1985). These hydroxyl radicals in turn can cause cellular damage via protein oxidation, lipid peroxidation, and nucleic acid cleavage (Santoro and Thiele, 1997). The ability of free copper to perform such deleterious chemical reactions is greatly enhanced when it is in its Cu(I) oxidation state, and the relatively high concentrations of glutathione in the cytoplasm of cells thus increase the potential for free copper ions to cause such damage (Tietze, 1969; Halliwell and Gutteridge, 1984, 1985).

Because copper is essential for life and can be highly toxic, it is critical that organisms possess mechanisms that allow them to acquire sufficient amounts for essential biochemical reactions, yet simultaneously, prevent its accumulation to toxic levels (Eide, 1998; Pena *et al.*, 1999). In this regard, cells have evolved a dual strategy to protect against copper toxicity by (1) tightly regulating its entry into the cytoplasm via the membrane-bound copper transporters (Dancis *et al.*, 1994a,b; Radisky *et al.*, 1997) and (2) expressing detoxification and scavenging proteins such as the metallothioneins that effectively act as a sink for free copper and other transition metal ions (Thiele, 1988; Szczypka and Thiele, 1989). Although intracellular copper sequestration by metallothionein dramatically limits the exposure of free copper to the cytoplasm, it is not responsible for directed copper delivery to organelles or to the various copper-containing proteins. The enzymes that use copper as a cofactor must therefore somehow acquire it in the face of this strict regulation of copper import and in the presence of cytoplasmic housekeeping molecules with a high capacity for copper chelation.

During the past several years, knowledge of how cells accomplish this has increased substantially with the discovery of a class of molecules called "copper chaperones" (Culotta *et al.*, 1997). The term copper chaperone is distinct from a "molecular chaperone," which assists in the folding of protein molecules (Bukau *et al.*, 2000), and is instead derived from the fact that copper chaperone molecules "escort" reactive copper by acquiring it (directly or indirectly) from the membrane-bound copper transporters, protecting it from housekeeping and scavenging molecules (and the cellular environment from it), and delivering and inserting it into target proteins, thereby activating them (Valentine and Gralla, 1997; O'Halloran and Culotta, 2000; Rosenzweig and O'Halloran, 2000). An important feature of copper chaperone proteins is that they recognize their cognate target molecules via specific protein–protein interactions and cannot substitute for each other across copper delivery pathways (Culotta *et al.*, 1997).

Currently, three such chaperone-mediated copper delivery pathways that are highly conserved between yeast and humans have been characterized (Fig. 1). Three small yeast copper-binding proteins, Atx1 (Hah1 or Atox1 in humans) (Lin and Culotta, 1995; Klomp *et al.*, 1997; Hung *et al.*, 1998), CCS (hCCS in humans) (Culotta *et al.*, 1997; Casareno *et al.*, 1998), and Cox17 (hCox17 in humans) (Glerum *et al.*, 1996a; Amaravadi *et al.*, 1997), are identified in high-affinity copper mobilization by delivering copper to late Golgi secretory compartments, to cytosolic Cu,ZnSOD, and

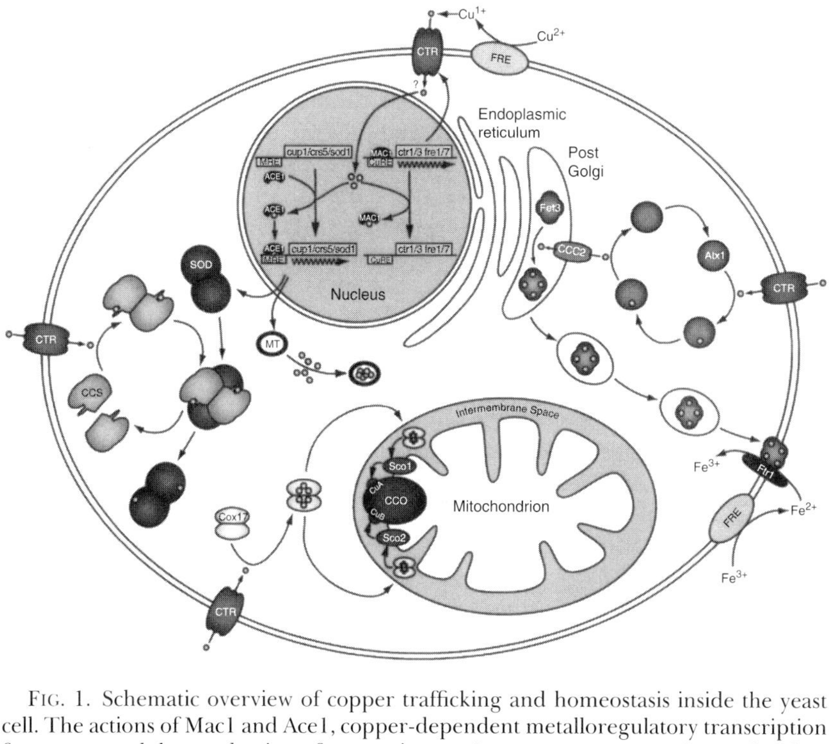


FIG. 1. Schematic overview of copper trafficking and homeostasis inside the yeast cell. The actions of Mac1 and Ace1, copper-dependent metalloregulatory transcription factors, control the production of copper import [copper transporter (Ctr) and reductase (Fre)] and detoxification/sequestration [metallothionein (MT)] machineries, respectively. Three chaperone-mediated delivery pathways are shown. Atx1 shuttles Cu(I) to the secretory pathway P-type ATPase Ccc2 (right). CCS delivers Cu(I) to the cytoplasmic enzyme copper–zinc superoxide dismutase (SOD) (left). Cox17 shuttles Cu(I) to cytochrome *c* oxidase (CCO) in the mitochondria (bottom). Mitochondrial proteins Sco1 and Sco2 may also play a role in copper delivery to the CuA and CuB sites of CCO. Copper metabolism and iron metabolism are linked through the actions of Fet3, a copper-containing ferroxidase required to bring iron into the cell (lower right) (see text).

to mitochondria, respectively. In addition to yeast and humans, copper chaperone homologues have been identified in bacteria, plants, and higher eukaryotes. Although fundamentally interesting in their own right, the importance of understanding the molecular details of how copper chaperones traffic copper ion within cells is underscored by findings that defects in genes involved in copper metabolism cause the human disorders Wilson's disease (Bull *et al.*, 1993), Menkes syndrome (Vulpe *et al.*, 1993), and iron deficiency anemia linked to copper deficiency (Vulpe *et al.*, 1999), and they might possibly play a role in the etiology of Alzheimer's disease (Huang *et al.*, 1999), prion diseases (Viles *et al.*, 1999), and familial amyotrophic lateral sclerosis (FALS) (Deng *et al.*, 1993; Rosen *et al.*, 1993).

While this chapter serves as a review of copper ion homeostasis and trafficking within cells, it touches only briefly on the regulation of intracellular copper concentrations at the level of uptake (the copper transporters) and sequestration (the metallothioneins) and instead focuses mainly on the current state of knowledge of the protein factors that perform the critical role of directed copper delivery to proteins and enzymes that use it as a cofactor. In particular, the past 2 years have produced a wealth of three-dimensional information on these molecules, setting the stage for a detailed understanding of the molecular mechanism(s) and determinants of specificity of copper transfer from the copper chaperones to their cognate target proteins. The future looks exciting for research on copper chaperones, as understanding the atomic details of copper transfer in these systems might serve as the starting point for the design of therapies for a variety of diseases resulting from defects in copper ion homeostasis and trafficking.

### *B. Regulation of Copper Uptake and Intracellular Copper Levels*

Studies in *Saccharomyces cerevisiae* have proven to be extremely powerful in the identification of components of the copper homeostatic machinery and, further, have provided fundamental information from which a comprehensive mechanistic understanding of copper homeostasis and trafficking in eukaryotic cells has begun to emerge. Clever genetic screens performed under conditions in which copper is either limiting (nutritional levels) or in excess (toxic levels) have led to the identification of many of the genes responsible for both copper uptake from the surrounding environment and copper sequestration and detoxification within the cytoplasm (Thiele, 1988, 1992; Culotta *et al.*, 1994; Dancis *et al.*, 1994a, b; Knight *et al.*, 1996). In-depth reviews describing the copper-sensing transcription factors, the gene products they regulate, and the interplay between copper and iron homeostasis are found elsewhere in this volume, and the brief description here is intended to give only the context within which the copper chaperone proteins are known to function.

*1. Copper Import Machinery*

At the nutritional level, copper ions are actively acquired by the yeast cell. Although in an aerobic environment copper is typically encountered in its Cu(II) state, it is transported into the cytoplasm by the transmembrane high-affinity copper transporters Ctr1 and Ctr3 either during or subsequent to its reduction to Cu(I) by the membrane-bound metalloreductases Fre1 and Fre7 [which also reduce Fe(III) to Fe(II)] (Lin and Kosman, 1990; Hassett and Kosman, 1995; Georgatsou *et al.*, 1997; Labbe and Thiele, 1999). The deletion of *ctr1* and *ctr3* genes results in a dramatic reduction in cell growth because of copper deprivation, and such yeast exhibit a number of phenotypes that can be overcome by the addition of exogenous copper, including respiratory defects due to lack of copper incorporation into cytochrome *c* oxidase, sensitivity to oxidative stress due to lack of copper incorporation into copper–zinc superoxide dismutase, and iron starvation due to lack of copper incorporation into the iron transport machinery (see below) (Dancis *et al.*, 1994a, b; Knight *et al.*, 1996). Ctr1 is a highly glycosylated, 406-amino-acid protein that oligomerizes in the plasma membrane (Dancis *et al.*, 1994a,b). Ctr3, a 241-amino-acid polypeptide and a trimer in the plasma membrane, was discovered because overexpression of its gene suppressed the copper starvation phenotypes observed in a *ctr1*Δ yeast strain (Knight *et al.*, 1996; Pena *et al.*, 1999). Although dissimilar in sequence, Ctr1 and Ctr3 have two or three predicted membrane-spanning motifs each, with Ctr1 containing eight copies of the consensus sequence Met-$X_2$-Met-X-Met in its extracellular domain, while Ctr3 possesses 11 cysteine residues, of which three pairs occur in a C-C or C-$X_2$-C motif (Dancis *et al.*, 1994a,b; Zhou and Gitschier, 1997). Because Cu(I) has different ligand preferences than Cu(II), and because methionine and cysteine are both excellent Cu(I) ligands, it has been postulated that the reduction to Cu(I) by Fre1/Fre7 might partially determine the specificity of the copper transport process by these proteins (Labbe and Thiele, 1999).

The *ctr1* gene was initially identified as a high-affinity copper transporter indirectly, through genetic selection for mutants defective in iron uptake (Askwith *et al.*, 1994; Dancis *et al.*, 1994a,b). There is a strict requirement for copper in order to bring iron into the cell because copper is the cofactor for the protein encoded by the *fet3* gene, a transmembrane ferroxidase that oxidizes Fe(II) to Fe(III) prior to or concurrent with its translocation across the plasma membrane by the iron permease Ftr1 (Fig. 1). Fet3, the yeast homologue of human ceruloplasmin (Askwith *et al.*, 1994; Blackburn *et al.*, 2000), is loaded with copper in the post-Golgi compartment through the actions of Ccc2, the yeast P-type copper-transporting ATPase homologous to the human Menkes (ATP7A) and Wilson's (ATP7B) disease proteins (Bull *et al.*, 1993; Vulpe *et al.*, 1993; Yuan *et al.*, 1995), and Atx1, the yeast copper chaperone homologous to human Hah1 (see below) (Lin and Culotta, 1995; Klomp *et al.*, 1997). If insufficient copper gains entry to the cytoplasm and ultimately to the *trans*-Golgi compartment through mutations in *ctr1*, *ctr3*, *ccc2*, or *atx1*, then Fet3 is not fully loaded with copper, and upon its translocation to the plasma membrane, it is unable to perform the oxidation of Fe(II) requisite for its transport. Mutations in *fet3* itself result in iron deficiency phenotypes almost identical to those observed for mutations in the *ctr1* gene (Dancis *et al.*, 1994a,b). The link between copper and iron homeostasis was delineated through the observation that *ctr1* mutants are defective in the uptake of *both* copper and iron. *fet3* mutants, on the other hand, are defective only in iron uptake as evidenced by the fact that deficiency in iron uptake in the *ctr1* mutants, but not the *fet3* mutants, was overcome by cell growth in the presence of elevated concentrations of copper (Dancis *et al.*, 1994a,b; Knight *et al.*, 1996).

Monitoring yeast cells for iron deficiency through insufficient copper loading into Fet3 has proven to be a convenient assay for proper copper trafficking through the secretory pathway. It has been used repeatedly in studies of Ccc2, the P-type ATPase responsible for copper translocation into the lumen of the endoplasmic reticulum (Yuan *et al.*, 1995, 1997), in studies of Atx1 and Hah1, the copper chaperones that shuttle copper ion to Ccc2 and the Menkes and Wilson's proteins, respectively (Klomp *et al.*, 1997; Lin *et al.*, 1997; Hung *et al.*, 1998) (see below), and in complementation assays used to ascribe function to higher eukaryotic gene products (Kampfenkel *et al.*, 1995; Payne *et al.*, 1998; Forbes *et al.*, 1999). For example, *copt1*, a putative copper transporter from *Arabidopsis thaliana*, and *hctr1*, a putative human copper transporter homologue gene, were both identified by their ability to functionally replace *ctr1* in maintaining cellular copper and iron homeostasis in yeast (Kampfenkel *et al.*, 1995; Zhou and Gitschier, 1997). Similarly, *ctr2*, a putative low-affinity copper transporter gene, was identified in yeast through its similarity to *copt1* and by its ability to complement the copper and iron deficiency of a *ctr1*Δ*ctr3*Δ strain (Kampfenkel *et al.*, 1995; Radisky and Kaplan, 1999).

### 2. *Copper Sequestration/Detoxification Machinery*

Under conditions where copper is in excess in the surrounding medium, the yeast cell's perspective shifts from one of active acquisition of copper ion to one of protecting the cytoplasm from its toxic effects. To accomplish this, *Sa. cerevisae* produce small cysteine-rich polypeptides with repeated C-X-X-C or C-X-C sequence motifs termed metallothioneins (MTs), encoded by the *cup1* and *crs5* genes. MTs are effective in copper ion detoxification because they bind it tightly within polymetallic thiolate bond clusters, shielding it from the cytoplasm and preventing it from performing Fenton-type reactions that can damage proteins, nucleic acids, and lipids (Fridovich, 1978; Halliwell and Gutteridge, 1984; Santoro and Thiele, 1997). *cup1Δ* yeast exhibit extreme sensitivity to copper salts, in line with its role in copper detoxification/sequestration (Hamer *et al.*, 1985; Ecker *et al.*, 1986). Figure 2a shows the three-dimensional structure of *Sa. cerevisae* metallothionein derived from the *cup1* gene complexed with seven Cu(I) ions as determined by nuclear magnetic resonance (NMR) (Peterson *et al.*, 1996). Although *cup1* MT binds up to 7 Cu(I) ions, *crs5* MT has been found to bind 11–12 Cu(I) ions (Jensen *et al.*, 1996). The role that metallothioneins play is likely more complex than simply acting as a copper ion sink, as they have been demonstrated to bind a variety of essential as well as nonessential metal ions (Karin, 1985).

*3. Regulation of Synthesis and Degradation of Copper Homeostasis Machinery*

The essential yet toxic nature of copper dictates that the synthesis of its import and detoxification machinery must be tightly regulated in order (1) to ensure that enough copper is available in the cell to drive critical biochemical processes, (2) to prevent its accumulation to toxic levels, and (3) to prevent the cell from wasting resources by producing proteins that are not needed under a given environmental condition. This copper-mediated regulation occurs both at the transcriptional and at the posttranslational levels. The transcription of copper import and copper detoxification/sequestration genes is reciprocally regulated by two so-called "copper sensor" proteins, the Cu-dependent metalloregulatory transcription factors Mac1 and Ace1 (Fig. 1, nucleus). The nutritional copper sensor, Mac1 (metal-binding activator), is a 417-amino-acid polypeptide that possesses an N-terminal DNA-binding domain with a zinc-finger motif, a *trans*-activation domain, and two repeated elements (REPs) near the C-terminus, containing a (C-X-C-$X_4$-C-X-C-$X_2$- C-$X_2$-His) sequence motif. These REPs, which likely bind copper ion, are absolutely required to regulate the copper-sensing function of Mac1, which in turn modulates its DNA-binding and *trans*-activation functions (Jensen and Winge, 1998; Labbe and Thiele, 1999; Serpe *et al.*, 1999). The toxic copper sensor Ace1 (activation of *cup1* expression, known also as Cup2) is a 225-amino-acid polypeptide that possesses an N-terminal DNA-binding domain and that cooperatively binds Cu(I) via specific cysteine residues to form a tetracopper cluster, thus modulating its

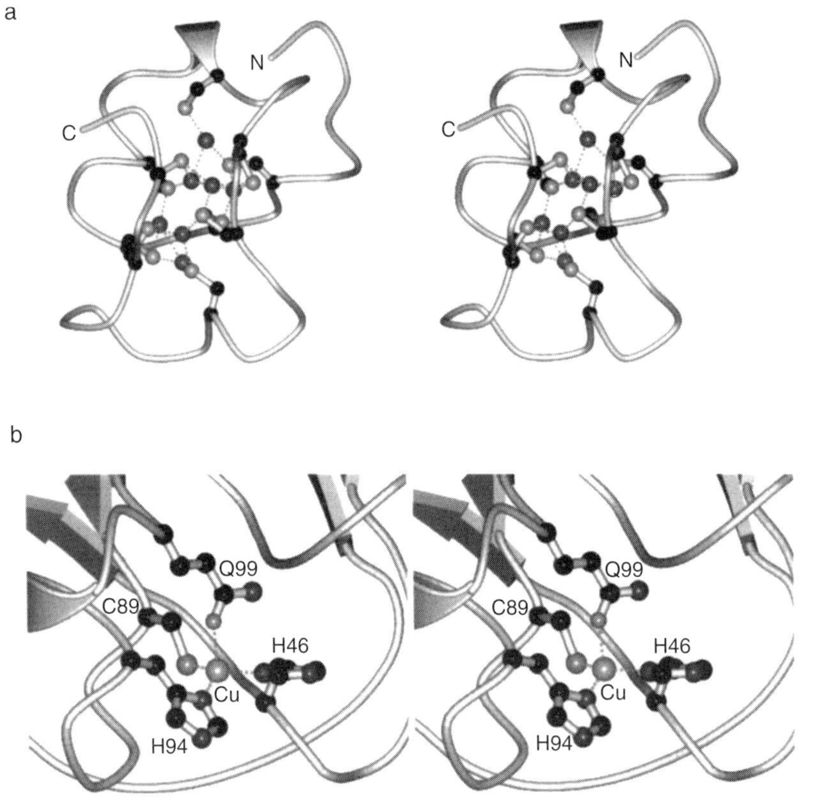


FIG. 2. Proteins that bind Cu(I). (a) *Saccharomyces cerevisiae* metallothionein (Cup1, pdb code 1aqr). Cup1 binds up to seven Cu(I) ions (medium gray spheres) using 10 cysteine sulfur atoms (light spheres) in a polythiolate cluster (Peterson *et al.*, 1996). All bonds shorter than 2.8 Å are shown as dotted lines. (b) *Cucumis sativus* stellacyanin (pdb code 1jer). Both Cu(I) and Cu(II) are bound by a pseudo-trigonal planar arrangement of $(His)_2Cys$ residues with an axial Gln ligand (Hart *et al.*, 1996). In other cupredoxins such as plastocyanin, a Met residue is the axial ligand (Adman, 1991).

DNA-binding activity (Furst *et al.*, 1988; Thiele, 1988; Szczypka and Thiele, 1989; Dameron *et al.*, 1991). As outlined below, Mac1 and Ace1 work cooperatively and reciprocally, sensing the amount of copper in the cytoplasmic environment and turning on and off the transcription of genes encoding the copper import and copper detoxification/sequestration proteins as appropriate.

In the case where copper is limiting in the surrounding medium, footprinting and DNA microarray studies reveal that Mac1 binds to *cis*-acting promoter elements (Cu response elements, or CuREs) upstream of the copper import machinery genes *ctr1*, *ctr3*, *fre1*, and *fre7*, activating their transcription (Fig. 1) (Labbe *et al.*, 1997; Gross *et al.*, 2000). Under these same conditions, Ace1 fails to bind to copper or to metal response elements (MREs) on the promoters of *cup1*, *crs5*, and *sod1* genes encoding components of the copper sequestration/detoxification and antioxidant machinery and their transcription is not activated (Gralla *et al.*, 1991; Zhou and Thiele, 1993). This is appropriate, because under copper-limiting conditions the cell needs import machinery to acquire it for vital enzymes and does not need the detoxification/sequestration machinery. Conversely, when copper is present in the surrounding medium above nutritional levels (in excess), Mac1 binds copper, undergoes a conformational change, and is released from the CuREs, thereby turning off the production of copper import proteins (Graden and Winge, 1997). Under these same conditions, Ace1 cooperatively binds Cu(I), undergoes a conformational change, and binds to the MREs, thus turning on the detoxification and antioxidant machinery genes *cup1*, *crs5*, and *sod1*, as well as elements of the iron transport machinery, *fet3* and *ftr1* (Zhou and Thiele, 1993; Labbe *et al.*, 1997; Gross *et al.*, 2000). The copper-sensing function of Mac1 is very sensitive, with half-maximal repression of the transcription of the genes encoding the copper import machinery occurring at concentrations of approximately ~20 nM (Joshi *et al.*, 1999). Deletion of the *mac1* gene results in copper ion starvation phenotypes similar to those associated with deletions in the *ctr1* and *ctr3* genes, and in fact, transcription of *ctr1* and *ctr3* genes is undetectable in *mac1*Δ strains (Jungmann *et al.*, 1993; Labbe *et al.*, 1997). The vital role of Ace1 in sensing toxic copper levels is emphasized by the observation that *ace1*Δ yeast, like *cup1*Δ strains, are extremely sensitive to elevated copper ion concentrations (Hu *et al.*, 1990).

At the posttranslational level, copper concentrations between 0.1 and 1.0 μM cause Ctr1 to be internalized in a fashion requiring the endocytosis machinery (Ooi *et al.*, 1996). It remains unclear whether this form of regulation plays a role in copper delivery to the cell's interior or whether it strictly provides a mechanism to reduce copper ion toxicity by reducing the number of active copper transporters on the cell surface. It has been suggested that because Ctr1 endocytosis occurs at copper concentrations much lower than the $K_m$ for copper transport, it may indeed play a role in copper delivery inside the cell (Lin *et al.*, 1997; Labbe and Thiele, 1999). Tangential support for this concept comes from the observation that the membrane-bound Menkes and Wilson's copper-transporting P-type ATPases move from the *trans*-Golgi network to the plasma membrane and endosomal compartments, respectively, in response to elevated copper levels, presumably so they can pump excess copper ion out of the cytoplasm (Petris *et al.*, 1996; Hung *et al.*, 1997). At higher copper concentrations (~10 μM), both Ctr1 and Mac1 are rapidly and specifically degraded. Ctr1 degradation occurs at the plasma membrane in a manner independent of the endocytosis machinery. The exact mechanism of Mac1 degradation remains unknown. It has been postulated that copper-dependent proteases, or proteases that recognize conformational changes induced by copper binding to low-affinity sites on the Ctr1 and Mac1 molecules might be responsible for the elimination of these proteins under conditions of copper excess (Ooi *et al.*, 1996; Zhu *et al.*, 1998). Thus, the copper-dependent degradation of Mac1 and Ctr1 is likely to serve as an important and effective cellular defense mechanism to minimize copper ion toxicity under conditions of copper ion excess (Zhu *et al.*, 1998).

### C. *The Need for Copper Chaperones*

The intricate and sensitive interplay between Mac1 and Ace1 ensures that the production of copper import and copper detoxification/sequestration machineries is tightly controlled and balanced. This in turn maintains a relatively steady-state total copper concentration in the cytoplasm over a range of copper ion concentrations in the surrounding medium. Until recently, it was assumed that proteins that use copper ion as a cofactor, such as Cu,ZnSOD (SOD1), acquire it from the intracellular copper pool through passive diffusion. At first glance, this assumption seems sound, as Cu,ZnSOD binds copper ions *in vitro* with a dissociation constant on the order of $10^{-15}$ M, and, using elemental analysis through inductively coupled plasma-atomic emission spectroscopy, O'Halloran and colleagues estimate the *total* copper concentration of an unstressed yeast cell to be 70 μ*M* (see Valentine and Pantoliano, 1981; Bertini *et al.*, 1998; Lippard, 1999; Rae *et al.*, 1999). Surprisingly, Culotta, Gitlin, and colleagues found that despite this seemingly plentiful supply of total copper ion in the cytoplasm under normal conditions, a protein factor encoded by the yeast *lys7* gene (later to be called the yeast copper chaperone for SOD1, or yCCS) is *absolutely* required for copper insertion into apo-SOD1 *in vivo*. *lys7*Δ yeast produce normal levels of the SOD1 polypeptide, but fail to incorporate copper into the protein and are thus devoid of superoxide scavenging activity and demonstrate phenotypes nearly identical to those of *sod1*Δ strains (see Culotta *et al.*, 1997). The requirement for yCCS is eliminated, however, when ambient copper levels are elevated (Culotta *et al.*, 1997; Rae *et al.*, 1999). O'Halloran and colleagues used this information, coupled with a series of *in vivo* and *in vitro* experiments, to estimate that in an unstressed yeast cell the total cytoplasmic concentration of water-bound copper ions [Cu(I) or Cu(II)] is less than $10^{-18}$ M, a concentration that corresponds to less than one free copper ion per cell. They use these calculations to explain the inability of apo-SOD1 to acquire copper *in* *vivo* by passive diffusion under normal growth conditions (see Rae *et al.*, 1999). It is interesting to note, however, that Mac1 senses cytoplasmic copper ion levels and appropriately regulates the transcription of the high-affinity copper import machinery. Mac1 must, therefore, be capable of acquiring labile copper ion from a source other than the limited aqueous cytoplasmic copper ion pool. Nonetheless, the observations described above led to the realization that the cytoplasmic milieu has an overcapacity for copper chelation and sequestration through nonspecific small-molecule interactions such as glutathione, through vesicular sites for copper concentration and storage, and through the induction of the metallothioneins discussed above (O'Halloran and Culotta, 2000). The role of the copper chaperone is thus to acquire copper in the face of this overcapacity for copper chelation and to guide it to the enzymes that need it for their function and that are otherwise unable to acquire it. In this context, copper chaperones might be considered as enzymes, lowering the kinetic barrier for copper transfer between unknown copper donor sites (possibly the membrane-bound transporters) and their target proteins (Huffman and O'Halloran, 2000). The following three sections describe what is known about copper chaperones that function in copper delivery to the secretory compartments (Atx1, Hah1), to cytosolic Cu,ZnSOD (CCS), and to the mitochondria (Cox17), with emphasis on their structural biology where appropriate.

## II. Copper Chaperones of the Atx1-like Family

### A. *Genetics and Chemistry*

#### 1. *Identification of Yeast Atx1 as a Copper Chaperone*

The *Sa. cerevisiae atx1* (antioxidant) gene encodes a 73-amino-acid polypeptide containing the sequence motif MXCXXC previously observed in the yeast secretory pathway P-type ATPase Ccc2 (Fu *et al.*, 1995), in its human homologues, the Menkes (ATP7A) and Wilson's (ATP7B) proteins (Bull *et al.*, 1993; Vulpe *et al.*, 1993), and in the bacterial mercury detoxification protein MerP (Sahlman and Skarfstad, 1993; Lin and Culotta, 1995; Morby *et al.*, 1995). As shown in Section II, B, the two cysteine residues of the MXCXXC motif in MerP bind Hg(II) ion in a nearly linear fashion, which in turn suggests that Atx1 and other proteins containing this motif may bind their metal ions in a similar mode (Lin *et al.*, 1997; Steele and Opella, 1997).

Culotta and colleagues first discovered the *atx1* gene in genetic screens and identified it as a multicopy suppressor of an oxygen-sensitive phenotype of *sod1*Δ *sod2*Δ mutant *Sa. cerevisiae* (see Lin and Culotta, 1995). The ability of Atx1 to rescue SOD1-deficient yeast is observed when the *atx1* copy number is high, but not when Atx1 is present at normal physiological levels (Portnoy *et al.*, 1999). Overexpression of *atx1* is incapable of compensating for SOD1, however, when cells are deprived of copper ion uptake either through the actions of bathocuproinedisulfonic acid (BCS), a copper-specific chelator known to deplete medium copper, or through deletion of the high-affinity copper transporter gene *ctr1* (Dancis *et al.*, 1994a,b; Lin and Culotta, 1995). Overexpression of an Atx1 protein with C15S, C18S mutations in the MXCXXC motif fails to suppress the *sod1Δ* phenotype, highlighting the importance of these residues and the likely requirement for copper ion binding by Atx1 in its observed antioxidant activity (Lin, 1997). Whether or not the antioxidant activity of Atx1 is physiologically relevant in *Sa. cerevisiae* when compared to SOD1 remains to be determined (Lin and Culotta, 1995; Portnoy *et al.*, 1999; Rosenzweig and O'Halloran, 2000).

A physiological role for Atx1 in copper trafficking was first suggested when it was localized to the cytosol through cell fractionation experiments, and, when it was observed that *atx1Δ* yeast, and those overexpressing the Atx1 C15S, C18S double mutation in the MXCXXC sequence motif are defective in the high-affinity uptake of iron (Lin, 1997; Lin *et al.*, 1997). The metabolic defects in iron metabolism due to these mutations in the *atx1* gene, that is, the failure to incorporate copper properly into the multicopper oxidase Fet3, are reversed by copper treatment in a way analogous to that observed for yeast cells lacking either the high-affinity copper transporter Ctr1 or the copper-transporting P-type ATPase Ccc2 in the secretory pathway (Askwith *et al.*, 1994; Dancis *et al.*, 1994a,b; Yuan *et al.*, 1995). Based on these observations, it was proposed that Atx1 might function as a freely diffusible copper shuttle involved in the trafficking of copper ion from Ctr1 to Ccc2 as shown schematically in Fig. 1 (Lin *et al.*, 1997). In support of this idea, overexpression of *atx1* is incapable of suppressing the iron deficiency of a *ccc2Δ* mutant. Conversely, overexpression of *ccc2* suppresses the iron dependence of an *atx1Δ* mutant, suggesting that Ccc2 functions downstream of Atx1 (Lin *et al.*, 1997). Further support for the role of Atx1 in iron homeostasis and in delivery of copper to the secretory pathway and ultimately to Fet3 comes from the observation that the *atx1* gene, like the *ccc2* and *fet3* genes, is regulated by the iron-sensing transactivation factor Aft1 and not the copper-sensing transactivator Mac1 (Lin *et al.*, 1997).

The copper chaperone function of Atx1 was definitively demonstrated by O'Halloran and colleagues in 1997 when they showed using electron paramagnetic resonance (EPR), X-ray absorption near edge structure (XANES), and extended X-ray absorption fine structure (EXAFS), methods that Atx1 specifically binds Cu(I) in a mixture of two- and three-sulfur ligand geometries even in the presence of a 20-fold excess of competing thiols (see Pufahl *et al.*, 1997). The observed two-sulfur ligand coordination of Cu(I) occurring at a bond distance of approximately 2.25 Å likely arises from intramolecular Cu(I) binding by Cys-15 and Cys-18 of the MXCXXC motif, while the third sulfur ligand observed at a bond distance of approximately 2.40 Å could come either from an exogenous thiol or from a cysteine residue of the MXCXXC motif in an interaction with another Atx1 molecule (Pufahl *et al.*, 1997). The observation of this type of two- and three-sulfur Cu(I) coordination in these studies was unprecedented because copper cysteinate proteins that stabilize Cu(I) were previously known to form either polynuclear metal thiolate clusters as observed in Ace1 and metallothionein (Fig. 2a) or a constrained $(His)_2Cys$ coordination geometry as observed in mononuclear blue copper proteins, representatives of which include plastocyanin and stellacyanin (Fig. 2b) (Adman, 1991; Hart *et al.*, 1996). Significantly, yeast two-hybrid analyses demonstrated that Atx1 *directly* associates with the MXCXXC-containing cytoplasmic domains of the vesicular Ccc2 P-type ATPase protein *in vivo*. Addition of the copper chelator BCS abrogates Atx1/Ccc2 interaction, indicating that their association is dependent on the presence of copper ion. These results are consistent with a copper transfer mechanism where Atx1 contacts the homologous domains in Ccc2 via protein–protein interactions and delivers the copper ion through a metal-bridged intermediate (Pufahl *et al.*, 1997) (see below).

In subsequent studies, Huffman and O'Halloran (2000) demonstrated that Cu(I)Atx1 *directly* donates copper to the first N-terminal Atx1-like domain of Ccc2 (Ccc2a) in a reversible manner *in vitro*, with a $K_{\text{exchange}} = 1.4 \pm 0.2$ ( $\sim 0.2$ kcal/mol), suggesting that vectorial delivery of copper ion by Atx1 to Ccc2a is not based solely on a significantly higher copper affinity of the target Ccc2a domain. The attainment of this equilibrium is rapid, with complete partitioning of copper between Atx1 and Ccc2a occurring in less than 1 min. In addition, the attainment of this equilibrium is unaffected by a 50-fold excess of glutathione, indicating that Atx1 is capable of protecting Cu(I) from other potentially abundant Cu(I) chelators found in the intracellular milieu (Huffman and O'Halloran, 2000). This result also demonstrates that Atx1 can release its Cu(I) cargo from its MXCXXC motif to a similar target motif, but the low value of $K_{\text{exchange}}$ measured *in vitro* might at first glance predict that the copper flux provided to the secretory pathway copper transporters is low. This led to the suggestion that copper transfer from the Atx1 chaperone to its Ccc2 target is not driven by thermodynamics, but rather that Atx1 acts as an enzyme, lowering the kinetic barrier for copper transfer along specific reaction coordinates (Huffman and O'Halloran, 2000). In this context, Atx1 would catalyze the equilibration of copper between unknown copper donor sites, possibly on the Ctr1/Ctr3 copper import machinery, and specific target sites on Ccc2. ATP hydrolysis by the Ccc2 P-type ATPase subsequently drives the transport of copper ion across the post-Golgi lipid bilayer into a separate thermodynamic compartment, which presumably contains a higher concentration of hydrated copper ions than does the cytoplasm and where the multicopper oxidases and other apo-proteins in the secretory compartment can obtain their copper ion cofactor, possibly through passive diffusion (Huffman and O'Halloran, 2000; O'Halloran and Culotta, 2000).

As mentioned above, two-hybrid studies demonstrated that Atx1 and Ccc2a interact *in vivo* in a copper-dependent fashion (Pufahl *et al.*, 1997). Although the presence of copper ion is necessary for the Atx1/Ccc2a interaction to occur, other residues on Atx1 that direct the specificity of the copper transfer interaction are also required. For example, the lysine patches represented by K24,28 and K61,62 of Atx1 were found to be critical for the delivery of copper ions to Ccc2. Converting these basic residues to their acidic counterparts, aspartic and glutamic acid, resulted in Atx1 molecules severely crippled in their capacity to deliver copper ion to Ccc2 and to Fet3 (Portnoy *et al.*, 1999).

Taken together, the above data suggest a stepwise mechanism of Cu(I) transfer from Atx1 to Ccc2a as illustrated in Fig. 3 (Brown *et al.*, 1991; Pufahl *et al.*, 1997; Huffman and O'Halloran, 2000). When the Cu(I) center can adopt a two-coordinate ligand geometry as in Atx1, the Cu(I) ion is subject to attack by the sulfur ligands of the MXCXXC motif

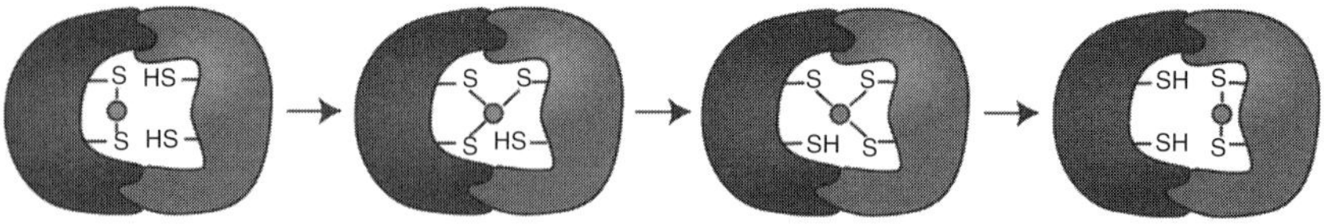


FIG. 3. Schematic representation of a proposed "bucket brigade" stepwise mechanism of Cu(I) transfer from Atx1 (dark) to Ccc2a, the N-terminal domain of the P-type ATPase Ccc2 (light) (Pufahl *et al.*, 1997). Atx1 and Ccc2a associate through electrostatic attraction and specific protein–protein interactions (see text). The Cu(I) ion, initially coordinated by the two cysteine residues of the Atx1 MXCXXC motif (left), is attacked by a cysteine residue of the Ccc2a MXCXXC motif, forming a bond of similar strength to those of the original two ligands (second from left). Rapid ligand exchange follows through a series of two- and three-coordinate intermediates (second from right). The copper ion is eventually transferred to the two cysteine residues of the target Ccc2a MXCXXC motif (right). Atx1 then presumably dissociates from Ccc2a to perform another cycle of copper ion delivery.

in the recipient protein, making a bond of similar strength to those of the original two metal ligands. This type of association can be further stabilized by electrostatic protein–protein interactions between positively charged residues on the Atx1 molecule and negatively charged residues on the Ccc2 copper receptor domain(s). When all three metal–ligand bonds are of similar energy, rapid ligand exchange can be observed. The intermolecular three-coordinate intermediate can subsequently lose one of the original ligands to give a new two-coordinate center in the target domain. The structural basis explaining the above observations and a structure-based mechanism for copper transfer from the Atx1-like chaperones to their targets are described below.

*2. Identification of Atx1 Homologues*

As indicated in Fig. 4 and Table I, the search for homologues to Atx1 has been fruitful, with discoveries in organisms ranging from bacteria to higher eukaryotes. Quickly following the discovery of *Sac. cerevisiae* Atx1, Gitlin and colleagues identified its 47% identical, 68-amino-acid human homologue, Hah1 (human Atx homologue 1) or Atox1, by screening a genomic human liver cDNA library with degenerate oligonucleotides corresponding to conserved regions of yeast Atx1 (see Klomp *et al.*, 1997). The *hah1* gene exists in humans as a single copy in the haploid genome, and RNA blot analyses show that *hah1* mRNA is present in all tissues tested (approximately 20), including high levels in the central nervous system (Klomp *et al.*, 1997). From this point forward, genetic and biochemical experiments on *hah1* and *atx1* and their gene products overlap almost as much as their homologies. *In vitro* studies using $^{64}$Cu demonstrate that Hah1 directly binds Cu(I) via Cys-12 and Cys-15, the two cysteines of its MXCXXC motif, and that copper binding is abrogated in a C12G, C15G double mutant (Hung *et al.*, 1998). *In vivo*, overexpression of wild-type *hah1* suppresses the oxygen-sensitive phenotype of *sod1*Δ yeast and restores copper delivery to Ccc2 and Fet3 in *atx1*Δ yeast, permitting growth on iron-depleted medium (Klomp *et al.*, 1997). Overexpression of *hah1* genes encoding mutations at Cys-12 and Cys-15 in *atx1*Δ yeast resulted in strains deficient in copper loading of Fet3 (Hung *et al.*, 1998). Together, these results strongly suggest that Hah1 is the functional homologue of Atx1 and, further, that the Menkes (ATP7A) and Wilson's (ATP7B) proteins are functional human homologues of *Sa. cerevisiae* Ccc2 (Klomp *et al.*, 1997). In this scenario, copper is transported into the cell by hCtr1 and shuttled by Hah1 to ceruloplasmin, the human homologue of Fet3, through the actions of the Menkes and Wilson's disease proteins (Bull *et al.*, 1993; Vulpe *et al.*, 1993). In support of this, expression of the wild-type Menkes P-type ATPase complemented the defects in copper loading to Fet3 in *ccc2*Δ yeast strains (Payne *et al.*,

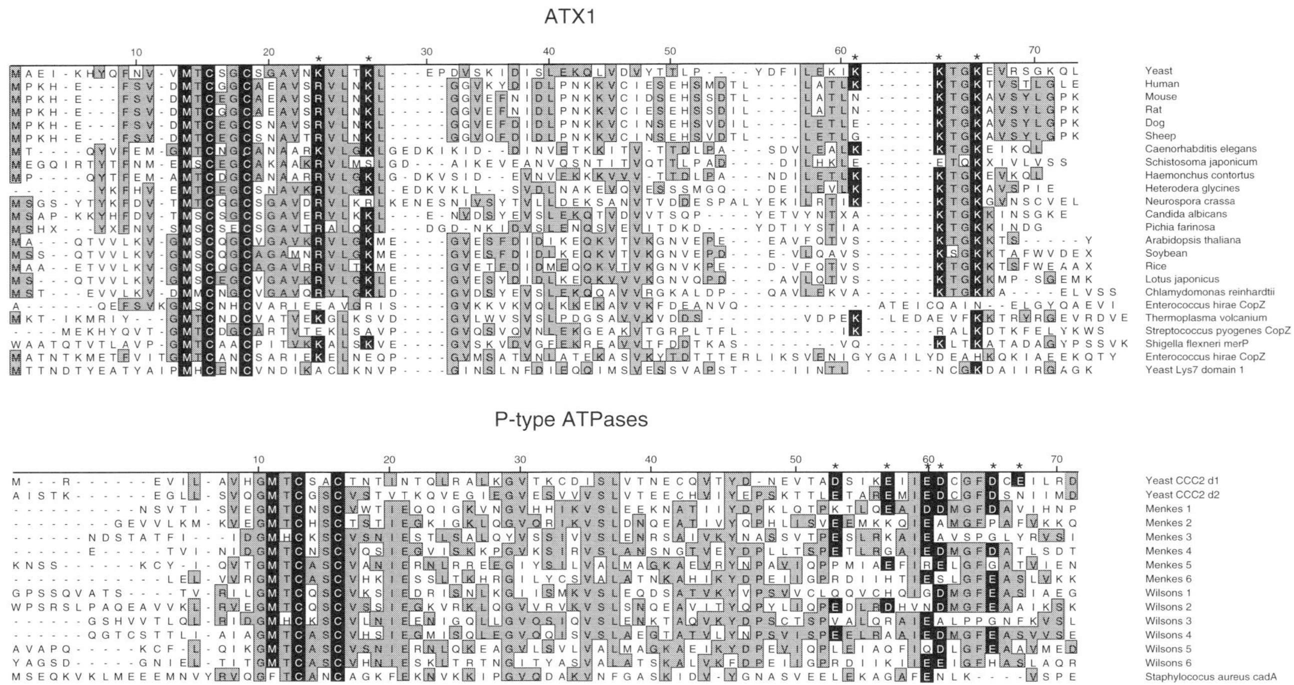


FIG. 4. Multiple sequence alignment of Atx1 and N-terminal domains of the P-type ATPases using the CLUSTAL method (Higgins and Sharp, 1989). Sequence numbering corresponds to that of the yeast proteins. (Top) The copper chaperones are shown and (bottom) the N-terminal domains of the target P-type ATPases are shown. The sulfur-containing components of the MXCXXC motifs are boxed in black between residues 10 and 20. Residues thought to be important in the electrostatic recognition between chaperones and target domains are boxed in black and labeled with an asterisk (see text).

TABLE I
*Sequence References for Atx1-like Copper Chaperones and Their Target Domains*

| Atx1 | Reference |
|---|---|
| Yeast (*Saccharomyces cerevisiae*) | Lin and Culotta, 1995 |
| Human (*Homo sapiens*) | Klomp *et al.*, 1997 |
| Mouse (*Mus musculus*) | Hamza *et al.*, 2000 |
| Rat (*Rattus norvegicus*) | Kelner *et al.*, 2000 |
| Dog (*Canis familiaris*) | Nanji and Cox, 1999 |
| Sheep (*Ovis ovaries*) | Lockhart and Mercer, 2000 |
| *Caenorhabditis elegans* | Wakabayashi *et al.*, 1998 |
| *Schistosoma japonicum* | Fan and Brindley, 1998 |
| *Haemonchus contortus* | Blaxter *et al.*, 2000a |
| *Heterodera glycines* | McCarter *et al.*, 1999 |
| *Neurospora crassa* | Nelson *et al.*, 1997 |
| *Candida albicans* | Souciet *et al.*, 2000 |
| *Pichia farinosa* | de Montigny *et al.*, 2000 |
| *Arabidopsis thaliana* | Himelblau *et al.*, 1998 |
| Soybean (*Glycine max*) | Amasino *et al.*, 1999 |
| Rice (*Oryza sativa*) | Mira and Penarrubia, 1999 |
| *Lotus japonicus* | Poulsen and Poedenphant, 2000 |
| *Chlamydomonas reinhardtii* | La Fontaine and Merchant, 2000b |
| *Thermoplasma volcanium* | Kawashima *et al.*, 1999 |
| *Streptococcus pyogenes* CopZ | Ferretti *et al.*, 2001 |
| *Shigella flexneri* MerP | Misra *et al.*, 1984 |
| *Enterococcus hirae* CopA and CopZ | Odermatt and Solioz, 1995 |
| **P-type ATPases** | **Reference** |
| Yeast (*Sa. cerevisiae*) Ccc2 | Fu *et al.*, 1995 |
| Human (*H. sapiens*) Menkes protein | Vulpe *et al.*, 1993 |
| Human (*H. sapiens*) Wilson protein | Bull *et al.*, 1993 |
| *Staphylococcus aureus* CadA | Wang and Novick, 1987 |

1998). In a fashion analogous to that observed for Atx1 and Ccc2, direct copper-dependent interactions between Hah1 and some of the N-terminal domains of the Menkes and Wilson's disease proteins were demonstrated both *in vitro* and *in vivo* using affinity chromatography, two-hybrid, and coimmunoprecipitation experiments (Hamza *et al.*, 1999; Larin *et al.*, 1999). Interestingly, when the coimmunoprecipitation experiments were repeated on Wilson's proteins harboring one of three disease-associated mutations (G85V, L492S, and G591D) in one of its N-terminal Atx1-like domains, a marked attenuation of the interaction with Hah1 was observed, suggesting that impaired copper delivery by Hah1 might account for the molecular basis of Wilson's disease in patients with these particular mutations (Hamza *et al.*, 1999). The Wilson P-type ATPase, Hah1, and copper are all abundant in brain tissue, and thus, a role for Hah1 in inherited neurodegenerative diseases is predicted (Rosenzweig and O'Halloran, 2000).

### *B. Structural Biology*

#### *1. Overall Protein Fold and Location of the Metal-Binding MXCXXC Motif*

Over the past several years, three-dimensional (3-D) studies of the chaperones of the Atx1-like family and their target domains on the P-type ATPases of the secretory pathway have dramatically increased our understanding about how these proteins function in copper binding and transfer. Table II provides a list of proteins of the Atx1-like family whose 3-D structures are known. Figure 5 illustrates the similarity in the overall folding topology and the general location of the conserved MXCXXC motifs in these proteins (Steele and Opella, 1997; Gitschier *et al.*, 1998; Rosenzweig *et al.*, 1999; Wimmer *et al.*, 1999; Wernimont *et al.*, 2000; Banci *et al.*, 2001). The conserved stable fold is described as "ferrodoxin-like" with a βαββαβ topology, where the two α-helices are superimposed on a four-stranded anti-parallel β-sheet (Hubbard *et al.*, 1997; Steele and Opella, 1997). The metal-binding site (boxed in Fig. 5) is on the periphery of the molecule, located at the first turn of the protein at the junction of β-strand 1 and α-helix 1. The observed accessibility of the metal-binding site is consistent with a metal transfer function to a target protein. As predicted by sequence analysis, the target N-terminal domains of the P-type ATPases, which are tethered to the post-Golgi membrane, also exhibit this folding motif and solvent-accessible metal-binding site. Now that the structures of the copper chaperone proteins (Atx1 and Hah1) and domains of their target proteins (Ccc2a and Mnk4) are in hand, they can be reconciled with the body of existing biochemical data to provide insight into the molecular determinants of the copper transfer process. The following describes the structures of Atx1, Ccc2a, Hah1, and Mnk4 proteins followed by a general model of copper ion transfer from chaperone to target.

#### *2. Atx1/Ccc2a*

The first structure of a copper chaperone, that of Atx1, was determined by Rosenzweig and colleagues in 1999 using single-crystal X-ray diffraction methods. Oxidized apo- and Hg(II)-bound forms were elucidated

TABLE II
*Three-Dimensional Structures of Members of the ATX1-like Family*

| Protein | Function | State | PDB | Method | Reference |
|---|---|---|---|---|---|
| MerP | Bacterial mercury detoxification protein | Reduced apo, | 1afi | NMR | Steele and Opella, 1997 |
| | | oxidized apo, | 2hqi | NMR | Qian *et al.*, 1998 |
| | | Hg(I) | 1afj | NMR | Steele and Opella, 1997 |
| CopZ | Bacterial copper chaperone to CopY repressor | Reduced apo, | 1cpz | NMR | Wimmer *et al.*, 1999 |
| | | Cu(I) | N/A | NMR | Wimmer *et al.*, 1999 |
| Mnk4 | Human copper receptor, secretory pathway | Reduced apo, | 1aw0 | NMR | Gitschier *et al.*, 1998 |
| | | Ag(I) | 2aw0 | NMR | Gitschier *et al.*, 1998 |
| Ccc2a | Yeast copper receptor, secretory pathway | Reduced apo, | 1fvq | NMR | Banci *et al.*, 2001 |
| | | Cu(I) | 1fvs | NMR | Banci *et al.*, 2001 |
| Atx1 | Yeast copper chaperone to P-type ATPase Ccc2 | Reduced apo, | 1fes | NMR | Arnesano *et al.*, 2001 |
| | | Cu(I) | 1fd8 | NMR | Arnesano *et al.*, 2001 |
| Atx1 | Yeast copper chaperone to | Oxidized apo (1.20 Å) | 1cc7 | X-ray | Rosenzweig *et al.*, 1999 |
| | P-type ATPase Ccc2 | Hg(II) (1.02 Å) | 1cc8 | X-ray | Rosenzweig *et al.*, 1999 |
| Hah1 | Human copper chaperone to Menkes, Wilson's P-type ATPases | Cu(I) (1.80 Å), | 1fee | X-ray | Wernimont *et al.*, 2000 |
| | | Hg(II) (1.75 Å), | 1fe4 | X-ray | Wernimont *et al.*, 2000 |
| | | Cd(I) (1.75 Å) | 1fe0 | X-ray | Wernimont *et al.*, 2000 |

at 1.20 and 1.02 Å resolution, respectively (see below) (Rosenzweig *et al.*, 1999). In these same studies, crystallization of Cu(I)Atx1 was attempted, but the loop containing the metal-binding MXCXXC motif was consistently disordered and copper ions were not visible, even when the crystals were grown anaerobically to prevent the oxidation of Cu(I) to Cu(II). It is possible that in Cu(I)Atx1, the conformation of the metal-binding loop is incompatible with the crystal lattice, as there exists a crystal contact with a symmetry-related molecule in the metal-binding loop region (Rosenzweig *et al.*, 1999). Although Hg(II) has been demonstrated to serve as a structurally useful model for low-coordination-number Cu(I) sites in a variety of proteins (Utschig *et al.*, 1995, 1997), it was initially unknown whether it could substitute for Cu(I) in metal transfer from Atx1 to Ccc2a. To test this, an *in vitro* metal transfer assay was developed to determine whether Hg(II)Atx1 is a functionally competent model for Cu(I)Atx1. When incubated with apo-Ccc2a, either Cu(I)Atx1 or

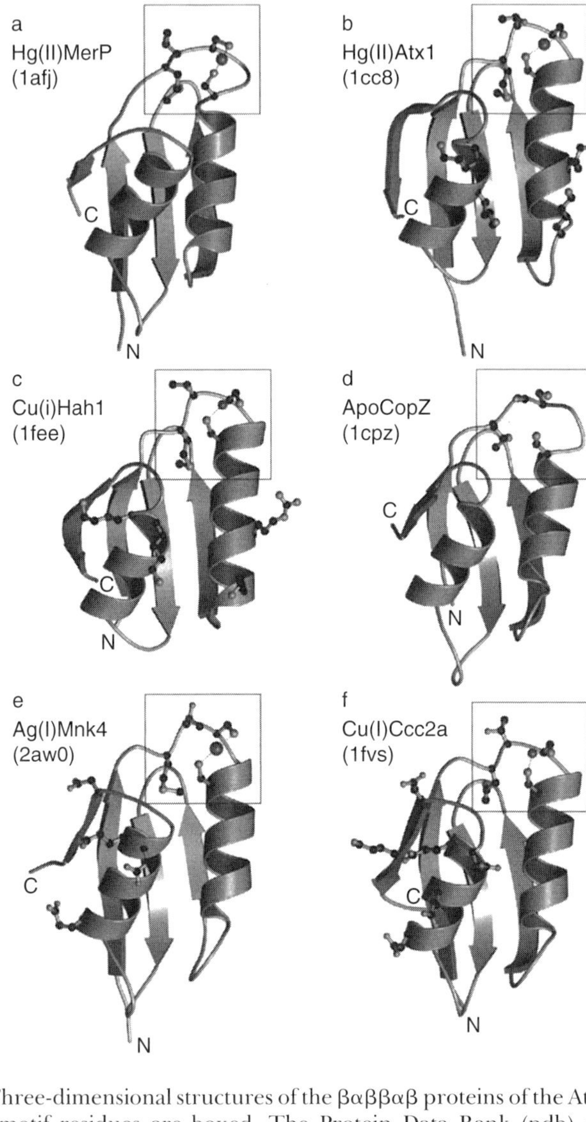


FIG. 5. Three-dimensional structures of the βαββαβ proteins of the Atx1-like family. MXCXXC motif residues are boxed. The Protein Data Bank (pdb) code for each structure is in parentheses. (a) NMR structure of *Shigella flexneri* Hg(II)MerP (Steele and Opella, 1997). (b) X-ray structure of *Saccharomyces cerevisiae* Hg(II)Atx1. K24, K28, K59, and K62, side chains important in the recognition of the Ccc2a target domain, are shown outside of the box (see text) (Rosenzweig *et al.*, 1999). (c) X-ray structure of human Cu(I)Hah1. R21, K25, K56, and K57, side chains important in the recognition of the fourth N-terminal domain of the Menkes protein, are shown outside of the box (Wernimont *et al.*, 2000). (d) NMR structure of *Enterococcus hirae* apoCopZ (Wimmer *et al.*, 1999). (e) NMR structure of human Ag(I)Mnk4, the fourth domain of the

Hg(II)Atx1 was able to donate a significant amount of metal to Ccc2a, suggesting both that Atx1 can transfer metal ions to its biological partner *in vitro* and that Hg(II) behaves in a similar fashion as Cu(I) in these experiments (Rosenzweig *et al.*, 1999; Huffman and O'Halloran, 2000).

The very-high-resolution structure determination of Hg(II)Atx1 is notable from a crystallographic standpoint because it is one of the largest structures to be determined using direct methods, a technique traditionally used to solve the so-called "phase problem" in small-molecule crystallography (Miller *et al.*, 1994; Rosenzweig *et al.*, 1999). As seen in Figs. 5a and 5b, the Atx1 molecule is small and compact in structure, with overall dimensions of approximately 24 × 27 × 36 Å, and possesses a folding topology and mode of Hg(II) binding similar to those of the previously determined NMR structure of the periplasmic mercury-binding protein MerP (Steele and Opella, 1997). The methionine side chain of the MXCXXC motif in Atx1 (and all subsequently determined Atx1-like proteins) participates in the formation of the hydrophobic core of the molecule (Fig. 5) and does not appear to play a role in metal binding (Rosenzweig *et al.*, 1999).

Figure 6a shows the MTCSGC motif in the Hg(II)-bound form of Atx1. The Hg(II) ion is bicoordinate, with Cys-15 and Cys-18 sulfur atoms acting as ligands at distances of 2.33 and 2.34 Å, respectively. The coordination is nearly linear, with a S–Hg(II)–S bond angle of ~167°. This observed mode of Hg(II) binding is consistent with previous interpretations of EXAFS and $^{199}$Hg NMR spectroscopic data for Hg(II)Atx1 and is similar to the coordination of Hg(II) in MerP and Ag(I) in the Menkes protein domain 4 (Fig. 5e) (Pufahl *et al.*, 1997; Steele and Opella, 1997; Gitschier *et al.*, 1998). Figure 6a also shows that the side chain oxygen atom of Thr-14 is fairly close to the mercury atom at a distance of approximately 3 Å, approaching the distance expected for a secondary bonding interaction (Wright *et al.*, 1990). The observed Hg–S bond distances agree well, however, with known two-coordinate complexes in model compounds, suggesting that the Thr-14 side chain oxygen atom does not contribute significantly to the binding of Hg(II) (Watton *et al.*, 1990; Wright *et al.*, 1990; Utschig *et al.*, 1995; Rosenzweig *et al.*, 1999). A role for Thr-14 in the recognition of its target protein Ccc2a is described below.

---

Menkes protein. E55, D62, D63, and D67, side chains important in the recognition of Hah1, are shown outside of the box (Gitschier *et al.*, 1998). (f) NMR structure of *Sa. cerevisiae* Cu(I)Ccc2a, the N-terminal domain of the P-type ATPase Ccc2. D53, E57, E60, D61, D65, and E67, residues important in the recognition of Atx1, are shown outside the box (see text) (Banci *et al.*, 2001). This figure and all subsequent protein structure renderings were created by the programs POV-Ray (POV-Team, 1997), Molscript (Kraulis, 1991), and/or Bobscript (Esnouf, 1999).

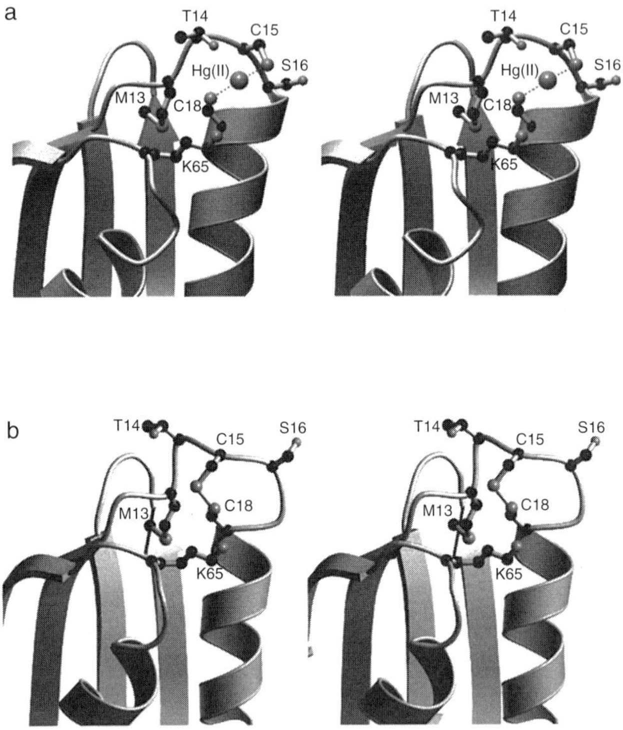


FIG. 6. The MTCSGC motif in Hg(II)-bound and oxidized apo forms of Atx1 [pdb codes 1cc8 and 1cc7, respectively (Rosenzweig *et al.*, 1999)]. (a) Hg(II)Atx1. Met13 participates in the formation of the hydrophobic core of the molecule. The side chain oxygen of Thr-14 is ~3 Å from the Hg(II) ion. Cys-15 and Cys-18 coordinate Hg(II) at a distance of ~2.3 Å. Lys-65, a residue thought to be important in the capture and release of copper, is also shown (see text). (b) Oxidized apoAtx1. Cys-15 and Cys-18 form a disulfide bond. The MTCSGC loop is somewhat conformationally flexible, as evidenced in the altered positions of Thr-14, Cys-15, and Ser-16.

Figure 6b shows the metal-binding loop in oxidized apoAtx1. Because the cell is a reducing environment, reduced apoAtx1 is postulated to be involved in metal binding and the oxidized apo structure may not be physiologically relevant. Importantly, however, the oxidized apo-Atx1 structure does demonstrate that the metal-binding loop is conformationally flexible. Relative to the Hg(II)Atx1 structure, the Cα of Cys-15 moves approximately 4 Å to allow the formation of a disulfide bond with Cys-18. This shift is concomitant with a rearrangement of Thr-14, Ser-16, and Gly-17. Ser-16 exhibits the largest shift in position, with a 5.6 Å difference in Cα position relative to the Hg(II)Atx1 structure. Except for this difference in the conformation of the metal-binding loop, the rest of the Atx1 protein is essentially unchanged in the two crystal forms (Rosenzweig *et al.*, 1999). Because yeast two-hybrid studies demonstrate that Atx1 and Ccc2a interact in a Cu(I)-dependent fashion, this conformational flexibility of the metal-binding loop may play a role in target recognition (see below).

Subsequent to the crystallographic work on Atx1, the solution structures of the reduced apo- and Cu(I)-bound forms of Atx1 were elucidated by NMR spectroscopy under an $N_2$ atmosphere (Arnesano *et al.*, 2001). A superposition of these mean energy-minimized structures is shown in Fig. 7a. Consistent with what is observed in the crystal structures of oxidized apo- and the Hg(II)-bound forms of Atx1, the significant changes observed in the two NMR structures manifest themselves mainly in the region of the metal-binding loop. The largest movement is demonstrated by Cys-18 in the reduced apoAtx1 structure, the Cα of which moves ~5.9 Å from its position in the copper-bound protein. In the Cu(I)Atx1 family of conformers (of which there are 35), the copper ion is coordinated by Cys-15 and Cys-18 with a S–Cu(I)–S angle of $120° \pm 40°$. This deviation from linearity suggests that the Cu(I) coordination number may not be fixed and that Cu(I) might prefer a coordination number greater than 2 (Arnesano *et al.*, 2001). As mentioned previously, EXAFS studies on Cu(I)Atx1 indicated that a third sulfur ligand from either a buffer thiol molecule or a protein side chain can also interact with the Cu(I) center (Pufahl *et al.*, 1997). Other than the two cysteine ligands, the closest approach of a protein atom to the Cu(I) center in the NMR structure comes from the side chain of Lys-65 at a distance of about 5 Å. Thus, at the time, the identity of the third ligand in the EXAFS data remained unresolved (Pufahl *et al.*, 1997; Arnesano *et al.*, 2001).

Figure 7b shows a superposition of the Hg(II)Atx1 (X-ray) and the mean energy-minimized Cu(I)Atx1 (NMR) structures (Rosenzweig *et al.*, 1999; Arnesano *et al.*, 2001). Cys-15 demonstrates the largest change in position in the two structures, with its Cα atom in Hg(II)Atx1 ~3 Å away from its position in Cu(I)Atx1. In these two structures, the side chain of Lys-65 has a different position with respect to the metal ion and its Cys ligands. In the Hg(I)Atx1 crystal structure depicted in Fig. 6a, the Nξ atom of Lys-65 is closer to the sulfur of Cys-18 than to the sulfur of Cys-15, while in the mean energy-minimized Cu(I)Atx1 solution structure shown in Fig. 7a, the Nξ atom of Lys-65 is very close to the sulfur of Cys-15 and is farther from Cys-18. This leads to the suggestion that the positive charge on Lys-65 can stabilize the overall negative charge resulting from binding Cu(I) to two cysteinate anions, while in the Hg(II)Atx1 protein, the charge is neutralized by the Hg(II) ion itself. Finally, Fig. 7b illustrates that this variation in conformation of the metal-binding loop results in a significantly different position of the metal ion in

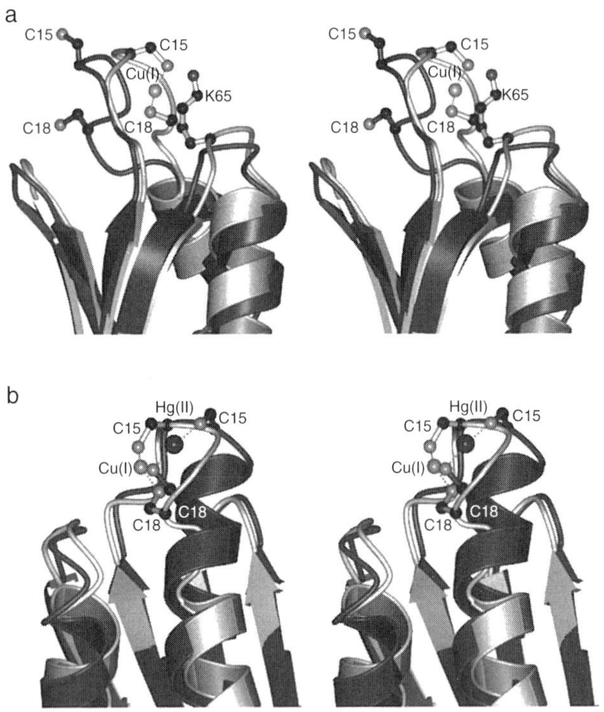


FIG. 7. The conformational flexibility of the MTCSGC-containing loop of Atx1. (a) Superposition of the mean energy-minimized NMR structures of reduced apo (dark gray) and Cu(I)-bound (light gray) forms of Atx1 (Amasino *et al.*, 1999). The reduced apoAtx1 structure presumably represents the conformation of the metal-binding loop before or after copper binding and transfer. The Cu(I)Atx1 structure shows that upon copper binding, it is shielded from the solvent, consistent with the chaperone function of the protein. The side chain nitrogen of Lys-65 is ~5 Å from the Cu(I) ion and may stabilize the net negative charge induced on Cu(I) by the two negatively charged thiolates of Cys-15 and Cys-18. (b) Superposition of the X-ray structure of Hg(II)Atx1 (dark gray) and the NMR structure of Cu(I)Atx1 (light gray) (Amasino *et al.*, 1999; Rosenzweig *et al.*, 1999). The difference in position of the Cu(I) and Hg(II) ions may represent conformations relevant to copper ion transport and delivery, respectively (see text and Fig. 9).

that Hg(II) is closer to the surface of the protein than is Cu(I) (Arnesano *et al.*, 2001). Both conformations likely represent positions of copper in Atx1 due to the flexibility of the metal-binding loop, which in turn is likely to be important in the metal transfer mechanism (see below).

Figure 5b shows that basic residues Lys-24, Lys-28, Lys-59, and Lys-62 in the Atx1 structure, residues implicated in two-hybrid analyses as being important in recognition of the target molecule Ccc2a, cluster on one face of the molecule, creating a positively charged surface. Recently, the NMR structures of reduced apo- and Cu(I)-bound Ccc2a were elucidated (Banci *et al.*, 2001). The Cu(I)Ccc2a structure shows that the copper ion is coordinated by Cys-13 and Cys-16 residues of its MXCXXC motif, with a S–Cu(I)–S angle of $120^\circ \pm 29^\circ$. These values are similar to those observed in the NMR structure of Cu(I)Atx1 (Arnesano *et al.*, 2001; Banci *et al.*, 2001). As illustrated in Fig. 5f, acidic residues Asp-53, Glu-57, Glu-60, Asp-61, Asp-65, and Glu-67 cluster on one face of the Ccc2a molecule, forming a potentially complementary negatively charged surface to the positively charged surface found on Atx1. These electrostatic surfaces are discussed further in Section II, C below.

*3. Hah1/Mnk4*

Subsequent to the determination of the Atx1 structure by X-ray crystallography, the structure of its human homologue, Hah1, was determined by single-crystal X-ray diffraction methods in the presence of Cu(I) (1.8 Å), Hg(II) (1.75 Å), and Cd(II) (1.75 Å) (Wernimont *et al.*, 2000). The unexpected result was that in each case, the crystallographic asymmetric unit contained two Hah1 molecules bridged by a single metal ion. As indicated from their superposition in Fig. 8a, these three structures are very similar, with root mean square (RMS) differences in atomic positions for backbone atoms in the range 0.2–0.6 Å for all possible monomer–monomer comparisions. In Cd(II)Hah1, the cadmium ion is tetrahedrally coordinated by four cysteine residues, Cys-12 and Cys-15 of the MXCXXC motif from each Hah1 monomer, at distances between 2.4 and 2.5 Å. These distances are in line with what is observed for tetrahedrally coordinated Cd(II) ions in the structure of the $Cd_5Zn_2$ form of rat metallothionein-2 (Robbins *et al.*, 1991). In Hg(II)Hah1, the metal-binding site demonstrates distorted tetrahedral geometry with Hg–S distances of 2.3, 2.5, 2.5, and 2.8 Å. The first three distances are well within the expected range for three–coordinate Hg(II) metal-binding sites, whereas the 2.8-Å Hg–S distance is too long for a covalent bond and likely corresponds to a secondary bonding interaction (Watton *et al.*, 1990; Wright *et al.*, 1990). The CuHah1 structure is the only X-ray structure of a MXCXXC-containing chaperone or target domain with copper bound in this motif. As mentioned above, crystallographic attempts to visualize Cu(I) binding to Atx1 were unsuccessful, resulting in a disordered metal-binding loop (Rosenzweig *et al.*, 1999). By contrast, the presence of a well-defined copper-binding site in CuHah1 can likely be attributed to two factors. First, anaerobic crystal growth was essential. CuHah1 crystals grown aerobically or those grown anaerobically but not immediately frozen resulted in structures with disordered metal-binding

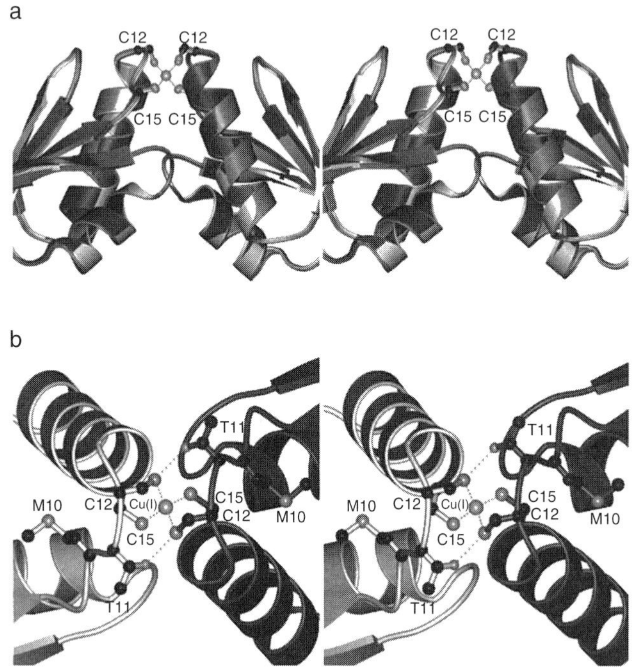


FIG. 8. The metal-binding site of Hah1. (a) Superposition of Cu(I) (light gray), Hg(II) (medium gray), and Cd(II) (dark gray) bound forms of human Hah1 as determined by X-ray crystallography [pdb codes 1fee, 1fe4, and 1fe0, respectively (Wernimont *et al.*, 2000) ]. The structures are very similar, and for clarity, the Cu(I) ion is the only one shown. (b) The copper-binding site of Cu(I)Hah1. The cysteine sulfur atoms of the MTCGGC motif from each monomer bind Cu(I) in a distorted tetrahedral geometry with bond distances of ~2.3 Å. The Cu(I)-bridged dimer is stabilized by hydrogen bonds between the side chain oxygen of Thr-11 in one monomer to the side chain sulfur atom of Cys-12 in the opposing monomer (see text).

loops and multiple positions for the copper ion. Second, simultaneous copper coordination by two Hah1 molecules probably contributes to its stabilization (Wernimont *et al.*, 2000).

Figure 8b shows the copper-binding site of CuHah1 in more detail. The cysteine sulfur atoms form a distorted tetrahedral geometry with Cu–S distances of 2.3, 2.3, 2.3, and 2.4 Å. Three-coordinate copper sites in Cu(I)-thiolate cluster-containing proteins such as copper metallothionein (Fig. 2a) demonstrate Cu–S distances between 2.2 and 2.3 Å (Pickering *et al.*, 1993). At 1.8-Å resolution, however, there is sufficient uncertainty in the metal–ligand distances that it is difficult to unambiguously describe the copper-binding site as being either three- or four-coordinate. From an energetic standpoint, however, a four-coordinate Cu(I) complex is unfavorable because it would generate a net −3 charge on the copper center. As shown in Fig. 8b, the Hah1 structures suggest an unanticipated role for the Thr-14 residue of the MTCXXC motif (and by extension, serine residues of those proteins with MSCXXC) in the copper chaperones and their target domains on the P-type ATPases. In each structure, the sulfur atom of Cys-12 is hydrogen bonded to the side chain oxygen of Thr-11 on the opposite protein molecule. Such a hydrogen bonding network can affect the structural and chemical properties of metal centers (Karlin *et al.*, 1998). This intermolecular hydrogen bonding network, along with the bound metal ion, provides the key forces that hold the two Hah1 molecules together, consistent with the observation that Atx1 and Hah1 interact with their target domains in two-hybrid experiments in a copper-dependent fashion (Pufahl *et al.*, 1997; Hamza *et al.*, 1999; Larin *et al.*, 1999). As mentioned previously, X-ray absorption spectroscopy indicated that Cu(I)Atx1 existed as a mixture of two- and three-coordinate species, with two sulfur ligands at a distance of approximately 2.25 Å, and a third ligand, most likely sulfur, at a distance of 2.40 Å (Pufahl *et al.*, 1997). At the time, it was suggested that the third ligand might come from an exogenous thiol present in the buffer or from a neighboring Atx1 molecule. The structure of CuHah1 strongly suggests that the third ligand likely comes from a second Atx1 molecule (Wernimont *et al.*, 2000).

Figure 5c shows that basic residues Arg-21, Lys-25, Lys-56, and Lys-57 in the Hah1 structure cluster on one face of the molecule in a mode analogous to that found in Atx1, creating a positively charged surface. The NMR structure of the Ag(I)-bound form of the fourth N-terminal domain of the Menkes P-type ATPase, the target of Hah1, shows that the silver ion is coordinated by Cys-14 and Cys-17 residues of its MXCXXC motif, with a S–Ag(I)–S angle restrained to be approximately linear (Gitschier *et al.*, 1998). As illustrated in Fig. 5e, acidic residues Glu-55, Asp-62, Asp-63, and Asp-67 cluster on one face of the Mnk4 molecule, forming a potentially complementary negatively charged surface to the positively charged surface found on Hah1. These electrostatic surfaces are discussed further in the structure-based mechanism described in the next section.

### C. *Metal Transfer Mechanism*

Based on the wealth of both *in vivo* and *in vitro* biochemical experiments described in Section II, A, a diffusion-driven or "bucket brigade" mechanism of metal ion transfer between MXCXXC-motifs on the copper chaperone proteins and their target domains was predicted (Pufahl *et al.*,

1997; Huffman and O'Halloran, 2000). This mechanism, illustrated schematically in Fig. 3, proposes that the Atx1 and Hah1 copper chaperones recognize their P-type ATPase target domains through protein–protein interactions, bringing their respective MXCXXC motifs into proximity. The Cu(I) ion, initially held nearly linearly by the chaperone cysteine thiolates, is attacked by a cysteine thiolate from the target domain, forming a bond of similar strength to those of the original two ligands. This intermediate is presumably reflected in two-hybrid studies demonstrating that chaperone and target proteins associate in a copper-dependent fashion (Pufahl *et al.*, 1997; Hamza *et al.*, 1999; Larin *et al.*, 1999). Rapid ligand exchange follows through a series of two- and three-coordinate intermediates, with the Cu(I) ion subsequently being shuttled to the target domain followed by dissociation of the apo-chaperone so that it can cycle to acquire additional copper ion (presumably) from the membrane copper transporter. The 3-D structures of Atx1, Ccc2a, Hah1, and Mnk4 now provide the means to allow visualization of copper ion transfer at the molecular level.

The following structure-based mechanism is predicated on the assumption that the Cu(I)Hah1 structure shown in Fig. 8a and 8b represents that of the three-coordinate copper-bridged intermediate shown schematically in Fig. 3. Using this Cu(I)Hah1 model as a template, representative chaperone and target protein domains are structurally aligned to each half of this template in order to envision copper ion movement from one to the other (see below). The copper delivery cycle begins when reduced apoAtx1 or apoHah1 acquires copper ion, possibly from the membrane-bound Ctr copper transporters. Figure 7a is representative of the conformations of the MTCXXC metal-binding loop before and after copper binding. After copper is bound by the chaperone, it adopts a two-coordinate Cu(I) geometry where the exposure of the copper ion to the cytoplasm is restricted, preventing the occurrence of unwanted Fenton chemistry and protecting the copper ion from competing exogenous ligands. As illustrated in Fig. 7b, when the Cu(I)Hah1 or Cu(I)Atx1 molecule begins to dock with its target protein, the copper ion moves from its protected environment [represented by Cu(I)Atx1] to the more solvent-exposed conformation [represented by Hg(II)Atx1] to prepare for its transfer.

Figures 8a and 8b show that, upon docking, a cysteine residue from the target protein forms a third primary bond to the copper ion. This intermediate conformation is further stabilized by a network of hydrogen bonds shown in Fig. 8b, where the threonine side chain of the MTCXXC motif in each protein makes a hydrogen bond to the N-terminal cysteine residue of the MTCXXC motif of its cognate molecule. A series of two- and three-coordinate intermediates now rapidly forms and dissipates, leaving the copper ion coordinated by the two Cys residues of the target protein, a conformation similar in structure to that demonstrated by the Ag(I)Mnk4 protein (Fig. 5e).

The entire proposed copper transfer mechanism, with the copper ion moving from right to left, is visualized in Fig. 9a (see color insert). The copper chaperone, in a conformation such that it is ready to donate the copper ion, is represented by the Hg(II)Atx1 structure in orange. It is structurally aligned to monomer B of the Cu(I)Hah1 template (yellow) with a RMS deviation of 0.94 Å for 261 target pair backbone atoms. On target molecule ligand attack, the copper moves from its two-coordinate position represented by the Hg(II) ion in Fig. 9a to the three-coordinate intermediate position represented by the Cu(I) ion. After decay of the three-coordinate intermediate, the copper ion is transferred to the target protein, represented by the structure of Ag(I)Mnk4 (blue) that has been structurally aligned to monomer A of the template with a RMS deviation of 1.3 Å for 254 target pair backbone atoms. Note that Arg-21 and Lys-25 from monomer B of Hah1 can form ion pairs with Asp-63 and Asp-67 of Mnk4. These residues likely play an important role in determining the specificity of Hah1/Mnk4 interaction. Similarly, as shown in Fig. 9b (see color insert), when the Ccc2a (green, left) and Hg(II)Atx1 (orange, right) structures are aligned to the template (not shown), there exists electrostatic complementarity between basic residues of Atx1 (Lys-24, Lys-28, Lys-59, and Lys-62) with acidic residues of Ccc2a (Glu-57, Glu-60, Asp-61, and Asp-65). Importantly, the steps depicted in Fig. 9a are likely applicable to intramolecular copper ion transfer between individual N-terminal Atx1-like domains in the Ccc2, Menkes, and Wilson P-type ATPases prior to copper ion translocation across the post-Golgi membrane into the secretory compartment. The structure-based copper transfer mechanism described above is also consistent with metal ion specificity studies that indicate that Hah1, when incubated with Cd(II), does not interact with and deliver cadmium to the N-terminal domains of the Wilson protein (Larin *et al.*, 1999). In these *in vitro* experiments, the Cd(II)Hah1 dimer is likely to be very stable, thus preventing its dissociation and docking with the Wilson's protein and precluding metal ion transfer. On the other hand, Cu(I) and Hg(II)Hah1 do interact with and deliver metal to the N-terminal domains of the Wilson's protein. These observations may be explained by the fact that Cu(I) and Hg(II) do not readily form four primary bonds to thiolates and are likely in equilibrium, thus allowing docking with a target P-type ATPase domain (Larin *et al.*, 1999; Wernimont *et al.*, 2000).

### *D. Bacterial Homologue*

In the gram-positive bacteria *Enterococcus hirae*, copper homeostasis is maintained by the *cop* operon, consisting of the *copA*, *copB*, *copY*, and *copZ* genes (Odermatt and Solioz, 1995; Wunderli-Ye and Solioz, 1999b). CopA and CopB are copper-transporting P-type ATPases analogous to those found in higher eukaryotes. CopA is believed to serve in the uptake of Cu(I) under conditions in which copper is limiting, while CopB is thought to function in the transport of copper out of the bacterial cell when intracellular copper levels become elevated (Odermatt and Solioz, 1995; Wunderli-Ye and Solioz, 1999a). CopY is a 53-kDa Zn(II)-loaded homodimeric protein that regulates expression of the *cop* operon by binding to the promoter in the absence of copper. In the presence of copper, the protein releases the promoter (Strausak and Solioz, 1997; Cobine *et al.*, 1999). Dameron and colleagues demonstrated that Cu(I)CopZ molecules can donate copper to Zn(II)CopY, displacing the bound Zn(II) and releasing CopY from the promoter DNA, thereby inducing the *cop* operon. The displacement of Zn(II) by Cu(I) suggests that the metal ions bind to the same thiolate-binding site within CopY (see Cobine *et al.*, 1999). Thus, CopZ is the first copper chaperone to be shown to deliver an inducer to a repressor molecule (Harrison *et al.*, 2000).

The NMR structure of apoCopZ determined by Wuthrich and colleagues is shown in Fig. 5d (see Wimmer *et al.*, 1999). Not surprisingly, the overall structure is similar to that observed for Mnk4, MerP, Atx1, Hah1, and Ccc2a, revealing the βαββαβ fold and the solvent-exposed MXCXXC motif. The NMR studies indicate that the loop containing the MXCXXC motif is quite mobile in the absence of metal. Addition of copper salts, however, leads to protein aggregation, and the metal-binding site could not be resolved due to signal loss in the MXCXXC loop region (Wimmer *et al.*, 1999). An additional function of CopZ may be to donate Cu(I) to CopB under elevated copper conditions or to receive it from CopA under conditions in which copper is limiting in the medium. In light of the significant sequence homology among Ccc2, CopA, and CopB, and the structural similarity between CopZ and the Atx1-like copper chaperone molecules, a metal transfer mechanism from the membrane-bound transporters to or from CopZ similar to that shown in Fig. 3 is possible. The exact role of CopZ in copper ion homeostasis awaits further experimentation.

## III. Copper Chaperones for Copper–Zinc Superoxide Dismutase

### A. *Genetics and Chemistry*

#### 1. *Discovery of the CCS Family*

The studies presented in Sections I and II above firmly establish the need for a copper chaperone to deliver the reactive metal to the enzymes in the secretory pathway. In parallel studies, Tzagoloff and co-workers identified an additional pathway involving the soluble yeast factor Cox17, a protein demonstrated to deliver copper ions to cytochrome oxidase in the mitochondria (see Section IV below) (see Glerum *et al.*, 1996a,b). Given the observation that different chaperones deliver copper ion to different thermodynamic compartments and to different target proteins, Culotta, Gitlin, and colleagues asked whether cytoplasmic enzymes that use copper ion as cofactor might also require copper chaperones for their activation. Using this concept as a working hypothesis, they predicted that insertion of copper into Cu,ZnSOD would involve a specific metal carrier protein (see below) (Culotta *et al.*, 1997).

Cu,ZnSOD is a 32-kDa homodimeric protein in the cytoplasm of eukaryotic and bacterial cells that catalyzes the disproportionation of superoxide into dioxygen and hydrogen peroxide ($2O_2^- + 2H^+ \rightarrow O_2 + H_2O_2$) through redox cycling of its catalytic copper ion (Fridovich, 1989; Valentine *et al.*, 1999). Higher organisms produce superoxide anion ($O_2^-$) as an occasional by-product during the one-electron reduction of dioxygen that occurs in respiration and photosynthesis (Davies, 1995; Richter *et al.*, 1995). Cells must therefore have ways to regulate superoxide concentrations as excess levels can inactivate enzymes containing iron–sulfur clusters and can lead to the formation of highly oxidizing species such as hydroxyl radical, which can be damaging to other cellular constituents (Valentine *et al.*, 1999). In higher mammals, Cu,ZnSOD is particularly abundant in red blood cells and in neurons. Regarding the latter, it constitutes ~1% of the mass of spinal tissue protein (see possible role in familial ALS below in Section III, D) (Pardo *et al.*, 1995; Wong *et al.*, 1995). 3-D structures of eukaryotic Cu,ZnSOD proteins from multiple species have been determined by X-ray crystallographic methods and have been demonstrated to be quite similar (Tainer *et al.*, 1982; Kitagawa *et al.*, 1991; Djinovic *et al.*, 1992; Djinovic-Carugo *et al.*, 1996; Ogihara *et al.*, 1996; Hart *et al.*, 1999). As shown in Fig. 10a, each monomer of the SOD1 dimer binds one catalytic copper and one structurally important zinc ion and displays the Greek key β-barrel-folding topology (Tainer *et al.*, 1982). Figure 10b shows the mode of Cu(I) binding in the yeast Cu,ZnSOD, where it is coordinated in a pseudo-trigonal-planar geometry by His-46, His-48, and His-120 (Ogihara *et al.*, 1996; Hart *et al.*, 1999). A series of high-resolution crystal structures of yeast CuZnSOD reaction intermediates offers a structure-based mechanism that enumerates the multiple movements occurring in the active site of the enzyme during the reaction cycle (Hart *et al.*, 1999).

*sod1*Δ strains of *Sa. cerevisiae* are oxygen sensitive and are auxotrophic for methionine and lysine when grown in air (Bermingham-McDonogh *et al.*, 1988; Chang and Kosman, 1990; Gralla and Valentine, 1991).

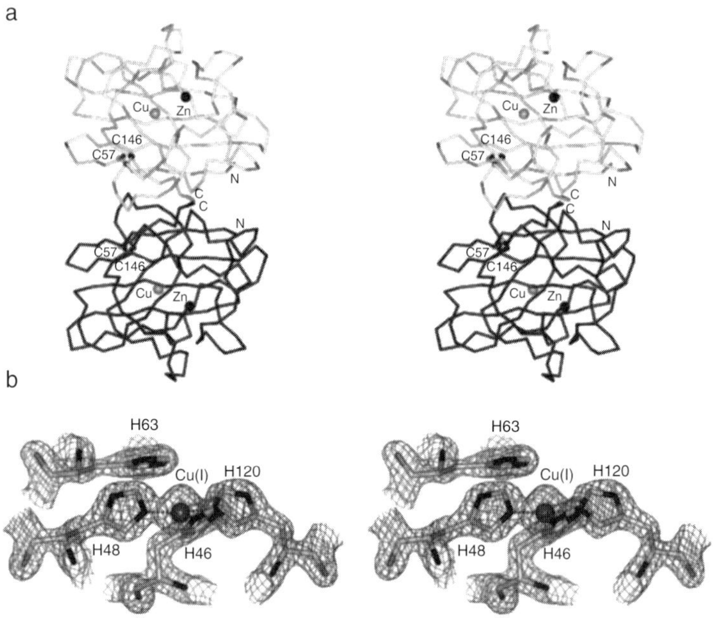


FIG. 10. The backbone structure of yeast copper–zinc superoxide dismutase and its copper-binding site [pdb code 2jcw (Hart *et al.*, 1999)]. (a) The ySOD1 homodimer. The molecular twofold axis runs horizontally in the plane of the page between the monomeric subunits (light and dark gray, respectively). The copper ion (light sphere), zinc ion (dark sphere), and disulfide bond between Cys-57 and Cys-146 (ball and stick) are shown in each subunit. (b) The pseudo-trigonal planar Cu(I)-binding site superimposed on $\sigma_A$-weighted electron density with coefficients $2mF_o\text{-}DF_c$ contoured at $1.1\sigma$. On oxidation to Cu(II), the copper ion becomes four-coordinate, moving toward and binding to His-63 in a distorted square-planar geometry.

Culotta, Gitlin, and colleagues surmised that yeast strains harboring a mutation in a gene for a putative copper chaperone for SOD1 (CCS) would exhibit phenotypes similar to those of *sod1*Δ strains. They noticed that mutations in the previously discovered *lys7* gene yielded yeast that are also auxotrophic for lysine and methionine when grown in air, supporting the idea that SOD1 and Lys7 might be functionally related (see Horecka *et al.*, 1995; Culotta *et al.*, 1997). The metabolic defects associated with both *sod1*Δ and *lys7*Δ null mutations are overcome when the yeast strains are grown anaerobically or when $CuSO_4$ is added to the growth medium (Chang *et al.*, 1991; Culotta *et al.*, 1997; Gamonet and Lauquin, 1998). Overexpession of the *lys7* gene does not, however, complement *sod1*Δ yeast, suggesting that Lys7 does not possess superoxide dismutase activity (Culotta *et al.*, 1997).

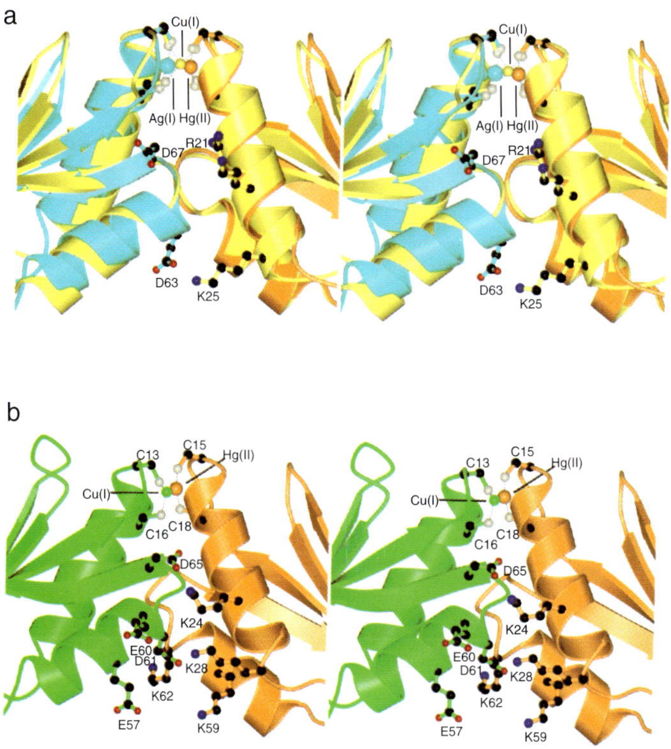


FIG. 9. Structure-based copper ion transfer between proteins of the Atx1-like family (see text for details). The structural alignments were performed using a modified version of the program ALIGN, with the Cu(I)Hah1 structure serving as the template (Satow *et al.*, 1986). (a) Copper ion transfer occurs from right to left. The Hg(II)Atx1 structure (orange) represents the copper chaperone with its metal-binding loop in a conformation appropriate to begin the transfer process. The Cu(I)Hah1 structure (yellow) represents the copper-bridged intermediate between chaperone and target protein. This intermediate is also represented schematically in Fig. 3. The Ag(I)Mnk4 structure (blue) represents the target domain after the copper transfer process is completed. The docking of the Hah1 chaperone to the Mnk4 target protein is likely facilitated by the electrostatic interactions between Hah1 residues Arg-21 and Lys-25 (yellow, right) and Mnk4 residues Asp-63 and Asp-67 (blue, left). (b) Electrostatic recognition of the target yeast Ccc2a domain (green, left) by the yeast Atx1 copper chaperone (orange, right). The Cu(I)Ccc2a and Hg(II)Atx1 structures were aligned to the Cu(I)Hah1 template as in (a). The docking of Atx1 to Ccc2a is likely facilitated by electrostatic interactions between the positively charged Lys-25, Lys-28, Lys-59, and Lys-62 side chains of Atx1 with the negatively charged Glu-57, Glu-60, Asp-61, and Asp-65 side chains of Ccc2a (see text).

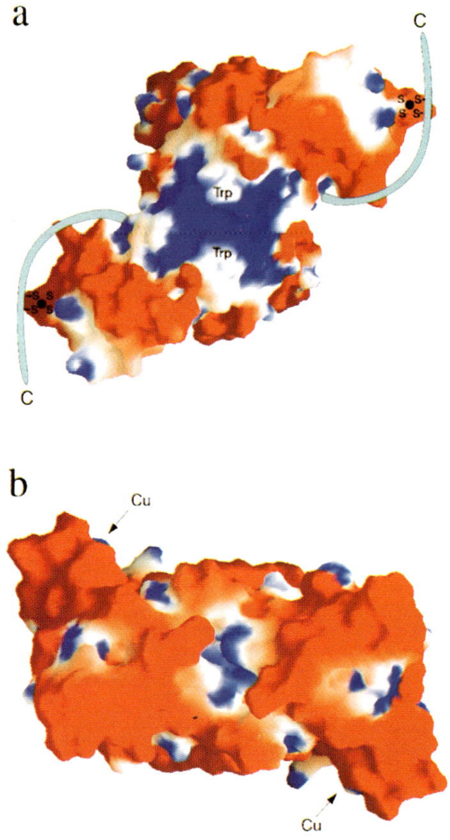


FIG. 17. Dimer of dimers model of yCCS/ySOD1 association and copper delivery. (a) Molecular surface map of the electrostatic potential of the yCCS dimer calculated using GRASP (Nicholls *et al.*, 1991). The blue and red contours are calculated at ± 2.5 kT and represent positive and negative electrostatic potentials, respectively. yCCS residues Arg-188 and Arg-217 shown in Fig. 14 are mainly responsible for the positively charged region shown here. The position of Trp-183 is indicated. Domain 3 of yCCS is represented as a stylized extended tail. The cysteine residues of the domain 3 CXC motif and the cysteine residues of the domain 1 MXCXXC motif bind copper in a mode similar to that illustrated in Fig. 16. (b) Molecular surface map of the electrostatic potential of the ySOD1 dimer calculated using GRASP (Nicholls *et al.*, 1991) as in (a). The copper-binding sites on ySOD1 are accessed in the plane of the page as indicated by arrows. The ySOD1 and yCCS surfaces shown are the putative docking surfaces, and the ySOD1 molecule must be rotated by 180° around the horizontal axis in the plane of the page to dock with the yCCS dimer shown in (a). Note that this 180° rotation puts the ySOD1 copper-binding sites in position to face domains 1 and 3 of yCCS such that their respective copper-binding sites are in proximity. Thus, in this model, the residues at the yCCS dimer interface serve not in target recognition, but to create the putative ySOD1 electrostatic interaction site on the surface of the protein (see text).

*In vivo* data supporting the hypothesis that Lys7 is the copper chaperone for SOD comes from the fact that yeast cells containing a *lys7*Δ null mutation produce normal levels of SOD1, but cannot incorporate copper into the enzyme, thereby rendering it inactive as a superoxide scavenger. Expression of *lys7* or its human counterpart in *trans* complements the *lys7*Δ null mutation and restores the production of holo Cu,ZnSOD. Furthermore, incorporation of $^{64}$Cu into SOD1 occurs only when Lys7 or its human homologue is present *in vivo*, showing that Lys7 (also termed yeast CCS or yCCS) or its human homologue (hCCS) is both necessary and sufficient for copper incorporation into SOD1 (Culotta *et al.*, 1997; Gamonet and Lauquin, 1998). CCS is specific for its SOD1 target. It is not needed for copper delivery to cytochrome *c* oxidase in the mitochondria as evidenced by the fact that *lys7*Δ null yeast strains are not defective in electron transport through the respiratory chain, nor does CCS deliver copper to the secretory pathway, as *lys7*Δ cells are found to be fully functional in the delivery of $^{64}$Cu to Fet3. Conversely, overexpression of the *atx1* gene could not complement the SOD1 copper-loading defect of *lys7*Δ mutants, further reinforcing the specificity of yCCS and of the copper chaperones of the mitochondrial and secretory pathways (Culotta *et al.*, 1997). A mouse *CCS* gene has also been discovered to be essential for *in vivo* copper incorporation into Cu,ZnSOD in transgenic mice. Although mice with targeted disruption of both CCS alleles are viable (which was itself a surprise to many) and possess normal levels of SOD1 protein, they reveal marked reductions in SOD1 activity when compared with control littermates (Wong *et al.*, 2000). Metabolic labeling with $^{64}$Cu showed that the loss of SOD1 activity in the *lys7*Δ null mice is the result of impaired copper incorporation into SOD1 and that this effect was specific because no abnormalities were seen in copper uptake, distribution, or incorporation into other copper-containing enzymes (Wong *et al.*, 2000).

O'Halloran and colleagues set out to determine whether yCCS is sufficient for direct and specific activation of apo-ySOD1 *in vitro* (see Rae *et al.*, 1999). Cu(I)yCCS and Cu(I)glutathione (GSH) both demonstrate the ability to activate zinc-loaded ySOD1 protein as evidenced by the standard kinetic assay using cytochrome *c* (Fridovich, 1985). Free copper in either the Cu(I) [$Cu(CH_3CN)_4PF_6$] or the Cu(II) ($CuSO_4$) state also activates zinc-loaded ySOD1 under similar assay conditions. As is observed in *in vivo* studies, yCCS is unnecessary for activation of ySOD1 *in vitro* if a sufficient pool of copper is available to the enzyme. In contrast, when the concentration of free copper is limited in the assay solution by the addition of the strong copper chelator BCS ($K_d \sim 10^{-20}$ M), purified Cu(I)yCCS, but not Cu(I)GSH, can activate Zn-loaded SOD1. As expected, $CuSO_4$ activation of zinc-loaded ySOD1 is abrogated in the presence of the Cu(II) chelator

EDTA ($K_d \sim 1.6 \times 10^{-19}$M) (Rae *et al.*, 1999). The primary conclusion derived from these studies is that in the presence of copper scavengers, Cu(I)yCCS activates ySOD1 but Cu(I)GSH and $CuSO_4$ cannot. Importantly, the copper insertion event likely occurs with direct transfer of the metal ion from yCCS to the ySOD1, as copper ions that might possibly dissociate from the yCCS or GSH donor molecules are rapidly sequestered by the BCS competitor molecule, which is present in 20-fold excess over the copper donor molecules under these assay conditions. This in turn suggests that specific protein–protein interactions are involved in the recognition of ySOD1 by yCCS (Rae *et al.*, 1999).

*2. Domain Architecture of CCS*

Figure 11 and Table III show a sequence alignment of known and putative CCS proteins that have been identified in *Saccharomyces cerevisiae* (yeast), (Horecka *et al.*, 1995, *Mus muscalus* (mouse), (Bartnikas *et al.*, 2000), *Homo sapiens* (human), (Culotta *et al.*, 1997), *Rattus norvegicus* (rat) (Hiromura *et al.*, 2000), *Drosophila melanogaster* (fruit fly) (Kirby and Phillips, 2001), *Schizosaccharomyces pombe* (Devlin *et al.*, 1996), *A. thaliana* (Kliebenstein *et al.*, 1998), and *Lycopersicon esculentum* (tomato) (Nersissian and Valentine, 1999). With the exception of the putative CCS protein from *Sc. pombe*, the largest of the copper chaperones identified to date are yCCS and hCCS, 249- (27.3 kDa) and 274- (29 kDa) amino-acid proteins, respectively. Each polypeptide contains three distinct domains (Schmidt *et al.*, 1999a). Figure 11 shows that the N-terminal domain, domain 1 (D1), possesses the MXCXXC copper-binding motif present in the Atx1-like family of proteins, including the copper-transporting ATPases. Domain 2 (D2) demonstrates strong sequence similarity to the SOD1 monomer and, in the case of hCCS, all but one of the metal-binding ligands found in SOD1 are retained. The C-terminal domain 3 (D3) does not share homology with other known proteins, but its sequence, which includes a potential copper-binding CXC motif, is highly conserved among the copper chaperones for SOD1 in all species studied to date. Yeast CCS and human CCS share 28% overall sequence identity and retain the same domain organization (Culotta *et al.*, 1997; Casareno *et al.*, 1998; Gamonet and Lauquin, 1998).

Culotta and colleagues conducted *in vitro* and *in vivo* structure–function analyses of yCCS in an effort to assign functions to the three distinct domains (see Schmidt *et al.*, 1999a). The invariant MXCXXC metal-binding motif within domain I (the N-terminal $\sim$70 amino acids) and the homology it shares with several other metallochaperones including Atx1 (36% identity), MerP (24% identity), and the CopA copper-transporting ATPase (34% identity) immediately led to the suggestion that this domain might be responsible for the insertion of copper ion into SOD1 (Gamonet and Lauquin, 1998; Falconi *et al.*, 1999;

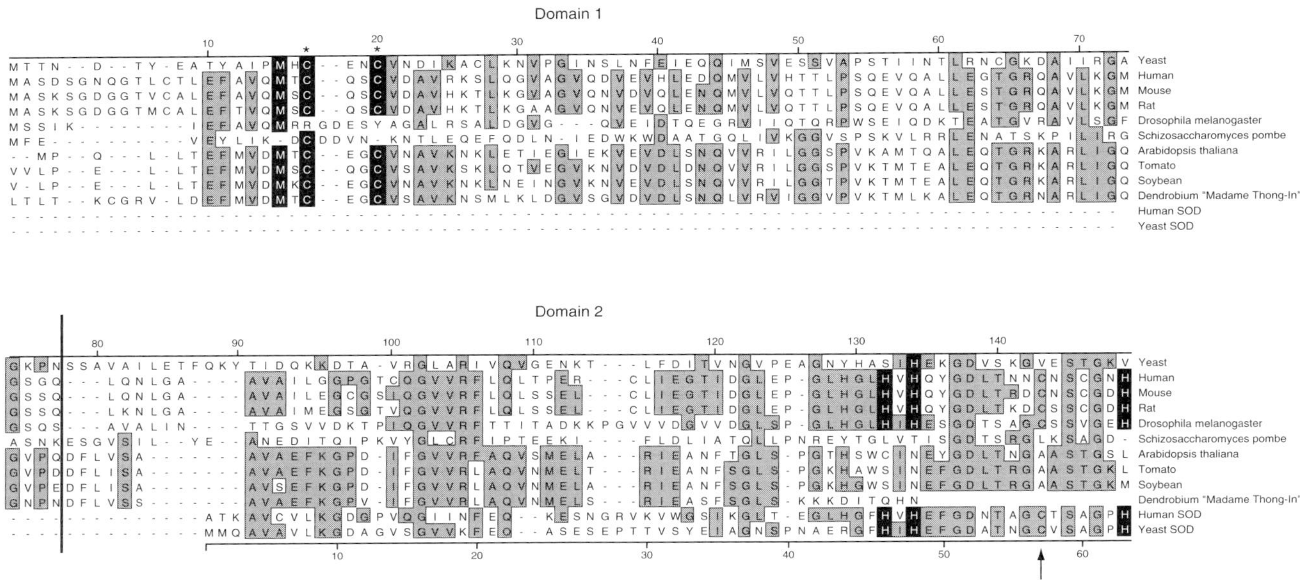


FIG. 11. *(continues)*

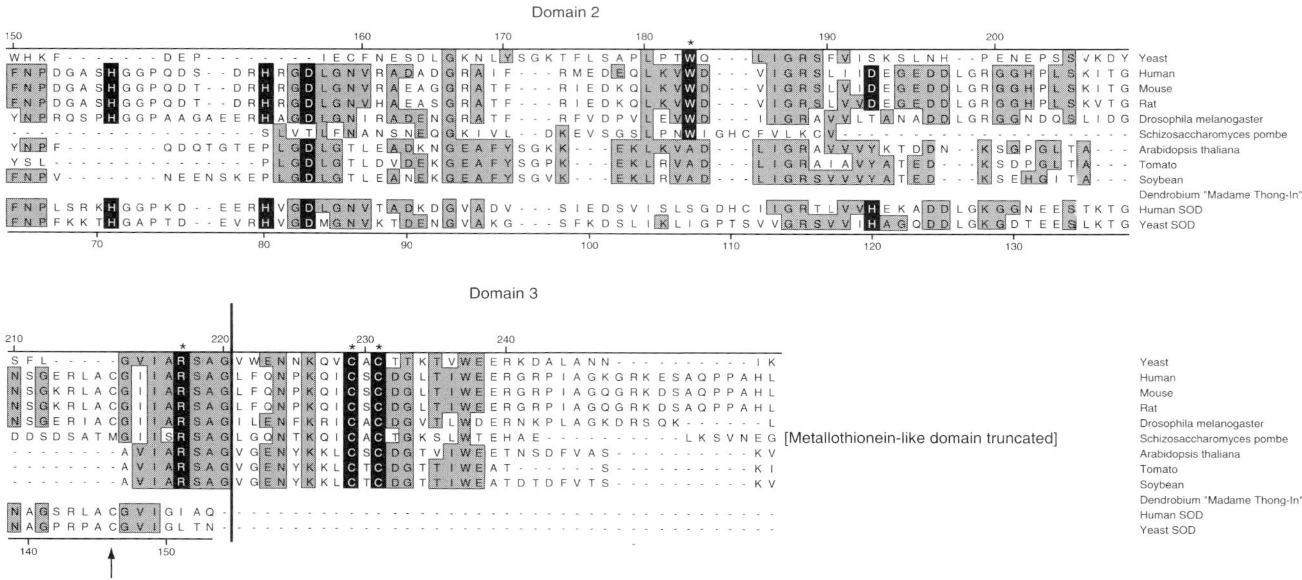


FIG. 11. Multiple sequence alignment of the copper chaperones for SOD1 using the CLUSTAL method (see text) (Higgins and Sharp, 1989). Sequence numbering corresponds to that of the yeast protein. Domain 1 cysteine residues believed to be important in uptake under copper-limiting conditions and domain 3 cysteine residues believed to be important in copper delivery to SOD1 are boxed in black and labeled with asterisks. Residues in domain 2 that correspond to copper and zinc ligands in yeast and human SOD1 are boxed in black. Residues in domain 2 postulated to be involved in SOD1 target recognition (Trp-183 and Arg-217) are boxed in black and labeled with asterisks. The vertical arrows underneath domain 2 represent the positions of cysteine residues that make the disulfide bond in yeast and human SOD1.

TABLE III
*Sequence References for Copper Chaperones for SOD1*

| CCS | Reference |
|---|---|
| Yeast (*Saccharomyces cerevisiae*) | Horecka *et al.*, 1995 |
| Human (*Homo sapiens*) | Culotta *et al.*, 1997 |
| Mouse (*Mus musculus*) | Bartnikas *et al.*, 2000 |
| Rat (*Rattus norvegicus*) | Hiromura *et al.*, 2000 |
| *Drosophila melanogaster* | Kirby and Phillips, 2001 |
| *Schizosaccharomyces pombe* | Devlin *et al.*, 1996 |
| *Arabidopsis thaliana* | Kliebenstein *et al.*, 1998 |
| Tomato (*Lycopersicon esculentum*) | Nersissian and Valentine, 1999 |
| Soybean (*Glycine max*) | Nersissian and Valentine, 2000 |
| *Dendrobium* "Madame Thong-In" | Yu and Goh, 1998 |
| Yeast (*Sa. cerevisiae*) SOD1 | Bermingham-McDonogh *et al.*, 1988 |
| Human (*H. sapiens*) SOD1 | Barra *et al.*, 1980 |

Schmidt *et al.*, 1999a). A series of yCCS truncation mutants expressed in a *lys7*Δ (*yCCS*Δ) null strain provide insight into the functions of the three domains. Yeast strains expressing polypeptides spanning either D1 alone or D1D2 alone do not complement the *lys7*Δ phenotype, whereas those expressing a polypeptide spanning D2D3 do retain copper delivery activity as long as copper is not limiting in the surrounding medium. When the growth medium contains the Cu(I) chelator BCS, the D2D3 protein does not complement the *lys7*Δ phenotype, whereas full-length yCCS does. D1 is thus not *absolutely* necessary for CCS activity. When D1 and the D2D3 constructs are expressed together *in trans* under copper-limiting conditions, however, CCS activity is regained, leading to the suggestion that D1 likely works in concert with D2D3 to incorporate copper into SOD1. Taken together these observations suggest that D1 of CCS functions to recruit copper under copper-limiting conditions, but is not necessary for target (SOD1) recognition or direct insertion of the copper into SOD1 (Schmidt *et al.*, 1999a). Interestingly, co-overexpression of Atx1 and D2D3, as well as overexpression of a chimeric Atx1–D2D3 construct, gives similar results as the expression of D2D3 alone, indicating the action of D1 is specific and that other Atx1-like domains cannot functionally substitute for it (Schmidt *et al.*, 1999a).

Domain 2 (amino acids 71–218 in yeast) of the CCS proteins bears striking homology to the SOD1 target molecule, especially in the case of hCCS. In fact, hCCS was originally identified as a fourth superoxide dismutase, *sod4*, in the human genome due to its significant homology with *sod1*. hCCS D2 shares 47% sequence identity with the human SOD1 monomer, including identity at three of the four copper ligands and identity at all the zinc ligands, whereas yCCS shares only 26% identity with ySOD1 and retains only one copper and one zinc ligand. Due to its similarity to SOD1, D2 has been postulated to play a role in target recognition. As illustrated in the sequence alignment in Fig. 11, in hCCS one of the copper ligands found in SOD1, His-120, is replaced by an aspartic acid (Asp-201), converting the "copper site" of hCCS to one possessing the ligands found in the zinc-binding site. Not surprisingly, if Asp-201 is mutated to a histidine, hCCS gains a substantial amount of superoxide dismutase activity, yet retains its ability to incorporate copper into SOD1. Conversely, if His-120 of SOD1 is mutated to an aspartate, it loses its SOD activity. The presence of Asp-201 in hCCS likely limits possible self-oxidation of the chaperone or oxidation of its target SOD1 (Schmidt *et al.*, 1999b). Recently, the zinc ligands of hCCS were mutated to alanine and the protein was expressed in *lys7*Δ yeast. Yeast expressing H147A and D167A hCCS mutants exhibit retarded growth and reduced SOD1 activity. The proteins are also less soluble, suggesting that these residues, when mutated, may affect the tertiary structure of hCCS and result in impaired stability and interaction with SOD1 (Endo *et al.*, 2000).

The third and C-terminal domain (amino acids 219–249 in yeast) of CCS (D3) was found to be essential for copper delivery to SOD1 in the yCCS truncation experiments mentioned above (Schmidt *et al.*, 1999a). Although it shares little homology with any known protein, it is the most highly conserved region among the copper chaperones for SOD1 with ~50% identity between yCCS and hCCS. Both cysteines in the invariant CXC motif of D3 are essential for copper chaperone activity, as expression of full-length yCCS proteins with either of these cysteine residues converted to serine fails to complement the *lys7*Δ null mutation (Schmidt *et al.*, 1999b). A peptide corresponding to D3 of hCCS binds copper *in vitro* in a copper to peptide stoichiometry of 0.53:1, most likely forming a copper-bridged peptide dimer (Schmidt *et al.*, 1999a). Based on these results, D3 is thought to be responsible for the direct transfer of copper to the copper-binding site located within SOD1.

*3. Interactions with SOD1*

The striking sequence homology of yCCS/hCCS and SOD1 and the results of the *in vitro* experiments examining copper transfer from yCCS to ySOD1 described above strongly suggest that a direct interaction between the chaperone and its target occurs, perhaps through the second domain of CCS. Direct interaction between hCCS and hSOD1 was demonstrated *in vitro* by glutathione *S*-transferase (GST) pull-down assays using either GST–CCS or GST–SOD1 attached to agarose beads. The same results were obtained in the presence or absence of reducing agent or in the presence of excess copper. hCCS D2D3 and FALS-associated hSOD1 mutants (see Section III, D) also exhibit interaction with hSOD1 (Casareno *et al.*, 1998). *In vivo*, coimmunoprecipitation of hCCS and hSOD1 in cell lysates confirms the interaction (Casareno *et al.*, 1998; Corson *et al.*, 1998). Immunofluorescence studies of both yCCS and hCCS reveal localization patterns similar to that of SOD1 in yeast and in mammalian liver cells (Culotta *et al.*, 1997; Casareno *et al.*, 1998). In a yeast interaction mating system, yCCS and ySOD1 physically interact *in vivo* (Schmidt *et al.*, 2000). Importantly, mutations corresponding to residues at the putative dimer interface of yCCS or at the dimer interface of ySOD1 abrogate this interaction. Neither yCCS D2 alone nor the D1D2 alone, however, was sufficient to direct interactions with ySOD1 (see below). yCCS D2D3 does physically interact with the target, and mutation of the cysteines in the CXC motif to serines does not block the interaction. Unlike the case with the Atx1 family of copper chaperone/target interactions, the interactions described above are not dependent on copper availability (Schmidt *et al.*, 2000). Activation of ySOD1 by yCCS in an *in vivo* assay for copper incorporation into SOD1 demonstrates that yCCS inserts copper into a preexisting pool of ySOD1 dimers and that this reaction occurs within minutes in the absence of new protein synthesis or protein unfolding by the major heat shock proteins. These data are consistent with a model in which prefolded dimers of SOD1 serve as substrate for the yCCS molecule (Schmidt *et al.*, 2000).

To help ascertain possible ways in which SOD1 interacts with CCS, the oligomeric state of CCS molecules under physiological conditions was examined. Analytical gel filtration studies indicate that apo-yCCS migrates as a monomer (29,000 ± 2,000 molecular weight) under physiological buffer conditions containing reducing agent, whereas copper-loaded yCCS migrates as a mixture of monomers and dimers (54,000 ± 4000 molecular weight) under the same conditions (Schmidt *et al.*, 1999a). Hart and colleagues used analytical ultracentrifugation sedimentation velocity and sedimentation equilibrium experiments to demonstrate that yCCS D2 alone is a monomer in solution both in the presence and in the absence of reducing agent (see Hall *et al.*, 2000). This inability of isolated D2 to form a dimer in solution likely accounts for the failure of polypeptides spanning D2 or D1D2 alone to interact with SOD1 in the two-hybrid analyses performed previously (Schmidt *et al.*, 2000). Moreover, sedimentation velocity and equilibrium analyses of full-length apo-yCCS reveal that it is a monomer in the presence of the reducing agent TCEP [tris(2-carboxyethyl)phosphine, a non-thiol-containing compound that reduces disulfide bonds but does not absorb at 280 nm] and that it exists in a monomer/dimer equilibrium ($K_d \simeq 3.6\times 10^{-6}$ M) in the absence of TCEP. Since both apo-yCCS and Cu(I)yCCS can exist as mixtures of monomers and dimers in solution, it was suggested that the unreduced apo-yCCS 3-D conformation may mimic that of the metal-loaded conformation. This could conceivably occur through disulfide bond formation from cysteine residues in the MXCXXC motif of D1 with cysteine residues of the CXC motif in D3. Such a conformation would be similar in local structure to one where Cu(I) bridges these domains. It was further suggested that an allosteric conformational change likely occurs upon metal binding that facilitates yCCS dimerization, perhaps as a prerequisite for interaction with SOD1 (Hall *et al.*, 2000) (see Section III, C below). Recently, using analytical gel filtration, Cu-free hCCS D2 and Cu-free full-length hCCS have been found to migrate as dimers (33,200 and 61,100 molecular weights, respectively) under non-reducing conditions (Rae *et al.*, 2001). In light of the experiments discussed thus far, two modes for the interaction of CCS with Cu,ZnSOD have been proposed: the heterodimer model and the dimer of dimers model (see below).

*4. Metal Binding by CCS*

Because the CCS molecules deliver copper ions to zinc-loaded SOD1, the mode(s) of copper binding by these proteins is of interest and can shed light on chaperone function. Valentine and colleagues examined the metal-binding properties of hCCS and tomato CCS (tCCS) molecules using cobalt as a spectroscopic probe (see Zhu *et al.*, 2000). tCCS was selected for comparative study with hCCS because it appears to lack the SOD-like metal-binding sites in D2, and it contains only four cysteine residues: two in the D1 MXCXXC motif and two in the D3 CXC motif. In contrast, hCCS contains nine cysteine residues: three in D1, four in D2, and two in D3. Based on the similarity in sequence between hCCS D2 and SOD1, two of the cysteine residues in hCCS D2 likely form a disulfide bridge (see below). Addition of Co(II) to purified hCCS and tCCS samples resulted in overall conformational changes as monitored by circular dichroism. Titration with Co(II) gave clear end-points of one Co(II)/tCCS protein and two Co(II)/hCCS proteins. In both tCCS and hCCS, Co(II) was observed to bind to three or four cysteine ligands in a tetrahedral geometry. An additional Co(II) binds to hCCS with a geometry similar to that found in the zinc site of Cu,ZnSOD, suggesting that it is coordinated at the putative zinc site of D2. Because tCCS contains only four cysteine residues, two in D1 and two in D3, the circular dichroism and electronic absorption spectroscopic data indicate that the metal is likely bound between D1 and D3 in these CCS molecules (Zhu *et al.*, 2000). Figure 12a is a schematic diagram that illustrates the conclusions of these spectroscopic metal-binding studies of hCCS with Co(II).

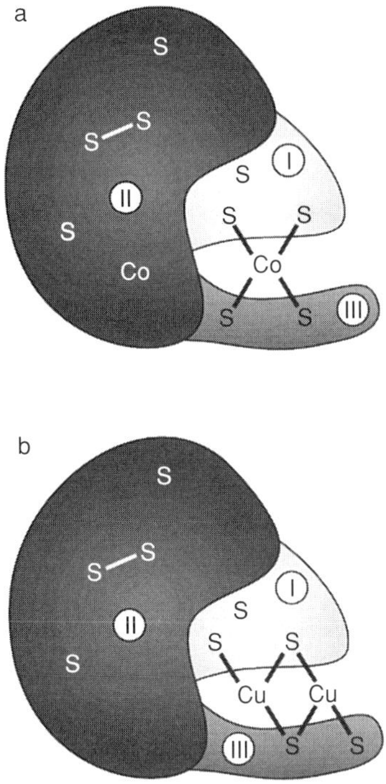


FIG. 12. Schematic diagram of metal binding by human CCS. hCCS domains 1, 2, and 3 are labeled with roman numerals. Cysteine residues are designated as S. The disulfide bond in domain 2 is indicated by S–S. (a) Cobalt binding to hCCS. Electronic absorption spectra indicate that two equivalents of Co(II) bind per hCCS monomer, one through three or four cysteine residues in a tetrahedral geometry, and one with a geometry similar to that found in the zinc site of SOD1 (see text) (Zhu *et al.*, 2000). (b) Copper binding to hCCS. XAS indicates that two Cu(I) ions bind per hCCS monomer in a sulfur-only liganding environment, with an additional heavy atom scatterer peak suggesting the presence of a $\mu_2$-bridged dicopper cluster (Eisses *et al.*, 2000).

A similar cooperation between D1 and D3 is observed in X-ray absorption spectroscopy (XAS) experiments performed by Blackburn and colleagues, in which reduced hCCS binds two Cu(I) ions in a sulfur-only liganding environment (see Eisses *et al.*, 2000). Interestingly, the presence of an additional heavy atom scatterer peak suggests that two Cu(I) ions bind in this environment in a $\mu_2$-bridged dicopper(I) cluster in which each copper is coordinated by three cysteine sulfur ligands. A possible geometry for such a site is shown in Fig. 12b. Rae *et al.* (2001) recently performed biochemical experiments to further determine the effects of copper binding on hCCS. When copper loading was attempted on hCCS domain 2 alone, no copper was found to be incorporated into the protein. However, domain 3 is cleaved more slowly in trypsin digests when hCCS is copper loaded than in the absence of copper, suggesting that domain 3 participates in and is stabilized by copper binding.

Several issues regarding CCS copper transfer remain to be resolved. It is unknown whether CCS adopts only one or several copper-bound modes or conformations. Though the spectroscopic studies show copper binding involves only domains 1 and 3, each domain may be able to function independently, forming the Cu(I)-bridged D1/D3 complex at only one stage of the copper delivery process. Since the mode of interaction between SOD1 and CCS remains to be fully resolved (see Section III, C), it is unclear how far the copper-binding site of CCS is from the copper site of SOD1 in the heterocomplex. It is also unknown whether CCS docking to SOD1 may allosterically induce the transfer of copper from CCS to SOD1 or whether another signal such as phosphorylation of CCS might trigger transfer (Falconi *et al.*, 1999; Hall *et al.*, 2000).

### *B. Structural Biology*

In 1999 and 2000, the understanding of how CCS molecules function in copper ion recruitment, SOD1 target recognition, and copper ion delivery was enhanced by the elucidation of the 3-D structures of three different CCS polypeptides. yCCS domain 2 (Hall *et al.*, 2000), full length apo-yCCS (Lamb *et al.*, 1999), and zinc-loaded hCCS domain 2 (Lamb *et al.*, 2000b) structures were all determined using single-crystal X-ray diffraction methods. yCCS domain 2 (yCCS-D2) was determined to 1.55-Å resolution using multiple-isomorphous replacement heavy-atom derivative methods by Hart and colleagues (pdb code 1ej8) (see Hall *et al.*, 2000). The yCCS-D2 protein is a flattened eight-stranded Greek key β-barrel with approximate dimensions of 46 Å × 25 Å × 16 Å. Figure 13a shows that the disk-shaped molecule possess one flat surface opposed by a saddle-shaped cleft. As indicated in Fig. 11, the yCCS-D2 protein is 26% identical in sequence to that of ySOD1, and thus it is not surprising that the overall Greek key β-barrel folding topology is conserved between the two proteins. The yCC2-D2 monomer structurally aligns with a monomer of ySOD1 with a rms deviation of 1.5 Å for 407 backbone atom target pairs. Figure 13b shows a superposition of yCCS-D2 with that of a monomer of ySOD1 determined at 1.7-Å resolution (Ogihara *et al.*, 1996).

The primary differences in the yCCS-D2 and ySOD1 β-barrel structures are as follows: (1) Residues 64–81 in ySOD1 that correspond to the so-called "zinc loop" that contains zinc ligands His-71 and His-80 are deleted in yCCS-D2. (2) The loop formed by residues 122–143 in ySOD1,

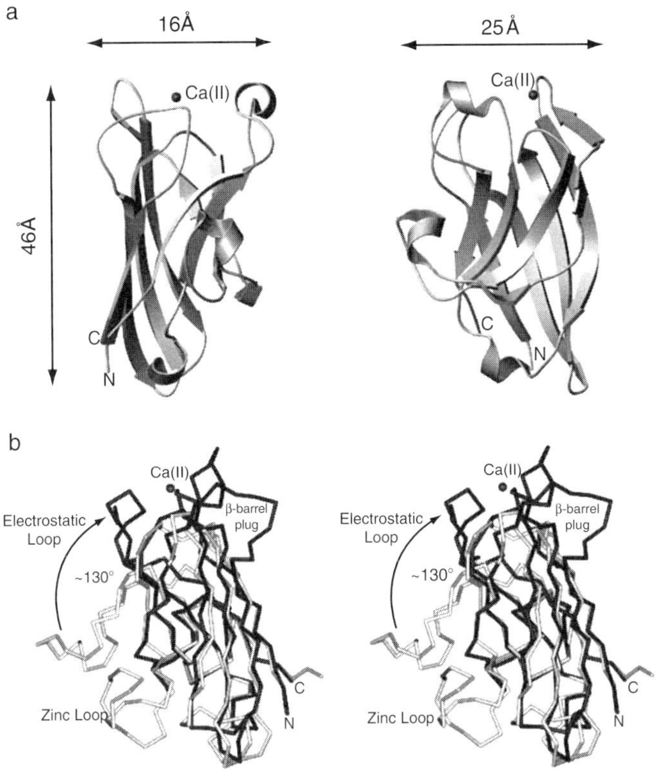


FIG. 13. X-ray crystal structure of yeast CCS domain 2 [pdb code 1ej8 (Hall *et al.*, 2000)]. (a) Overall fold and dimensions of the monomeric yCCS-D2 protein. The right image looks into the saddle-shaped cleft (see text). The left image is rotated counterclockwise (looking from the top) approximately 90° in the plane of the page relative to the right image. The position of the bound Ca(II) ion is indicated. (b) Superposition of the yCCS-D2 structure (black) on a monomer of yeast SOD1 (light gray) [pdb code 2jcw (Hart *et al.*, 1999)]. The region corresponding to the zinc loop in ySOD1 is absent in yCCS-D2. The region corresponding to the electrostatic loop in ySOD1 is rotated approximately 130° in yCCS-D2 into a conformation that extends the β-barrel. The positions of the bound Ca(II) ion and the β-barrel plug are indicated (see text).

known as the "electrostatic loop" or the "active site lid loop" (Tainer *et al.*, 1982), is truncated and rotated by about 130° in the yCCS-D2 structure (Fig. 13b). The loss of the zinc loop and the altered conformation of the electrostatic loop are responsible for the formation of the saddle shape seen in Fig. 13a and for the formation of a β-barrel in yCCS-D2 that is approximately 13 Å longer than that found in yeast SOD1. (3) The apolar β-barrel "plug" interactions found in all known Cu,ZnSODs (Deng *et al.*,

1993; Hart *et al.*, 1999) are replaced in yCCS-D2 by a series of completely new interactions that are predominantly electrostatic, hydrogen bonded, and water mediated. (4) A single Ca(II) ion is coordinated at the apex of this elongated yCCS-D2 β-barrel, linking three of the four loops together via direct interactions with the protein or through a network of bound water molecules (Hall *et al.*, 2000). (5) Somewhat surprisingly, the yCCS-D2 structure is monomeric in solution and in the crystal, while all known eukaryotic SOD1 proteins are dimeric (Hall *et al.*, 2000).

As shown in Fig. 14a, multiwavelength anomalous diffraction X-ray crystallographic analysis of full-length apo-yCCS to 1.8-Å resolution by Rosenzweig and colleagues reveals both D1 and D2 components of the holo-yCCS protein (pdb code 1qup) (see Lamb *et al.*, 1999). The CXC-containing C-terminal 27 amino acids (D3) that are absolutely required for metal ion delivery to SOD1 are disordered in the crystal and could not be modeled. Consistent with the 34% sequence identity shared between them (see Fig. 4), yCCS-D1 displays considerable structural homology to Atx1. As is observed in the oxidized apo-Atx1 crystal structure (Fig. 6b), the two cysteine residues of the yCCS-D1 MXCXXC motif also form a disulfide bond (Lamb *et al.*, 1999; Rosenzweig *et al.*, 1999). This led to the suggestion that the conformation of yCCS-D1 may change upon copper binding as it does upon mercury binding in the Hg(II)Atx1 structure (Fig. 6 and 7) with Cys-17, Cys-20, and His-16 of the MXCXXC motif the likely candidates for copper binding (Lamb *et al.*, 1999). The sequence alignment in Fig. 4 shows that yCCS-D1 lacks many of the conserved lysine residues clustered near the copper site of Atx1 that are known to be involved in Cu(I)Atx1 recognition of Ccc2a. The loss of these positively charged residues by yCCS-D1 may help explain the observation that co-overexpression of Atx1 and yCCS-D2D3 proteins, or overexpression of a chimeric Atx1-yCCS-D2D3 protein, cannot functionally substitute for full-length yCCS in yeast under copper-limiting conditions (Schmidt *et al.*, 1999a). However, the lysine residue thought to be important for metal capture and/or delivery to the target protein is conserved between Atx1 (Lys-65) and yCCS (Lys-66) (Lamb *et al.*, 1999; Arnesano *et al.*, 2001).

As shown in Fig. 14b, and in contrast to the monomeric structure of yCCS-D2 shown in Fig. 13a, full-length apo-yCCS displays a D2-mediated dimer in the crystallographic asymmetric unit (Lamb *et al.*, 1999; Hall *et al.*, 2000). The full-length apo-yCCS dimer interface resembles that of ySOD1 in overall size and buried surface area, and about 64% of the amino acids found at the yCCS dimer interface are conserved relative to ySOD1 (Ogihara *et al.*, 1996; Lamb *et al.*, 1999). Notable differences between the full-length apo-yCCS and ySOD1 interfaces, however, include the substitution of charged residues in yCCS (Lys-136, Arg-217) for

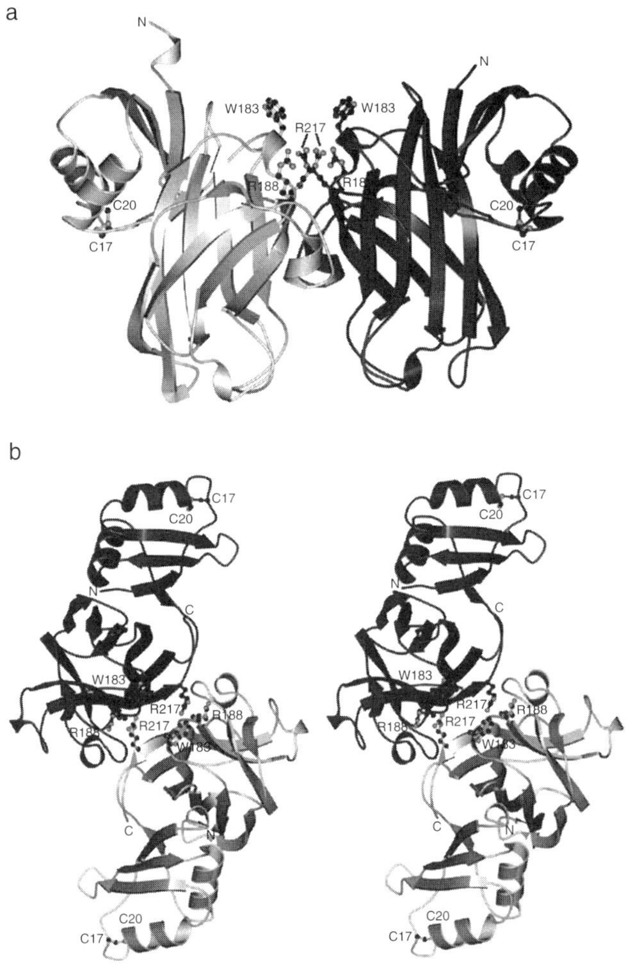


FIG. 14. X-ray crystal structure of full-length yeast CCS [pdb code 1qup (Lamb *et al.*, 1999)]. (a) One monomer of yCCS is in light gray and the other is in dark gray. The cysteine residues of the MXCXXC motif in domain 1 are labeled and form a disulfide bond in each subunit. Amino acid side chains that are important in the formation of the positive patch at the dimer interface (Arg-188 and Arg-217) and the solvent-exposed Trp-183 residues of loop 6 at the center of this patch are shown in ball-and-stick representation. Domain 3 is not visible in the crystal structure (see text). (b) Stereo view of the image in (a) rotated 90° in the horizontal plane of the page and then 90° counterclockwise around an axis perpendicular to the page. The side chains that form the putative ySOD1 interaction surface are represented as ball-and-stick. The cysteine residues of the domain 1 MXCXXC motif are also represented in ball-and-stick.

hydrophobic residues in yeast SOD1 (Phe-50, Leu-151) and the presence of Trp-183 on loop 6, which is shortened relative to ySOD1. As shown in Fig. 14b, these two positively charged amino acid substitutions are translated into four local 3-D changes because of the twofold molecular symmetry at the dimer interface, and they dramatically alter the electrostatic character of this region relative to ySOD1 (see below). Interestingly, two sulfate ions are bound at the yCCS dimer interface by symmetry-related residues Lys-136 and Arg-188 (not shown), apparently stabilizing the homodimeric structure in the crystal. The presence of the sulfate ions adds between six and eight favorable hydrogen-bonding interactions in dimeric yCCS as compared with dimeric ySOD1. Although the significance of this observation remains unclear, sulfate or phosphate ions present in the cytoplasm may facilitate yCCS dimer formation *in vivo* (Hall *et al.*, 2000).

Finally, as shown in Fig. 15, the crystal structure of domain 2 of human CCS (hCCS-D2) (pdb code 1do5) has been determined to 2.75 Å resolution through molecular replacement methods, using the yeast SOD1 dimer as a search model (pdb code 1sdy) (Djinovic *et al.*, 1992; Lamb *et al.*, 2000a). Although the full-length hCCS protein was set up in crystallization trials, the crystals contained only hCCS-D2 alone, suggesting that proteolysis of domains 1 and 3 occurred in the hanging drop during the time frame of the crystallization experiment. hCCS-D2 forms a homodimer in the crystal that in overall appearance very closely resembles the

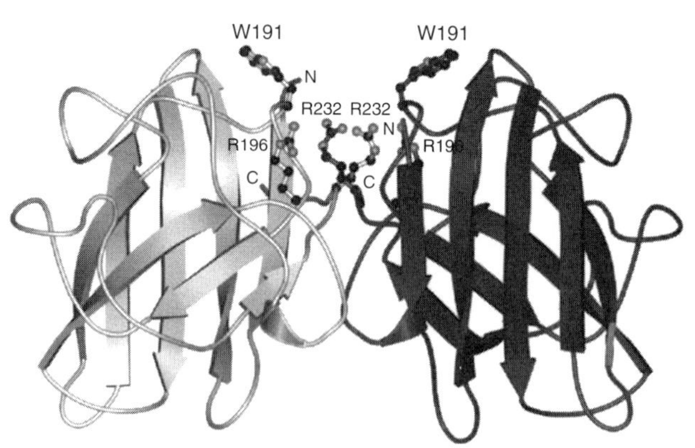


FIG. 15. X-ray crystal structure of human CCS domain 2 [pdb code 1do5 (Lamb *et al.*, 2000b)]. One monomer of hCCS-D2 is in light gray and the other is in dark gray. Amino acid side chains that are important in the formation of the positive patch at the dimer interface (Arg-196 and Arg-232) and the solvent-exposed Trp-191 residues of loop 6 at the center of this patch are shown in ball-and-stick representation. Domains 1 and 3 are not present (see text).

hSOD1 homodimer. Similar associative interactions occur at the dimer interfaces of both hCCS-D2 and full-length apo-yCCS, where the same four main chain hydrogen bonds and a number of hydrophobic contacts are found (Lamb *et al.*, 2000a). As shown in the sequence alignment in Fig. 11, hCCS-D2 retains almost all of the structural characteristics of its target SOD1 protein and binds one Zn(II) ion per monomer in the zinc-binding site. His-120, a copper ligand in both yeast and human SOD1, is replaced by an aspartic acid (Asp-201) in hCCS. This led Desideri and colleagues to predict that Zn(II) might bind because the zinc site in SOD1 also consists of a $(His)_3Asp$ ligand set (see Falconi *et al.*, 1999). The putative copper site of hCCS-D2, however, is not occupied by a metal ion, but rather, the electron density suggests that a water molecule resides in that site (Lamb *et al.*, 2000a). As mentioned previously, the inability to bind copper at this site likely protects against self-oxidation and against oxidation of its target SOD1 (Schmidt *et al.*, 1999a). The zinc and electrostatic loop elements remain structurally intact, although many of the residues important for SOD1 catalysis are absent, such as the charged residues of the electrostatic loop of SOD1 that are exchanged for hydrophobic amino acids in hCCS. One exception is Arg-224 in hCCS, a residue that is analogous to Arg-143 in both ySOD1 and hSOD1. This residue functions in those enzymes as an electrostatic "sink" to pull superoxide into the catalytic site (Getzoff *et al.*, 1992). The fact that Arg-224 is conserved in hCCS likely accounts for the fact that D201H hCCS mutant can function as a superoxide dismutase, whereas the absence of the other charged residues of the electrostatic loop likely explains why this mutant is not as active as the hSOD1 molecule (Schmidt *et al.*, 1999a; Lamb *et al.*, 2000a). Notably, loop 6 of hCCS-D2 exhibits significant sequence identity to the same region of yCCS-D2. Relative to SOD1, loop 6 is shorter in both CCS proteins and possesses a solvent-exposed tryptophan residue (Trp-191 in hCCS, Trp-183 in yCCS). The shortening of loop 6 causes two arginine residues (Arg-196 and Arg-232 in hCCS and Arg-188 and Arg-217 in yCCS) to become more solvent exposed, creating a positive patch on the protein's surface (see below). Because loop 6 is the only structural element unique to domain 2 of the CCS proteins, it may be important in metal delivery, either by facilitating interactions with SOD1 or, perhaps, by working with domain 3 to incorporate copper (Lamb *et al.*, 2000a).

### C. *Metal Transfer Mechanisms*

#### 1. *Heterodimer Model*

In light of the significant sequence similarity between domain 2 of the CCS proteins and their target SOD1 molecules, particularly at the homodimer interface, an attractive hypothesis is that CCS recognizes its target SOD1 protein through this surface, forming a CCS/SOD1 heterodimer (Casareno *et al.*, 1998; Falconi *et al.*, 1999; Lamb *et al.*, 1999, 2000a, 2000b; Schmidt *et al.*, 1999b; Rae *et al.*, 2001). The formation of such a heterodimeric CCS/SOD1 complex appears possible, as structural models obtained by superimposing domain 2 of yCCS or hCCS onto one of the monomers of its respective target SOD1 homodimer reveal no critical steric interference that would preclude such an interaction. The amount of buried surface area in the CCS/SOD1 heterodimer interfaces in these modeling studies is similar to that observed buried in the CCS or SOD1 homodimers themselves (Falconi *et al.*, 1999; Lamb *et al.*, 1999, 2000b; Hall *et al.*, 2000). Figure 16 shows the yCCS/ySOD1 heterodimer model generated as described above, illustrating the position of the yCCS MXCXXC motif in D1 relative to that of the copper-binding site in ySOD1. Although domain 3 is not visible in the crystal structure, spectroscopic studies mentioned above and the previously described roles of domain 1 in copper ion uptake and of domain 3 in copper ion delivery suggest that they act together in the copper transfer process. D3 is thus stylistically rendered such that the cysteine residues of the CXC motif and the cysteine residues of the D1 MXCXXC motif bind Cu(I) in a

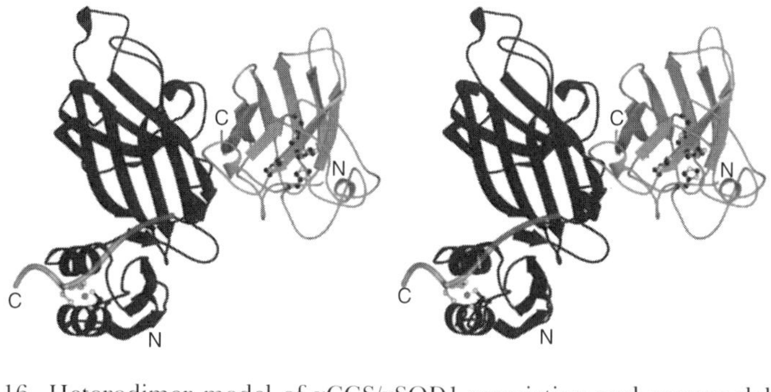


FIG. 16. Heterodimer model of yCCS/ySOD1 association and copper delivery. A monomer of yCCS is shown in dark gray and a monomer of ySOD1 is shown in light gray. The ligands of the ySOD1 copper site are shown as ball-and-stick. Domain 3 of yCCS containing the CXC motif is rendered stylistically as a transparent light gray tube. The Cu(I) ion is modeled in a geometry similar to that shown in the Cu(I)Hah1 structure shown in Fig. 8, with two cysteine residues coming from the domain 3 CXC motif and two cysteine residues coming from the domain 1 MXCXXC motif. This geometry is also similar to that suggested by the hCCS Co(II) electronic absorption binding studies shown schematically in Fig. 12a. In this model, copper ion delivery to ySOD1 occurs though movement of yCCS domain 3 from the position shown to the ySOD1 copper-binding site (~40 Å). Thus, the residues yCCS at the heterodimer interface are postulated to serve in ySOD1 target recognition (see text).

conformation analogous to that observed in Cu(I)Hah1 (Fig. 8a). When copper ion is limiting in the environment, the CXC motif of D3 would presumably acquire Cu(I) from the MXCXXC motif in a fashion analogous to the way Ccc2a is thought to acquire Cu(I) from Atx1, through a series of two- and three-coordinate intermediates (Fig. 3). For metal delivery, the CXC motif would presumably "swing" over to deposit the copper into the vacant SOD1 copper-binding site (Falconi *et al.*, 1999). This mode of switch-like copper translocation is illustrated schematically in the top row of Fig. 18, where copper ion delivery requires that the CXC residues move approximately 40 Å (Poulos, 1999; Hall *et al.*, 2000; Rae *et al.*, 2001).

*2. Dimer of Dimers Model*

As shown in Figs. 14 and 15, both full-length yCCS and hCCS proteins are dimeric in their crystal structures (Lamb *et al.*, 1999, 2000b). Their target molecules, ySOD1 and hSOD1, are also very stable homodimers (Valentine and Pantoliano, 1981; Djinovic *et al.*, 1992; Parge *et al.*, 1992; Deng *et al.*, 1993; Ogihara *et al.*, 1996; Poulos, 1999). Inspection of the electrostatic surface potentials of the yCCS and hCCS homodimers and of the ySOD1 and hSOD1 homodimers and the 3-D arrangements of the metal-binding sites in the CCS and SOD1 proteins together suggest a possible mechanism of CCS/SOD1 association where a dimer of yCCS or hCCS interacts with a dimer of SOD1 in such a way that symmetry-related domains 1 and 3 of the CCS molecules could be close in space to the symmetry-related copper-binding sites of homodimeric SOD1. As illustrated in Fig. 17, the yCCS dimeric interface has a positively charged nature relative to yeast SOD1 due to the substitution of charged residues for their hydrophobic counterparts, most notably Lys-136 and Arg-217 in yCCS for Phe-50 and Leu-151 in ySOD1. As mentioned previously, these two substitutions translate into four local charge changes due to the twofold symmetry at the yCCS dimer interface. These residues, coupled with the shortening of loop 6 resulting in the exposure of Arg-188 (a residue conserved in ySOD1), are responsible for six solvent-exposed positive charges at the yCCS dimer interface. This positively charged surface patch is also present in hCCS (Lamb *et al.*, 2000a). As shown in Figs. 14 and 15, it is notable that Trp-183 in yCCS (Trp-191 in hCCS) sits at the heart of the symmetrical dimer interface surface and is completely solvent exposed. The symmetry-related tryptophan indole rings face one another (but are not within stacking distance) and appear poised for interaction with another protein, possibly via stacking interactions.

The dimer of dimers model of yCCS/ySOD1 interaction and copper delivery illustrated in Fig. 17 (see color insert) can be summarized as follows: The positively charged (blue) yCCS dimer interface surface features (Fig. 17a) with dual solvent-exposed Trp-183 residues could serve as the platform upon which the predominantly negatively charged (red) SOD1 dimer will interact (Fig. 17b). The negatively charged ySOD1 electrostatic surface shown in Fig. 17b is the putative docking surface. For interaction with a dimer of yCCS, the ySOD1 molecule would rotate approximately 180° around the horizontal axis in the plane of the page before interacting with the positively charged yCCS docking platform. The resultant ySOD1 rotation and docking to yCCS put both of the copper-binding sites of ySOD near both of the MXCXXC and CXC metal-binding sites of yCCS-D1 and yCCS-D3 (represented stylistically). Under copper-limiting conditions, the CXC motif of yCCS-D3 can acquire copper from the MXCXXC motif of yCCS-D1 and, via a small movement or conformational change, insert the copper ion into the high-affinity copper-binding site of ySOD1. yCCS-D3/ySOD1 contacts could be the "switch" that triggers copper movement from one to the other. This concept is represented schematically in the bottom row of Fig. 18.

### 3. *Heterodimer versus Dimer of Dimers Model*

As mentioned, the heterodimer model of CCS/SOD1 association and copper transfer makes use of a common structural scaffold and a partially conserved dimer interface to determine the specificity of interaction. This model requires that at some point both the SOD1 and the CCS homodimers dissociate to monomers and subsequently reassociate as a single heterodimer or a pair of heterodimers. O'Halloran and colleagues propose two potential modes through which this could occur (see Rae *et al.*, 2001). As illustrated in the top row of Fig. 18, the CCS and SOD1 homodimers might simply dissociate into their individual monomeric subunits. These individual CCS and SOD1 monomeric subunits may subsequently encounter each other and recombine to form heterodimers, undergo copper ion transfer, again dissociate into monomeric subunits, and finally recombine to form their respective homodimeric forms. Alternatively, as shown in the second row of Fig. 18, a dimer of CCS and a dimer of SOD1 might associate and undergo a "monomer rearrangement" mechanism where two heterodimers are formed in a loosely associated tetrameric complex, followed by copper ion transfer and another rearrangement to again yield their respective homodimeric forms. The details of how the monomers would rearrange in this model, however, are unclear. In both of these models, the CXC motif of domain 3 cycles between a conformation where it receives copper from the MXCXXC motif of domain 1 and a conformation corresponding to its copper delivery function to SOD1. Based on the model shown in Fig. 16, this requires that the CXC motif move approximately 40 Å during this process (Lamb *et al.*, 1999; Poulos, 1999; Hall *et al.*, 2000). The distance

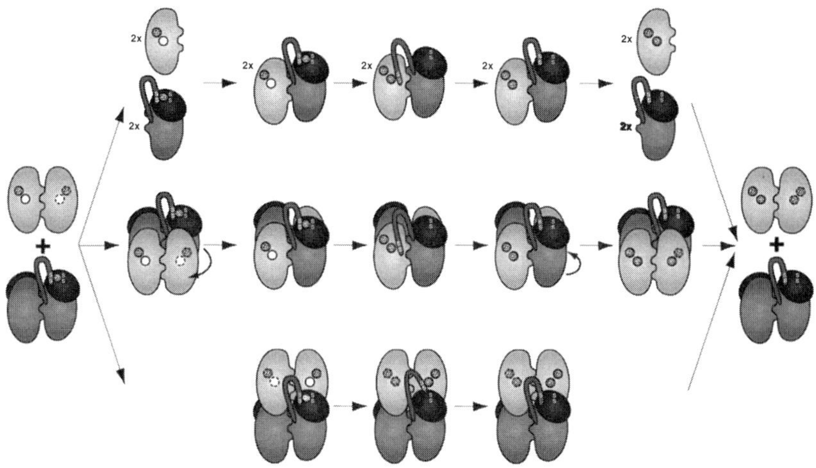


FIG. 18. Schematic representation of possible copper ion transfer mechanisms between hCCS and hSOD1 (see text). CCS and zinc-loaded SOD1 (apoSOD1) homodimers are shown at the left in dark gray and light gray, respectively. Dashed circles represent metal-binding sites on the side of the molecule away from the reader, while solid circles represent metal-binding sites facing toward the reader. In all cases, SOD1 begins devoid of copper (left) and ends replete with copper (right). (Top) Cu(I)CCS and SOD1, initially homodimers, dissociate into monomers. These different monomeric subunits subsequently encounter each other in the cytoplasm to form heterodimers, undergo copper ion transfer as depicted in the heterodimer model in Fig. 16, and again dissociate into their monomeric subunits. Finally, the apoCCS and Cu(I)SOD1 monomeric subunits encounter a self-subunit to re-form homodimers. (Middle) Cu(I)CCS and SOD1 homodimers associate to form a heterotetramer. Subunit "swapping" occurs in this heterotetrameric complex to form two heterodimers, followed by copper ion transfer as depicted in the heterodimer model in Fig. 16. After copper transfer, another subunit "swapping" event takes place to re-form the respective homodimers followed by dissociation of the homodimers from the heterotetrameric complex. (Bottom) CCS and SOD1 homodimers do not dissociate or swap, but encounter each other through the surface features described in the legend to Fig. 17 such that the empty copper-binding sites on SOD1 are immediately adjacent to domains 1 and 3 of the CCS molecules. Copper transfer occurs though domain 3 of CCS, followed by separation of the CCS and SOD1 homodimers.

from the C-terminus of yCCS-D2 to the CXC motif of yCCS-D3 is sufficient to span the distance in this cycling movement as long as domain 3 is in a relatively extended conformation (Rae *et al.*, 2001).

While this mode of copper ion delivery from the chaperone to SOD1 is attractive due to the similarities between the CCS and the SOD1 dimer interfaces, it is interesting to note that SOD1 exists in manyfold excess relative to the copper chaperone *in vivo* (Rae *et al.*, 1999; Rothstein *et al.*,

1999). This means that nascent SOD1, unless captured by a monomer of CCS immediately as it is translated, can form homodimers prior to receiving copper ion. The native SOD1 homodimer is extremely stable, exhibiting very tight binding between subunits even under harsh conditions such as high concentration of denaturant and low pH (Valentine and Pantoliano, 1981; Poulos, 1999). For example, when copper and zinc are removed, the SOD1 homodimer has been observed to dissociate to its monomeric subunits, but only after the addition of 7 M guanidine HCl at pH 5.0 (Hartz and Deutsch, 1972). In light of this intrinsic stability of the SOD1 homodimer, it may be unlikely that a monomer of CCS could effectively compete with and displace a monomer of dimeric SOD1 or that SOD1 would dissociate to monomers so that it would be available to encounter a monomer of CCS as depicted in the top two rows of Fig. 18 (Hall *et al.*, 2000). A counterargument proposed is that, although the interface between monomers of SOD1 is very strong in a thermodynamic sense, exchange of monomers between dimeric proteins can be kinetically facile if the two have similar association constants (Rae *et al.*, 2001). In this context, hCCS is 47% identical to hSOD1 and always exists as a dimer in solution and, therefore, the strength of the dimer interfaces in the hCCS and hSOD1 molecules is likely to be similar. While this is true for the human proteins, it is likely not the case for yCCS and ySOD1, as yCCS is observed to be a dimer in solution only when copper-loaded or under elevated concentrations under nonreducing conditions (Schmidt *et al.*, 1999a; Hall *et al.*, 2000).

In an effort to assess the validity of the heterodimer model, Rosenzweig and colleagues sought to isolate a species corresponding to the 43-kDa molecular mass of a yCCS/ySOD1 heterodimeric complex (see Lamb *et al.*, 2000a). They reasoned that if formed, such a complex is likely to be transient to facilitate copper transfer and not hinder the catalytic function of SOD1. To "trap" the putative metal delivery complex, a H48F mutant of ySOD1 that is incapable of binding copper in the copper-binding site was generated by site-directed mutagenesis. Phenylalanine was chosen because it is unlikely to coordinate copper ion and because it fills the cavity vacated by the histidine residue. Apo-yCCS and Cu-yCCS were mixed with wild-type and H48F ySOD1 in the presence and absence of zinc and allowed to incubate overnight. The resulting mixtures were run on an analytical gel filtration column. The peaks from the gel filtration column were subsequently analyzed using dynamic light scattering. If the sample was deemed monodisperse, analytical ultracentrifugation and chemical crosslinking experiments were carried out to try to ascertain the molecular mass of the species in solution. In all cases, in the absence of zinc, the samples were polydisperse. When the analytical ultracentrifugation sedimentation equilibrium experiments were performed on the monodisperse samples, the equilibrium data were fitted directly to a single ideal species model. The primary conclusion drawn from these experiments was that in the presence of zinc, either apo-yCCS or Cu-yCCS could form a heterodimer with H48F ySOD1, but not with wild-type ySOD1. The putative apo-yCCS/H48F-SOD1 heterodimer was assigned a molecular mass of 42,000 ± 3500 while the putative Cu-yCCS/H48F-SOD1 heterodimer was assigned a molecular mass of 38,100 ± 3100 (Lamb *et al.*, 2000a).

Although technically difficult to perform, the results of the experiments described above suggest that a yCCS/ySOD1 heterodimeric species can form *in vitro*. Rosenzweig and colleagues summarize their findings by asserting that copper insertion into SOD1 is likely to occur via the heterodimeric complex for the following reasons: (1) SOD1 activation by Cu-yCCS requires zinc, and zinc promotes heterodimer formation. (2) Mutations at the dimer interfaces of either CCS or SOD1 abrogate SOD1 activation in yeast cells. (3) The heterodimer formed with wild-type SOD1 is less stable than that formed with the H48F mutant, consistent with a transient docked complex that dissociates after copper transfer (See Lamb *et al.*, 2000a; Rosenzweig and O'Halloran, 2000)].

The reasons given above in support of the heterodimer model of CCS/SOD1 association and copper delivery can also be used, however, to support the dimer of dimers model. For example, the presence of zinc would both strengthen the association between monomers in the ySOD1, hSOD1, and hCCS homodimers and cause these molecules to undergo a conformational change in the zinc subloop upon zinc binding that could facilitate a dimer/dimer interaction. Although mutations at the dimer interfaces of either CCS or SOD1 would prevent the formation of CCS/SOD1 heterodimers, they would also prevent homodimerization of these molecules, and thus, this observation cannot distinguish between the two models. Finally, just as a wild-type ySOD1/yCCS complex is suggested to be transient in the above studies, the fact that higher order oligomers were not detected cannot preclude such an interaction. Clearly, a great deal of future effort will be required to elucidate the molecular basis of chaperone-assisted copper transfer in solution *in vitro*.

As shown in the bottom row of Fig. 18, the advantages of the dimer of dimers model over the heterodimer model are several: (1) There is no need to disrupt the very stable SOD1 homodimer. (2) Instead of the 40-Å movement by the CXC motif of domain 3 to the copper-binding site of SOD1, the distance would be much shorter, minimizing the possibility of copper loss. The movement of copper from lower (MXCXXC) to higher (Cu,ZnSOD) affinity sites via domain 3 would occur in a very confined space and in a rapid time due to this proximity. (3) Copper can be delivered to both copper sites of dimeric SOD1 simultaneously. (4) Practically all of the biochemical and genetic data compiled at this writing are completely compatible with a dimer/dimer mechanism of CCS/SOD1 association and copper delivery (Hall *et al.*, 2000). Although the dimer of dimers model seems to possess advantages over the heterodimer model of CCS/SOD1 association and copper delivery, it must be stressed that both are possible. More experiments are necessary to delineate which of the two models, or whether some as yet undefined model, is correct. This promises to be an active area of research in the years to come.

### *D. CCS and Familial Amyotrophic Lateral Sclerosis*

To date, over 90 different single-site mutations in hSOD1 have been individually linked to the inherited (familial) form of the neurodegenerative disease amyotrophic lateral sclerosis (Rosen *et al.*, 1993). Because aberrant copper binding in FALS-associated SOD1 mutants has been proposed to mediate the toxic gain-of-function that underlies FALS, a comprehensive understanding of how SOD1 obtains its copper by hCCS may be particularly important in determining the molecular basis for the cause of this disease (Pardo *et al.*, 1995; Rabizadeh *et al.*, 1995; Ripps *et al.*, Wong *et al.*, 1995; Gurney *et al.*, 1996; Reaume *et al.*, 1996; Culotta *et al.*, 1997; Valentine *et al.*, 1999). Furthermore, since aberrant copper chemistry is a possible mechanism by which FALS-associated mutant SODs gain their toxic function, and since practically all the FALS mutant SOD1s tested so far bind copper *in vivo*, attenuating the incorporation of copper into SOD1 by CCS could be important as a potential therapeutic avenue. Thus, exploring CCS and SOD1 structure and function will not only lend invaluable fundamental information as to the molecular basis for copper transfer to SOD1, it may also lead to the development of inhibitors of CCS that can be used in the treatment of FALS.

## IV. Copper Chaperones for Cytochrome *c* Oxidase

### *A. Genetics and Chemistry*

Cox17, an 8.1-kDa cysteine-rich protein, was the first copper chaperone to be identified. *Saccharomyces cerevisiae* harboring mutations in *cox17* are respiratory deficient, a phenotype resulting from their inability to assemble a functional cytochrome *c* oxidase complex (Glerum *et al.*, 1996a). *cox17* mutant yeast are, however, able to express all the subunits of the cytochrome *c* oxidase complex, indicating that the lesion must lie in a posttranslational step that is essential for assembly of the functional complex in the mitochondrial membrane. Unlike other cytochrome *c* oxidase complex assembly-deficient phenotypes, however, the respiratory deficiency in *cox17* mutants can be overcome with increased levels of exogenous copper, indicating that Cox17 is involved in copper delivery to the mitochondrial compartment (Glerum *et al.*, 1996a). In support of this, cell fractionation and Western blotting experiments indicate that Cox17 is largely localized to the inner membrane space of the mitochondrion, with 60% of the protein accumulating in that part of the organelle and the remaining 40% being localized in the cytosol (Glerum *et al.*, 1996a; Beers *et al.*, 1997). Interestingly, Cox17 does not have a classic mitochondrial import sequence, suggesting that it must be internalized by an alternative mechanism. Its small size, 69 amino acids in yeast and only 62 or 63 amino acids in mammals, may allow it to pass through the mitochondrial pores by diffusion and thus enter into the intermembrane space (Neupert, 1997). It is also possible that Cox17 enters via a pathway similar to that used by cytochrome *c*, the apo form of which is able to reversibly traverse the outer mitochondrial membrane using a process that may or may not be assisted by a protein component (Neupert, 1997).

Cytochrome *c* oxidase, the terminal oxidase in cellular respiration, is an essential enzyme that carries out the four-electron reduction of oxygen to water. The mammalian enzyme complex consists of at least 13 different polypeptide subunits and a number of prosthetic groups including iron, copper, magnesium, zinc, and heme (Saraste, 1990; Tsukihara *et al.*, 1995). Three of the subunits, CoxI, CoxII, and CoxIII, constitute the catalytic core of the enzyme and are encoded in the mitochondrial genome. This catalytic core contains the two copper-binding sites, termed CuA and CuB. The CuA site is a binuclear copper center located in CoxII near the surface of the complex and adjacent to the inner membrane space of the mitochondrion (Tsukihara *et al.*, 1995). The CuB site binds a single copper atom and is located in CoxI closer to the lumen of the mitochondrion, placing it deeper in the complex (Tsukihara *et al.*, 1995). Because the two copper-containing components of the enzyme are synthesized in the mitochondrion itself, copper ions must be brought into the mitochondrion from the cytosol and subsequently inserted into the assembling complex.

The overall assembly of cytochrome *c* oxidase on the inner mitochondrial membrane is controlled by a large number of nuclear encoded genes (Tzagoloff and Dieckmann, 1990). Four of these genes, *sco1, sco2, cox11*, and *cox17*, encode proteins that appear to be involved in copper incorporation into the catalytic core of the enzyme, though precisely which one(s) is (are) responsible for insertion of copper into the complex remains unclear (Horvath *et al.*, 2000). Sco1 is anchored in the inner mitochondrial membrane and is essential for the accumulation of CoxI and CoxII subunits as well as the proper assembly of the cytochrome *c* oxidase complex in the membrane (Schulze and Rodel, 1989; Krummeck and Rodel, 1990; Buchwald *et al.*, 1991). Sco1 is proposed to be directly involved in the insertion of copper into cytochrome *c* oxidase (Glerum *et al.*, 1996b; Petruzzella *et al.*, 1998; Horvath *et al.*, 2000). The Sco1 protein contains a CXXXC motif that, if mutated, leads to respiratory deficiency in yeast (Rentzsch *et al.*, 1999). *sco1*Δ yeast are also respiratory deficient, but unlike *cox17*Δ yeast, the respiratory defects cannot be overcome by increasing the exogenous copper concentrations, suggesting that it functions downstream of Cox17 (Glerum *et al.*, 1996a). Recent evidence suggests that Cox17 may be specifically responsible for providing copper to the CuA site in CoxII, and it may thus also be a copper-trafficking protein, but because it is not freely diffusible, it is not technically classified as a copper chaperone. The precise function of transmembrane protein Sco2 remains unclear, although it is able to rescue the original Cox17 respiratory-deficient mutant in the presence of copper in the growth medium, suggesting that it also plays a role in the activation of cytochrome *c* oxidase (Glerum *et al.*, 1996b). Yeast two-hybrid experiments indicate that Cox17 interacts with both Sco1 and Sco2 (Uetz and Hughes, 2000).

Subsequent to its discovery in yeast, and as illustrated in Fig. 19, homologues of Cox17 have been identified in human (Amaravadi *et al.*, 1997; Horvath *et al.*, 2000; Punter *et al.*, 2000), *Sa. cerevisiae* (yeast) (Johnston *et al.*, 1997), *Sus scrofa* (pig) (Chen *et al.*, 1997), *Mus musculus* (mouse) (Kako *et al.*, 2001), *R. norvegicus* (rat) (Kako *et al.*, 2000), *Gallus gallus* (chicken) (Burnside *et al.*, 2001), *Ophiophagus hannah* (king cobra) (Lee and Zhang, 2000), *Xenopus laevis* (African clawed toad) (Clifton *et al.*, 1999), *Danio rerio* (zebrafish) (Clark *et al.*, 1998), *D. melanogaster* (fruit fly) (Adams *et al.*, 2000), *Caenorhabditis elegans* (nematode) (Wilson *et al.*, 1994), *Necator americanus* (parasitic nematode) (Blaxter *et al.*, 2000b), *Chlamydomonas reinhardtii* (single-celled algae) (La Fontaine and Merchant, 2000a), and *Sc. pombe* (fission yeast) (Wood *et al.*, 2000) (see also Table IV).

The copper-binding capacity of Cox17 remains somewhat ambiguous. The cysteine content of the protein suggests that it should bind 2–3 mol eq of copper; however, the purified protein was found to contain only 0.3 mol/mol protein, suggesting that a substantial amount of copper is lost during purification (Beers *et al.*, 1997). Under oxygen-limiting conditions, the copper-binding capacity of the purified protein was found to be 1.8 mol/mol protein, suggesting that 2 copper atoms bind per protein molecule (Beers *et al.*, 1997). A similar result was obtained using a thrombin-cleaved GST fusion protein (Srinivasan *et al.*, 1998). Analysis of this protein suggested that it had a Gly-Ser dipeptide appended to the N-terminus, while metal analysis indicated that it contained $2 \pm 0.2$ copper atoms per molecule (Srinivasan *et al.*, 1998). Subsequent metal

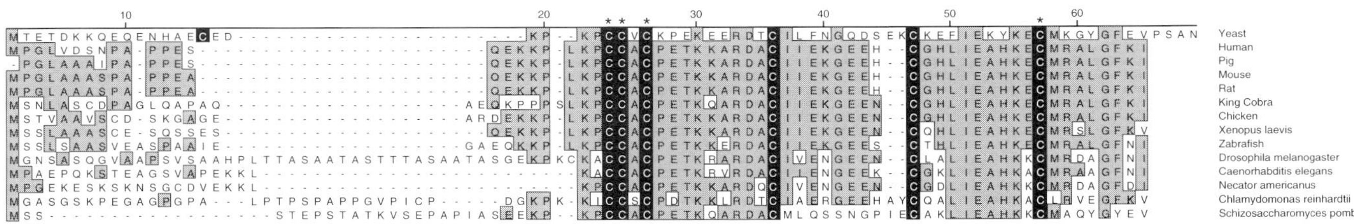


FIG. 19. Multiple sequence alignment of Cox17 proteins using the CLUSTAL method (Higgins and Sharp, 1989). Sequence numbering corresponds to that of the yeast protein. All cysteine residues in *Saccharomyces cerevisiae* Cox17 and the seven invariant cysteine residues across all species are boxed in black. Cysteine residues that if mutated result in respiratory-deficient yeast are labeled with asterisks.

TABLE IV
*Sequence References for the Copper Chaperones for Cytochrome c Oxidase*

| Cox 17 | Reference |
|---|---|
| Yeast (*Saccharomyces cerevisiae*) | Johnston *et al.*, 1997 |
| Human (*Homo sapiens*) | Punter *et al.*, 2000 |
| Pig (*Sus scrofa*) | Chen *et al.*, 1997 |
| Mouse (*Mus musculus*) | Kako *et al.*, 2001 |
| Rat (*Rattus norvegicus*) | Kako *et al.*, 2000 |
| King cobra (*Ophiophagus hannah*) | Lee and Zhang, 2000 |
| Chicken (*Gallus gallus*) | Burnside *et al.*, 2001 |
| *Xenopus laevis* | Clifton *et al.*, 1999 |
| Zebrafish (*Danio rerio*) | Clark *et al.*, 1998 |
| *Drosophila melanogaster* | Adams *et al.*, 2000 |
| *Caenorhabditis elegans* | Wilson *et al.*, 1994 |
| *Necator americanus* | Blaxter *et al.*, 2000b |
| *Chlamydomonas reinhardtii* | La Fontaine and Merchant, 2000a |
| *Schizosaccharomyces pombe* | Wood *et al.*, 2000 |

analysis of the native protein, however, suggests 3 copper atoms per molecule of Cox17 (Heaton *et al.*, 2000). New data demonstrate that the biophysical properties of the GST fusion protein and the native protein differ, which may account for the discrepancy in copper binding (Heaton *et al.*, 2001).

The nature of the bound copper in Cox17 was studied in some detail using the thrombin-cleaved GST fusion protein. The ultraviolet absorption spectrum of the metallated version of this protein reveals acid-labile transitions that are consistent with thiolate ligation of the copper atoms (Srinivasan *et al.*, 1998). The metallated Cox17 protein also shows luminescence around 570 nm, consistent with Cu(I) coordination in a solvent-shielded environment with a trigonal coordination geometry (McCleskey *et al.*, 1996). A similar emission is observed with Cup1 (Fig. 2a) and Ace1, which are also trigonally coordinated Cu(I) thiolate complexes (Andersson and Kurland, 1990). X-ray absorption near-edge spectroscopy also supports a trigonal Cu(I)-thiolate complex (Srinivasan *et al.*, 1998). Interestingly, six of the seven cysteines present in *Sa. cerevisiae* Cox17 are conserved in all the Cox17 homologues that have been sequenced to date (Fig. 19). A model for the binuclear Cu(I) cluster that is similar to the model that has been proposed for the binuclear Cu(I) cluster of human CCS is suggested (Fig. 12b) (Eisses *et al.*, 2000). Despite this potential similarity in copper binding, however, Cox17 and hCCS share little sequence homology.

A mutational analysis revealed which cysteine residues are important for copper binding to Cox17. Substitution of each cysteine residue individually reveals that only three, those forming the CCXC motif in Fig. 19, are essential for function, i.e., metallation of the cytochrome *c* oxidase complex. Replacement of any of these residues with serine results in respiratory-deficient cells (Heaton *et al.*, 2000). However, mutation of any one of these residues does not alter the metal-binding capacity of the protein, suggesting that other non-essential cysteine residues can substitute for copper binding, but cannot substitute for functionality (Heaton *et al.*, 2000). Mutation of any of the remaining, non-CCXC motif cysteine residues in the yeast protein has no effect on cell respiration or metal binding, nor do mutations appear to have a synergistic effect, since substitution of all four non-CCXC motif cysteines with serines had no effect on cytochrome *c* oxidase activity (Heaton *et al.*, 2000).

Interestingly, mutation of Cys-57 to serine has no effect on cytochrome *c* oxidase activity, even though it is the mutation of this residue to tyrosine that led to the initial identification of Cox17 (Glerum *et al.*, 1996a; Heaton *et al.*, 2000). Copper binding to Tyr-57 Cox17 is normal, but the protein fails to accumulate in the mitochondrion, leading to the respiratory defect. Tyrosine is less chemically related to cysteine than serine, and it has been suggested that this mutation may affect docking of the Cox17 protein with Sco1 (Heaton *et al.*, 2000). This suggestion is supported by the fact that overexpression of Sco1 can suppress the respiratory defect in Tyr-57 *cox17* mutants but not *cox17* deletion mutants (Glerum *et al.*, 1996b). Also, a respiratory defect is present when a hemagglutinin epitope tag is added to the C-terminus of Cox17, even though mitochondrial uptake and copper binding are normal, suggesting that the tag may affect the interaction of Cox17 with Sco1 or other protein partners (Heaton *et al.*, 2000).

### *B. Metal Transfer Mechanism*

Although it is unclear precisely how copper is transferred to cytochrome *c* oxidase, the following model for metallation of the complex is beginning to emerge. As indicated in Fig. 1, Cox17 might obtain its copper from the high-flux copper transporter Ctr1, located in the cytoplasmic membrane. At least two, and possibly three, copper ions are incorporated into each Cox17 molecule, and these are coordinated by at least three cysteine residues in an as yet undetermined polycopper cluster (Srinivasan *et al.*, 1998; Heaton *et al.*, 2000, 2001). Lacking a classic mitochondrial import sequence, but being very small in size, Cox17 is thought to enter the intermembrane space in the mitochondrion by diffusion through the porous outer mitochondrial membrane. Once in the inner membrane space where assembly of the cytochrome *c* oxidase complex takes place, copper is most likely transferred to CoxI and CoxII via intermediary proteins. Since the copper-binding sites of Sco1 and Cox17 are both oriented toward the inner membrane space, it has been proposed that Cox17 passes its copper to Sco1, a suggestion that is supported by the yeast two-hybrid data (Uetz and Hughes, 2000). From Sco1, the copper is transferred to the cytochrome *c* oxidase complex, and mutational analysis of the CXXXC Cu(I)-binding site in Sco1 supports this proposal, as mutation of this site abolishes metallation of cytochrome *c* oxidase (Rentzsch *et al.*, 1999). The CuA site of cytochrome *c* oxidase protrudes into the inner mitochondrial membrane space, and there is evidence to suggest that Sco1 directly metallates this site (Tsukihara *et al.*, 1995; Dickinson *et al.*, 2000). Metallation of the CuB site, which is located deeper in the complex toward the lumen side of the membrane, is known to require the presence of Cox11, a protein that is essential for the correct formation of the CuB site in CoxI (Hiser *et al.*, 2000). Cox11 is also essential for the accumulation of CoxI in yeast (Tzagoloff *et al.*, 1990). If Cox11 is the copper delivery protein for the CuB site, Cox17 might also deliver copper to this protein. The transmembrane protein Sco2, which is a homologue of Sco1, also plays a role in activation of cytochrome *c* oxidase complex, although its function remains unclear (Glerum *et al.*, 1996b). It is currently unknown whether Cox17 shuttles back and forth to and from the cytoplasmic membrane or whether it is degraded after performing its copper transfer function.

Considerable progress has been made toward the elucidation of the function of Cox17 since its identification 5 years ago. It has been shown to be a copper-containing protein that is essential for assembly of the cytochrome *c* oxidase complex and is present in both the mitochondria and the cytosol. These features are strongly indicative of a copper chaperone-like function. It appears that Cox17 interacts directly with Sco1 and Sco2, but the delineation of the remainder of the copper-trafficking pathway from the chaperone to the cytochrome *c* oxidase complex remains unclear. It is also unclear exactly how Cox17 binds copper, since the CCXC motif is unique to date. If it binds three copper atoms with only three cysteine residues, the coordination of the metal ions will be extremely interesting, and elucidation of the protein structure will undoubtedly yield new insights into copper transportation in the cell.

## V. Conclusions

Copper, a redox-active transition metal, is both a blessing and a curse for the living cell. The electronic properties that make it useful as a catalytic cofactor also render it quite toxic. The results of the wide variety of studies outlined in this chapter establish the existence of a new class of copper-containing proteins, the copper chaperones. Although they are not enzymes in the traditional sense, the work reviewed here demonstrates that they lower the activation barrier for Cu(I) transfer to specific protein-binding sites and that they do so in the presence of a cytoplasm with an impressive capacity for copper chelation. The target proteins are enzymes that exist in several distinct thermodynamic compartments. Over the past 5 years, a great deal of progress has been made toward understanding the genetic, biochemical, and structural aspects of the copper chaperone protein function. Deletion of a single copper chaperone appears to impair copper delivery only to its specific target without affecting the remaining copper proteins in the cell, highlighting the specificity of the copper delivery process. Important strides have been made in the determination of the 3-D structures of the individual copper chaperones of the Atx1 and CCS families, as well as their cognate target proteins. These structures provide the platform upon which a detailed understanding of the molecular determinants of target recognition and metal transfer will be assembled. The next big challenge in this regard is to elucidate the 3-D structures of the copper chaperone proteins in complex with their target molecules. This will be especially important in understanding the mechanistic aspects of the copper chaperones for superoxide dismutase, as most fundamental mechanistic issues remain unresolved. So far, there is little information as to where the copper chaperone proteins themselves become loaded with copper. This aspect of copper trafficking promises to be an area of active investigation in the very near future. Finally, the possibility that other transition metals such as nickel, iron, molybdenum, and manganese are also shuttled by metallochaperones may pave the way for intensified research in this area. It will be intriguing to see where we stand in terms of our knowledge of metallochaperones and the trafficking of transition metal ions 5 years from now.

## Acknowledgments

We thank Edie Gralla and Aram Nersissian for helpful discussions and Les Hall for help in assembling the final manuscript. Funding support from National Institutes of Health Grant NS39112, The Robert A. Welch Foundation Grant AQ-1399, and the ALS Association is gratefully acknowledged.

## References

Adams, M. D., Celniker, S. E., Gibbs, R. A., Rubin, G. M., and Venter, C. J. (2000). Unpublished. Genbank Accession No. AAF48421.

Adman, E. T. (1991). *Adv. Protein Chem.* **42**, 145–197.
Amaravadi, R., Glerum, D. M., and Tzagoloff, A. (1997). *Hum. Genet.* **99**, 329–333.
Amasino, R. M., Crowell, D., Mira, H., and Penarrubia, L. (1999). Unpublished. Genbank Accession No. AF198627.
Andersson, S. G., and Kurland, C. G. (1990). *Microbiol. Rev.* **54**, 198–210.
Arnesano, F., Banci, L., Bertini, I., Huffman, D. L., and O'Halloran, T. V. (2001). *Biochemistry* **40**, 1528–1539.
Askwith, C., Eide, D., Van Ho, A., Bernard, P. S., Li, L., Davis-Kaplan, S., Sipe, D. M., and Kaplan, J. (1994). *Cell* **76**, 403–410.
Banci, L., Bertini, I., Ciofi-Baffoni, S., Huffman, D. L., and O'Halloran, T. V. (2001). *J. Biol. Chem.* **276**, 8415–8426.
Barra, D., Martini, F., Bannister, J. V., Schinina, M. E., Rotilio, G., Bannister, W. H., and Bossa, F. (1980). *FEBS Lett.* **120**, 53–56.
Bartnikas, T. B., Waggoner, D. J., Casareno, R. L., Gaedigk, R., White, R. A., and Gitlin, J. D. (2000). *Mamm. Genome* **11**, 409–411.
Beers, J., Glerum, D. M., and Tzagoloff, A. (1997). *J. Biol. Chem.* **272**, 33191–33196.
Bermingham-McDonogh, O., Gralla, E. B., and Valentine, J. S. (1988). *Proc. Natl. Acad. Sci. USA* **85**, 4789–4793.
Bertini, I., Mangani, S., and Viezzoli, M. S. (1998). *Adv. Inorg. Chem.* **45**, 127–251.
Blackburn, N. J., Ralle, M., Hassett, R., and Kosman, D. J. (2000). *Biochemistry* **39**, 2316–2324.
Blaxter, M. L., Parkinson, J., Whitton, C., Daub, J., Guiliano, D., Hall, N., Quayle, M., and Barrell, B. (2000a). Unpublished. Genbank Accession No. BF060227.
Blaxter, M. L., Parkinson, J., Whitton, C., Daub, J., Guiliano, D., Hall, N., Quayle, M., and Barrell, B. (2000b). Unpublished. Genbank Accession No. BG467702.
Brown, N. L., Camakaris, J., Lee, B. T., Williams, T., Morby, A. P., Parkhill, J., and Rouch, D. A. (1991). *J. Cell. Biochem.* **46**, 106–114.
Buchwald, P., Krummeck, G., and Rodel, G. (1991). *Mol. Gen. Genet.* **229**, 413–420.
Bukau, B., Deuerling, E., Pfund, C., and Craig, E. A. (2000). *Cell* **101**, 119–122.
Bull, P. C., Thomas, G. R., Rommens, J. M., Forbes, J. R., and Cox, D. W. (1993). *Nat. Genet.* **5**, 327–337.
Burnside, J., Morgan, R. W., and Cogburn, L. A. (2001). Unpublished. Genbank Accession No. BG713534.
Capaldi, R. A. (1990). *Annu. Rev. Biochem.* **59**, 569–596.
Casareno, R. L., Waggoner, D., and Gitlin, J. D. (1998). *J. Biol. Chem.* **273**, 23625–23628.
Chang, E. C., Crawford, B. F., Hong, Z., Bilinski, T., and Kosman, D. J. (1991). *J. Biol. Chem.* **266**, 4417–4424.
Chang, E. C., and Kosman, D. J. (1990). *J. Bacteriol.* **172**, 1840–1845.
Chen, Z. W., Bergman, T., Ostenson, C. G., Efendic, S., Mutt, V., and Jornvall, H. (1997). *Eur. J. Biochem.* **249**, 518–522.
Clark, M., Johnson, S. L., Lehrach, H., Lee, R., Li, F., Marra, M., Eddy, S., Hillier, L., Kucaba, T., Martin, J., Beck, C., Wylie, T., Underwood, K., Steptoe, M., Theising, B., Allen, M., Bowers, Y., Person, B., Swaller, T., Gibbons, M., Pape, D., Harvey, N., Schurk, R., Ritter, E., Kohn, S., Shin, T., Jackson, Y., Cardenas, M., McCann, R., Waterston, R., and Wilson, R. (1998). Unpublished. Genbank Accession No. BE201771.
Clifton, S., Johnson, S. L., Blumberg, B., Song, J., Hillier, L., Pape, D., Martin, J., Wylie, T., Underwood, K., Theising, B., Bowers, Y., Person, B., Gibbons, M., Harvey, N., Ritter, E., Jackson, Y., McCann, R., Waterston, R., and Wilson, R. (1999). Unpublished. Genbank Accession No. BF613133.
Cobine, P., Wickramasinghe, W. A., Harrison, M. D., Weber, T., Solioz, M., and Dameron, C. T. (1999). *FEBS Lett.* **445**, 27–30.

Corson, L. B., Strain, J. J., Culotta, V. C., and Cleveland, D. W. (1998). *Proc. Natl. Acad. Sci. USA* **95**, 6361–6366.

Culotta, V. C., Howard, W. R., and Liu, X. F. (1994). *J. Biol. Chem.* **272**, 23469–23472.

Culotta, V. C., Klomp, L. W., Strain, J., Casareno, R. L., Krems, B., and Gitlin, J. D. (1997). *J. Biol. Chem.* **272**, 23469–23472.

Dameron, C. T., Winge, D. R., George, G. N., Sansone, M., Hu, S., and Hamer, D. (1991). *Proc. Natl. Acad. Sci. USA* **88**, 6127–6131.

Dancis, A., Haile, D., Yuan, D. S., and Klausner, R. D. (1994a). *J. Biol. Chem.* **269**, 25660–25667.

Dancis, A., Yuan, D. S., Haile, D., Askwith, C., Eide, D., Moehle, C., Kaplan, J., and Klausner, R. D. (1994b). *Cell* **76**, 393–402.

Davies, K. J. (1995). *Biochem. Soc. Symp.* **61**, 1–31.

de Montigny, J., Spehner, C., Souciet, J., Tekaia, F., Dujon, B., Wincker, P., Artiguenave, F., and Potier, S. (2000). *FEBS Lett.* **487**, 87–90.

Deng, H. X., Hentati, A., Tainer, J. A., Iqbal, Z., Cayabyab, A., Hung, W. Y., Getzoff, E. D., Hu, P., Herzfeldt, B., Roos, R. P., *et al.* (1993). *Science* **261**, 1047–1051.

Devlin, K., Churcher, C. M., Barrell, B. G., Rajandream, M. A., and Walsh, S. V. (1996). Unpublished. Genbank Accession No. Z70043. SwissProt Accession No. Q10357.

Dickinson, E. K., Adams, D. L., Schon, E. A., and Glerum, D. M. (2000). *J. Biol. Chem.* **275**, 26780–26785.

Djinovic, K., Gatti, G., Coda, A., Antolini, L., Pelosi, G., Desideri, A., Falconi, M., Marmocchi, F., Rotilio, G., and Bolognesi, M. (1992). *J. Mol. Biol.* **225**, 791–809.

Djinovic-Carugo, K., Battistoni, A., Carri, M. T., Polticelli, F., Desideri, A., Rotilio, G., Coda, A., Wilson, K. S., and Bolognesi, M. (1996). *Acta Crystallogr. D Sect.* **52**, 176–188.

Ecker, D. J., Butt, T. R., Sternberg, E. J., Neeper, M. P., Debouck, C., Gorman, J. A., and Crooke, S. T. (1986). *J. Biol. Chem.* **261**, 16895–16900.

Eide, D. J. (1998). *Annu. Rev. Nutr.* **18**, 441–469.

Eisses, J. F., Stasser, J. P., Ralle, M., Kaplan, J. H., and Blackburn, N. J. (2000). *Biochemistry* **39**, 7337–7342.

Endo, T., Fujii, T., Sato, K., Taniguchi, N., and Fujii, J. (2000). *Biochem. Biophys. Res. Commun.* **276**, 999–1004.

Esnouf, R. M. (1999). *Acta Crystallogr.* Sect. D **55**, 938–940.

Falconi, M., Iovino, M., and Desideri, A. (1999). *Structure Fold. Des.* **7**, 903–908.

Fan, J., and Brindley, P. J. (1998). Unpublished. Genbank Accession No.AI168925.

Ferretti, J. J., McShan, W. M., Adjic, D., Savic, D., Savic, G., Lyon, K., Primeaux, C., Sezate, S. S., Surorov, A. N., Kenton, S., Lai, H., Lin, S., Qian, Y., Jia, H. G., Najar, F. Z., Ren, Q., Zhu, H., Song, L., White, J., Yuan, X., Clifton, S. W., Roe, B. A., and McLaughlin, R. E. (2001). Unpublished. Genbank Accession No.AE006600.

Fielden, E. M., and Rotilio, G. (1984). "Copper Proteins and Copper Enzymes"(Lontie, R., Ed.), Vol. II, pp. 27–61. CRC Press, Boca Raton, FL.

Forbes, J. R., Hsi, G., and Cox, D. W. (1999). *J. Biol. Chem.* **274**, 12408–12413.

Fridovich, I. (1978). *Science* **201**, 875–880.

Fridovich, I. (1985). "CRC Handbook of Methods for Oxygen Radical Research"(Greenwald, R. A., Ed.), Vol. 1, pp. 213–215. CRC Press, Boca Raton, FL.

Fridovich, I. (1989). *J. Biol. Chem.* **264**, 7761–7764.

Fu, D., Beeler, T. J., and Dunn, T. M. (1995). *Yeast* **11**, 283–292.

Furst, P., Hu, S., Hackett, R., and Hamer, D. (1988). *Cell* **55**, 705–717.

Gamonet, F., and Lauquin, G. J. (1998). *Eur. J. Biochem.* **251**, 716–723.

Georgatsou, E., Mavrogiannis, L. A., Fragiadakis, G. S., and Alexandraki, D. (1997). *J. Biol. Chem.* **272**, 13786–13792.

Getzoff, E. D., Cabelli, D. E., Fisher, C. L., Parge, H. E., Viezzoli, M. S., Banci, L., and Hallewell, R. A. (1992). *Nature* **358**, 347–351.

Gitschier, J., Moffat, B., Reilly, D., Wood, W. I., and Fairbrother, W. J. (1998). *Nat. Struct. Biol.* **5**, 47–54.
Glerum, D. M., Shtanko, A., and Tzagoloff, A. (1996a). *J. Biol. Chem.* **271**, 14504–14509.
Glerum, D. M., Shtanko, A., and Tzagoloff, A. (1996b). *J. Biol. Chem.* **271**, 20531–20535.
Graden, J. A., and Winge, D. R. (1997). *Proc. Natl. Acad. Sci. USA* **94**, 5550–5555.
Gralla, E. B., Thiele, D. J., Silar, P., and Valentine, J. S. (1991). *Proc. Natl. Acad. Sci. USA* **88**, 8558–8562.
Gralla, E. B., and Valentine, J. S. (1991). *J. Bacteriol.* **173**, 5918–5920.
Gross, C., Kelleher, M., Iyer, V. R., Brown, P. O., and Winge, D. R. (2000). *J. Biol. Chem.* **275**, 32310–32316.
Gurney, M. E., Cutting, F. B., Zhai, P., Andrus, P. K., and Hall, E. D. (1996). *Pathol. Biol.* **44**, 51–56.
Hall, L. T., Sanchez, R. J., Holloway, S. P., Zhu, H., Stine, J. E., Lyons, T. J., Demeler, B., Schirf, V., Hansen, J. C., Nersissian, A. M., Valentine, J. S., and Hart, P. J. (2000). *Biochemistry* **39**, 3611–3623.
Halliwell, B., and Gutteridge, J. M. (1984). *Lancet* **2**, 1095.
Halliwell, B., and Gutteridge, J. M. (1985). *Mol. Aspects Med.* **8**, 89–193.
Hamer, D. H., Thiele, D. J., and Lemontt, J. E. (1985). *Science* **228**, 685–690.
Hamza, I., Klomp, L. W., Gaedigk, R., White, R. A., and Gitlin, J. D. (2000). *Genomics* **63**, 294–297.
Hamza, I., Schaefer, M., Klomp, L. W., and Gitlin, J. D. (1999). *Proc. Natl. Acad. Sci. USA* **96**, 13363–13368.
Harrison, M. D., Jones, C. E., Solioz, M., and Dameron, C. T. (2000). *Trends Biochem. Sci.* **25**, 29–32.
Hart, P. J., Balbirnie, M. M., Ogihara, N. L., Nersissian, A. M., Weiss, M. S., Valentine, J. S., and Eisenberg, D. (1999). *Biochemistry* **38**, 2167–2178.
Hart, P. J., Nersissian, A. M., Herrmann, R. G., Nalbandyan, R. M., Valentine, J. S., and Eisenberg, D. (1996). *Protein Sci.* **5**, 2175–2183.
Hartz, J. W., and Deutsch, H. F. (1972). *J. Biol. Chem.* **247**, 7043–7050.
Hassett, R., and Kosman, D. J. (1995). *J. Biol. Chem.* **270**, 128–134.
Heaton, D., Nittis, T., Srinivasan, C., and Winge, D. R. (2000). *J. Biol. Chem.* **275**, 37582–37587.
Heaton, D. N., George, G. N., Garrison, G., and Winge, D. R. (2001). *Biochemistry* **40**, 743–751.
Higgins, D. G., and Sharp, P. M. (1989). *CABIOS* **5**, 151–153.
Himelblau, E., Mira, H., Lin, S. J., Culotta, V. C., Penarrubia, L., and Amasino, R. M. (1998). *Plant Physiol.* **117**, 1227–1234.
Hiromura, M., Chino, H., Sonoda, T., and Sakurai, H. (2000). *Biochem. Biophys. Res. Commun.* **275**, 394–400.
Hiser, L., Di Valentin, M., Hamer, A. G., and Hosler, J. P. (2000). *J. Biol. Chem.* **275**, 619–623.
Horecka, J., Kinsey, P. T., and Sprague, G. F., Jr. (1995). *Gene* **162**, 87–92.
Horvath, R., Lochmuller, H., Stucka, R., Yao, J., Shoubridge, E. A., Kim, S. H., Gerbitz, K. D., and Jaksch, M. (2000). *Biochem. Biophys. Res. Commun.* **276**, 530–533.
Hu, S., Furst, P., and Hamer, D. (1990). *New Biologist* **2**, 544–555.
Huang, X., Cuajungco, M. P., Atwood, C. S., Hartshorn, M. A., Tyndall, J. D., Hanson, G. R., Stokes, K. C., Leopold, M., Multhaup, G., Goldstein, L. E., Scarpa, R. C., Saunders, A. J., Lim, J., Moir, R. D., Glabe, C., Bowden, E. F., Masters, C. L., Fairlie, D. P., Tanzi, R. E., and Bush, A. I. (1999). *J. Biol. Chem.* **274**, 37111–37116.
Hubbard, T. J., Murzin, A. G., Brenner, S. E., and Chothia, C. (1997). *Nucleic Acids Res.* **25**, 236–239.
Huffman, D. L., and O'Halloran, T. V. (2000). *J. Biol. Chem.* **275**, 18611–18614.

Hung, I. H., Casareno, R. L., Labesse, G., Mathews, F. S., and Gitlin, J. D. (1998). *J. Biol. Chem.* **273**, 1749–1754.

Hung, I. H., Suzuki, M., Yamaguchi, Y., Yuan, D. S., Klausner, R. D., and Gitlin, J. D. (1997). *J. Biol. Chem.* **272**, 21461–21466.

Jensen, L. T., Howard, W. R., Strain, J. J., Winge, D. R., and Culotta, V. C. (1996). *J. Biol. Chem.* **271**, 18514–18519.

Jensen, L. T., and Winge, D. R. (1998). *EMBO J.* **17**, 5400–5408.

Johnston, M., Hillier, L., Riles, L., Albermann, K., Andre, B., Ansorge, W., Benes, V., Bruckner, M., Delius, H., Dubois, E., Dusterhoft, A., Entian, K. D., Floeth, M., Goffeau, A., Hebling, U., Heumann, K., Heuss-Neitzel, D., Hilbert, H., Hilger, F., Kleine, K., Kotter, P., Louis, E. J., Messenguy, F., Mewes, H. W., Hoheisel, J. D., *et al.* (1997). *Nature* **387**, 87–90.

Joshi, A., Serpe, M., and Kosman, D. J. (1999). *J. Biol. Chem.* **274**, 218–226.

Jungmann, J., Reins, H. A., Lee, J., Romeo, A., Hassett, R., Kosman, D., and Jentsch, S. (1993). *EMBO J.* **12**, 5051–5056.

Kako, K., Takahashi, Y., and Munekata, E. (2001). Unpublished. Genbank Accession No. AB047323.

Kako, K., Tsumori, K., Ohmasa, Y., Takahashi, Y., and Munekata, E. (2000). *Eur. J. Biochem.* **267**, 6699–6707.

Kampfenkel, K., Kushnir, S., Babiychuk, E., Inze, D., and Van Montagu, M. (1995). *J. Biol. Chem.* **270**, 28479–28486.

Karin, M. (1985). *Cell* **41**, 9–10.

Karlin, S., Zhu, Z. Y., and Karlin, K. D. (1998). *Biochemistry* **37**, 17726–17734.

Kawashima, T., Yamamoto, Y., Aramaki, H., Nunoshiba, T., Kawamoto, T., Watanabe, K., Yamazaki, M., Kanehori, K., Amano, N., Ohya, Y., Makino, K., and Suzuki, M. (1999). *Proc. Jpn. Acad.* **75**, 213–218.

Kelner, G. S., Lee, M., Clark, M. E., Maciejewski, D., McGrath, D., Rabizadeh, S., Lyons, T., Bredesen, D., Jenner, P., and Maki, R. A. (2000). *J. Biol. Chem.* **275**, 580–584.

Kirby, K., and Phillips, J. P. (2001). Unpublished. Genbank Accession No. AAK07691.

Kitagawa, Y., Tanaka, N., Hata, Y., Kusunoki, M., Lee, G. P., Katsube, Y., Asada, K., Aibara, S., and Morita, Y. (1991). *J. Biochem.* **109**, 477–485.

Kliebenstein, D. J., Saracco, S. A., and Last, R. L. (1998). Unpublished. Genbank Accession No. AF061517.

Klomp, L. W., Lin, S. J., Yuan, D. S., Klausner, R. D., Culotta, V. C., and Gitlin, J. D. (1997). *J. Biol. Chem.* **272**, 9221–9226.

Knight, S. A., Labbe, S., Kwon, L. F., Kosman, D. J., and Thiele, D. J. (1996). *Genes Dev.* **10**, 1917–1929.

Knowles, P. F., and Yadav, K. D. S. (1984). "Copper Proteins and Copper Enzymes"(Lontie, R., ed.), Vol. II, pp. 103–130. CRC Press, Boca Raton, FL.

Kraulis, P. J. (1991). *J. Appl. Crystallogr.* **24**, 946–950.

Krummeck, G., and Rodel, G. (1990). *Curr. Genet.* **18**, 13–15.

La Fontaine, S., and Merchant, S. (2000a). Unpublished. Genbank Accession No. AF280543.

La Fontaine, S., and Merchant, S. (2000b). Unpublished. Genebank Accession No. AF280056.

Labbe, S., and Thiele, D. J. (1999). *Trends Microbiol.* **7**, 500–505.

Labbe, S., Zhu, Z., and Thiele, D. J. (1997). *J. Biol. Chem.* **272**, 15951–15958.

Lamb, A. L., Torres, A. S., O'Halloran, T. V., and Rosenzweig, A. C. (2000a). *Biochemistry* **39**, 14720–14727.

Lamb, A. L., Wernimont, A. K., Pufahl, R. A., Culotta, V. C., O'Halloran, T. V., and Rosenzweig, A. C. (1999). *Nat. Struct. Biol.* **6**, 724–729.

Lamb, A. L., Wernimont, A. K., Pufahl, R. A., O'Halloran, T. V., and Rosenzweig, A. C. (2000b). *Biochemistry* **39**, 1589–1595.

Larin, D., Mekios, C., Das, K., Ross, B., Yang, A. S., and Gilliam, T. C. (1999). *J. Biol. Chem.* **274**, 28497–28504.
Lee, W., and Zhang, Y. (2000). Unpublished. Genbank Accession No.AF297035.
Lin, C. M., and Kosman, D. J. (1990). *J. Biol. Chem.* **265**, 9194–9200.
Lin, S. J. (1997). "Roles of the *Saccharomyces cerevisiae ATX1* and *ATX2* Genes in Oxygen Radical and Metal Ion Homeostasis." Ph.D. Thesis, John Hopkins University School of Public Health, Baltimore, MD.
Lin, S. J., and Culotta, V. C. (1995). *Proc. Natl. Acad. Sci. USA* **92**, 3784–3788.
Lin, S. J., Pufahl, R. A., Dancis, A., O'Halloran, T. V., and Culotta, V. C. (1997). *J. Biol. Chem.* **272**, 9215–9220.
Linder, M. C., and Goode, C. A. (1991). "Biochemistry of Copper." Plenum, New York.
Lippard, S. J. (1999). *Science* **284**, 748–749.
Ljones, T., and Skotland, T. (1984). "Copper Proteins and Copper Enzymes" (Lontie, R., Ed.), Vol. II, pp. 131–158. CRC Press, Boca Raton, FL.
Lockhart, P. J., and Mercer, J. F. (2000). *Biochim. Biophys. Acta* **1490**, 11–20.
McCarter, J., Clifton, S., Chiapelli, B., Pape, D., Martin, J., Wylie, T., Dante, M., Marra, M., Hillier, L., Kucaba, T., Theising, B., Bowers, Y., Gibbons, M., Ritter, E., Bennett, J., Franklin, C., Tsagareishvili, R., Ronko, I., Kennedy, S., Maguire, L., Beck, C., Underwood, K., Steptoe, M., Allen, M., Person, B., Swaller, T., Harvey, N., Schurk, R., Kohn, S., Shin, T., Jackson, Y., Cardenas, M., McCann, R., Waterston, R., and Wilson, R. (1999). Unpublished. Genbank Accession No. BF014007.
McCleskey, T. M., Mizoguchi, T. J., Richards, J. H., and Gray, H. B. (1996). *Inorg. Chem.* **35**, 3434–3435.
Miller, R., Gallo, S. M., Khalak, H. G., and Weeks, C. M. (1994). *J. Appl. Crystallogr.* **27**, 613–621.
Mira, H., and Penarrubia, L. (1999). Unpublished. Genbank Accession No. AF198626.
Misra, T. K., Brown, N. L., Fritzinger, D. C., Pridmore, R. D., Barnes, W. M., Haberstroh, L., and Silver, S. (1984). *Proc. Natl. Acad. Sci. USA* **81**, 5975–5979.
Morby, A. P., Hobman, J. L., and Brown, N. L. (1995). *Mol. Microbiol.* **17**, 25–35.
Nanji, M. S., and Cox, D. W. (1999). *Genomics* **62**, 108–112.
Nelson, M. A., Kang, S., Braun, E. L., Crawford, M. E., Dolan, P. L., Leonard, P. M., Mitchell, J., Armijo, A. M., Bean, L., Blueyes, E., Cushing, T., Errett, A., Fleharty, M., Gorman, M., Judson, K., Miller, R., Ortega, J., Pavlova, I., Perea, J., Todisco, S., Trujillo, R., Valentine, J., Wells, A., Werner-Washburne, M., Natvig, D. O., *et al.* (1997). *Fungal Genet. Biol.* **21**, 348–363.
Nersissian, A. M., and Valentine, J. S. (1999). Unpublished. Genbank Accession No. AAD12307.
Nersissian, A. M., and Valentine, J. S. (2000). Unpublished. Genbank Accession No. AF329816.
Neupert, W. (1997). *Annu. Rev. Biochem.* **66**, 863–917.
Nicholls, A., Sharp, K. A., and Honig, B. (1991). *Proteins* **11**, 281–296.
Odermatt, A., and Solioz, M. (1995). *J. Biol. Chem.* **270**, 4349–4354.
Ogihara, N. L., Parge, H. E., Hart, P. J., Weiss, M. S., Goto, J. J., Crane, B. R., Tsang, J., Slater, K., Roe, J. A., Valentine, J. S., Eisenberg, D., and Tainer, J. A. (1996). *Biochemistry* **35**, 2316–2321.
O'Halloran, T. V., and Culotta, V. C. (2000). *J. Biol. Chem.* **275**, 25057–25060.
Ooi, C. E., Rabinovich, E., Dancis, A., Bonifacino, J. S., and Klausner, R. D. (1996). *EMBO J.* **15**, 3515–3523.
Pardo, C. A., Xu, Z., Borchelt, D. R., Price, D. L., Sisodia, S. S., and Cleveland, D. W. (1995). *Proc. Natl. Acad. Sci. USA* **92**, 954–958.
Parge, H. E., Hallewell, R. A., and Tainer, J. A. (1992). *Proc. Natl. Acad. Sci. USA* **89**, 6109–6113.

Payne, A. S., Kelly, E. J., and Gitlin, J. D. (1998). *Proc. Natl. Acad. Sci. USA* **95**, 10854–10859.
Pena, M. M., Lee, J., and Thiele, D. J. (1999). *J. Nutr.* **129**, 1251–1260.
Peterson, C. W., Narula, S. S., and Armitage, I. M. (1996). *FEBS Lett.* **379**, 85–93.
Petris, M. J., Mercer, J. F., Culvenor, J. G., Lockhart, P., Gleeson, P. A., and Camakaris, J. (1996). *EMBO J.* **15**, 6084–6095.
Petruzzella, V., Tiranti, V., Fernandez, P., Ianna, P., Carrozzo, R., and Zeviani, M. (1998). *Genomics* **54**, 494–504.
Pickering, I. J., George, G. N., Dameron, C. T., Kurz, B., Winge, D. R., and Dance, I. G. (1993). *J. Am. Chem. Soc.* **115**, 9498–9505.
Portnoy, M. E., Rosenzweig, A. C., Rae, T., Huffman, D. L., O'Halloran, T. V., and Culotta, V. C. (1999). *J. Biol. Chem.* **274**, 15041–15045.
Poulos, T. L. (1999). *Nat. Struct. Biol.* **6**, 709–711.
Poulsen, C., and Poedenphant, L. (2000). Unpublished. Genbank Accession No. AW428839.
POV-Team (1997). Persistence of Vision Ray Tracer: POV-Team. Available at http://www.povray.org.
Pufahl, R. A., Singer, C. P., Peariso, K. L., Lin, S. J., Schmidt, P. J., Fahrni, C. J., Culotta, V. C., Penner-Hahn, J. E., and O'Halloran, T. V. (1997). *Science* **278**, 853–856.
Punter, F. A., Adams, D. L., and Glerum, D. M. (2000). *Hum. Genet.* **107**, 69–74.
Qian, H., Sahlman, L., Eriksson, P. O., Hambraeus, C., Edlund, U., and Sethson, I. (1998). *Biochemistry*, **37**, 9316–9322.
Rabizadeh, S., Gralla, E. B., Borchelt, D. R., Gwinn, R., Valentine, J. S., Sisodia, S., Wong, P., Lee, M., Hahn, H., and Bredesen, D. E. (1995). *Proc. Natl. Acad. Sci. USA* **92**, 3024–3028.
Radisky, D., and Kaplan, J. (1999). *J. Biol. Chem.* **274**, 4481–4484.
Radisky, D. C., Snyder, W. B., Emr, S. D., and Kaplan, J. (1997). *Proc. Natl. Acad. Sci. USA* **94**, 5662–5666.
Rae, T. D., Schmidt, P. J., Pufahl, R. A., Culotta, V. C., and O'Halloran, T. V. (1999). *Science* **284**, 805–808.
Rae, T. D., Torres, A. S., Pufahl, R. A., and O'Halloran, T. V. (2001). *J. Biol. Chem.* **276**, 5166–5176.
Reaume, A. G., Elliott, J. L., Hoffman, E. K., Kowall, N. W., Ferrante, R. J., Siwek, D. F., Wilcox, H. M., Flood, D. G., Beal, M. F., Brown, R. H., Jr., Scott, R. W., and Snider, W. D. (1996). *Nat. Genet.* **13**, 43–47.
Rentzsch, A., Krummeck-Weiss, G., Hofer, A., Bartuschka, A., Ostermann, K., and Rodel, G. (1999). *Curr. Genet.* **35**, 103–108.
Richter, C., Gogvadze, V., Laffranchi, R., Schlapbach, R., Schweizer, M., Suter, M., Walter, P., and Yaffee, M. (1995). *Biochim. Biophys. Acta* **1271**, 67–74.
Ripps, M. E., Huntley, G. W., Hof, P. R., Morrison, J. H., and Gordon, J. W. (1995). *Proc. Natl. Acad. Sci. USA* **92**, 689–693.
Robb, D. A. (1984). "Copper Proteins and Copper Enzymes" (Lontie, R., Ed.), Vol. II, pp. 207–240. CRC Press, Boca Raton, FL.
Robbins, A. H., McRee, D. E., Williamson, M., Collett, S. A., Xuong, N. H., Furey, W. F., Wang, B. C., and Stout, C. D. (1991). *J. Mol. Biol.* **221**, 1269–1293.
Rosen, D. R., Siddique, T., Patterson, D., Figlewicz, D. A., Sapp, P., Hentati, A., Donaldson, D., Goto, J., O'Regan, J. P., Deng, H. X., *et al.* (1993). *Nature* **362**, 59–62.
Rosenzweig, A. C., Huffman, D. L., Hou, M. Y., Wernimont, A. K., Pufahl, R. A., and O'Halloran, T. V. (1999). *Structure Fold. Des.* **7**, 605–617.
Rosenzweig, A. C., and O'Halloran, T. V. (2000). *Curr. Opin. Chem. Biol.* **4**, 140–147.
Rothstein, J. D., Dykes-Hoberg, M., Corson, L. B., Becker, M., Cleveland, D. W., Price, D. L., Culotta, V. C., and Wong, P. C. (1999). *J. Neurochem.* **72**, 422–429.
Ryden, L. (1984). "Copper Proteins and Copper Enzymes" (Lontie, R., Ed.), Vol. III, pp. 37–100. CRC Press, Boca Raton, FL.

Ryden, L. (1988). *Prog. Clin. Biol. Res.* **274**, 349–366.

Sahlman, L., and Skarfstad, E. G. (1993). *Biochem. Biophys. Res. Commun.* **196**, 583–588.

Santoro, N., and Thiele, D. J. (1997). Yeast stress responses. *In* "Molecular Biology Intelligence Unit (Unnumbered)" (Hohmann S., and Mager, W. H., Eds.), pp. 171–211. Chapman & Hall R. G. Landes, New York Austin.

Saraste, M. (1990). *Q. Rev. Biophys.* **23**, 331–366.

Satow, Y., Cohen, G. H., Padlan, E. A., and Davies, D. R. (1986). *J. Mol. Biol.* **190**, 593–604.

Schmidt, P. J., Kunst, C., and Culotta, V. C. (2000). *J. Biol. Chem.* **275**, 33771–33776.

Schmidt, P. J., Rae, T. D., Pufahl, R. A., Hamma, T., Strain, J., O'Halloran, T. V., and Culotta, V. C. (1999a). *J. Biol. Chem.* **274**, 23719–23725.

Schmidt, P. J., Ramos-Gomez, M., and Culotta, V. C. (1999b). *J. Biol. Chem.* **274**, 36952–36956.

Schulze, M., and Rodel, G. (1989). *Mol. Gen. Genet.* **216**, 37–43.

Serpe, M., Joshi, A., and Kosman, D. J. (1999). *J. Biol. Chem.* **274**, 29211–29219.

Solomon, E. I., and Lowery, M. D. (1993). *Science* **259**, 1575–1581.

Souciet, J., Aigle, M., Artiguenave, F., Blandin, G., Bolotin-Fukuhara, M., Bon, E., Brottier, P., Casaregola, S., de Montigny, J., Dujon, B., Durrens, P., Gaillardin, C., Lepingle, A., Llorente, B., Malpertuy, A., Neuveglise, C., Ozier-Kalogeropoulos, O., Potier, S., Saurin, W., Tekaia, F., Toffano-Nioche, C., Wesolowski-Louvel, M., Wincker, P., and Weissenbach, J. (2000). *FEBS Lett.* **487**, 3–12.

Srinivasan, C., Posewitz, M. C., George, G. N., and Winge, D. R. (1998). *Biochemistry* **37**, 7572–7577.

Steele, R. A., and Opella, S. J. (1997). *Biochemistry* **36**, 6885–6895.

Strausak, D., and Solioz, M. (1997). *J. Biol. Chem.* **272**, 8932–8936.

Szczypka, M. S., and Thiele, D. J. (1989). *Mol. Cell. Biol.* **9**, 421–429.

Tainer, J. A., Getzoff, E. D., Beem, K. M., Richardson, J. S., and Richardson, D. C. (1982). *J. Mol. Biol.* **160**, 181–217.

Thiele, D. J. (1988). *Mol. Cell. Biol.* **8**, 2745–2752.

Thiele, D. J. (1992). *Nucleic Acids Res.* **20**, 1183–1191.

Tietze, F. (1969). *Anal. Biochem.* **27**, 502–522.

Tsukihara, T., Aoyama, H., Yamashita, E., Tomizaki, T., Yamaguchi, H., Shinzawa-Itoh, K., Nakashima, R., Yaono, R., and Yoshikawa, S. (1995). *Science* **269**, 1069–1074.

Tzagoloff, A., Capitanio, N., Nobrega, M. P., and Gatti, D. (1990). *EMBO J.* **9**, 2759–2764.

Tzagoloff, A., and Dieckmann, C. L. (1990). *Microbiol. Rev.* **54**, 211–225.

Uetz, P., and Hughes, R. E. (2000). *Curr. Opin. Microbiol.* **3**, 303–308.

Underwood, E. J. (1977). "Trace Elements in Human and Animal Nutrition," 4th ed. Academic Press, New York.

Utschig, L. M., Baynard, T., Strong, C., and O'Halloran, T. V. (1997). *Inorg. Chem.* **36**, 2926–2927.

Utschig, L. M., Bryson, J. W., and O'Halloran, T. V. (1995). *Science* **268**, 380–385.

Valentine, J. S., and Gralla, E. B. (1997). *Science* **278**, 817–818.

Valentine, J. S., Hart, P. J., and Gralla, E. B. (1999). *In* "Copper Transport and Its Disorders: Molecular and Cellular Aspects" (Leone, A., and Mercer, J. F. B., Eds.), Kluwer Academic/Plenum, New York.

Valentine, J. S., and Pantoliano, M. W. (1981). *In* "Metal Ions in Biology" (Spiro, T. G., Ed.), Vol. 3, pp. 291–358. Wiley, New York.

Viles, J. H., Cohen, F. E., Prusiner, S. B., Goodin, D. B., Wright, P. E., and Dyson, H. J. (1999). *Proc. Natl. Acad. Sci. USA* **96**, 2042–2047.

Vulpe, C., Levinson, B., Whitney, S., Packman, S., and Gitschier, J. (1993). *Nat. Genet.* **3**, 7–13.

Vulpe, C. D., Kuo, Y. M., Murphy, T. L., Cowley, L., Askwith, C., Libina, N., Gitschier, J., and Anderson, G. J. (1999). *Nat. Genet.* **21**, 195–199.

Wakabayashi, T., Nakamura, N., Sambongi, Y., Wada, Y., Oka, T., and Futai, M. (1998). *FEBS Lett.* **440**, 141–146.
Wang, P. Z., and Novick, R. P. (1987). *J. Bacteriol.* **169**, 1763–1766.
Watton, S. P., Wright, G., MacDonnell, F. M., Bryson, W., Sabat, M., and O'Halloran, T. V. (1990). *J. Am. Chem. Soc.* **112**, 2824–2826.
Wernimont, A. K., Huffman, D. L., Lamb, A. L., O'Halloran, T. V., and Rosenzweig, A. C. (2000). *Nat. Struct. Biol.* **7**, 766–771.
Wilson, R., Ainscough, R., Anderson, K., Baynes, C., Berks, M., Bonfield, J., Burton, J., Connell, M., Copsey, T., Cooper, J., *et al.* (1994). *Nature* **368**, 32–38.
Wimmer, R., Herrmann, T., Solioz, M., and Wuthrich, K. (1999). *J. Biol. Chem.* **274**, 22597–22603.
Wong, P. C., Pardo, C. A., Borchelt, D. R., Lee, M. K., Copeland, N. G., Jenkins, N. A., Sisodia, S. S., Cleveland, D. W., and Price, D. L. (1995). *Neuron* **14**, 1105–1116.
Wong, P. C., Waggoner, D., Subramaniam, J. R., Tessarollo, L., Bartnikas, T. B., Culotta, V. C., Price, D. L., Rothstein, J., and Gitlin, J. D. (2000). *Proc. Natl. Acad. Sci. USA* **97**, 2886–2891.
Wood, V., Rajandream, M. A., Barrell, B. G., Devlin, K., and Churcher, C. M. (2000). Unpublished. Genbank Accession No. CAB75401.
Wright, J. W., Natan, M. J., MacDonnell, F. M., Ralston, D. M., and O'Halloran, T. V. (1990). *Prog. Inorg. Chem.* **38**, 323–412.
Wunderli-Ye, H., and Solioz, M. (1999a). *Adv. Exp. Med. Biol.* **448**, 255–264.
Wunderli-Ye, H., and Solioz, M. (1999b). *Biochem. Biophys. Res. Commun.* **259**, 443–449.
Yu, H., and Goh, C. J. (1998). Unpublished. Genbank Accession No. AAC79870.
Yuan, D. S., Dancis, A., and Klausner, R. D. (1997). *J. Biol. Chem.* **272**, 25787–25793.
Yuan, D. S., Stearman, R., Dancis, A., Dunn, T., Beeler, T., and Klausner, R. D. (1995). *Proc. Natl. Acad. Sci. USA* **92**, 2632–2636.
Zhou, B., and Gitschier, J. (1997). *Proc. Natl. Acad. Sci. USA* **94**, 7481–7486.
Zhou, P., and Thiele, D. J. (1993). *Genes Dev.* **7**, 1824–1835.
Zhu, H., Shipp, E., Sanchez, R. J., Liba, A., Stine, J. E., Hart, P. J., Gralla, E. B., Nersissian, A. M., and Valentine, J. S. (2000). *Biochemistry* **39**, 5413–5421.
Zhu, Z., Labbe, S., Pena, M. M., and Thiele, D. J. (1998). *J. Biol. Chem.* **273**, 1277–1280.

# FET3P, CERULOPLASMIN, AND THE ROLE OF COPPER IN IRON METABOLISM

BY DANIEL J. KOSMAN

Department of Biochemistry, School of Medicine and Biomedical Sciences, State University of New York, Buffalo, New York 14214



## I. COPPER PUMPS, FERROXIDASES, AND IRON HOMEOSTASIS IN EUKARYOTES

Eukaryotes, large and small, exhibit defects in iron homeostasis when in a copper-deficient nutritional condition or secondary to a defect in copper metabolism. This physiologic linkage between copper and iron is now well understood at the molecular level, at least in terms of the gene products that metabolically link these two essential metal nutrients. The central component in this linkage is a multicopper ferroxidase: ceruloplasmin or hephaestin, Fet3p or Fet5p, in mammals and the yeast *Saccharomyces cerevisiae*, respectively. However, each of these copper proteins relies on a copper ATPase found in the membrane of a specific vesicular compartment for the copper necessary for each protein's activation. The copper pumping that any one of these ATPases does may be critical to copper homeostasis as well—for copper excretion, for example. These pumps, in turn, rely on a protein—a copper chaperone—that ferries the copper from the plasma membrane copper permease through the cytosol to this vesicular compartment. The permease relies on a plasma membrane cuprireductase to supply it with the Cu(I) as substrate for uptake. Nonetheless, irrespective of whether or how a defect in any one of these enzymes, transporters, chaperones, or pumps may contribute to a dysfunction in copper handling, there most certainly will be a direct impact



*ADVANCES IN PROTEIN CHEMISTRY, Vol. 60*


0065-3233/02 $35.00

on the copper incorporation into one or more of these ferroxidases leading to a secondary effect on iron homeostasis. The copper ferroxidases are central to this secondary nutritional, metabolic, essentially epistatic relationship between copper and iron in eukaryotes. These copper ferroxidases are the focus of this chapter.

## II. Biologic Copper Sites and the Multicopper Oxidases

Copper proteins are classified on the basis of the "type" of copper site that they contain. There are three types of copper sites (Solomon and Lowery, 1993). They differ as to coordination number, type of ligand, and geometry; these differences in turn are the basis for these sites' electronic signatures and chemical (redox) activity. Type 1 Cu(II) exhibits a very strong absorbance at $\sim$600 nm($\varepsilon = 5000\ M^{-1}\ cm^{-1}$). This absorbance imparts a striking blue color to type 1 copper-containing proteins at concentrations $>$100 $\mu$M. This absorbance is due to a charge transfer transition from the cysteine sulfur ligand that characterizes the type 1 site to the Cu(II). This Cys-S $\pi$ to $Cu^{2+}d_{x^2-y^2}$ transition also places significant unpaired electron spin density on the sulfur rather than on the copper. As a result, the type 1 Cu(II) has a correspondingly small parallel electron spin-nuclear spin hyperfine coupling evident in the continuous wave electron paramagnetic resonance (cwEPR) spectrum [$A_{\parallel} = (43\text{–}95) \times 10^{-4}\ cm^{-1}$]. In addition to this Cys ligand, type 1 sites also have two histidine imidazole ligands. These three ligands typically describe a trigonal plane such that the geometry of the site overall can be described as distorted tetrahedral or distorted trigonal pyramid with the fourth ligand, if present, at the apex of this pyramid. However, it is the cysteine sulfur ligand common to all type 1 sites that dominates their electronic and chemical properties (Solomon and Lowery, 1993; Solomon *et al.*, 1996).

Type 2 copper sites are "normal" in that they exhibit only the weak (forbidden) d–d transitions typical of Cu(II). They are "nonblue." Not surprisingly, they lack the cysteine sulfur ligand found at all type 1 sites while most commonly containing only histidine imidazole coordination by the protein. The absence of significant electron spin transfer to the ligands at type 2 Cu(II) results in these sites having the copper hyperfine coupling typical of (pseudo)square-planar Cu(II) complexes containing nitrogenous and/or oxygenous ligands. $A_{\parallel}$ values for type 2 Cu(II) sites in proteins range from $140 \times 10^{-4}$ to $190 \times 10^{-4}\ cm^{-1}$. The wave function that describes the unpaired electron in type 1 Cu(II) (spin density higher at the S) results also in a smaller spin–orbit coupling in comparison to the wave function that describes the unpaired electron at type 2 Cu(II). With the electron spin density higher at the metal in this latter case, there is greater spin–orbit coupling. This difference results in a larger contribution of orbital angular momentum to the magnetization of an electron at a type 2 Cu(II) and thus in a stronger interaction with the magnetic field. This is reflected in a larger $g_{\parallel}$ value for type 2 sites, typically >2.2; in contrast, $g_{\parallel}$ values for type 1 Cu(II) sites are closer to the free electron value of 2.002.

Another feature that distinguishes type 2 from type 1 copper sites is that the former nearly always have at least one water molecule from solvent as one of the inner sphere ligands; many have two solvent-derived, exchangeable ligands. Type 1 Cu(II), in contrast, does not have a coordinated, solvent-derived ligand. This structural difference is directly linked to the different functions that the two sites have in the electron transfer reactions involving copper proteins. Known mechanisms involving type 1 sites are restricted to outer-sphere electron transfer processes while type 2 sites catalyze reactions that involve a direct coordination of an electron donor or acceptor. Thus, in the case of the multicopper oxidases, the type 1 copper is the site of entry of the electron from the one-electron reductant, while dioxygen—the electron acceptor or oxidant—is reduced at the type 2 copper site.

Type 3 copper is the third type of Cu(II) site in biology. A type 3 site has two distinguishing electronic properties. First is its relatively strong absorbance in the near-UV at ~330 nm($\varepsilon$ = 3–5000 $M^{-1}$ $cm^{-1}$). This transition is indicated by a shoulder on the much stronger absorbance at ~280 nm due to aromatic amino acid residues. Second, type 3 sites are diamagnetic despite the presence of Cu(II). The lack of a cwEPR spectrum, for example, is due to the fact that the type 3 Cu(II) site contains two copper atoms that are antiferromagnetically coupled through a bridging oxygen (-OH) atom. The absorbance at 330 nm is due to charge transfer from this ligand onto the copper atoms. This bridging ligand is the hallmark of type 3 copper sites, at least in their fully oxidized state; in other respects they vary as to the nature of the protein ligands although most contain histidine imidazole or tyrosine phenol ligands or both.

What distinguishes multicopper oxidases from other copper proteins is that they contain one each of these three types of copper site (Solomon and Lowery, 1993; Solomon *et al.*, 1996). Not only does this make them excellent models for *all* copper proteins, but because they have four redox-active metal ions, they also serve as paradigms for other enzymes that couple a one-electron reductant to a four-electron oxidant, most notably cytochrome *c* oxidase. Indeed, the three copper sites (and four copper atoms) in the multicopper oxidases play essentially equivalent roles in comparison to the two heme groups and two copper atoms in cytochrome *c* oxidase.

Despite large differences in their overall sequence and resulting structure, the multicopper oxidases contain signature sequences for these three types of copper ligand arrays. This sequence homology is evident in the alignments given in Fig. 1 (see color insert) for *Neurospora crassa* laccase (Lac); *Cucimus sativus* ascorbate oxidase (AO); and two ferroxidases, human ceruloplasmin (hCp) and Fet3p, from *Sa. cerevisiae*. Otherwise, these proteins share little sequence in common overall, although all are glycoproteins. On the other hand, within the principal structural domains of the type 1 sites in Lac, AO, and hCp as determined crystallographically, there is a high level of sequence similarity (49–64%) and identity (38–42%), suggesting that these proteins evolved from a common ancestor at least with respect to the domains that comprised the type 1 copper site ligands (Messerschmidt and Huber, 1990). Most reasonably this ancestral gene encoded a small "blue" copper protein like azurin or plastocyanin (Lerch and Germann, 1988; Rydén, 1988). However, these homology studies do not address the question of why hCp is a ferroxidase while Lac and AO are not. Physiologically, this remains the most critical question about structure–function relationships in this class of proteins.

Spectral data for the Fet3 protein illustrate the electronic properties of the three types of Cu(II) sites found in multicopper oxidases (Blackburn *et al.*, 2000; Hassett *et al.*, 1998). The cwEPR spectrum shown in Fig. 2 for

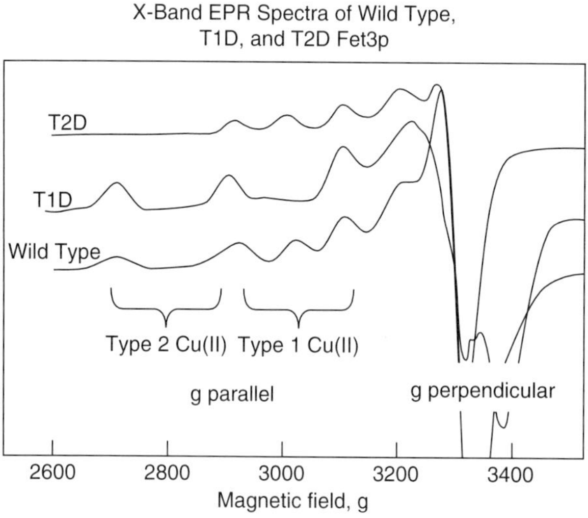


FIG. 2. Electron paramagnetic resonance spectra of wild-type and T1D and T2D mutants of Fet3p as indicated. Spectra were obtained at a microwave frequency of 9.5 GHz and 120 K. The samples were prepared in 25% v/v ethylene glycol/50 mM MES buffer, pH 6.0. The instrument settings were constant with values as follows: microwave power, 10 mw; modulation frequency, 100 kHz; modulation amplitude, 10 G; time constant, 0.02 s; sweep time, 60 s (Hassett *et al.*, 1998).

the wild-type protein has contributions from both the type 1 and the type Cu(II) electron Zeeman and nuclear spin interactions. The spectra for mutant proteins that lack either the type 2 copper (T2D) or the type 1 copper (T1D) exhibit the EPR spectrum of the remaining Cu(II) site only. These latter spectra show clearly the relatively smaller $A$ and $g$ values associated with the type 1 Cu(II) (seen in the T2D protein) in comparison to the type 2 Cu(II) (seen in the T1D protein) as noted above. The UV–visible absorbance spectra of this set of three Fet3 proteins are shown in Fig. 3.

Included in this latter set is a double mutant protein, a T1D/T2D protein that lacks both the type 1 and the type 2 copper atoms. Only the type 3 binuclear cluster contributes to the nonprotein absorbance in this protein, demonstrating that the shoulder at 330 nm is due to this cluster. This cluster also contributes a broad absorbance centered at 720 nm as the spectrum of this double mutant demonstrates. The absorbance of the wild-type protein at 608 nm is clearly due to the type 1 Cu(II) since it is seen only in protein forms that possess this site. The spin Hamiltonian and absorbance values for the copper sites in Fet3p are summarized in Table I. Additional properties of these copper site-depleted Fet3p mutant proteins are discussed below.

One of the significant differences between the multicopper oxidases as a group and the ferroxidases is that only the latter efficiently catalyze the

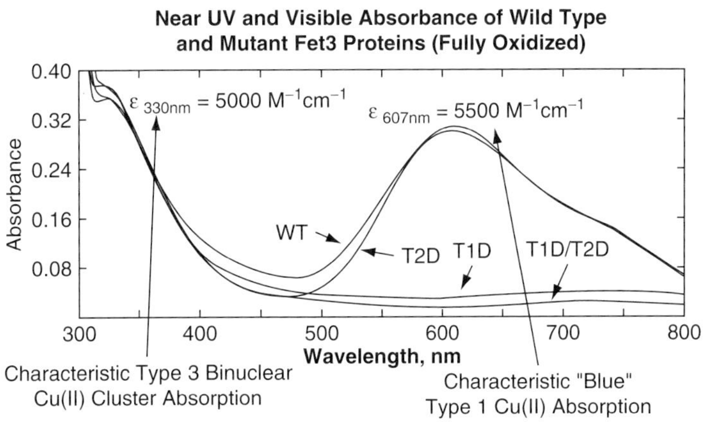


FIG. 3. Near-UV and visible absorbance spectra of wild-type, T1D, T2D, and T1D/T2D mutants of Fet3p proteins (fully oxidized) as indicated. All samples were prepared in 50 mM MES buffer, pH 6.0. Spectra were recorded following treatment of the samples with 0.5 mM hydrogen peroxide to ensure that all copper atoms present were in the cupric state (Hassett *et al.*, 1998).

TABLE I
*Spin Hamiltonian and Absorbance Parameters for the Copper Sites in Fet3p*[a]

| Copper site | $g_{\parallel}$ | $A_{\parallel}$, $\times 10^{-4}(\text{cm}^{-1})$ | $g\perp$ | $\lambda_{max}$(nm) | $\varepsilon(\text{M}^{-1}\text{cm}^{-1})$ |
|---|---|---|---|---|---|
| Type 1 | 2.19 | 89 | 2.05 | 608 | 5500 |
| Type 2 | 2.24 | 195 | 2.05 | Not resolvable | <100 |
| Type 3 | — | Diamagnetic | — | 330 | ~5000 |
| | | | | ~720 | ~300 |

[a]Data taken from Hassett *et al.* (1998), Blackburn *et al.* (2000), and Machonkin *et al.* (2001).

reaction shown in Eq. (1), commonly referred to as the ferroxidase reaction (Frieden and Osaki, 1974; Osaki *et al.*, 1966):

$$4\text{Fe(II)} + \text{O}_2 + 4\text{H}^+ \longrightarrow 4\text{Fe(III)} + 2\text{H}_2\text{O}. \quad (1)$$

Thus, only the "ferroxidases" exhibit significant activity toward Fe(II) as the reducing substrate while retaining appreciable activity toward most common organic reductants like hydroquinone, *o*-dianisidine, and *p*-phenylenediamine. This ferroxidase activity can be measured by the absorbance change at 315 nm due to the production of Fe(III). The change in molar absorbance at 315 nm for the formation of Fe(III) from Fe(II) is 2200 $\text{M}^{-1}\,\text{cm}^{-1}$ (Bonomi *et al.*, 1966). [The speciation of the "Fe(III)" formed in these experiments is undoubtedly complex, but, despite this complexity, this molar change in absorbance is relatively insensitive to biologically relevant conditions of pH and buffer composition.] Conversely, the reaction can be followed by measurement of $O_2$ consumption using an oxygen electrode. These two approaches were used to determine the kinetic constants for the ferroxidase reaction catalyzed by Fet3p (Hassett *et al.*, 1998). These experimental data are shown in Fig. 4. The fitted constants for this reaction are given in Fig. 4. Appropriately, the $V_{max}$ for Fe(III) production is four times the $V_{max}$ for $O_2$ consumption. This is consistent with the stoichiometry of the overall redox reaction [Eq. (1)]. For comparison, the $V_{max}$ for Fe(II) turnover by ceruloplasmin (Cp) is 550 $\text{min}^{-1}$ (Osaki, 1966). Possible bases for the somewhat greater activity of Cp towards Fe(II) in comparison to Fet3p are discussed below.

Alternatively, the amount of Fe(II) remaining can be determined by quenching the reaction with a Fe(II)-specific chelating agent, like bathophenanthroline disulfonate (BPS), which yields a strongly absorbing Fe(II) complex ($\varepsilon_{520\,\text{nm}} \sim 4 \times 10^4\ \text{M}^{-1}\ \text{cm}^{-1}$). There are two drawbacks to the use of this indirect means of assay. First, one is measuring a (small) change in the amount of substrate remaining against a large initial reading. Second, thermodynamically, the BPS added becomes part of the reaction

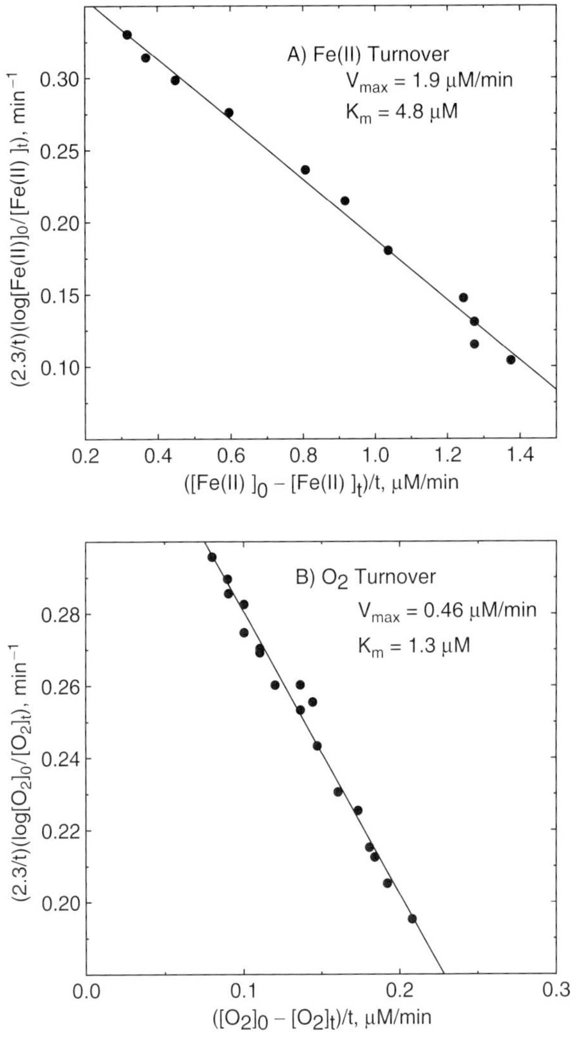


FIG. 4. Analysis of the kinetic constants of the ferroxidase reaction catalyzed by soluble Fet3p. Fe(II) oxidation (A) and $O_2$ consumption (B) were measured continuously and the residual substrate concentration was plotted with respect to time according to the integrated form of the Michealis–Menten equation as indicated in each panel. Fe(II) oxidation was followed by the appearance of Fe(III) at 315 nm while $O_2$ consumption was determined by the use of an $O_2$ electrode. The [Fet3p] = 0.2 μM in 0.1 M MES buffer, pH 6.0, at 25°C. The curve in each panel is a linear least-squares fit of the data to

$$2.3/t\{\log(S_0/[S_t])\} = (-1/K_m)\{([S]_0 - [S]_t/t\} + V_{\max}/K_m$$

computed using the kinetic constants given in the figure (Hassett *et al.*, 1998).

in that, given its $10^6$-fold greater affinity for Fe(II) than for Fe(III), it has the potential to drive the Fe(III) formed back to Fe(II) in the presence of any reducing equivalents available in the solution essentially independent of the reduction potential of these equivalents. The more direct assays used in Fig. 4 are strongly preferable to this BPS-based assay.

Another assay that has been used to demonstrate the ferroxidase reaction is based on one hypothesis for how Cp functions in iron homeostasis, namely, as a catalyst of the transfer of plasma Fe(II) to transferrin (Tf) generating Tf•Fe(III) (Frieden and Osaki, 1974; Osaki *et al.*, 1966). Fe(III)-loaded Tf exhibits an absorbance in the red, $\varepsilon_{460\,nm} = 4560\,M^{-1}\,cm^{-1}$. The $V_{max}$ for formation of Tf•Fe(III) catalyzed by Cp has been estimated at $12\,min^{-1}$, which is ~10-fold greater than the rate at which Tf•Fe(III) is produced in the absence of Cp, but at the same pH, [Fe(II)], and partial pressure of $O_2$. Although not analyzed so rigorously as above, the Fet3p-dependent Fe(III) loading of Tf has also been demonstrated (de Silva *et al.*, 1997). However, these studies have not determined the mechanism by which Cp activates the Fe(III) loading of *apo*-Tf. For example, there is no strong evidence that Cp and Tf interact thereby allowing for a channeling of the Fe(III) produced at the type 1 site in Cp to Tf. However, the kinetic data given above do indicate that the inherent Fe(II) turnover capacity of Cp is ~20-fold greater than its ability to facilitate the iron loading of Tf. Thus, even if no specific protein–protein interaction were involved in this iron loading, the Cp-dependent generation of Fe(III) alone would potentiate this process by providing more substrate—Fe(III)—for it. Nonetheless, the mechanistic relationship(s) between Cp ferroxidase and iron-loading activities remains an intriguing and critical question in the ceruloplasmin/iron trafficking story that is presented in more detail below.

## III. The Ferroxidases

### A. *A Historical Perspective*

Cartwright and Wintrobe and their co-workers suggested a link between copper deficiency and anemia in mammals 50 years ago (see Lahey *et al.*, 1952). Cartwright subsequently demonstrated that this copper-dependent anemia was unresponsive to iron supplementation but was corrected on administration of ceruloplasmin (see Lee *et al.*, 1968). The molecular basis of this link was indicated by Osaki and Friedan, who characterized the ferroxidase activity of ceruloplasmin and kinetically demonstrated that Cp could play a critical role in catalyzing trafficking of the potentially cytotoxic Fe(II) in the plasma to *apo*-Tf (see Frieden and

Osaki, 1974; Osaki, 1966; Osaki *et al.*, 1966). Using the kinetic values given above, they estimated that without the ferroxidase activity of Cp in the plasma 80% of the iron released from erythrocyte turnover would accumulate as non-Tf-bound Fe(II) and thereby would be unavailable for reabsorption by the reticuloendothelial system. Furthermore, this "free" Fe(II) could catalyze the formation of reactive oxygen species via the Fenton reaction. This, in turn, could lead to a subsequent organismal pathophysiology (Miyajima *et al.*, 1996; Nakano, 1993). This inference has been strikingly confirmed by research over the past 6 years in both yeasts and mammals; this research has directly tested the hypothesis that multicopper oxidase-dependent ferroxidase activity is essential to eukaryotic iron homeostasis (Askwith *et al.*, 1996; Harris *et al.*, 1995; 1998; Wessling-Resnick, 1999).

### *B. The Present*

Based on present sequence data, known or likely ferroxidase enzymes can be identified in several eukaryotes. These enzymes are listed in Table II. All are multicopper oxidases, by sequence homology at least. In mammals, they include ceruloplasmin and, most likely, hephaestin (Hp), although only mouse Hp (mHp) has been characterized at this time (Vulpe *et al.*, 1999). The alignments in Fig. 5A show that mHp is essentially

TABLE II
*Known and Likely Ferroxidase Enzymes*

| Kingdom/Order | Organism | Protein | NCBI Accession No. | Function evaluated |
|---|---|---|---|---|
| Fungi/Ascomycota | *Saccharomyces cerevisiae* | Fet3p | 6323703 | Y |
| | | Fet5p | 6321067 | |
| | *Schizosaccharomyces pombe* | Fio1p | 1067210 | Y |
| | *Candida albicans* | CaFet3p | 1684656 | N |
| | *Arxula adeninivorans* | Fet3 homologue | 7798835 | N |
| Metazoa/ Mammalia | *Mus musculus* | Ceruloplasmin | 6680997 | Y |
| | | Hephaestin | 6754180 | Y |
| | *Rattus norvegicus* | Ceruloplasmin | 6978695 | Y |
| | | Hephaestin | n.s.[a] | |
| | *Homo sapiens* | Ceruloplasmin | 1620909 | Y |
| | | Hephaestin | 7662254 | Y |

[a]n.s., not sequenced.

**A**

```
Cp            1 ----MKFLLLSTFIFLYSSLALARDKHYFIGITEAVWDYASG--TEEKKLISVDTEQSNF
Hephaestin    1 MKAGHLLWALLLMHSLWSIPTDGAIRNYYLGIQDMQWNYAPKGRNVITNQTLNNDTVASS

Cp           55 YLQNGPDRIGRKYKKALYFEYTDGTFSKTIDKPAWLGFLGPVIKAEVEDKVYVHLKNLAS
Hephaestin   61 FLKSGKNRIGSSYKKTVYKEYSDGTYTEEIAKPAWLGFLGPLLQAEVGDVILIHLKNFAS

Cp          115 RIYTFHAHGVTYTKEYEGAVYPDNTTDFQRADDKVLPGQQYVYVLHANE-PSPGEGDSNC
Hephaestin  121 RPYTIHPHGVFYEKDSEGSLYPDGSSGYLKADDSVPPGGSHVYNWSIPESHAPTEADPAC

Cp          174 VTRIYHSHVDAPKDIASGLIGPLILCKKGSLYKE---KEKNIDQEFVLMFSVVDENLSWY
Hephaestin  181 LTWIYHSHVDAPRDIATGLIGPLITCKRGTLDGNSPPQRKDVDHNFFLLFSVIDENLSWH

Cp          231 LEDNIKTFCSEPEKVDKDNEDFQESNRMYSINGYTFGSLPGLSMCAADRVKWYLFGMGNE
Hephaestin  241 LDDNIATYCSDPASVDKEDGAFQDSNRMHAINGFVFGNLPELSMCAQKHVAWHLFGMGNE

Cp          291 VDVHSAFFHGQALTSRNYQTDIINLFPATLIDAYMVAQNPGVWMLSCQNLNHLKAGLQAF
Hephaestin  301 IDVHTAFFHGQMLSIRGHHTDVANIFPATFVTAEMVPQKSGTWLISCEVNSHLRSGMQAF

Cp          351 FQVQDCNKPSSKDNIRGKHVRHYYIAAEEVIWNYAPSGIDIETEEKLTASGSDSGVFFEQ
Hephaestin  361 YKVDSCSMDPPVDQLTD-KVRQYFIQAHEIQWDYGPIGYDGRTGKSLREPGSGPDKYFQK

Cp          411 GATRIGGSYKKMAYREYTDGSFTNRKERGPDEEHLGILGPVIWAEVGDTIKVTFHNKGQH
Hephaestin  420 SSSRIGGTYWKVRYEAFQDETEQERVHQ-EEETHLGILGPVIRAEVGDTIQVVFYNRASQ

Cp          471 HLSIQPMGVSFTAENEGTYYGPPGASSQQAASHVAPKXTFTYEQTVPKEMGPTYADPVCL
Hephaestin  479 PFSIQPHGVFYEKNSEGTVYNDGTSHPKVAKSFEKVT----YYWTVPPHAGPTAQDPACL

Cp          531 SKMYYSAVDPTKDIFTGLIGPMKICKKGSLLADGRQKDVDKEFYLFPTVFDENESLLLDD
Hephaestin  535 TWMYFSAADPTRDTNSGLVGPLLVCKAGALGADGKQKGVDKEFFLLFTVFDENESWYNNA

Cp          591 NIRMFTHAPDQVDKEDEDFQESNKMHSMNGFMYGNQSWPHMCLGESIVWYLFSAGNEADV
Hephaestin  595 NQAAGMLDSRLLSEDVEGFQDSNRMHAINGFLFSNLPRLDMCKGDTVAWHLLGLGTETDV

Cp          651 HGIYFSGNTYLCKGEERDTANLFPHKSLTLLMNPDTKGTFDVECLTTDHYTGGMKQKYTV
Hephaestin  655 HGVMFEGNTVQLQGMRKGAVMLFPHTFVTAIMQPDNPGIFEIYCQAGSHREEGMQATYNV

Cp          711 NQCQRQFEDFT-VYLGERTYYVDAVEVEWDYSPSRAWEKELHHLQEQN-VSNVFLDKEEF
Hephaestin  715 SQCSSHQDSPRQHYQASRVYYIMAEEIEWDYCPDRSWELEWHNTSEKDSYGHVFLSNKDG

Cp          769 FIGSKYKKVVYRQFTDSSFREQVKRRAEEDEHLGILGPPIHANVGDKVKVVFKNMATRPY
Hephaestin  775 LLGSKYKKAVFREYTDGTFRIPRPRSGPE-EHLGILGPLIRGEVGDILTVVFKNKASRPY

Cp          829 SIHAHGVKTESS-TVVPTLPGEVATYTWQIPERSGAGREDSACIPWAYYSTVDRVKDLYS
Hephaestin  834 SIHAHGVLESNTGGPQAAEPGEVLTYQWNIPERSGPGPSDSACVSWIYYSAVDPIKDMYS

Cp          888 GLIGPLIVCRKSYVKVFSPKK--KMEFFLLFLVFDEMESWYLDDNIKTYSEHP-EKVNKD
Hephaestin  894 GLVGPLVICRNGILEPNGGRNDMDREFALLFLIFDENQSWYLKENIATYGPQESSHVNLK

Cp          945 NEEFLESNKMHAINGKMFGNLQGLTMHVKDEVNWYLMGMGNEIDLHTVHFHGHSFQYKHR
Hephaestin  954 DATFLESNKMHAINGKLYANLRGLTVYQGERVAWYMLAMGQDTDIHTVHFHAESFLYQNG

Cp         1005 GVYSSDVFDLFPGTYQTLEMFPQTPGTWLLHCHVTDHVHAGMATTYTVLPVEQETKSG--
Hephaestin 1014 QSYRADVVDLFPGTFEVVEMVASNPGTWLMHCHVTDHVHAGMETIFTVLSHEEHYSTMTT
                                                  T1/T3 Box
Cp              ------------------------------------------------------------
Hephaestin 1074 ITKEIGKAVILRDIGGDNVKMLGMNIPIKDVEILSSALIAICVLLLLIALALGGVVWYQH
                                                C-terminal membrane anchor
Cp              -------------------------
Hephaestin 1134 RQRKLRKRNRRSILDDSFKLLSLKQ
```

FIG. 5. Sequence alignments among known and/or likely ferroxidase proteins. (A) Alignment of the encoded sequences of mouse Cp (gi6680997) and hephaestin (gi4322374). (B) Alignment of encoded sequences of Fet3p (gi6323703) and Fet5p (gi6321067) from *Saccharomyces cerevisiae*; a Fet homologue from *Candida albicans* (gi684656); and Fio1p (gi1067210) from *Schizosaccharomyces pombe*. In both alignment sets, the copper ligands are indicated by boldface dots. In (A), the carboxyl-terminal, membrane-anchoring extension found in hephaestin and not in Cp is boxed. In (B), the four homology elements common to the fungal "ferroxidases" are denoted Boxes 1–4. Within two of these boxes, specific residues have been mutated. These are indicated

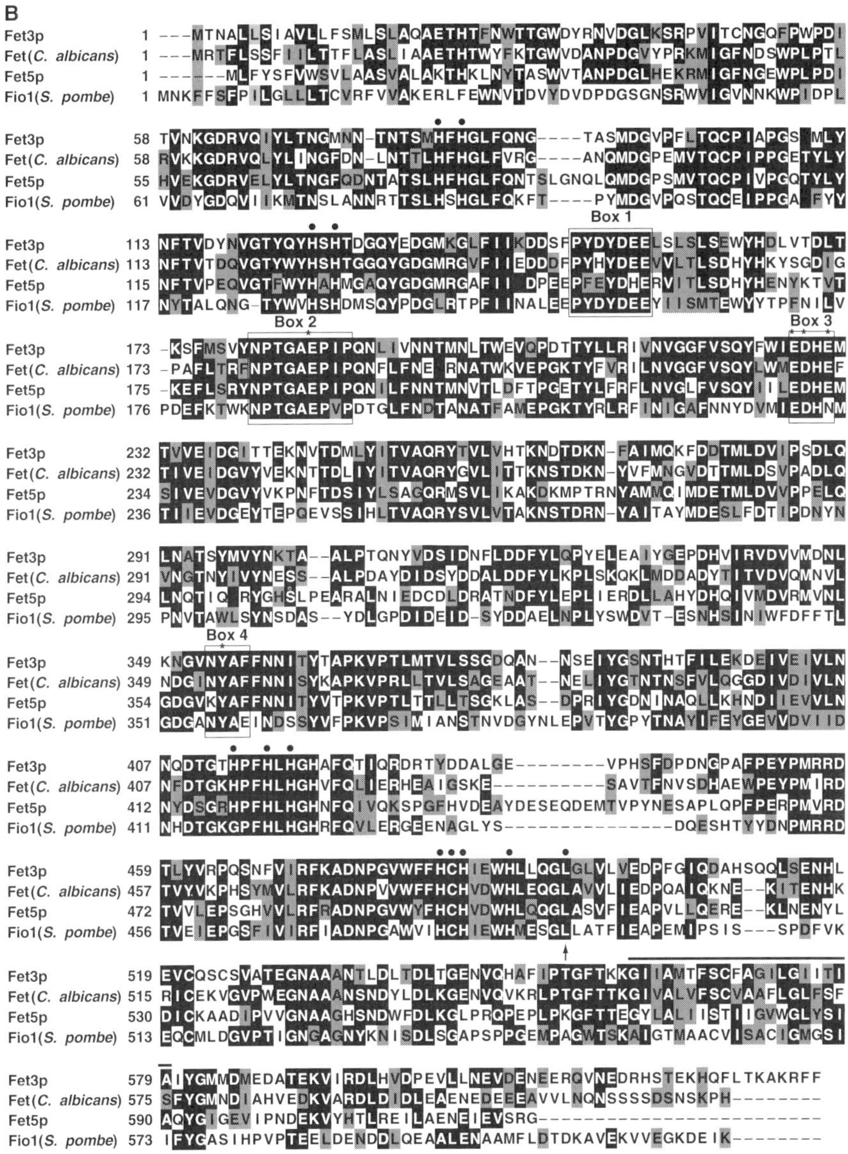


by an asterisk. The arrow below the Leu residue in the type 1 site of these fungal proteins shows that as a group this site is trigonal in this class of multicopper oxidase. The bar above residues 560–579 in Fet3p indicates the predicted carboxyl-terminal membrane-spanning element in this protein; the alignment shows that this element is likely conserved in Fet5p (a known type 1 membrane protein) and the Fet homologue from *Ca. albicans*, but is most likely absent in Fio1p. These sequences and others noted in the text can be retrieved from the Web site maintained by the National Center for Biotechnology Information using the accession numbers given (www.ncbi.nlm.nih.gov).

mouseCp with an 86-amino-acid carboxyl-terminal extension. This extention includes a predicted membrane-spanning element. This element is boxed in Fig. 5A. Fet3p (Askwith *et al.*, 1994; Blackburn *et al.*, 2000; de Silva *et al.*, 1995, 1997; Hassett *et al.*, 1998) and Fet5p (Spizzo *et al.*, 1997; Urbanowski and Piper, 1999) in *Sa. cerevisiae* represent the ferroxidases found in lower eukaryotes. In addition, Fet homologues are present in the genomes of the yeasts *Schizosaccharomyces pombe* (Askwith and Kaplan, 1997), *Candida albicans*, and *Arxula adeninivorans*. In Fig. 5B, the sequences of the first two proteins are aligned with those of Fet3p and Fet5p. In both alignment sets (Figs. 5A and 5B), the copper ligand motifs highlighted in Fig. 1 are marked with a boldface dot. Cursory inspection shows that aside from these copper-binding motifs, the Cp/Hp pair has little in common with the yeast proteins.

One can reasonably ask how is it possible to characterize a multicopper oxidase as a ferroxidase without testing for the requisite substrate specificity? One alternative to a direct enzymatic assay, which has been done for only (Cp) and Fet3p, is to demonstrate that a mutant cell or organism that does not produce the gene product exhibits a defect in iron homeostasis that could result from the absence of a putative ferroxidase activity. This has been done also only for Cp and Fet3p, although some data have linked Fet5p (Urbanowski and Piper, 1999) and Fio1p (Askwith and Kaplan, 1997) to iron metabolism as well. For the experimentally challenged, a more appealing approach would be to delineate sequence motifs that are specific to the ferroxidase activity. For example, one could propose that ferroxidases but not the other multicopper oxidases have amino acid residues that specifically bind the Fe(II) substrate and/or the Fe(III) product. Crystallographic evidence has implicated a cohort of D, E, and H residues that may serve as a substrate/product iron-binding site in hCp (Lindley *et al.*, 1997; Zaitseva *et al.*, 1996). Subsequent sequence alignments and homology modeling also suggested some possible residues of this type in other known or potential ferroxidases, principally sequences rich in D or E and some containing Y, as well (Buonaccorsi di Patti *et al.*, 1999; Murphy *et al.*, 1997). The former residues would make good ligands for Fe(II); the latter are more appropriate for Fe(III). Both types of residue could serve as part of the electron transfer path for the electron from the Fe(II) substrate to the type 1 Cu(II). Elements like these are found in the yeast ferroxidases and are identified in Fig. 5B (Boxes 1–4). The role of some residues in the ferroxidase reaction catalyzed by Fet3p has been tested by mutagenesis. Those residues tested in mutagenesis experiments are marked with an asterisk. The results of these experiments are reviewed below; in summary, however, these experiments have been negative or have provided only limited insight (Askwith and Kaplan, 1998; Bonaccorsi di Patti *et al.*, 2000). It is important to emphasize again that the yeast proteins are overall unlike the Cp/Hp pair and are therefore unlikely to possess the tertiary fold that results in the active site pocket seen in the hCp structure. Thus, although hCp and Fet3p are both multicopper oxidases, the ferroxidase activity that they share uniquely among this class of copper enzymes may represent an instance of functional convergence.

These putative ferroxidase motifs highlighted in Fig. 5B have not turned up in other genome databases, such as for *Drosophila* or *Caenorhabditis elegans*. Although flies express three proteins that contain all of the copper liganding motifs common to the multicopper oxidases as shown in Fig. 1 (see Table III) none contains any of the possible ferroxidase motifs indicated in Fig. 5B. The same is true for the F21D5.3 protein from *Cae. elegans* and the three proteins from *Arabidopsis* (Table III); all are clearly multicopper oxidases but are otherwise similar to neither the mammalian nor the yeast ferroxidase enzymes. Of course, any or all of these proteins may instead be functional Lac or AO homologues and play no role in the iron metabolism of their respective organisms. This certainly could be true of one or more of the *Arabidopsis* proteins, but is less likely for those from the Metazoans since there is no evidence for laccase-like activity in these higher eukaryotes, for example. Clearly, what physiologic role(s) they do play remains a completely open question.

One also should be aware of the cohort of proteins that provide copper resistance to bacteria that commonly are encoded by an extrachromosomal element. Sequence analysis indicates that one of these proteins is a multicopper oxidase since this member of the group contains the copper liganding motifs highlighted in Fig. 1. However, this member also contains a M/S-rich motif that is thought to be essential to the copper trafficking supported by a copper transporter like Ctrlp (*Sa. cerevisiae*) (Dancis *et al.*, 1994) or the CopA (Cha and Cooksey, 1991) and CopB

TABLE III
*Uncharacterized Eukaryotic Multicopper Oxidases*

| Kingdom/Order | Organism | NCBI Accession No. |
|---|---|---|
| Chromista/Kinetoplastida | *Leishmania major* | 7191095 |
| Plantae/Spermatophyta | *Arabidopsis thaliana* | 6759442 |
| | | 7546691 |
| | | 9955523 |
| Metazoa/Nematoda | *Caenorhabditis elegans* | 3876110 |
| Metazoa/Insecta | *Drosophila melanogaster* | 7293395 |
| | | 7297515 |
| | | 7301401 |

(Odermatt *et al.*, 1993) proteins of *Pseudomonas syringae* and *Enterococcus hirae*, respectively. An example of this sequence, arbitrarily chosen from *Xanthomonas campestris*, is

$$_{387}\text{MGMDHGGMSGMSMSASDSSDSSDSSNKPAMAMNM}_{420}.$$

This putative protein product (gi1073083) is termed a "copA" homologue although it does not have the CXXC motifs common to the gene products similar to the copA protein from *Escherichia coli* that is a known copper-translocating ATPase (Rensing *et al.*, 2000). In summary, bacteria also produce multicopper oxidases, and they potentially could be involved in metal metabolism. However, essentially no research has been reported that in any way tests this possible involvement.

### *C. Cell Locale and Biologic Function*

What is not evident in the alignments shown in Fig. 5 is the fact that many if not all organisms that produce one ferroxidase produce two genetically distinct forms. This is more directly indicated by the information in Tables II and III. For example, *Sa. cerevisiae* produces Fet3p and Fet5p while mammals produce ceruloplasmin and hephaestin. This is not an example of redundancy: in both cases, the two enzymes have distinct cell (organismal) locales and thus have specific and distinct physiologic functions. Although far from fully delineated, the roles that these four proteins play in iron metabolism in their respective organisms probably do broadly illustrate how ferroxidase activity functions in iron homeostasis. The fact that several genera do produce more than one multicopper oxidase (Table III) makes it more likely that at least one of these proteins in its respective organism does fulfill this function.

Protein structure determines cell locale and therefore proscribes, if not determines, physiologic function. All four proteins are synthesized in an immature form. Of course, all are synthesized as *apo* or copper-free proteins. In addition, all are glycosylated. Since glycosylation occurs in the endoplasmic reticulum (ER) and Golgi, it is not surprising that all four proteins carry an encoded signal sequence that targets the nascent polypeptide to the ER. [Note that in this chapter the numbering of amino acid residues in all cases is in reference to the translation product (the mRNA sequence), not the mature protein that lacks the encoded, but ultimately cleaved, amino-terminal signal recognition peptide.] On the other hand, ceruloplasmin is secreted and thus is carried along the default pathway in vesicular trafficking to the plasma membrane. This is comparable to Fet3p, which is found in the yeast cell membrane. The feature that distinguishes these two ferroxidases is the presence of a carboxyl-terminal, membrane-spanning domain in Fet3p that makes it a type 1 membrane protein. In contrast, the encoded Cp is a fully soluble protein with no membrane-spanning elements. Nonetheless, these two proteins follow a parallel trafficking pathway to the plasma membrane. Since Fet3p is contained within the vesicular membrane it is retained in the plasma membrane. As a lumenal, soluble protein, Cp normally is released from the cell.

However, there is one instance in which Cp is found tethered to the extracellular surface of cells and that is in the central nervous system (Patel and David, 1997; Patel *et al.*, 2000). This anchoring is due to the posttranslational attachment of a glycosylphosphatidylinositol—or GPI—unit to a carboxyl-terminal extension of the normal Cp sequence. This extension results from an alternative splice variant of the single Cp pre-mRNA that deletes 5 amino acid residues while encoding an additional 30. This alternative sequence encodes the signal for GPI anchor addition. In effect, this converts Cp into a "Fet3p" with regard to its cellular targeting and ultimate cell locale. Thus, astrocytes, for example, produce predominantly the membrane-anchored, Fet3p form of Cp, indicating that the pathophysiology due to iron deposition in the brain in Cp-deficiency states is most likely due to the lack of this ferroxidase activity on the surface of neural cells (Harris *et al.*, 1995, 1998; Nakano, 1993).

Fet5p (Spizzo *et al.*, 1997; Urbanowski and Piper, 1999) and hephaestin (Vulpe *et al.*, 1999) are both intracellular and are alike in that both have carboxyl-terminal membrane-spanning domains. In the case of hephaestin, this element is contained within an 86-amino-acid-residue extension of the homologous Cp sequence as noted above. Fet3p and Fet5p, in contrast, are closely homologous both in sequence and in size, including carboxyl-terminal membrane-spanning elements. This comparison between the yeast and human cell, and between these four ferroxidases, is illustrated in Fig. 6. In this context, Fio1p in *Sc. pombe* may not be a strict Fet3p homologue; it does not contain an apparent membrane-spanning domain in its carboxyl-terminal region (Fig. 5B). The fact that expression of *fio1* does not complement in a *fet3*Δ-containing strain may reflect this apparent structural difference between the two gene products (Askwith and Kaplan, 1997). Clearly, precise delineation of the role of Fio1p in iron metabolism in *Sc. pombe* remains to be achieved. Doing so would be important since *Sc. pombe* has been seen as evolutionarily closer to higher eukaryotes than is *Sa. cerevisiae*.

Fet5p and hephaestin are type 1 membrane proteins; both are found associated with vesicular compartments. Fet5p is localized to the vacuolar membrane in *Sa. cerevisiae* while hephaestin appears to be associated with a vesicular compartment within the villus cells of the small intestine. These two (apparent) ferroxidases share another similarity: neither is

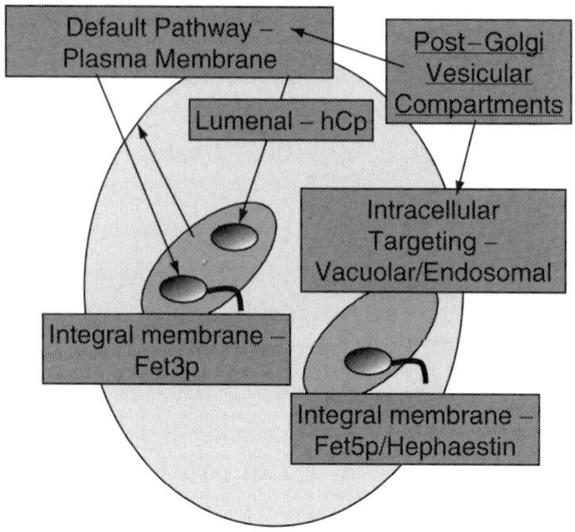


FIG. 6. Cartoon depicting the cell locale and targeting of Fet3p and Cp in comparison to Fet5p and hephaestin.

involved in the uptake of iron into their respective cells. In yeast, iron uptake from the environment is due to the Fet3p/Ftr1p plasma membrane complex (Stearman *et al.*, 1996). In mammals, uptake of iron from the lumen of the intestine is due to the action of the DMT1/DCT1/Nramp2 protein (Wessling-Resnick, 1999), which is strongly produced in the proximal duodenum and found in the mucosal (apical) surface (Tandy *et al.*, 2000; Trinder *et al.*, 2000). Nonetheless, hephaestin is required for maintenance of normal body levels of iron. This is evident in the pathology of the *sla* mouse. The mutation in this sex-linked disease maps to the hephaestin locus (Vulpe *et al.*, 1999). Despite a severe organismal microcytic hypochromic anemia, intestinal epithelial cells from the *sla* mouse exhibit a normal iron uptake; however, the iron taken into these cells apparently is not released into the circulation (Manis, 1970). This phenotype strongly suggests that hephaestin is required for the export of iron at the basolateral surface of the enterocyte. Preliminary data indicate that a DMT1 homologue, MTP1/Ferroportin/IREG1, is also involved in this cellular iron export into the circulation at the basolateral membrane (Abboud and Haile, 2000; Donovan *et al.*, 2000; McKie *et al.*, 2000). The $Fe^{2+}$ oxidation necessary prior to iron binding to circulating Tf certainly could be catalyzed by hephaestin, but this would require that hephaestin be physically "downstream" from $Fe^{2+}$ transport out of the enterocyte. Clearly, the mechanistic link between the $Fe^{2+}$ "exporter" and the ferroxidase in the intestinal epithelial cell remains a very open question. It should also be noted that DMT1 and MTP1 expression is not restricted to this cell type. Therefore, nontransferrin iron trafficking across cell barriers may occur in other tissues, as well. Of interest would be the possible role that the GPI-anchored form of ceruloplasmin might play in iron uptake and efflux, particularly in the central nervous system (Harris *et al.*, 1998, 1999; Patel and David, 1997; Patel *et al.*, 2000). Significantly, targeted gene disruption in mice has demonstrated a role for Cp in iron efflux, at least in reticuloendothelial cells (Kupffer cells in the liver, for example) (Harris *et al.*, 1999; Richardson, 1999). In the case of this cell type, Cp works from the extracellular side of the cell membrane to stimulate efflux; whether the Cp is free or cell-associated in some fashion is not known, however. In any event, much like Fet3p and the GPI-anchored form of Cp, this extracellular Cp appears to couple the oxidation of $Fe^{2+}$ to its targeting to a $Fe^{3+}$-specific protein: Ftr1p in the case of yeast iron uptake and Tf in the case of mammalian iron homeostasis.

The genetic disorder of iron trafficking in the *sla* mouse is strikingly similar to the defect in copper trafficking linked to the Menkes protein. Menkes disease, a genetic disorder that maps to the MNK locus in humans, ATP7A, is characterized by a systemic copper deficiency despite normal copper uptake from the lumen of the intestine (Harrison and Dameron, 1999). The MNK protein is a copper ATPase that pumps copper into a vesicular compartment within mammalian cells, including the intestinal epithelium. The copper that accumulates in these vesicles may be secreted from the cells by exocytosis. This is suggested by the observation that the MNK protein in cultured epithelial cells rapidly translocates from the *trans*-Golgi network to the plasma membrane with an increase in extracellular copper. It cycles back once the levels of extracellular copper are reduced (Petris *et al.*, 1996). Therefore, it is likely that the systemic copper deficiency in Menkes disease is due to the failure of copper to be secreted from the intestinal epithelium into the circulation in much the same fashion that iron is retained in the same tissue in the *sla* mouse. What is possible also is that in addition to targeting copper for export, the Menkes protein could be supplying copper for hephaestin, as well. If this is the case, then defects in the APT7A ATPase (the Menkes protein) would cause systemic deficiencies of both copper and iron.

As a single-cell organism, a yeast cell is not polar in an organ sense, apical versus basolateral, for example, as is the case of intestinal epithelial cells. As a free-living organism, yeast accumulates nutrients for future use in times of nutrient depletion. Thus, little copper or iron is secreted from the yeast cell. On the other hand, yeast must have a mechanism to mobilize and use stored nutrients. Iron is stored in *Sa. cerevisiae* in a vesicular compartment, perhaps a (the) vacuole, most likely as an Fe(III) polyphosphate (Raguzzi *et al.*, 1988). This iron phosphate is probably not significantly different chemically than the iron core found in mammalian ferritin. Also, iron most likely is reductively mobilized from this Fe(III) polymer much as iron is mobilized in ferritin. The fate of this Fe(II) thereby produced within the vacuole is similar to that of the Fe(II) produced at the plasma membrane by the ferrireductases/cuprireductases that support iron and copper uptake in yeast. This Fe(II) is oxidized by Fet5p, and the Fe(III) produced becomes substrate for the Fth1p iron permease that is coupled to Fet5p in the vacuolar membrane (Urbanowski and Piper, 1999). Thus, much as hephaestin makes iron available to the human organism by supporting its trafficking across the epithelial cell barrier in the intestine, Fet5p makes iron available to the yeast cell by supporting its export from the yeast vacuole.

## IV. Fet3 and Ftr1 in Iron Uptake in *Saccharomyces cerevisiae*: The Molecular Link between Copper and Iron Metabolism

Fet3p and Ftr1p most likely form a heterodimeric complex in the yeast plasma membrane (Stearman *et al.*, 1996). Fet5p and Fth1p form a homologous complex in the vacuolar membrane (Urbanowski and Piper, 1999). The evidence for this model is indirect in that a specific physical interaction between the two proteins has not been directly demonstrated. The lack of evidence for a similar functional interaction between Cp and Tf was noted above. Fet3p and Ftr1p do localize to the plasma membrane. However, the most compelling evidence for a physical interaction between the two is that this localization depends on the presence of both proteins; if one is not produced, then the other remains intracellular (Stearman *et al.*, 1996). Furthermore, in the absence of Ftr1p production, the Fet3 protein that is made remains in an enzymatically inactive form, presumably as the *apo*-protein.

The *apo*-Fet3p appears to obtain its prosthetic copper atoms in a post-Golgi vesicle. This is indicated by the fact that in yeast mutants lacking a component required for vesicular trafficking from the ER (e.g. *sec18*), or through the Golgi (*sec7*), or from the Golgi to a post-Golgi compartment (vacuolar sorting mutants, or *vps*), Fet3p is recovered only as an inactive, *apo*-protein (Yuan, *et al.*, 1997). In these mutants, Fet3p would accumulate in the donor compartment upstream from the subsequent, and blocked, vesicular trafficking event. However, this inactive protein can be activated *in vitro* simply by the addition of aqueous copper. Even the enzyme activity of the *apo*-Fet3p fractionated in a polyacrylamide gel can be recovered in this fashion. Consequently, PAGE analysis of membrane extracts from various *sec* and *vps* mutants demonstrated that except in *sec1* and *sec4* mutants, and some *vps* mutants, *apo*-Fet3p was present; this was shown by the oxidase activity present in gels incubated with copper sulfate prior to histochemical staining. Sec1p and Sec4p are involved in the targeting and fusion of secretory vesicles to the plasma membrane, while Vps proteins certainly have some specificity as to the type(s) of post-Golgi vesicles with which they associate. Therefore, it is not surprising that some *vps* mutants lacked active Fet3p while others exhibited normal Fet3p oxidase activity without the copper treatment.

Direct, *in situ* immunofluorescence studies also indicate that *apo*-Fet3p obtains its copper in a post-Golgi vesicle. These studies demonstrate that the copper ATPase in yeast, Ccc2p, which pumps copper into the lumen of this Fet3p-containing vesicle, colocalizes with the yeast chloride channel, Geflp (Gaxiola *et al.*, 1998). Gef1p-dependent transport of $Cl^-$ into the vacuolar lumen neutralizes the electrochemical potential-positive inside linked to the import of copper through Ccc2p and protons through the vacuolar $H^+$-ATPase, Vma1p (Davis-Kaplan *et al.*, 1998; Gaxiola *et al.*, 1998). This cocompartmentalization is indicated by functional studies, also, in that both Gef1p and Vma1p mutants exhibit the characteristic respiratory deficiency of an iron-starved yeast (Eide *et al.*, 1993; Greene *et al.*, 1993). As with the Ccc2p mutant, but not a Fet3p mutant, this respiration deficiency, scored as a growth defect on a non-fermentable carbon source such as glycerol or ethanol, can be rescued by growth in the presence of 100 μM copper. This physiologic suppression parallels the recovery of oxidase activity that follows treatment of *apo*-Fet3p-containing gels with aqueous copper (Davis-Kaplan *et al.*, 1998; Yuan *et al.*, 1997).

The uptake and trafficking of copper to Fet3p (and, perhaps, Fet5p) and the epistatic relationships between the gene products involved in this pathway are illustrated in Fig. 7. What this cartoon emphasizes is that iron homeostasis in yeast requires the activities of four proteins that either transport copper or are activated by copper. The former are the copper permease, Ctr1 p (Dancis *et al.*, 1994); the copper chaperone, Atx1p (Lin *et al.*, 1997); and the copper ATPase, Ccc2p (Yuan *et al.*, 1995). Fet3p and Fet5p are the copper-activated proteins that ultimately are required for the normal trafficking of iron into and within the yeast cell. Loss of function in any one of the former proteins is correctable by addition of sufficient nutritional copper so as to bypass the specific step supported by the particular gene product. Loss of Fet3p function, however, is not correctable in this fashion, consistent with the fact that active Fet3p is the end-product of this copper delivery pathway. Each one of these yeast gene products is found in human cells: hCtr1p (Zhou and Gitschier, 1997), HAH1p (Klomp *et al.*, 1997), the Menkes protein (Payne and Gitlin, 1998), hCp, and hephaestin. In yeast, deletion of the gene encoding any one of the copper-trafficking proteins results in an organismal iron

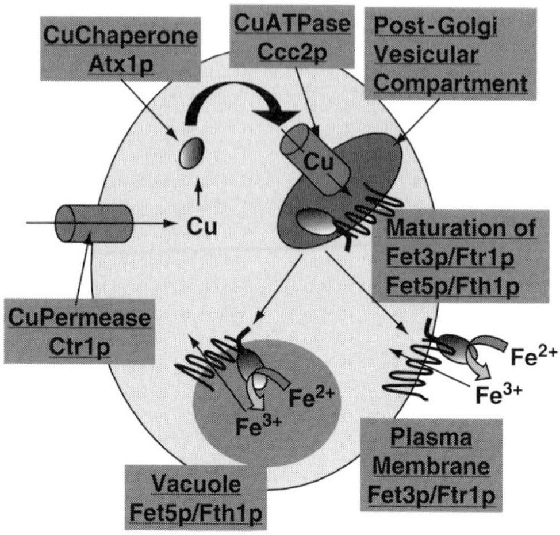


FIG. 7. Cartoon depicting the functional and epistatic relationship between copper and iron uptake and trafficking in yeast. The handling of iron by eukaryotic cells is ultimately dependent on a copper ferroxidase. In yeast, either copper deficiency or loss of function in any of the copper-handling proteins indicated in the cartoon causes an iron deficiency that is correctable readily by extranutritional copper but not by iron. Loss of function in Fet3p or Ftr1p, however, is correctable by neither metal ion.

deficiency that is due to a loss of iron uptake and that results in a loss of respiratory capacity: the cells cannot grow on a nonfermentable carbon source. Each of these deletions is complemented by expression of the human gene encoding the human protein counterpart. Thus, when it comes to the interrelationship between copper and iron handling, there is essentially no difference between yeasts and humans.

## V. FERROXIDASE STRUCTURE: hCp AND Fet3p

hCp is the only ferroxidase whose structure has been determined crystallographically (Lindley *et al.*, 1997; Zaitseva *et al.*, 1996). Lindley *et al.* (1997) have published a 3.1-Å map (PDB Accession No. 1KCW). This structure showed that as in Lac (1A65) (Ducros *et al.*, 1998) and AO (1AOZ) (Messerschmidt *et al.*, 1992a, 1992b, 1993), the two other multicopper oxidases with crystallographically determined structures, the type 2 and type 3 copper atoms in hCp are trigonally arranged with an atom-to-atom spacing of ca. 3.5Å. This "trinuclear" cluster is, in turn, ca. 13 Å from the type 1 copper atom. This arrangement of the three copper sites is diagrammed in Fig. 8. As noted previously, the type 1 Cu(II) is the

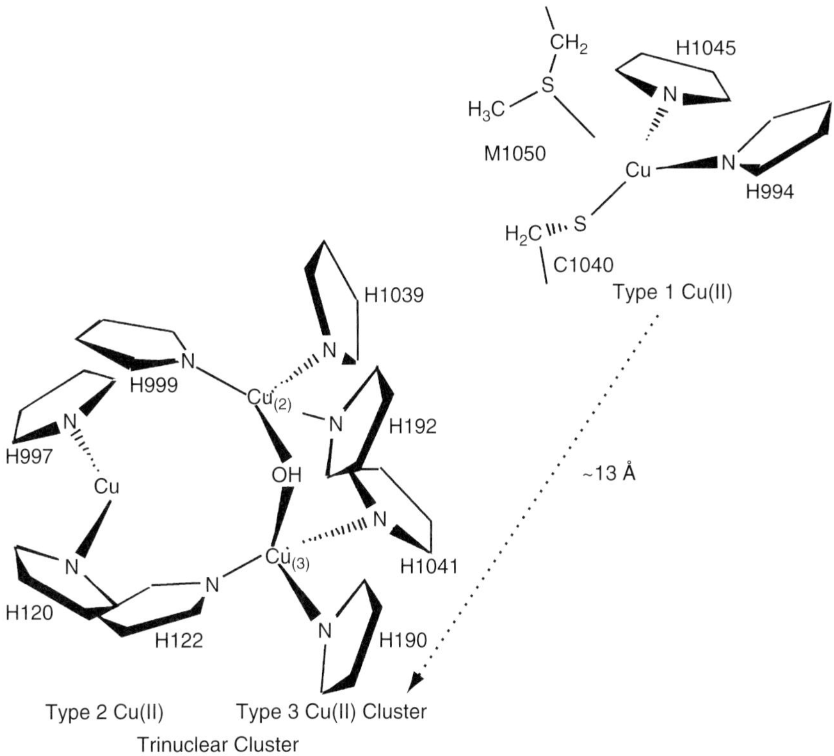


FIG. 8. Spatial relationship between the three copper sites in a multicopper oxidase. The structure (and residue numbering) is a representation of these sites in hCp. The trinuclear cluster is a near-isosceles triangle, ~3.4 Å on a side. The notation $Cu_{(2)}$ and $Cu_{(3)}$ is taken from Messerschmidt *et al.* (1992a) and corresponds to the notation Type 3″ and Type 3′, respectively, used by Zaitseva *et al.* (1996). Oxygen is thought to bind between the type 2 copper and $Cu_{(2)}$.

redox partner of the reducing substrate. The trinuclear cluster, in turn, is responsible for the 4 one-electron transfers to dioxygen that result in the production of two water molecules (Shin *et al.*, 1996). Although the details of these two redox reactions are not fully delineated, their general characteristics are quite well understood.

As is discussed immediately below, the structures of the type 1 sites in hCp, Lac, and AO—and in Fet3p—differ in some details; nonetheless, the electron transfer reactions at this site in the four proteins are all outer sphere in nature (Solomon *et al.*, 1996). This mechanism is dictated by the fact that type 1 sites, as a class, do not have exchangeable, solvent-accessible inner coordination sites to which a reductive ligand could bind. On the other hand, the relationship between the copper atom and the solvent-accessible surface of the protein does vary among these proteins. For example, the type 1 copper in hCp is completely buried (Zaitseva *et al.*, 1996) while this site in *Coprinus cinereus* Lac is much closer to the surface (Ducros *et al.*, 1998). This difference is apparent in Fig. 9 (see color insert), which displays Connelly surfaces adjacent to the type 1 sites in the two proteins. Connelly surfaces, in effect, show the solvent surface of a macromolecule. The green-hued surface patch in the Lac structure (Fig. 9, left) is due to the $N^{\varepsilon}$ of H494; clearly, the edge of this copper ligand is strongly exposed to solvent. In contrast, the corresponding histidine in hCp, H1045, is buried in the protein and completely shielded from the solvent (Fig. 9, right). In both cases, the electron transfer from substrate is outer sphere; however, given the striking difference in surface accessibility the precise mechanism (route) for how the electron gets to the Cu(II) may be quite different in the two proteins.

Although the structure of Fet3p is not known, ESEEM data suggest that its type 1 copper has a solvent accessibility that is intermediate between the situation in hCp and Lac (Aznar *et al.*, 2002). ESEEM—electron spin-echo envelope modulation—is sensitive to spin interactions that are weak compared to the interaction of the electron spin magnetization with the instrument magnetic field or with those fields due to strongly interacting nuclei. Those interactions are detected in the cwEPR spectrum. Thus, ESEEM that is due to the weak interaction ($\sim$1 MHz at a field of 3 kG) of the distal, noncoordinating N in a histidine imidazole found at a Cu(II) site in a protein can be resolved easily (Cornelius *et al.*, 1990). The ESEEM spectra of the unpaired electron at the type 2 Cu(II) (in the T1D mutant) and the type 1 Cu(II) (in the T2D mutant) are shown in Fig. 10 (Aznar *et al.*, 2002). The depth and period of modulation of the electron spin magnetization can be correlated to the number of equatorially coordinated histidine imidazoles at a Cu(II) site. The modulation depth of the ESEEM spectrum due to the type 2 Cu(II) is best ascribed to a single, equatorially coordinated histidine imidazole at this site. The absence of strong $^{14}N$ modulation at the type 1 Cu(II) is consistent with the dominant effect of the coordinated cysteine sulfur and the significant spin delocalization onto that ligand (Solomon and Lowery, 1993; Solomon *et al.*, 1996).

The spectra in Fig. 10 were taken in $H_2O$ and therefore have the potential to show a modulation due to the $^1H$ of any water that might be coordinated to or magnetically "near" the type 2 Cu(II), as well. This modulation is well resolved in the type 2 Cu(II) spectrum but is essentially absent in the type 1 Cu(II) spectrum. In order to better resolve this modulation, however, one takes spectra in $^1H_2O$ and $^2H_2O$ and calculates the ratio of the experimental modulation patterns: in effect, the pattern

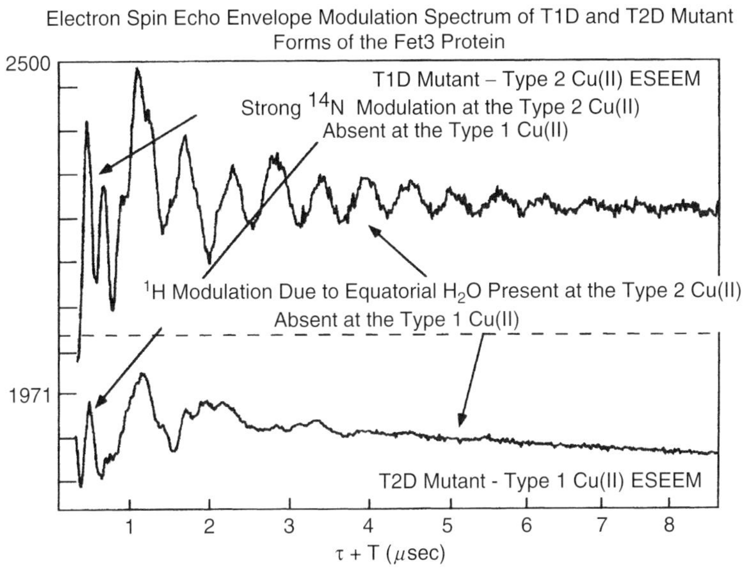


FIG. 10. ESEEM spectra for T1D and T2D Fet3p. The upper spectrum of the type 2 Cu(II) site (T1D mutant) was collected at 2690 G with $\tau = 260$ ns. The lower spectrum of the type 1 Cu(II) site (T2D mutant) was collected at 2825 G with $\tau = 250$ ns. The measurement conditions were as follows: microwave frequency, 8.80 GHz; microwave power, 43 dB; sample temperature, 4.2 K. The sample was 0.5 mM Fet3p in MES buffer, pH 6.0, containing 25% ethylene glycol (Aznar *et al.*, 2002).

in deuterium oxide provides the negative control for water proton modulation (Hegg *et al.*, 1999). Again, the strength and pattern of this ratioed modulation can be interpreted in terms of how water might be coordinated to the Cu(II) or how it might magnetically interact with the Cu(II). This is illustrated in the theoretical traces given in Fig. 11 for one equatorial water, or for one axial water, or for ambient water only ("outer-sphere" water). These three traces can be compared to the experimental trace for the ratioed envelope for the type 1 site in Fet3p. This experimental trace can be best modeled by a Cu(II) that has a "half-shell" of ambient water only, with no directly coordinated water molecules. The fit that is shown is for a model that has this "hemisphere" of water molecules at 3.75 Å from the type 1 Cu(II). In comparison to the known structures of Lac and hCp (Fig. 9), this result indicates that in Fet3p the $N^{\varepsilon}$ of His-489 (corresponding to His-494 and His-1045 in Lac and hCp, respectively) would be just at the protein surface and contributing little to the Connelly surface. On the other hand, the H$\varepsilon$2 hydrogen at His-489 would be solvent exposed and H-bonded to water. Further ESEEM

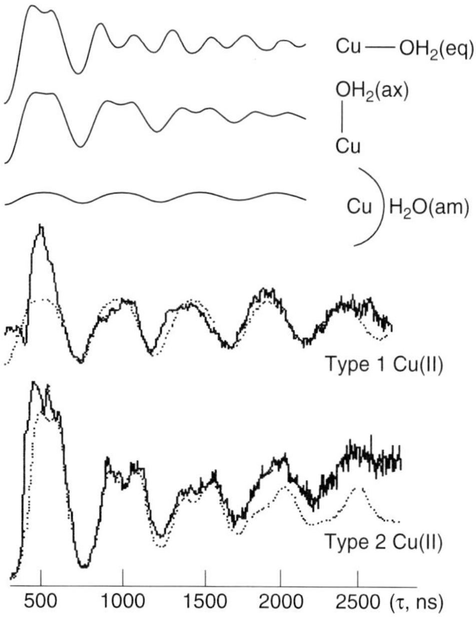


FIG. 11. Simulated and experimental two-pulse $^2H_2O$ (solvent water) ESEEM spectra. Theoretical ESEEM spectra for equatorial, axial, and ambient water are calculated as indicated. These can be compared to the experimental envelopes for the Fet3p type 1 and type 2 Cu(II) sites (solid lines) and the simulations for these envelopes assuming for the type 1 copper, only ambient water, and for the type 2 copper, a combination of one equatorial, one axial, and ambient water (dotted lines) (Aznar *et al.*, 2002).

analysis was consistent with this inference. The type 1 Cu(II) sites in the three proteins are directly compared in the illustrations shown in Fig. 12.

In addition to the relative degree of solvent exposure represented in these models, the difference in coordination can also be seen. Lac (A) and Fet3p (C) have three-coordinate, trigonal type 1 Cu(II) sites while hCp (B) has a four-coordinate, distorted tetrahedral structure. Type 1 copper sites in general exhibit this divergence, which can also be seen in the sequence alignments in Fig. 1: AO, like hCp, has the thioether of Met as a fourth ligand. What is interesting is that although this ligand does modulate the properties of type 1 Cu(II) to some extent, other features, such as protein-induced distortion, overall are much stronger determinants of type 1 copper structure and reactivity (Malmström, 1994; Solomon *et al.*, 1996).

The ratioed envelope for the type 2 Cu(II) spectrum is best fit with a combination of one equatorial water, one axial water, and a distribution of

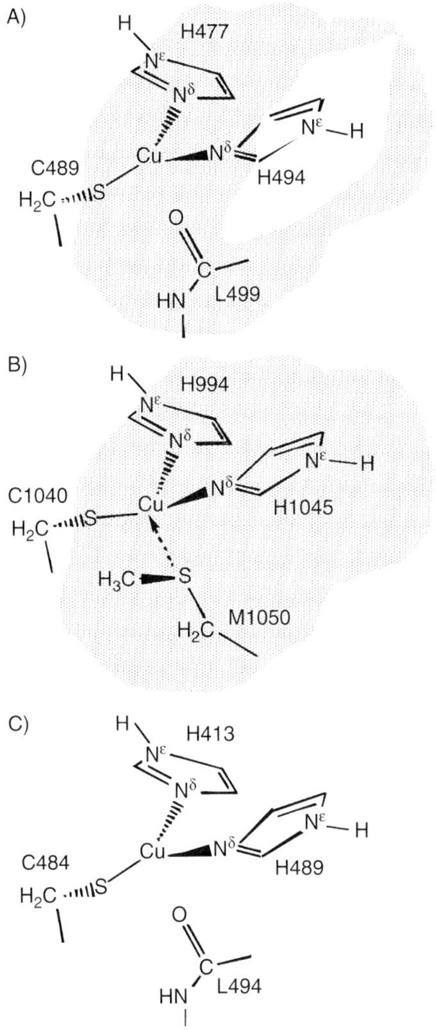


FIG. 12. Model of the type 1 Cu(II) structures in Lac, hCp, and Fet3p. The solvent accessibility is indicated by the shading. As shown in Fig. 9, His-494 in Lac is strongly solvent exposed while in hCp, His-1045 is fully buried within the protein. ESEEM analysis (Fig. 11) suggests that this residue in Fet3p, His-489, is at or near the surface of the protein, more similar to Lac than to hCp.

ambient water at a radius of 4–8 Å (Fig. 11). Together with the Fourier transform shown in Fig. 10 that is attributable to a single equatorially coordinated histidine, the water modulation analysis suggests a structure

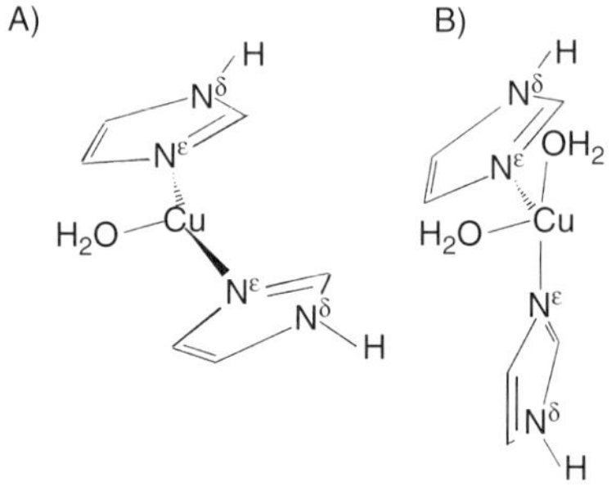


FIG. 13. Alternative type 2 Cu(II) coordination. (A) The structure of the type 2 site in AO and hCp is represented in which the two His ligands define a single, presumably equatorial plane. An equatorial $H_2O$ is indicated by the fact that anions, including peroxide, the two-electron oxygen reduction intermediate, bind equatorially in AO (Messerschimidt, *et al.*, 1993). (B) The type 2 Cu(II) coordination proposed for Fet3p. ESEEM indicates that this site has only one equatorial His and one equatorial and one axial water. The other His at this site may be axial since it does not contribute to the N modulation in the ESEEM pattern (Aznar *et al.*, 2002).

of the type 2 Cu(II) site in Fet3p that is significantly different from what crystallographic analysis has indicated for the type 2 sites in hCp (Lindley *et al.*, 1997; Zaitseva *et al.*, 1996) and AO (Messerschmidt *et al.*, 1992a, 1993). This site in these proteins appears to have two equatorial histidine imidazoles and one equatorial water ligand in an essentially trigonal planar complex. This is illustrated in Fig. 13A. In contrast, Fet3p appears to have only one equatorial histidine imidazole and one equatorial and one axial water. In as much as the type 2 site in Fet3p has both conserved histidines (His-81 and His-416) it is possible that the second of these is an axial ligand at this site and therefore does not contribute to the modulation pattern. This model is presented in Fig. 13B.

## VI. Ferroxidase Reaction

### A. *Linking Reaction to Iron Metabolism*

*In vitro* experiments with Cp and Fet3p have shown definitively that both enzymes have a strong substrate specificity toward Fe(II) that is not shared by the other well-characterized multicopper oxidases. Furthermore, several independent analyses have demonstrated that the formation of Tf•Fe(III) in a mixture of *apo*-Tf and Fe(II) is strongly catalyzed by Cp. At the other end of the experimental spectrum are the well-established correlations between organismal abnormalities in iron handling in

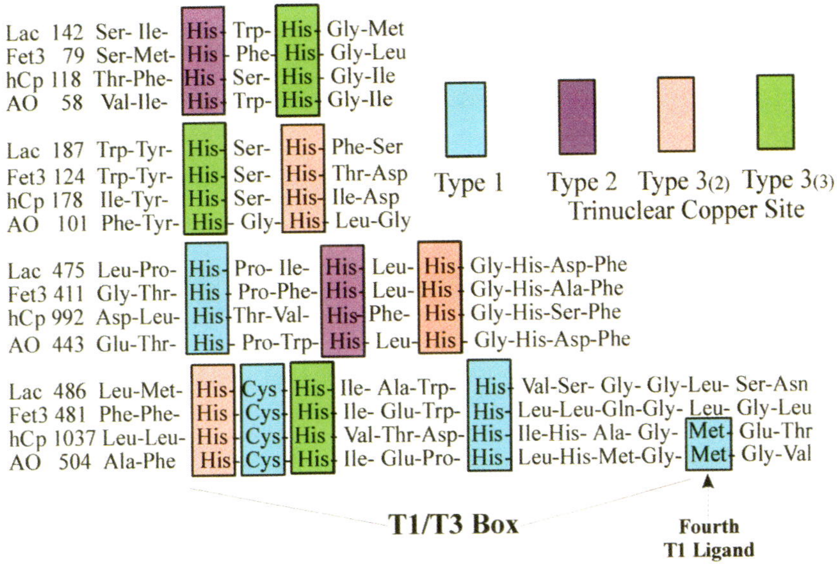


FIG. 1. Sequence alignments of the copper-binding motifs in multicopper oxidases. The sequences are for *Neurospora crassa* laccase (Lac); *Cucimus sativus* ascorbate oxidase (AO); and two ferroxidases, hCp and Fet3p, from *Saccharomyces cerevisiae*. The key indicates the ligand assignments to the three copper sites found in these proteins. The sequence numbering is referenced to the ATG that encodes the start codon in the translated mRNA and not to the carboxyl-terminal residue of the processed, mature protein. Note that all of these proteins are glycoproteins and are targeted to the endoplasmic reticulum. Thus, the nascent polypeptide contains a canonical signal recognition sequence that is cleaved as part of the proteins' overall processing, which includes the posttranslational insertion of the four active site copper atoms.

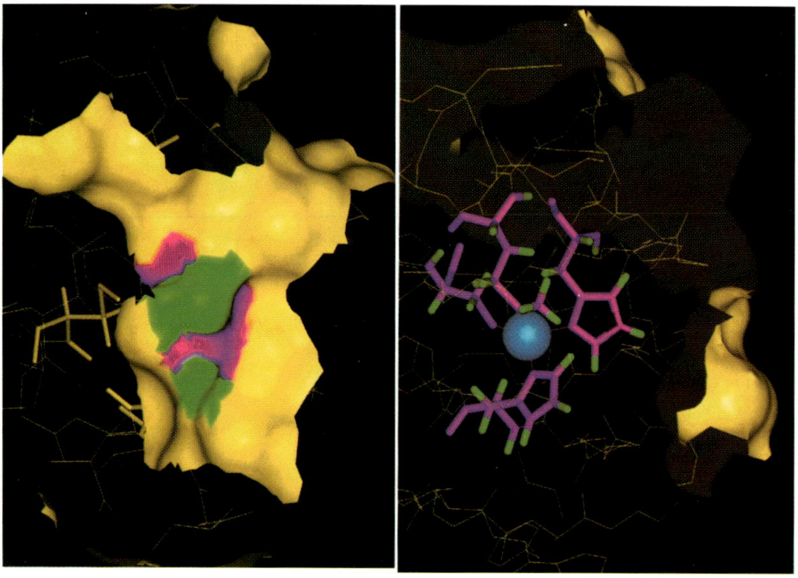

FIG. 9. The Connelly surface at the type 1 copper sites in *Coprinus cinereus* Lac (left) and hCp (right). These surfaces were generated using InsightII software from Molecular Dynamics. The structure files came from the PDB; the accession numbers are as follows: for Lac, 1A65, and for hCp, 1KCW. The type 1 site in hCp is completely shielded from solvent; H1045 (to the right), which is the closest ligand to the surface, is at least 7 Å from it. In contrast, the corresponding His ligand in Lac, His-494, is strongly solvent exposed. The olive green shading in the right panel is due to the $N^{\varepsilon}$ of this residue and the magenta indicates the adjacent ring carbons. ESSEM data indicate that this site in Fet3p is more similar to Lac than to hCp (see text and Fig. 12).

eukaryotes and genetic, nutritional, or pathologic defects in copper metallobiochemistry in those organisms. Furthermore, in many of these latter instances, a ferroxidase can be directly implicated. For example, yeast that do not produce an active Fet3p cannot take up iron, while humans that lack Cp exhibit a variety of tissue damage that can be ascribed to the presence of free, redox-active iron. However compelling these observations are in support of the hypothesis that ferroxidases are essential to iron homeostasis, the fact remains that our knowledge of the mechanism(s) that underlies this essentially is extremely limited.

Consider the observation noted above that Cp "catalyzes" the formation of Tf•Fe(III). With the rate of this process ca. 1/40 that of the Cp-catalyzed ferroxidase reaction, at the least one can say that whatever the mechanism is, it is not rate-limited by the formation of Fe(III) per se. Conceivably, the increased rate of Tf•Fe(III) formation in the presence of Cp could be solely due to an increase in the (local) concentration of Fe(III). However, there is no indication from the progress curves in Tf•Fe(III) formation that the rate of this process increases following a lag in this putative accumulation of Fe(III). In fact, there is an inherent limiting feature of a mechanism that has the Fe(III) accumulating from the Cp reaction as the basis for the Cp catalysis of the appearance of Tf•Fe(III). There is a strongly competing reaction, namely, the hydrolysis of Fe(III) at the neutral or slightly alkaline pH that is common to most biological systems. Frieden and co-workers have documented this limitation and concluded that the most efficient catalysis, as measured by the rate of Tf•Fe(III) formation and by the fractional saturation of the Tf produced, was obtained under conditions that, in fact, would limit the *accumulation* of Cp-generated Fe(III) or its hydrolysis (see Chidambaram *et al.*, 1983). This analysis, however, did not provide any insight to the molecular details of the Tf•Fe(III) formation. What is true is that under the conditions of [Fe(II)] and [$O_2$] in the plasma (~1 and 100 μM, respectively), the auto-oxidation of Fe(II) would be at most 10% the rate at which Tf•Fe(III) formation occurs in the presence of Cp; that is, the scavenging of Fe(II) by Tf in the absence of Cp would be strongly rate-limited by the generation of Fe(III). The Cp reaction is not limited in this fashion since the $K_m$ values for the two reactants, Fe(II) and $O_2$, are both ca. 1 μM and, therefore, Cp in the plasma would be at ~50% of full activity with respect to the ferroxidase reaction (Osaki, 1966; Osaki *et al.*, 1966). Even this fractional activity would be 20-fold greater than the maximum rate at which Cp drives the formation of Tf•Fe(III). In summary, in the overall reaction Fe(II) +$O_2$+ Tf, the Cp–ferroxidase step is *kinetically* competent. Kinetics, however, can only rule a step in or out; it cannot prove that the step actually happens, nor can the simple kinetic experiment establish the mechanism of a step even if its presence has been established by other means.

How the Fet3p ferroxidase reaction is linked to iron uptake in yeast is similarly ill defined. On the other hand, the essentiality of the former to the latter process is more rigorously established in that iron uptake requires that the Fet3p in the plasma membrane, presumably in some association with the Ftr1p permease, be an enzymatically active protein. In this regard, the yeast system has been invaluable in establishing a more concrete link between copper ferroxidase activity and biologic function in iron handling. It is a simple exercise in yeast to generate and produce *in vivo* mutant proteins to test the role a protein plays in some cell process and/or to determine the role of a particular amino acid residue in that protein's biologic function. A number of Fet3p mutants have been generated and subsequently studied *in vivo* and *in vitro* in both membrane-bound (Askwith and Kaplan, 1998) and soluble forms (Aznar *et al.*, 2002; Blackburn *et al.*, 2000; Bonaccorsi di Patti *et al.*, 2000; Hassett *et al.*, 1998; Machonkin *et al.*, 2002). These mutagenesis experiments have adopted one or both of two different strategies: (1) substitution of residues that are likely ligands to one of the copper atoms, thus generating copper-depleted forms of the protein, and (2) substitution of residues that could play a role in the binding of Fe(II) and/or the (outer sphere) transfer of the electron from Fe(II) to the type 1 Cu(II). The first type of mutant would be inactive since all four copper atoms are required for substrate turnover while the second type of mutant could show some altered kinetic behavior without a complete loss of ferroxidase activity.

Some of the functional characteristics of these mutant proteins are reviewed below. However, irrespective of other outcomes of these experiments, they uniformly demonstrate that if the Fet3p in the plasma membrane is inactive as a ferroxidase, the cell exhibits no Ftr1p-dependent iron uptake. To appreciate the significance of this result, one needs to appreciate the steps involved in iron (and copper) uptake in yeast (Askwith *et al.*, 1996; Hassett and Kosman, 1995; Kosman, 1993; Wessling-Resnick, 1999). This is diagrammed in Fig. 14.

The first key element in the uptake of these two metal ions is that the substrate is the lower valent state species, Cu(I) in the case of copper and Fe(II) in the case of iron (Dancis *et al.*, 1990, 1992; Hassett and Kosman, 1995; Kosman, 1993). Normally, these reduced valence species are provided by the action of plasma membrane metal reductases, an activity in yeast provided predominantly by the product of the *FRE1* gene (Dancis *et al.*, 1992). However, Fe(II) [or Cu(I)] provided exogenously to the cell is equally competent for uptake and, in most experimental regimes, is added directly or generated *in situ* by the addition of a strong reductant like ascorbate or dithionite. Cu(I) is the direct substrate for uptake, through the Ctr1p copper permease in most yeast strains (Dancis *et al.*, 1994). However, the presence of Fe(II), *although required*, alone is not

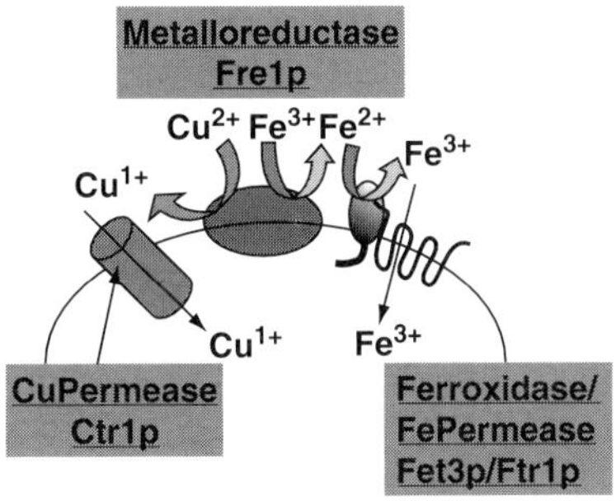


FIG. 14. Redox cycling in the uptake of copper and iron. The lower valent state species is substrate for uptake of copper and iron. The system in the yeast *Saccharomyces cerevisiae* is diagrammed. The Fre1 protein reduces environmental $Cu^{2+}$ and $Fe^{3+}$. The cuprous ion is substrate for the copper permease, Ctr1p. $Fe^{2+}$ is substrate for Fet3p; its oxidation to $Fe^{3+}$ is an obligate step in iron uptake through Ftr1p. Exogenous ferric iron is not taken up by yeast cells unless it is cycled through the ferrireduction–ferroxidation reactions catalyzed by Fre1p and Fet3p.

sufficient for the uptake of iron. This Fe(II), whether generated by the Fre1 protein or by the action of ascorbate on exogenous Fe(III) or added as ferrous ammonium sulfate, must undergo an obligate oxidation (ferroxidation) by Fet3p prior to iron transport into the cell. If the Fet3p in the membrane is inactive, there is no iron uptake. Most importantly, exogenous Fe(III) is *not* competent for uptake; that is, while one can bypass the Fre1p reaction by adding Fe(II), one cannot productively bypass the Fet3p ferroxidase reaction by adding Fe(III). In short, iron uptake through Ftr1p is *coupled* to the Fet3p ferroxidase reaction in the strictest sense of that word.

A variety of kinetic data support this tight coupling. This model requires, for example, that iron uptake is $O_2$-dependent since, according to this model, uptake is downstream from the ferroxidase reaction. This is the case as is shown in Fig. 15, which gives the $[O_2]$ dependence of $^{59}Fe$ uptake. This kinetic analysis also gives the $K_m$ for oxygen in uptake. The $V_{max}$ for uptake can also be obtained. These kinetic constants can be compared to the kinetic constants for the ferroxidase reaction catalyzed by Fet3p *in vitro*. The data for this analysis are shown in Fig. 4. Table IV summarizes all of the relevant kinetic constants for iron and $O_2$ in either the Fet3p ferroxidase reaction *in vitro* or the Fet3p-dependent uptake of iron *in vivo*.

These constants indicate that the ferroxidase reaction catalyzed by Fet3p might be rate-limiting in iron uptake. The value of $k_{cat}$ for $^{59}Fe$ uptake, 36 $min^{-1}$, is based on the assumption that there are 1000

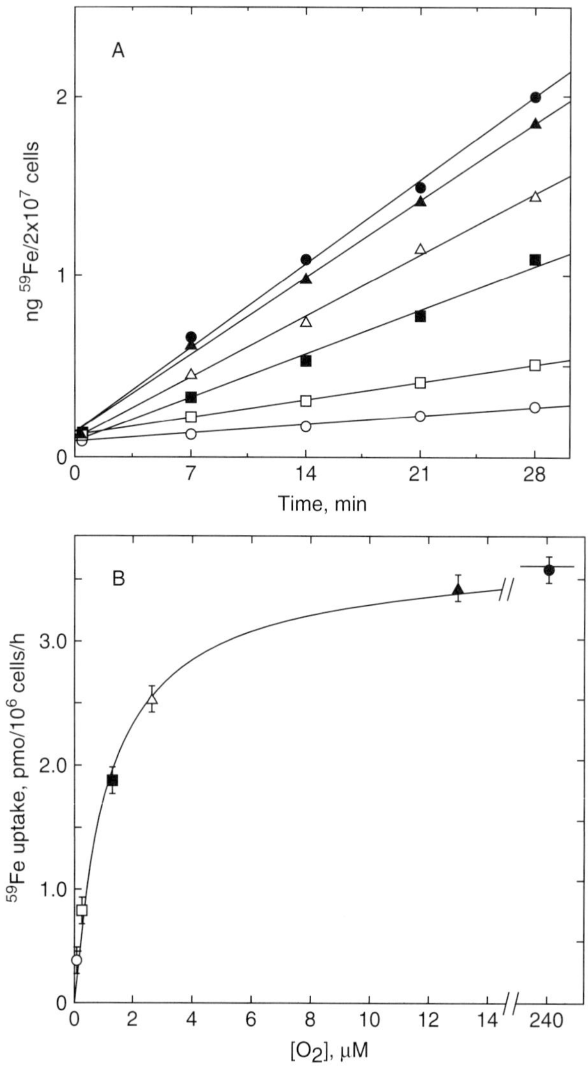


FIG. 15. Oxygen dependence of iron uptake in yeast. (A) Progress curves for $^{59}$Fe uptake by wild-type yeast strain DEY1457 are shown. The individual uptake curves are obtained at different concentrations of $O_2$ ranging from 1.2 to 240 μM. (B) The uptake velocities derived from the traces in (A) are plotted versus the $[O_2]$; the line is the theoretical fit of the data to the Michaelis–Menten equation using the fitted constants given in Table IV.

TABLE IV
*Kinetic Constants for the Fet3p Ferroxidase Reaction in Vitro and* $^{59}$Fe *Uptake in Vivo*

| Reaction | $K_m(O_2)$ (μM) | $K_m(Fe^{2+})$ (μM) | $k_{cat}$ (min$^{-1}$) |
|---|---|---|---|
| Fet3p ferroxidation | 1.3 | 4.8 | 9.5 |
| $^{59}$Fe uptake | 1.2 | 5[a] | 36[b] |

[a]Value taken from Dancis *et al.* (1990).
[b]Value calculated from $V_{max}$ for $^{59}$Fe uptake found at limiting $[O_2]$ (Fig. 15B) of 3.6 pmol Fe/$10^6$ cells/h, assuming there are $10^3$ Fet3p•Ftr1p complexes/cell.

Fet3p/Ftr1p complexes in the yeast plasma membrane. (Uptake is most commonly reported on a per cell basis and so provides "normal" $V_{max}$ values. Converting these to "molar" values requires an assumption about the normality of the transporter in the cell membrane.) This estimate has no experimental basis, but it is not likely to be more than an order of magnitude in error. Thus, at the least the $k_{cat}$ values for ferroxidation and uptake are very similar. Furthermore, there is a good correspondence between the $K_m$ values for oxygen and iron in the two kinetic processes, i.e., the $K_m$ values associated with iron uptake are comparable to those values found for the ferroxidase reaction alone. With respect to oxygen, in particular, this result is consistent with a sequential process in which the $O_2$ utilization step generates as a product the substrate for subsequent iron transport. However, although somewhat more rigorous than the results described above for the Cp, Tf system, these data are no more definitive as to how these two processes are mechanistically linked.

Structure–function studies on the Ftr1 protein are an equally logical experimental approach since the permease is a partner in this apparently coupled process. Such data for Ftr1p are very limited, however. Sequence analysis suggests that Ftr1p has six transmembrane-spanning domains; the topology of the protein in the yeast plasma membrane could be as pictured in Fig. 16 (top) (Stearman *et al.*, 1996). Fet3p is included in this diagram. In this orientation, the carboxyl-terminal domain in Ftr1p is on the extracellular surface, on the same side of the membrane as the Fet3p active site, as indicated. It should be noted that the *relative* orientation of Fet3p and Ftr1p shown in the top panel of Fig. 16 is the opposite of what has been indicated for Fet5p and Fth1p; the proposed membrane topology of this heterodimeric complex is shown in the bottom panel of Fig. 16 (Urbanowski and Piper, 1999). Keeping in mind that the lumen of an intracellular compartment is topologically continuous with the extracellular milieu, one can appreciate that the permease component has a

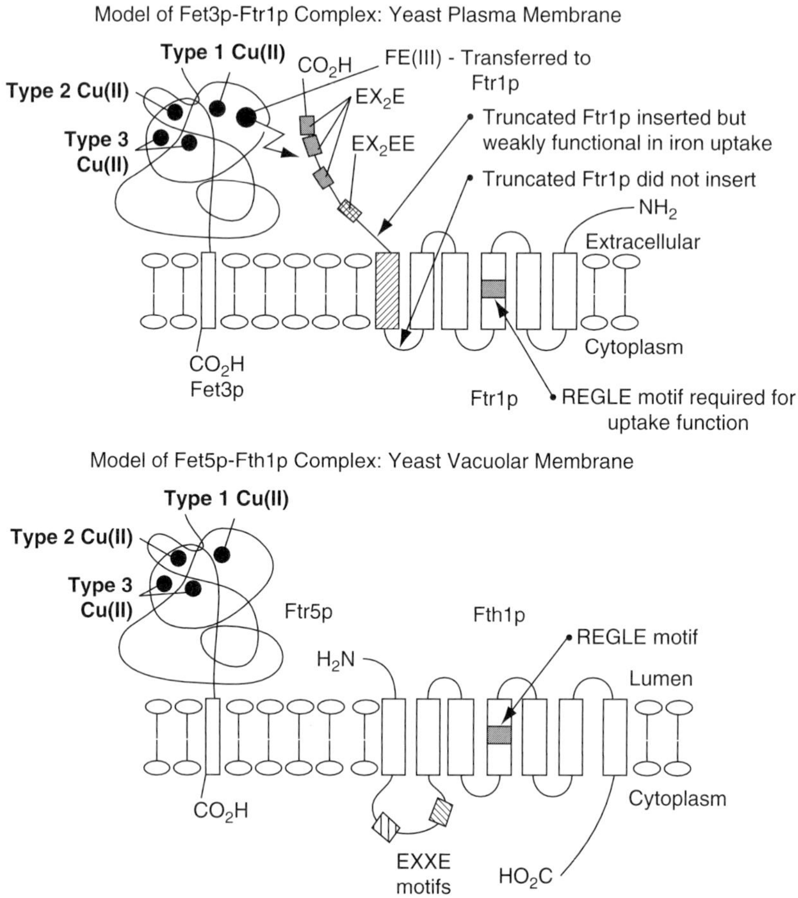


FIG. 16. Models for the membrane topology and orientation for the Fet3p/Ftr1p and Fet5p/Fth1p complexes in yeast membranes. (Top) Fet3p/Ftr1p are pictured in the yeast plasma membrane oriented to the extracellular space. The (R)EXXE motifs that may be involved in iron uptake and the transmembrane domain required for correct assembly are indicated. These roles have been suggested by mutagenesis studies (Stearman *et al.*, 1996). (Bottom) Model of Fet5p/Fth1p complex proposed for the membrane of the yeast vacuole oriented to the lumen (inside) of the vacuole. In this model, the EXXE motifs in Fth1p are not oriented toward the ferroxidase site on Fet5p (Urbanowski and Piper, 1999).

somewhat different predicted topology in the two cases. An experiment has indicated that the carboxyl-terminal domain of Fth1p has a cytoplasmic orientation in the vacuolar membrane as shown in the model (Urbanowski and Piper, 1999). No comparable experiment(s) has established the precise orientation of Ftr1p in the plasma membrane, nor has the actual domain structure of either protein been experimentally established.

However, Stearman *et al.* (1996) did generate two C-terminal truncations of Ftr1p that were informative. Protein that lacked the carboxyl-terminal residues 278–404 failed to insert into the plasma membrane (see Fig. 16, top). This protein would lack what is seen in the diagram as constituting the sixth membrane-spanning domain; its failure to be correctly targeted is consistent with this structural assignment. In contrast, a truncated protein that lacked residues 334–404 was targeted to the membrane; however, the iron uptake supported by this mutant protein was less than 10% of the rate supported by wild-type Ftr1p. This extreme part of the carboxyl-terminal domain is predicted to be entirely in the extracellular space as indicated in Fig. 16 (top). Inspection of the Ftr1p C-terminal sequence shows that it contains four EXXE repeats. The latter mutant contained none of these repeats, suggesting that they play some role in the trafficking of iron during uptake. Given their extracellular orientation, they could play a specific role in the transfer of the Fet3p Fe(III) product to the domain(s) in Ftr1p that is directly involved in the passage of the iron across the plasma membrane. These motifs are roughly conserved in the *Sc. pombe* Ftr1p homologue, Fip1p, in that Fip1p has one DXXD, one EXXE, and one EXXD element in the corresponding carboxyl-terminal region (Askwith and Kaplan, 1997). It may be significant that in contrast Fth1p has no such motifs in this region of its primary structure. It does have elements like these but as noted in Fig. 16 (bottom), they do not align with those in Ftr1p and Fip1p. Obviously, how ferroxidation is linked to iron permeation remains to be experimentally established. Nonetheless, it is clear that the yeast system will be relatively amenable to the necessary systematic, quantitative tests of possible mechanisms.

### B. *Structure and Reactivity*

#### 1. *Structural Studies: hCp*

Crystallographic analysis has provided us with a detailed structure of hCp; on the other hand, essentially all of the structure–function analyses have been done on Fet3p. Also, except for the copper site structural homology, the two proteins are quite different. hCp is composed of six plastocyanin-like domains (plastocyanin is a type 1 copper-containing protein) that are arranged in a trigonal array (Zaitseva *et al.*, 1996). One result of this domain replication is a conformational fold that produces a distinct, negatively charged patch on the protein surface adjacent to the catalytically active type 1 Cu(II). This copper atom is in domain 6. (Domains 2 and 4 contain type 1-like copper sites that do not participate in the ferroxidase reaction.) Lindley *et al.* (1997) have proposed that this patch contains two ligand arrays essential to the ferroxidase reaction. First, they interpreted a region of electron density adjacent to the type 1 Cu(II) as a "labile" copper-binding site that includes residues E291, E954, H959, and D1044. (As noted, this numbering is referenced to the mRNA sequence that includes the signal sequence cleaved during processing.) They postulate that Fe(II) binds at this site and is oxidized in an outer-sphere electron transfer to the type 1 Cu(II). H959 and/or D1044 could provide the path for this process since both are in contact with the two histidine imidazoles that are ligands to the type 1 copper. Immediately adjacent to and on the solvent side of this labile site is a region in which additional electron density is observed when the crystals are soaked in Fe(III). The authors refer to this as the Fe(III)-holding site for the ferric ion product of the ferroxidase reaction. The ligands at this putative Fe(III) site are not well delineated. However, D249, E950, E951, and E976 are in the vicinity of this electron density. Also, the electron density due to E954 shifts away from the labile site toward this "holding" site, suggesting that this side chain could play a role in the channeling of the Fe(III) from one site to the other. As appealing as this model is, it is based on fairly limited structural data. Systematic structure–function studies have yet to be done to test this model, which is illustrated in Fig. 17.

### 2. *Structure–Function Studies: Fet3p*

In contrast, structure–function studies on Fet3p have been reported, but, in most respects, are of little relevance to the model that Lindley *et al.* (1997) have proposed for the iron channeling in hCp. Fet3p most certainly does not have the trimeric structure that Cp has and thus cannot have the ligand arrays that Lindley proposes for the labile and holding sites. In both sites, the ligands come from at least two of the six domains. There is nothing in the Fet3p sequence to indicate that Fet3p has a tertiary fold of this nature. On the other hand, Fet3p most reasonably has one or more ligand arrays that play a similar function. What the crystallographic data on hCp and their interpretation suggest is that these arrays would be composed of similar amino acid side chain types, specifically D or E, and H, residues. Indeed, Askwith and Kaplan (1998) based their mutagenesis studies on the very reasonable assumption that such residues would provide this type of metal ion coordination. They prepared several Fet3p mutants and tested them for their ability to support iron uptake *in vivo* and for their ferroxidase activity *in vitro* as visualized by a histochemical enzyme stain of membrane extracts fractionated in polyacrylamide gels. Among these mutants were E227A, D228A, and E230A. These are found in "ferroxidase" Box 3 (Fig. 5B) and, except for the latter residue in Fio1p, are fully conserved in all the yeast ferroxidases identified by sequence analysis to date. Based on the hCp

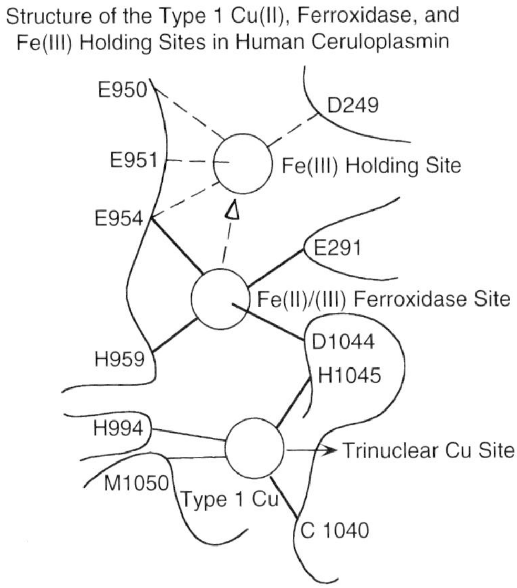


FIG. 17. Model of the type 1 Cu(II), ferroxidase, and Fe(III) holding sites in hCp. $Fe^{2+}$ oxidation is proposed to occur with the metal bound at a site adjacent to the type 1 copper. Crystallographic data suggest that E291, E954, H959, and D1044 are ligands to the iron at this site (Lindley *et al.*, 1997). The $Fe^{3+}$ produced is suggested to then migrate to a "holding site" that is closer to the protein surface. The ligation at this site is not resolved crystallographically; the residues indicated are potential ligands.

structure, all were predicted to play a role in the binding of Fe(II) or the electron transfer reaction or both (Buonaccorsi di Patti *et al.*, 1999; Murphy *et al.*, 1997). In fact, none of these mutants constructed by Askwith and Kaplan (1998) exhibited any significant loss of ferroxidase activity in this qualitative histochemical assay, and all supported a normal rate of iron uptake. Possibly, a more rigorous kinetic analysis would reveal some deficit in enzymatic activity in one or more of these mutant proteins. However, in light of their normal physiologic activity, it is highly unlikely that any one of these residues plays a critical role in the structure and function of Fet3p.

One of the two systems that has been developed to produce soluble, recombinant Fet3p involves expression of *FET3* in insect cells using the standard Baculovirus system (Bonaccorsi di Patti *et al.*, 2000). The protein produced localizes to the cell membrane from which it can be released in soluble form by mild trypsin digestion. The protein released is in a copper-free, *apo* form and must be incubated with copper *in vitro* to attain enzymatic activity. In this system, two mutants were constructed based on modeling studies that compared hCp to Fet3p (Buonaccorsi di Patti *et al.*, 1999). These studies suggested that Fet3p residues E185 and Y354 could be involved in iron binding. The inclusion of Y354 in this set is somewhat surprising inasmuch as Y would preferentially bind Fe(III), not Fe(II). On the other hand, both residues are fully conserved among the yeast ferroxidase homologues (see Fig. 5B, Box 2 and Box 4). In any event, only the E185A mutant exhibited strikingly different kinetics with Fe(II) as substrate. For this mutant, $V_{max}$ was reduced 2.5-fold while $K_m$ for Fe(II) was increased 6-fold. The result of these two changes was a 16-fold decrease in $V_{max}/K_m$ for Fe(II) turnover. This result is ambiguous since these kinetic constants can reflect the contributions of several discrete steps in the overall enzymatic reaction. Among these are the inter- and intramolecular electron transfer steps, the electron transfers to $O_2$, and the release of $H_2O$. Therefore, it is not possible to conclude from this type of experiment that a particular amino acid residue is playing a specific role in only one of these steps. Nonetheless, mutation at E185 does alter the ferroxidase activity of Fet3p. Therefore, this residue may be a candidate for one of the ligands involved in iron binding and/or electron transfer into the type 1 Cu(II). These initial data indicate that its role in these processes should be thoroughly delineated. As noted, E185 is found within one of the highly conserved motifs that appear to *distinguish* the yeast ferroxidases from those found in other eukaryotes, e.g., flies and mammals. Therefore, it may have a specific role in the ferroxidase reaction of Fet3p and there may not be a functionally homologous residue in hCp.

### *3. Copper Site Structural Studies: Fet3p*

What all of these proteins have in common, however, are the ligand arrays for the four copper atoms. A yeast system has been used to produce recombinant Fet3p that lacks one or more of these copper atoms as a result of substitution for one of the liganding side chains (Aznar *et al.*, 2002; Blackburn *et al.*, 2000; Hassett *et al.*, 1998; Machonkin *et al.*, 2001). This system is based on a truncation of the *FET3* gene at nucleotide +1666 (at amino acid residue 555); a nucleotide sequence encoding the FLAG epitope was appended at that point. The result of this manipulation was the production of a Fet3p that lacked its carboxyl-terminal, membrane-spanning domain. Instead, it was epitopetagged at its C-terminus. Lacking its membrane domain, this protein was secreted into the growth medium rather than remaining tethered to the plasma membrane. This strategy is illustrated in Fig. 18. The protein produced was easily recovered from the growth medium by ion-exchange chromatography. The yield of pure protein was 5–10 mg/liter depending on the precise protein species being produced. This system has two significant

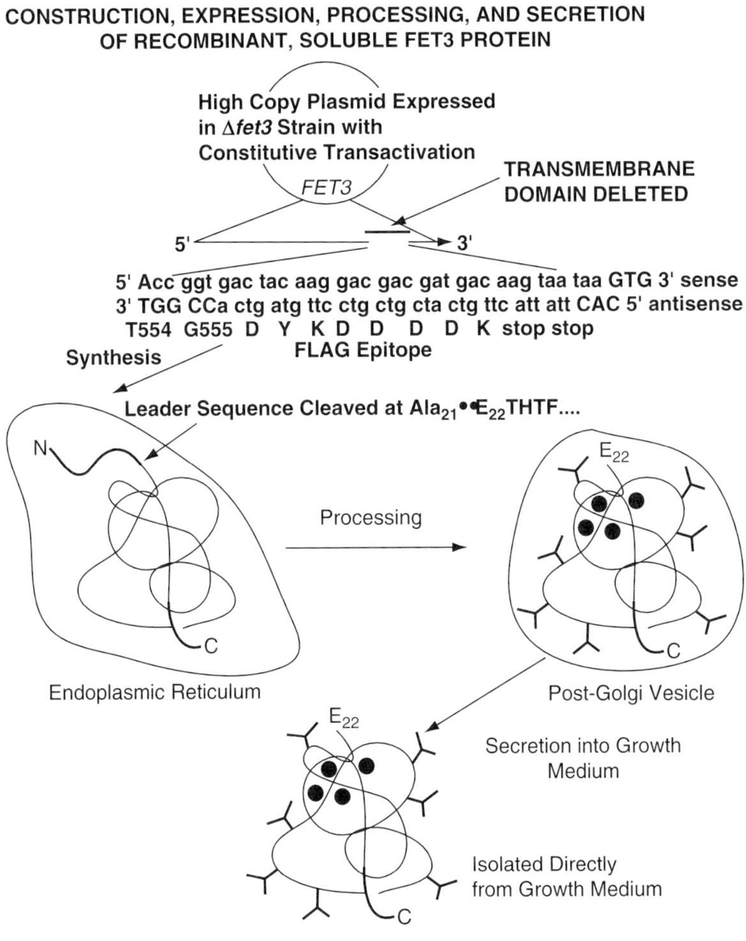


FIG. 18. Strategy for construction, expression, processing, and secretion of a recombinant, soluble Fet3 protein in yeast (Hassett *et al.*, 1998).

advantages over the Baculovirus system. First, the protein is recovered directly from the growth medium without protease treatment. More importantly, the protein is processed normally within the yeast cell and therefore contains all of its copper prosthetic groups. The kinetic data shown in Fig. 4 and summarized in Table IV were obtained with this recombinant protein without any additional copper treatment. This system has been exploited to produce a family of mutant Fet3 proteins. Thus, T1D, T2D, and T1D/T2D mutants have been produced, where T1D means "type 1 depleted." The mutations that resulted in the loss of copper binding to the three copper sites are illustrated in Fig. 19. This mutagenesis strategy allowed for the isolation of each of the copper sites in turn for either spectroscopic or kinetic study independent of the contributions from one or more of the other sites. For example, the EPR data shown in Fig. 2 and the ESEEM data shown in Figs. 10 and 11 for the type 1 and type 2 Cu(II) could be obtained only because in these samples the type 2 (T2D) and type 1 Cu(II) (T1D) sites were deleted, allowing for EPR study of the remaining Cu(II).

These copper site mutants have been particularly useful in two different types of experiments. First, the T2D protein allows for investigation of the electron transfer to the type 1 Cu(II) in the absence of turnover. This is because in the multicopper oxidase reaction, electron transfer from the type 1 copper—as Cu(I)—to the trinuclear cluster where $O_2$ is reduced requires the type 2 Cu(II) (Solomon and Lowery, 1993; Solomon

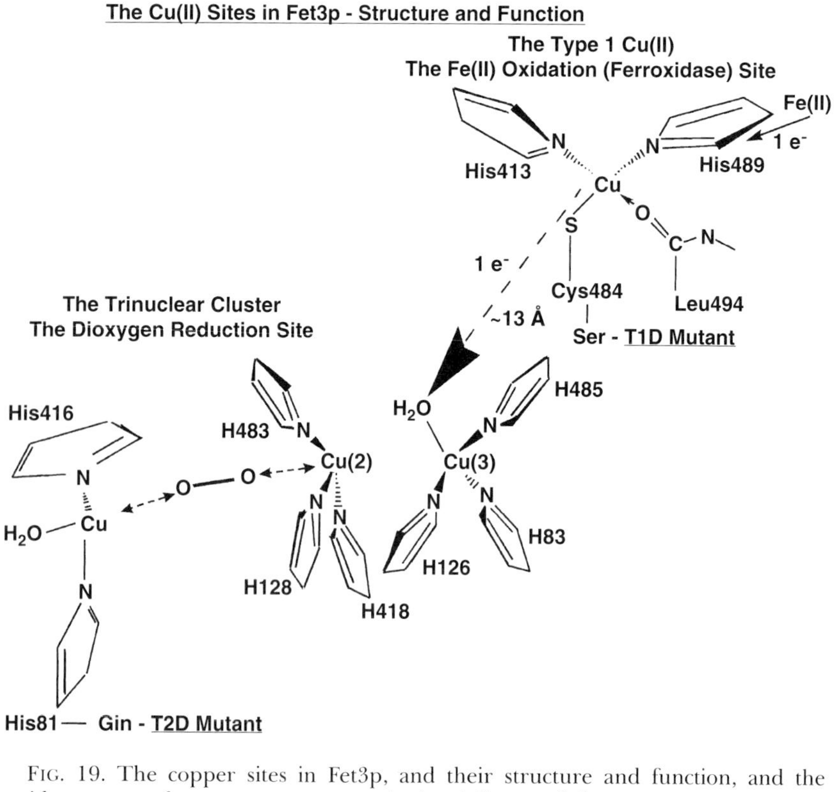


FIG. 19. The copper sites in Fet3p, and their structure and function, and the residues mutated to generate copper-depleted forms of the enzyme (Aznar *et al.*, 2002; Blackburn *et al.*, 2000; Hassett *et al.*, 1998; Machonkin *et al.*, 2001).

*et al.*, 1996). Thus, addition of Fe(II) or other substrate (*o*-dianisidine, *p*-phenylenediamine, hydroquinone) to the fully oxidized T2D protein results in the reduction of the type 1 Cu(II) without further turnover (Blackburn *et al.*, 2000). This reaction is shown in Fig. 20A with Fe(II) as

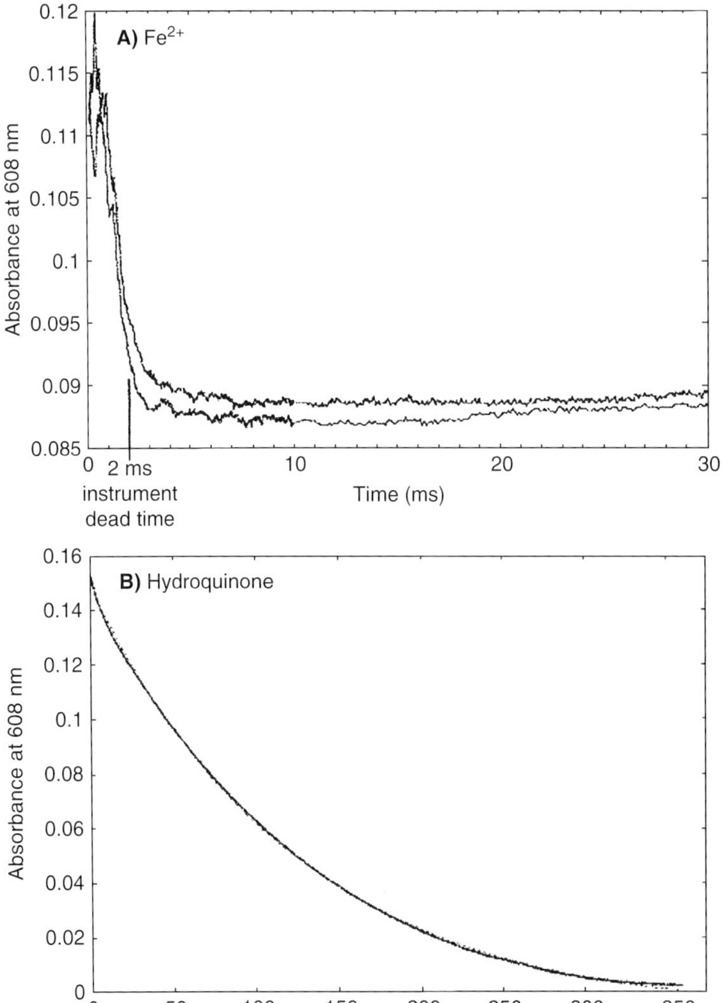


FIG. 20. Reduction of the type 1 Cu(II) in Fet3p by $Fe^{2+}$ (A) and hydroquinone (B). The reduction is followed by the 608-nm absorbance due to the type 1 Cu(II). The [Fet3p] was 35 μM in MES, pH 6.5, with $[Fe^{2+}] = 243$ μM and [hydroquinone] = 565 μM (Machonkin *et al.*, 2001).

substrate and in Fig. 20B with hydroquinone as substrate (Machonkin *et al.*, 2001). The results demonstrate clearly the strong selectivity that Fet3p has toward Fe(II) in comparison to hydroquinone. The reaction with Fe(II) is complete within the instrument dead time ($\sim$2 ms); therefore, the $k_{obs}$ must be $>1200 s^{-1}$. Under similar conditions of reactant concentrations, the $k_{obs}$ for reduction of the type 1 Cu(II) in Fet3p by hydroquinone was 0.008 $s^{-1}$. The large $k_{obs}$ for the reaction with Fe(II) suggests that electron transfer takes place within a Fet3p Fe(II) substrate complex although it does not require it.

Using a stopped-flow technique, Machonkin *et al.* (2001) measured the rate constant for the Fet3p–hydroquinone reaction and compared it to the type 1 reduction rates for hCp and Lac. The purpose of this comparison was to establish any possible correlation between solvent accessibility of the type 1 sites in these proteins (e.g., see comparisons shown in Figs. 9 and 12) and their reactivity. The second-order rate constants for the reaction of hCp and Fet3p with hydroquinone were $6.1 \times 10^4$ and $3.5 \times 10^5 M^{-1}s^{-1}$, respectively. The value for the *Co. cinereus* Lac reaction was $>10^7 M^{-1}s^{-1}$. This pattern of reactivity parallels the solvent accessibility of the type 1 site in the three proteins in that this site in hCp is completely buried while one edge of this site in *Co. cinereus* is fully solvent accessible. As suggested by the ESEEM data, the Fet3p type 1 Cu(II) falls in between these two extremes (Aznar *et al.*, 2002).

There was no similar correlation between reactivity toward Fe(II) and solvent exposure. Fet3p and hCp exhibited similar rates of type 1 Cu(II) reduction by Fe(II) ($k_{obs} > 1200 s^{-1}$) while the rate with *Co. cinereus* Lac was $>23 s^{-1}$. In other words, laccases can use Fe(II) as substrate but have no better than 1–2% of this activity in comparison to Fet3p and hCp. In addition, they are at least 100-fold better than the ferroxidases in the turnover of bulky organic reductants. Combining the structure and reactivity features of these proteins indicates that the type 1 sites in the ferroxidases are less accessible to these large reductants and at the same time possess specificity elements that support the recognition and binding of Fe(II) as substrate. As outlined above, some of these elements may have been identified in hCp; they remain uncharacterized in Fet3p.

Theoretically, the redox potential of the type 1 Cu(II) could provide the specificity toward a particular reductive substrate in comparison to another. Type 1 copper sites do exhibit a remarkable variability in reduction potential, from 240 mV in nitrite reductase (LaCroix *et al.*, 1996; Suzuki *et al.*, 1994) to >1 V for a noncatalytic type 1 copper in hCp (Machonkin *et al.*, 1998). This variability is certainly due in part to the presence of the methionine ligand found in the four-coordinate type 1 site (as in hCp and AO) that is absent in Lac and Fet3p, for example (cf. Figs. 1 and 12). Mutagenesis studies indicate that this ligand "tunes" the $E^{\circ\prime}$ down by ~100 mV (Xu *et al.*, 1999). On the other hand, the ferroxidase-active, type 1 sites in Fet3p and hCp have essentially identical reduction potentials—427 and ~450 mV, respectively—despite the fact that the latter has Met coordination while the former does not (Machonkin *et al.*, 2000; Machonkin and Solomon, in press). In summary, available data do not suggest that a simple linear free energy relationship links substrate specificity to type 1 reduction potential in the multicopper oxidases, nor does the presence or absence of a Met ligand predict the $E^{\circ}$. Instead, ferroxidase activity appears to be an acquired trait due to protein elements that, in addition to the copper ligands and environment, directly modulate the spectral and redox properties of the type 1 copper atom itself.

One recombinant Fet3p mutant is unique among multicopper oxidase species and has been particularly informative about the structure of the type 3 binuclear cluster in these species. This is the TlD/T2D double mutant that contains only this type 3 site (Blackburn *et al.*, 2000). EXAFS analysis of this protein contains contributions from electron ejection and scattering from only the type 3 copper atoms and thus provides direct structural information about this cluster. The K-edge XAS spectrum for this mutant in its oxidized and reduced states is shown in Fig. 21. The oxidized sample has a nearly featureless edge with a midpoint energy of 8990 eV typical of tetragonally distorted type 2 Cu(II) centers, i.e., those with predominantly histidine imidazole coordination. The reduced type 3 cluster exhibited a pronounced shoulder at 8984 eV just below the

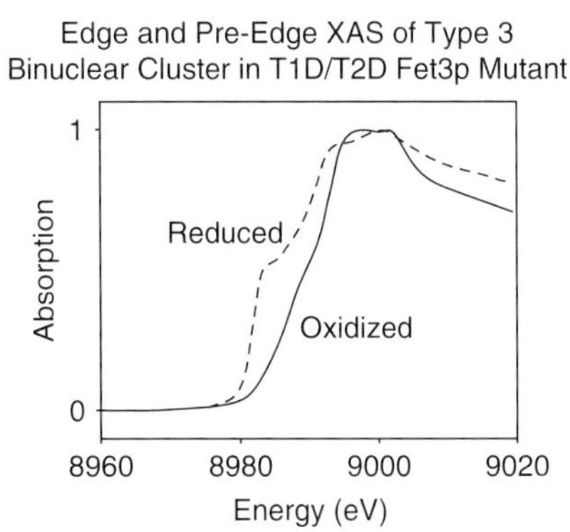


FIG. 21. Edge and pre-edge XAS of the binuclear type 3 copper site in a Fet3p double mutant. The TlD/T2D Fet3p sample was 0.5 mM in MES buffer, pH 6.0, at 11–14 K. The data were collected at the Standford Synchrotron Radiation Laboratory (Blackburn *et al.*, 2000). The XAS spectra for the oxidized and reduced type 3 sites are indicated.

midpoint energy of 8987 eV. The shoulder is characteristic of a three-coordinate site with a significant doming of the Cu(I) out of the trigonal plane, presumably toward a more weakly bound fourth ligand.

The EXAFS analysis of the reduced and the oxidized T1D/T2D mutant protein and of the T1D mutant that had an intact trinuclear center gave a rather complete picture of the coordination at the type 3 cluster. The best fit of the EXAFS data in all cases included the coordination of three histidine imidazoles to each of the two type 3 coppers consistent with the K-edge results. Furthermore, the bridging oxygen ligand that electronically couples the two Cu(II) ions in the oxidized state was strongly evident in the data. Also evident was the scattering that gave a Cu–Cu distance in the cluster of 3.33 Å. The fitted data for the oxidized cluster are indicated in the structure shown in Fig. 22A.

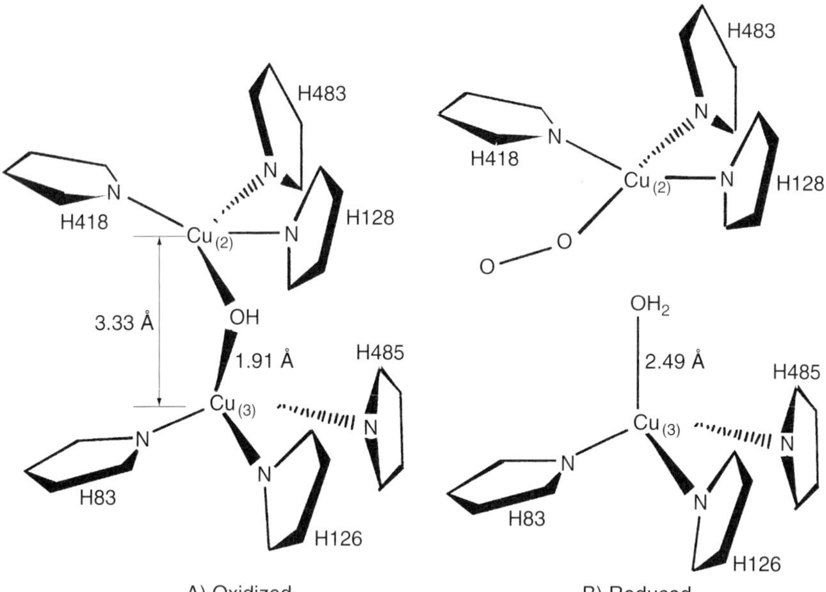


FIG. 22. Model for the coordination changes at the binuclear copper site in Fet3p on reduction. This model is based on the fitting of the EXAFS data from the oxidized and reduced forms of T1D and T1D/T2D Fet3p. This fitting indicated three significant structural differences on cluster reduction: (1) loss of the bridging oxygen ligand; (2) separation of the two copper atoms; (3) appearance of a nonbridging, O/N ligand with a relatively long bond to one of the copper atoms. Also pictured in the model for the reduced state is the dioxygen liganded to $Cu_{(2)}$ as has been proposed based on a variety of spectral and kinetic data. This model suggests that the proposed water at $Cu_{(3)}$ resulting from protonation of the bridging (–OH) on reduction could serve as an acid catalyst of the reduction of the bound $O_2$ (Blackburn *et al.*, 2000).

The analysis of the data for the reduced protein was slightly more ambiguous. This ambiguity followed from the fact that the best fit required an additional single oxygen or nitrogen first-shell atom at 2.49 Å distributed between the two copper atoms. This was at a considerably longer distance than the bridging O in the oxidized cluster (2.49 Å versus 1.91 Å; see Figs. 22A and 22B). Also, the Cu–Cu distance in the reduced cluster was greater since there was no evidence of Cu backscattering in the data. One interpretation of these data is given in Fig. 22B. In the model, the two copper atoms move apart upon reduction. The oxygen bridge is disrupted and the putative hydroxide is protonated, thus making cluster reduction electroneutral. (As presented in Fig. 22, the formal charge of both the oxidized and the reduced states is +2; this is arbitrary, of course, since it depends on how one writes the state of protonation/ionization of the bridging oxygen/hydroxide.) In this model, the two Cu(I)ions could now serve different roles in $O_2$ binding and reduction. The open apical coordination site on $Cu_{(2)}$ (see Fig. 22 for numbering; see also Fig. 8) would be one binding site for the dioxygen molecule; the other site, tethering the O–O to the protein, would be the type 2 Cu(I), as has been indicated by several studies (cf. Shin *et al.*, 1996). The other type 3 Cu(I), in addition to supplying one of the electrons for dioxygen reduction, serves as part of an acid catalyst, thereby also supplying a proton(s) to the reduced oxygen species, $O_2^{2-}$ and $HO^{\cdot}$. There certainly needs to be proton transfer coincident with electron transfer to the dioxygen. The EXAFS data on the T1D/T2D Fet3p are the first to suggest a model for this part of dioxygen reduction at the trinuclear cluster of a multicopper oxidase.

## VII. Convergence of Structural and Cell Biology in Iron Metabolism

There really is no experimental limitation to establishing a complete understanding of how the ferroxidases Fet3p and hCp, and their congeners in the same and other organisms, work, in both molecular and physiologic terms. Within the next few years, most likely a high-resolution structure of Fet3p will become available, as will data on structure–function studies on mutant forms of recombinant hCp. The current gaps in knowledge will have been filled. There will be data on the electrophysiology of Fe(II) oxidation by Fet3p and iron uptake through Ftr1p in a heterologous eukaryotic system, e.g., *Xenopus* oocytes or transfected Caco-2 cells. These studies would test models for the coupling mechanism that links Fe(II) oxidation to iron uptake. The cellular role of hephaestin in human iron metabolism and that of Fet5p/Fth1p in yeast will have been established. Additional ferroxidases will have been identified, their genes cloned, and their cell functions characterized. The links between ferroxidases and the $Fe^{2+}$ transporters—those in both the plasma membrane and intracellular membranes—in the various cell types in mammals will become clearer. And in the background of these quantitative, structural, and mechanistic studies will be the musings about the evolution of the multicopper oxidase that had ferroxidase activity. Copper and iron, of course, are intimately associated with aerobic metabolism, although copper and iron enzymes are not confined to obligate aerobes. Although the ferroxidase reaction is explicitly an aerobic process, taking Fet3p as an example ($K_m$ for $O_2 = 1\,\mu M$, equivalent to a partial pressure under a 0.1% $O_2$ atmosphere) it can proceed under microaerobic conditions. These conditions do not obtain for yeast under most laboratory conditions, but they certainly do for hCp since free $O_2$ in the plasma is strongly buffered by hemoglobin, and they certainly would have obtained in the early stages of geologic aerobiosis. Certainly, one of the early events in aerobiosis would have been evolution of the copper- and iron-requiring processes of respiration that enormously extended the energy-production capacity of facultative aerobes. Mechanisms for supplying iron to cells to support these respiratory functions would have been part of this early evolution. In this view, the ferroxidases may have been the first of the multicopper oxidases and may have been selected for specifically as a component of iron uptake in these first eukaryotes. There is an irony in this view in that the target for this cellular iron, whose uptake required the action of a copper ferroxidase, was itself a copper- and iron-dependent enzyme—cytochrome *c* oxidase—and both enzymes were dioxygen reductases. On the other hand, this may not be irony, but a very reasonable evolutionary trick. After all, both enzymes are also multinuclear metallo-oxidases with surprising homologies despite their disparate overall structures and cofactors. Reasonably, they are part of the same adaptive stream; perhaps they also part of the same evolutionary tree. This question, too, may be brought closer to resolution in these next few, exciting years of research on ferroxidases and iron homeostasis.

## Acknowledgments

A breakfast with Dr. Daniel Yuan resulted in a collaboration on Fet3p from *Saccharomyces cerevisiae*, a collaboration that was made possible by Dr. Yuan's pioneering efforts. This collaboration has included the laboratories of Drs. Ninian Blackburn, John McCracken, and Edward Solomon. Members of these labs who have participated in this collaboration are Martina Ralle, Constantino Aznar, Timothy Machonkin, Liliana Quintanar, and Amy Palmer. All members of my own lab contributed to this work, most particularly Richard Hassett, Scott Severance, and Kimberly Prohaska. Of course, little good happens in the Kosman lab without the critical assistance of Annette Romeo, and our work on Fet3p and the iron–copper connection owes much to her efforts. The support for this work by the NIH (DK53820) is gratefully acknowledged.

## References


Abboud, S., and Haile, D. J. (2000). A novel mammalian iron-regulated protein involved in intracellular iron metabolism. *J. Biol. Chem.* **275**, 19906–9912.

Askwith, C., Eide, D., Van Ho, A., Bernard, P. S., Li, L., Davis-Kaplan, S., Sipe, D. M., and Kaplan, J. (1994). The *FET3* gene of *S. cerevisiae* encodes a multicopper oxidase required for ferrous iron uptake. *Cell* **76**, 403–410.

Askwith, C., and Kaplan, J. (1997). An oxidase-permease-based iron transport system in *Schizosaccharomyces pombe* and its expression in *Saccharomyces cerevisiae*. *J. Biol. Chem.* **272**, 401–405.

Askwith, C. C., de Silva, D., and Kaplan, J. (1996). Molecular biology of iron acquisition in *Saccharomyces cerevisiae*. *Mol. Microbiol.* **20**, 27–34.

Askwith, C. C., and Kaplan, J. (1998). Site-directed mutagenesis of the yeast multicopper oxidase Fet3p. *J. Biol. Chem.* **273**, 22415–22419.

Aznar, C. P., Kosman, D. J., Hassett, R., Yuan, D., and McCracken, J. (2002). Pulsed EPR studies on the type 1 and type 2 Cu(II) sites in the yeast multicopper oxidase, Fet3p. *Biochemistry*, in press.

Blackburn, N. J., Ralle, M., Hassett, R., and Kosman, D. J. (2000). Spectroscopic analysis of the trinuclear cluster in the Fet3 protein from yeast, a multinuclear copper oxidase. *Biochemistry* **39**, 2316–2324.

Bonaccorsi di Patti, M. C., Felice, M. R., Camuti, A. P., Lania, A., and Musci, G. (2000). The essential role of Glu-185 and Tyr- 354 residues in the ferroxidase activity of *Saccharomyces cerevisiae* Fet3. *FEBS Lett.* **472**, 283–286.

Bonomi, F., Kurtz, D. M., Jr., and Cui, X. (1996). Ferroxidase activity of recombinant *Desulfovibrio vulgaris* rubrerythrin. *J. Biol. Inorg. Chem.* **1**, 67–72.

Buonaccorsi di Patti, M. C., Pascarella, S., Catalucci, D., and Calabrese, L. (1999). Homology modeling of the multicopper oxidase Fet3 gives new insights in the mechanism of iron transport in yeast. *Protein Eng.* **12**, 895–897.

Cha, J., and Cooksey, D. A. (1991). Copper resistance in *Pseudomonas syringae* mediated by periplasmic and outer membrane proteins. *Proc. Natl. Acad. Sci. USA* **88**, 8915–8919.

Chidambaram, M. V., Barnes, G., and Frieden, E. (1983). Ceruloplasmin and the reactions forming diferric transferrin. *FEBS Lett.* **159**, 137–140.

Cornelius, J. B., McCracken, J., Clarkson, R. B., and Belford, R. L. (1990). Electron spin echo envelope modulation angle selection studies of axial pyridine coordination to copper(II) benzoylacetonate. *J. Phys. Chem.* **94**, 6977–6982.

Dancis, A., Klausner, R. D., Hinnebusch, A. G., and Barriocanal, J. G. (1990). Genetic evidence that ferric reductase is required for iron uptake in *Saccharomyces cerevisiae*. *Mol. Cell. Biol.* **10**, 2294–2301.

Dancis, A., Roman, D. G., Anderson, G. J., Hinnebusch, A. G., and Klausner, R. D. (1992). Ferric reductase of *Saccharomyces cerevisiae*: Molecular characterization, role in iron uptake and transcriptional control by iron. *Proc. Natl. Acad. Sci. USA* **89**, 3869–3873.

Dancis, A., Yuan, D. S., Haile, D., Askwith, C., Eide, D., Moehle, C., Kaplan, J., and Klausner, R. D. (1994). Molecular characterization of a copper transport protein in *S. cerevisiae*: An unexpected role for copper in iron transport. *Cell* **76**, 393–402.

Davis-Kaplan, S. R., Askwith, C. C., Bengtzen, A. C., Radisky, D., and Kaplan, J. (1998). Chloride is an allosteric effector of copper assembly for the yeast multicopper oxidase Fet3p: An unexpected role for intracellular chloride channels. *Proc. Natl. Acad. Sci. USA* **95**, 13641–13645.

de Silva, D., Davis-Kaplan, S., Fergestad, J., and Kaplan, J. (1997). Purification and characterization of Fet3 protein, a yeast homologue of ceruloplasmin. *J. Biol. Chem.* **272**, 14208–14213.

de Silva, D. M., Askwith, C. C., Eide, D., and Kaplan, J. (1995). The *FET3* gene product required for high affinity iron transport in yeast is a cell surface ferroxidase. *J. Biol. Chem.* **270**, 1098–1101.

Donovan, A., Brownlie, A., Zhou, Y., Shepard, J., Pratt, S. J., Moynihan, J., Paw, B. H., Drejer, A., Barut, B., Zapata, A., Law, T., Brugnara, C., Lux, S. E., Pinkus, G. S., Pinkus, J. L., Kingsley, P. D., Palis, J., Fleming, M. D., Andrews, N. C., and Zon, L. I. (2000). Positional cloning of zebrafish ferroportin1 identifies a conserved vertebrate iron exporter. *Nature* **403**, 776–781.

Ducros, V., Brzozowski, A. M., Wilson, K. S., Brown, S. H., Ostergaard, P., Schneider, P., Yaver, D. S., Pedersen, A. H., and Davies, G. J. (1998). Crystal structure of the type 2 Cu depleted laccase from *Coprinus cinereus* at 2.2 Å resolution. *Nat. Struct. Biol.* **5**, 310–316.

Eide, D. J., Bridgham, J. T., Zhao, Z., and Mattoon, J. R. (1993). The vacuolar $H^+$-ATPase of *Saccharomyces cerevisiae* is required for efficient copper detoxification, mitochondrial function, and iron metabolism. *Mol. Gen. Genet.* **241**, 447–456.

Frieden, E., and Osaki, S. (1974). Ferroxidases and ferrireductases: Their role in iron metabolism. *Adv. Exp. Med. Biol.* **48**, 235–265.

Gaxiola, R. A., Yuan, D. S., Klausner, R. D., and Fink, G. R. (1998). The yeast CLC chloride channel functions in cation homeostasis. *Proc. Natl. Acad. Sci. USA* **95**, 4046–4050.

Greene, J. R., Brown, N. H., DiDomenico, B. J., Kaplan, J., and Eide, D. J. (1993). The *GEF1* gene of *Saccharomyces cerevisiae* encodes an integral membrane protein: Mutations in which have effects on respiration and iron-limited growth. *Mol. Gen. Genet.* **241**, 542–553.

Harris, Z. L., Durley, A. P., Man, T. K., and Gitlin, J. D. (1999). Targeted gene disruption reveals an essential role for ceruloplasmin in cellular iron efflux. *Proc. Natl. Acad. Sci. USA* **96**, 10812–10817.

Harris, Z. L., Klomp, L. W. J., and Gitlin, J. D. (1998). Aceruloplasminemia: An inherited neurodegenerative disease with impairment of iron homeostasis. *Am. J. Clin. Nutr.* **67**, 972S–977S.

Harris, Z. L., Takahashi, Y., Miyajima, H., Serizawa, M., Macgillivray, R. T. A., and Gitlin, J. D. (1995). Aceruloplasminemia: Molecular characterization of this disorder of iron metabolism. *Proc. Natl. Acad. Sci. USA* **92**, 2539–2543.

Harrison, M. D., and Dameron, C. T. (1999). Molecular mechanisms of copper metabolism and the role of the Menkes disease protein. *J. Biochem. Mol. Toxicol.* **13**, 93–106.

Hassett, R., and Kosman, D. J. (1995). Evidence for Cu(II) reduction as a component of copper uptake by *Saccharomyces cerevisiae*. *J. Biol. Chem.* **270**, 128–134.

Hassett, R. F., Yuan, D. S., and Kosman, D. J. (1998). Spectral and kinetic properties of the Fet3 protein from *Saccharomyces cerevisiae*, a multinuclear copper ferroxidase enzyme. *J. Biol. Chem.* **273**, 23274–23282.

Hegg, E. L., Whiting, A. K., Saari, R. E., McCracken, J., Hausinger, R. P., and Que, L. J. (1999). Herbicide-degrading α-keto acid-dependent enzyme TfdA: Metal coordination environment and mechanistic insights. *Biochemistry* **38**, 16714–16726.

Klomp, L. W. J., Lin, S.-J., Klausner, R. D., Yuan, D. S., Culotta, V. C., and Gitlin, J. D. (1997). Identification and functional expression of *HAH1*, a novel human gene involved in copper homeostasis. *J. Biol. Chem.* **272**, 9221–9226.

Kosman, D. J. (1993). Transition metal ion uptake in yeasts and filamentous fungi. *In* "Metal Ions in Fungi" (Winklemann, G., and Winge, D., Eds.), pp. 1–38. Dekker, New York.

LaCroix, L. B., Shadle, S. E., Wang, Y., Averill, B. A., Hedman, B., Hodgson, K. O., and Solomon, J. S. (1996). *J. Am. Chem. Soc.* **118**, 7775–7768.

Lahey, M. E., Gubler, M. S., Chase, G. E., Cartwright, G. E., and Wintrobe, M. M. (1952). Studies on copper metabolism. II. Hematologic manifestations of copper deficiency in swine. *Blood* **7**, 1053–1065.

Lee, G. R., Nacht, S., Lukens, J. N., and Cartwright, G. E. (1968). Iron metabolism in copper-deficient swine. *J. Clin. Invest.* **47**, 2058–2069.

Lerch, K., and German, U. A. (1988). Evolutionary relationships among copper proteins containing coupled binuclear copper sites. *In* "Oxidases and Related Redox Systems" (K. T. S., Mason, H. S., and Morrison, M., Eds.), pp. 331–348. A. R. Liss, New York.

Lin, S.-J., Pufahl, R. A., Dancis, A., O'Halloran, T. V., and Culotta, V. C. (1997). A role for the *Saccharomyces cerevisiae ATX1* gene in copper trafficking and iron transport. *J. Biol. Chem.* **272**, 9215–9220.

Lindley, P. F., Card, G., Zaitseva, I., Zaitsev, V., Reinhammar, B., Selin-Lindgren, E., and Yoshida, K. (1997). An X-ray structural study of human ceruloplasmin in relation to ferroxidase activity. *J. Biol. Inorg. Chem.* **2**, 454–463.

Machonkin, T. E., Quintanar, L., Palmer, A. E., Hassett, R., Severance, S., Kosman, D. J., and Solomon, E. I. (2001). Spectroscopy and reactivity of the type 1 copper site in Fet3p from *Saccharomyces cerevisiae*: Correlation of structure with reactivity in the multicopper oxidases. *J. Am. Chem. Soc.* **123**, 5507–5517.

Machonkin, T. E., and Solomon, E. I. "The thermodynamics, kinetics, and molecular mechanism of intramolecular electron transfer in human ceruloplasmin." *J. Am. Chem. Soc.* **122**, 12547–12560.

Machonkin, T. E., Zhang, H. H., Hedman, B., Hodgson, K. O., and Solomon, E. I. (1998). Spectroscopic and magnetic studies of human ceruloplasmin: Identification of a redox-inactive reduced type 1 copper site. *Biochemistry* **37**, 9570–9578.

Malmström, B. G. (1994). Rack-induced bonding in blue-copper proteins. *Eur. J. Biochem.* **223**, 711–718.

Manis, J. (1970). Active transport of iron by intestine: Selective defect in the mouse. *Nature* **227**, 385–386.

McKie, A. T., Marciani, P., Rolfs, A., Brennan, K., Wehr, K., Barrow, D., Miret, S., Bomford, A., Peters, T. J., Farzaneh, F., Hediger, M. A., Hentze, M. W., and Simpson, R. J. (2000). A novel duodenal iron-regulated transporter, IREG1, implicated in the basolateral transfer of iron to the circulation. *Mol. Cell* **5**, 299–309.

Messerschmidt, A., and Huber, R. (1990). The blue oxidases, ascorbate oxidase, laccase and ceruloplasmin. Modeling and structural relationships. *Eur. J. Biochem.* **187**, 341–352.

Messerschmidt, A., Luecke, H., and Huber, R. (1993). X-ray structures and mechanistic implications of three functional derivatives of ascorbate oxidase from zucchini. *J. Mol. Biol* **230**, 997–1014.

Messerschmidt, A., Ladenstein, R., Huber, R., Bolognesi, M., Avigliano, L., Petruzzelli, R., Rossi, A., and Finazzi-Agró, A. (1992a). Refined cystal structure of ascorbate oxidase at 1.9 Å resolution. *J. Mol. Biol.* **224**, 179–205.

Messerschmidt, A., Steigemann, W., Huber, R., Lang, G., and Kroneck, P. M. H. (1992b). X-ray crystallographic characterization of type 2 depleted ascorbate oxidase from zucchini. *Eur. J. Biochem.* **209**, 597–602.

Miyajima, H., Takahashi, Y., Serizawa, M., Kaneko, E., and Gitlin, J. D. (1996). Increased plasma lipid peroxidation in patients with aceruloplasminemia. *Free Rad. Biol. Med.* **20**, 757–760.

Murphy, M. E., Lindley, P. F., and Adman, E. T. (1997). *Protein Sci.* **6**, 761–770.

Nakano, M. (1993). A possible mechanism of iron neurotoxicity. *Eur. Neurol.* **33**, 44–51.

Odermatt, A., Suter, H., Krapt, R., and Solioz, M. (1993). Primary structure of two P-type ATPases involved in copper homeostasis in *Enterococcus hirae*. *J. Biol. Chem.* **268**, 12775–12779.

Osaki, S. (1966). Kinetic studies of ferrous ion oxidation with crystalline human ferroxidase (ceruloplasmin). *J. Biol. Chem.* **241**, 5053–5059.

Osaki, S., Johnson, D. A., and Frieden, E. (1966). The possible significance of the ferrous oxidase activity of ceruloplasmin in normal human serum. *J. Biol. Chem.* **241**, 2746–2751.

Patel, B. N., and David, S. (1997). A novel glycosylphosphatidylinositol-anchored form of ceruloplasmin is expressed by mammalian astrocytes. *J. Biol. Chem.* **272**, 20185–20190.

Patel, B. N., Dunn, R. J., and David, S. (2000). Alternative RNA splicing generates a glycosylphosphatidylinositol-anchored form of ceruloplasmin in mammalian brain. *J. Biol. Chem.* **275**, 4305–4310.

Payne, A. S., and Gitlin, J. D. (1998). Functional expression of the Menkes disease protein reveals common biochemical mechanisms among the copper-transporting P-type ATPases. *J. Biol. Chem.* **273**, 3765–3770.

Petris, M., Mercer, J., Culvenor, J., Lockhart, P., Glesson, P., and Camakaris, J. (1996). Ligand-regulated transport of the Menkes copper P-type ATPase efflux pump from the Golgi apparatus to the plasma membrane: A novel mechanism of regulated trafficking. *EMBO J.* **15**, 101–112.

Raguzzi, F., Lesuisse, E., and Crichton, R. R. (1988). Iron storage in *Saccharomyces cerevisiae*. *FEBS Lett.* **231**, 253–258.

Rensing, C., Fan, B., Sharma, R., Mitra, B., and Rosen, B. P. (2000). CopA: An *Escherichia coli* Cu(I)-translocating P-type ATPase. *Proc. Natl. Acad. Sci. USA* **97**, 652–656.

Richardson, D. R. (1999). Role of ceruloplasmin and ascorbate in cellular iron release. *J. Lab. Clin. Med.* **134**, 454–465.

Rydén, L. (1988). Evolution of blue copper proteins. *In* "Oxidases and Related Redox Systems" (K. T. S., Mason, H. S., and Morrison, M., Eds.), pp. 349–366. A. R. Liss, New York.

Shin, W., Sundaram, U. M., Cole, J. L., Zhang, H. H., Hedman, G., Hodgson, K. O., and Solomon, E. I. (1996). Chemical and spectroscopic definition of the peroxide-level intermediate in the multicopper oxidases: Relevance to the catalytic mechanism of dioxygen reduction to water. *J. Am. Chem. Soc.* **118**, 3202–3215.

Solomon, E. I., and Lowery, M. D. (1993). Electronic structure contributions to function in bioinorganic chemistry. *Science* **259**, 1575–1581.

Solomon, E. I., Penfield, K. W., Gewirth, A. A., Lowery, M. D., Shadle, S. E., Guckert, J. A., and LaCroix, L. B. (1996). Electronic structure of the oxidized and reduced blue copper sites: Contributions to the electron transfer pathway, reduction potential, and geometry. *Inorg. Chim. Acta* **243**, 67–78.

Solomon, E. I., Sundaram, U. M., and Machonkin, T. E. (1996). Multicopper oxidases and oxygenases. *Chem. Rev.* **96**, 2563–2605.

Spizzo, T., Byersdorfer, C., Duesterhoeft, S., and Eide, D. (1997). The yeast FET5 gene encodes a FET3-related multicopper oxidase implicated in iron transport. *Mol. Gen. Genet.* **256**, 547–556.

Stearman, R., Yuan, D. S., Yamaguchi-Iwai, Y., Klausner, R. D., and Dancis, A. (1996). A permease–oxidase complex involved in high-affinity iron uptake in yeast. *Science* **271**, 1552–1557.

Suzuki, S., Kohzuma, T., Deligeer, A., Yamaguchi, K., Nakamura, N., Shidara, S., Kobayashi, K., and Tagawa, S. (1994). *J. Am. Chem. Soc.* **116**, 11145–11146.

Tandy, S., Williams, M., Leggett, A., Lopez-Jimenez, M., Dedes, M., Ramesh, B., Srai, S. K., and Sharp, P. (2000). Nramp2 expression is associated with pH-dependent iron uptake across the apical membrane of human intestinal Caco-2 cells. *J. Biol. Chem.* **275**, 1023–1029.

Trinder, D., Oates, P. S., Thomas, C., Sadleir, J., and Morgan, E. H. (2000). Localisation of divalent metal transporter 1 (DMT1) to the microvillus membrane of rat duodenal enterocytes in iron deficiency, but to hepatocytes in-iron overload [See comments]. *Gut* **46**, 270–276.

Urbanowski, J. L., and Piper, R. C. (1999). The iron transporter Fth1p forms a complex with the Fet5 iron oxidase and resides on the vacuolar membrane. *J. Biol. Chem.* **274**, 38061–38070.

Vulpe, C. D., Kuo, Y. M., Murphy, T. L., Cowley, L., Askwith, C., Libina, N., Gitschier, J., and Anderson, G. J. (1999). Hephaestin, a ceruloplasmin homologue implicated in intestinal iron transport, is defective in the *sla* mouse. *Nat. Genet.* **21**, 195–199.

Wessling-Resnick, M. (1999). Biochemistry of iron uptake. *Crit. Rev. Biochem. Mol. Biol.* **34**, 285–314.

Xu, F., Palmer, A. E., Yaver, D. S., Berka, R. M., Gambetta, G. A., Brown, S. H., and Solomon, E. I. (1999). Targeted mutations in *Trametes villosa* laccase. Axial perturbations of the T1 copper. *J. Biol. Chem.* **274**, 12372–12375.

Yuan, D. S., Dancis, A., and Klausner, R. D. (1997). Restriction of copper export in *Saccharomyces cerevisiae* to a late Golgi or post-Golgi compartment in the secretory pathway. *J. Biol. Chem.* **272**, 25787–25793.

Yuan, D. S., Stearman, R., Dancis, A., Dunn, T., Beeler, T., and Klausner, R. D. (1995). The Menkes/Wilson disease gene homologue in yeast provides copper to a ceruloplasmin-like oxidase required for iron uptake. *Proc. Natl. Acad. Sci. USA* **92**, 2632–2636.

Zaitseva, I., Zaitsev, V., Card, G., Moshkov, K., Bax, B., Ralph, A., and Lindley, P. (1996). The X-ray structure of human ceruloplasmin at 3.1 Å: Nature of the copper centres. *J. Biol. Inorg. Chem.* **1**, 15–23.

Zhou, B., and Gitschier, J. (1997). hCTR1: A human gene for copper uptake identified by complementation in yeast. *Proc. Natl. Acad. Sci. USA* **94**, 7481–7486.

# BLUE COPPER-BINDING DOMAINS

BY ARAM M. NERSISSIAN AND ERIC L. SHIPP

Department of Chemistry and Biochemistry, University of California, Los Angeles, Los Angeles, California 90095



## I. Introduction

Blue copper proteins were among the first proteins to be isolated. Their intense blue color made them particularly attractive targets for researchers, and examples such as laccase, stellacyanin, and plastocyanin are among the best-characterized metalloproteins known. Now we are entering the "genomic era" and new prospects for identification of novel members of previously characterized protein families have opened. In the case of blue copper proteins, the availability of such data has allowed us to define a larger protein family based on the "blue-copper-binding" (BCB) domain, a structurally conserved 90- to 150-amino-acid sequence module. Most, but not all, BCB domains house a single copper ion tightly bound to the polypeptide in a fashion known as blue or type 1. Other,



ADVANCES IN PROTEIN CHEMISTRY, Vol. 60


0065-3233/02 $35.00

more recently identified domains bind copper differently or not at all. Genes encoding such sequences occur in all domains of living organisms, Bacteria (both Archaebacteria and Eubacteria) and Eukaryota, that inhabit a whole spectrum of environments. These genes produce either single BCB domain proteins or more complex, multidomain proteins composed exclusively of two, three, or six BCB domains. In addition, there are genes in which the BCB domain is fused with unrelated sequence domains.

The purpose of this chapter is twofold. We provide a comprehensive inventory of the occurrence and distribution of BCB domain-containing proteins based on analysis of the genomic and EST sequence data currently available, and we propose a classification system. Some analysis of codon usage for conserved amino acids involved in copper binding will be used to trace the evolutionary history of BCB domains within a single genome. In the second, major, portion of this chapter, the structural and physical characteristics of each kind of BCB domain protein will be summarized. Perhaps surprisingly for proteins that are so well studied biophysically, the biological functions of many of the canonical blue copper proteins are not known. In recent years some progress has been made in this direction, and it will be mentioned where it is relevant.

## II. Four Classes of BCB Domain-Containing Proteins

Close inspection of currently available sequences of proteins carrying BCB domains clearly indicated that they can be classified into four major classes, which are described below. This classification is based on their ability to bind copper and the specific features of their domain organization. Members of the first three classes harbor single or multiple type 1 blue copper-binding sites, while members of the fourth class do not appear to bind copper. Domain organizations of the precursors of all currently known protein families that contain a BCB domain are shown in Fig. 1.

1. The first class is cupredoxins—single-domain blue copper proteins composed of only one BCB domain. These proteins include plastocyanin, azurin, pseudoazurin, amicyanin, auracyanins, rusticyanin, halocyanin, and sulfocyanin (see Section IV). Plantacyanin of the phytocyanin family (Section V), subunit II of the cytochrome *c* oxidase, and the recently characterized nitrosocyanin also fall into this class. The last two are single BCB domain polypeptides closely related structurally to cupredoxins, but harboring, respectively, a binuclear copper site known as CuA and a novel type of copper-binding site called "red" (see Sections IX and X).

2. The second class consists of multidomain blue copper proteins composed of exclusively two or more BCB domains and includes nitrite reductase (Section IV, E), multicopper blue oxidases such as laccase, ascorbate oxidase, ceruloplasmin, and hephaestin (Section VII), and some sequences found in extreme halophilic archaea (see Section V, E).

3. The third class consists of proteins that are composed of one or more BCB domains fused to a sequence domain(s) characteristic of evolutionarily unrelated protein families. Such a mosaic domain organization has been found in the phytocyanin protein family, stellacyanins, uclacyanins, and the recently characterized dicyanins (Section V); in blood coagulation factor VIII (Section VIII); and in nitrous oxide reductase (Section IX).

The functions of the above three classes of proteins are directly related to the protein-bound copper ions. In those cases where functions have been unequivocally established, they are either electron storage and transfer or redox catalysis. Representative proteins from each of the above-mentioned classes have been purified directly from their natural sources and extensively characterized both structurally and biophysically. In addition, there are well-established protocols for their expression and purification from different heterologous systems. [Factor VIII is a special case for which blue copper binding has not been experimentally demonstrated, although at least two such sites can be identified in its amino acid sequence (Section VIII).]

4. The fourth class includes proteins that have BCB domains that lack any obvious copper-binding site. These proteins must, therefore, carry out a radically different function(s), other than that of redox chemistry. Overall domain organization can be similar to any of the above-mentioned three classes. Proteins belonging to this class are early nodulins and dinodulins of the phytocyanin family (Section V), blood coagulation factor V (Section VIII), plant sequences labeled as "pectinesterases" (Section VII, A), ephrin ligands, which are structural homologues of phytocyanins and are involved in cell-to-cell signaling in metazoa (Section VI), and subunit II of bacterial quinol oxidases. The class 4 proteins appear to be involved in the cell-to-cell communications (except factor V) that orchestrate cell differentiation and other morphogenic processes. In plants, they also participate in the progression of pathogenesis and symbiosis. DNA sequence analysis based on codon usage for the functionally important amino acids indicates that at least early nodulins are likely to have evolved from corresponding blue copper proteins (see Section V, D).

BCB domain-containing proteins are most abundant in plants, which are also the only organisms that contain genes encoding proteins of all four classes. In the haploid genome of *Arabidopsis thaliana*, the complete

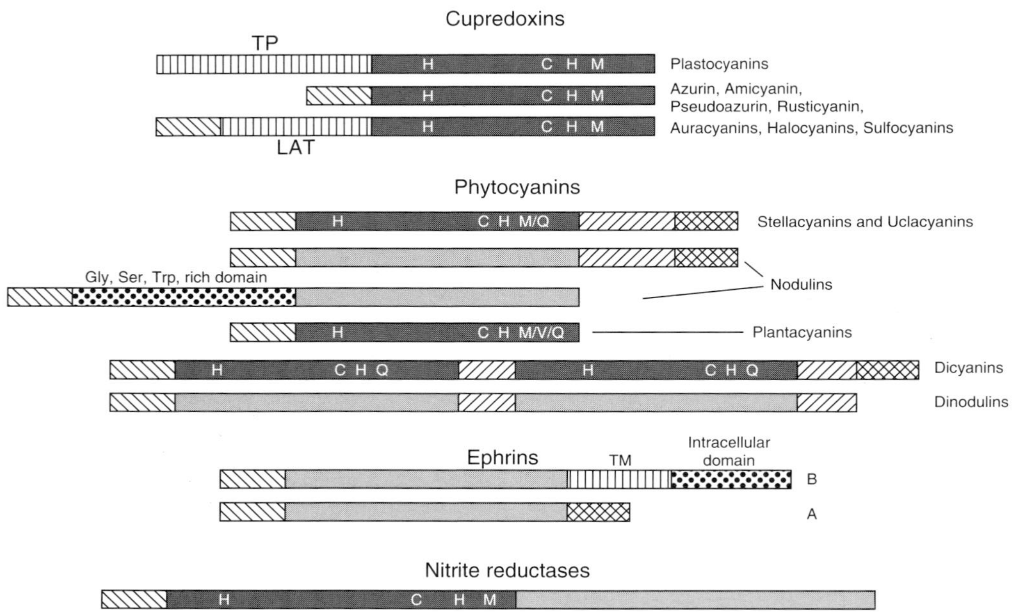
Cupredoxins
TP
H
C H M
Plastocyanins
Azurin, Amicyanin,
Pseudoazurin, Rusticyanin,
Auracyanins, Halocyanins, Sulfocyanins
LAT
Phytocyanins
C H M/Q
Stellacyanins and Uclacyanins
Nodulins
Gly, Ser, Trp, rich domain
C H M/V/Q
Plantacyanins
C H Q
Dicyanins
Dinodulins
Ephrins
TM
Intracellular domain
B
A
Nitrite reductases

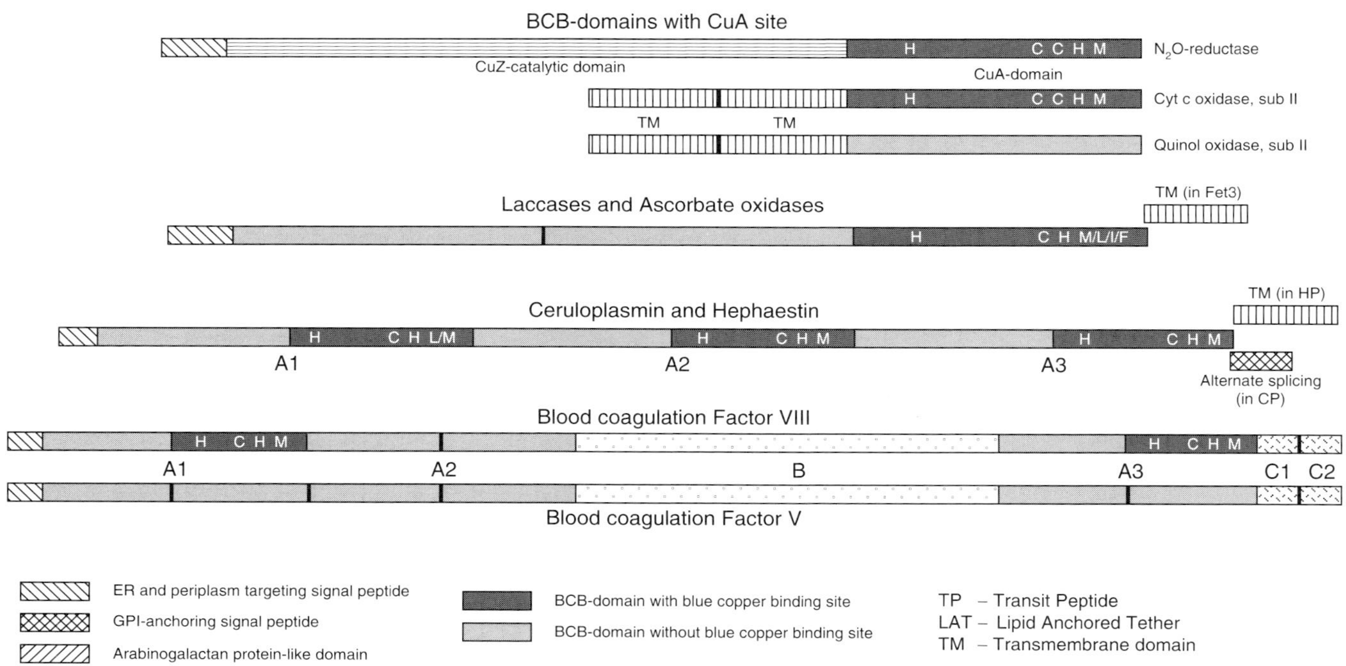


FIG. 1. Domain organizations of the precursors of all known major families of proteins containing BCB domains.

sequence of which has been recently determined (The *Arabidopsis* Genome Initiative, 2000), at least 84 genes encode polypeptides in which such domains can be identified: 2 plastocyanins, 3 ascorbate oxidases, 42 phytocyanins (14 with a type 1 copper site and 28 without), 18 laccases, and 19 putative pectinesterases. The pectinesterases one composed of three BCB domains similar to laccases and ascorbate oxidases, but lack any obvious copper-binding site. cDNA sequences corresponding to most of these genes can be also identified in expressed sequence tag (EST) databases, suggesting that they are transcriptionally active. In contrast, other organisms house only a handful, 2 to 13, of such genes. Analysis of the sequences released to the GenBank database (at the URL http://www.ncbi.nlm.nih.gov) as of May 2001 revealed that many bacteria have at least one gene encoding a cupredoxin and also a gene encoding a laccase-like protein. In some fungi, laccases constitute a multigene family composed of at least 3 different genes, while 3 to 7 genes encoding putative cupredoxins can be identified in extreme haloalkaliphilic archaeons. To date, 13 such proteins have been identified in vertebrates: 2 multicopper oxidases, ceruloplasmin and hephaestin, each composed of six BCB domains; blood coagulation factors V and VIII with a mosaic domain organization containing six BCB domains; 5 ephrin-A and 3 ephrin-B ligands for Eph receptor tyrosine kinases containing a single BCB domain, which is a structural homologue of plant phytocyanins; and the mitochondrial genome-encoded single BCB domain protein, subunit II of the cytochrome c oxidase containing a CuA site. Table I summarizes all information necessary to extract sequence information for those genes from the GenBank database.

The large number of BCB domain proteins in plants may be explained by the phenomenon of genome duplication, believed to occur widely in the plant kingdom, as well as by different lateral gene transfers (The *Arabidopsis* Genome Initiative, 2000; Vision *et al.*, 2000). Individual segments derived from different chromosomes of *Arabidopsis* have been noted to display similar gene order and contents, indicating that they originated from common ancestral segments through large-scale, genome-wide duplication events, possibly polyploidy. Much of the *Arabidopsis* genome is internally duplicated, with more than 100 duplicated blocks with seven or more open reading frames having been identified. This duplication is hypothesized to have originated through a single polyploidy event estimated to have occurred 112 million years ago (Vision *et al.*, 2000). It is believed that *Arabidopsis* once had a tetraploid genome, a situation commonly seen in many contemporary plant species. After these duplication events, the chromosomes "collapsed"; this was followed by their random rearrangements and stabilization, giving rise to the five haploid chromosomes seen in contemporary *Arabidopsis*.

TABLE I
*GenBank Accession Numbers and Names of the Genes and cDNAs Encoding BCB Domain-Containing Proteins in Arabidopsis, Human, and Archaeon Halobacterium sp. NRC-1 Genomes*

| GenBank Accession No. (gene/cDNA) | Name | Chromosome | Gene ID |
|---|---|---|---|
| | *Arabidopsis thaliana* | | |
| Plantacyanin | | | |
| AC004138/U76297 | AtPNC | 2 | At2g02850/T17M13.2 |
| Stellacyanins | | | |
| Z15058/AF296825 | BCB, AtSTC1 | 5 | At5g20230/F5024 (*bcb*) |
| AC003105 | AtSTC2 | 2 | At2g26720/F18A8.9 |
| AF077407 | AtSTC3 | 5 | At5g26330/F9D12.16 |
| AB022219 | AtSTC4 | 3 | MKP6.25 |
| AC005311 | AtSTC5 | 2 | At2g31050/T16B12.14 |
| Uclacyanins | | | |
| AC005700/U76298 | AtUCC1 | 2 | At2g32300/T32F6.18 |
| AC003672/U76299 | AtUCC2 | 2 | At2g44790/F16B22.32 |
| AL163852/AF039404 | AtUCC3 | 3 | At3g60280/F27H5.70 |
| AL163852 | AtUCC4 | 3 | At3g60270/F27H5.60 |
| AL163912 | AtUCC5 | 5 | Does not have ID (complementary/nt 79,996–80,705) |
| AP000381 | AtUCC6 | 3 | At3g27200/K17E12.2 |
| AC006551 | AtUCC7 | 1 | At1g22480/F12K8.17 |
| AC067754 | AtUCC8 | 1 | At1g72230/T9N14.17 |
| Early nodulins | | | |
| AB007644 | AtEN1 | 5 | At5g53870/K19P17.3 |
| AL021749 | AtEN2 | 4 | At4g28360/F2009.30 |
| AB013396 | AtEN3 | 5 | At5g57920/MT120.18 |
| AC010793 | AtEN4 | 1 | At1g79800/F20B17.22 |
| AC016041 | AtEN5 | 1 | At1g48940/F27J15.27 |
| AC005964 | AtEN6 | 5 | At5g25090/T11H3.100 |
| AC006585 | AtEN7 | 2 | At2g25060/F27C12.2 |
| AC009519 | AtEN8 | 1 | At1g64640/F1N19.2 |
| AF160182 | AtEN9 | 4 | At4g30590/F17I123 |
| AL163817 | AtEN10 | 5 | At5g14350/F18022.140 |
| AL034567 | AtEN11 | 4 | At4g32490/F8B4.190 |
| AL035602 | AtEN12 | 4 | At4g27520/T29A15.10 |
| AL049607 | AtEN13 | 4 | At4g31840/F11C18.40 |
| AP000410 | AtEN14 | 3 | At3g20570/K10D20.11 |
| AP001303 | AtEN15 | 3 | At3g18590/K24M9.8 |

(*continues*)

TABLE I *continued*

| GenBank Accession No. (gene/cDNA) | Name | Chromosome | Gene ID |
|---|---|---|---|
| AC005170 | AtEN16 | 2 | At2g23990/T29E15.19 |
| AC005623 | AtEN17 | 2 | Does not have ID (nt 35,145–35,726) |
| AC008261 | AtEN18 | 3 | At3g01070/T4P13.25 |
| AC034106 | AtEN19 | 1 | At1g17800/F2H15.3 |
| AL049640/X97206 | AtEN20 | 4 | At4g12880/T20K18.230 |
| AL353993/U77721 | AtEN21 | 5 | At5g15350/F8M21.240 |
| AC006932/U76300 | AtEN22 | 1 | At1g08500/T27G7.18 |
| AC006438 | AtEN23 | 2 | At2g15770/F19G14.23 |
| AC006438 | AtEN24 | 2 | At2g15780/F19G14.22 |
| AL035521 | AtEN25 | 4 | At4g34300/F10M10.70 |
| AL031032 | AtEN26 | 4 | At4g33930/F17I5.120 |
| Dinodulin | | | |
| AC007915 | AtDN1 | 1 | F27F5.14 |
| AL132966 | AtDN2 | 3 | At3g53330/F4P12.30 |
| Plastocyanins | | | |
| AC009978/M20937 | AtPC1 | 1 | At1g76100/T23E18.3 |
| AC026234/M98456 | AtPC2 | 1 | At1g20340/F14O10.6 |
| Ascorbate oxidases | | | |
| AL022605 | AtA01 | 4 | At4g39830 |
| AC069325/AB004798 | AtA02 | 5 | At5g21100 |
| AC069325 | AtA03 | 5 | Does not have ID (complementary/nt 39,303–41,710 |
| Laccases | | | |
| AB005240 | AtLC1 | 5 | At5g03260/MOK16.17 |
| AC003028 | AtLC2 | 2 | At2g38080/F16M14.1 |
| AL137189 | AtLC3 | 5 | At5g01190/F7J8.170 |
| AB015475 | AtLC4 | 5 | At5g60020/MMN10.27 |
| AC005315 | AtLC5 | 2 | At2g29130/T9I4.21 |
| AC034107 | AtLC6 | 1 | At1g18140/T10F20.14 |
| AB010692 | AtLC7 | 5 | At5g05390/K18I23.20 |
| AC007020 | AtLC8 | 2 | At2g40370/T3G21.14 |
| AC002338 | AtLC9 | 2 | At2g30210/T9D9.2 |
| AL163652 | AtLC10 | 5 | At5g07130/T28J14.70 |
| AB017064 | AtLC11 | 5 | At5g48100/MDN11.18 |
| AL391712 | AtLC12 | 5 | At5g09360/T5E8.160 |
| AC006418 | AtLC13 | 2 | At2g46570/F13A10.10 |

(*continues*)

TABLE I *continued*

| GenBank Accession No. (gene/cDNA) | Name | Chromosome | Gene ID |
|---|---|---|---|
| AL137189 | AtLC14 | 5 | At5g01040/F7J8.20 |
| AL137189 | AtLC15 | 5 | At5g01050/F7J8.30 |
| AC011436 | AtLC16 | 3 | At3g09220/F3L24.9 |
| AC016972 | AtLC17 | 1 | At1g71040/F23N20.3 |
| AF000657 | AtLC18 | 1 | At1g23010/F19G10.5 |
| Putative "pectinesterases" | | | |
| AC004482 | AtPE1 | 2 | At2g23630/F27L4.18 |
| AL161590 | AtPE2 | 4 | At4g37160 |
| AB010700 | AtPE3 | 5 | At5g66920/MUD21.18 |
| AC008046 | AtPE4 | 1 | At1g41830/F5A13.5 |
| AC009978 | AtPE5 | 1 | At1g76160/T23E18.10 |
| AC013482 | AtPE6 | 1 | At1g21850/T26F17.6 |
| AC013482 | AtPE7 | 1 | At1g21860/T26F17.7 |
| AL022140 | AtPE8 | 4 | At4g22010 |
| AL035524 | AtPE9 | 4 | At4g28090/T13J8.200 |
| AL035539 | AtPE10 | 4 | At4g38420/F22I13.190 |
| AC005223 | AtPE11 | 1 | At1g55570/T5A14.1 |
| AP000603 | AtPE12 | 3 | At3g13390/MRP15.2 |
| AC005223 | AtPE13 | 1 | At1g55560/T5A14.2 |
| AP000603 | AtPE14 | 3 | At3g13400/MRP15.3 |
| AB018109 | AtPE15 | 5 | At5g51480/K17N15.3 |
| AL035396 | AtPE16 | 4 | At4g25240/F24A6.80 |
| AL049730 | AtPE17 | 4 | At4g12420 |
| AB020745 | AtPE18 | 5 | At5g48450/MJE7.8 |
| AC006434 | AtPE19 | 1 | At1g75790/F10A5.2 |
| | *Homo sapiens* | | |
| M13699 | Ceruloplasmin | 3 | |
| AF148860 | Hephaestin | *X* | |
| Z99572/M14335 | Factor V | 1 | |
| K01740/M14113/ X01179 | Factor VIII | *X* | |
| M57730 | Ephrin A1 | 1 | |
| AC004258/AJ007292 | Ephrin A2 | 19 | |
| L37360 | Ephrin A3 | 1 | |
| U14188 | Ephrin A4 | 1 | |
| AC008822 AC008952/U26403 | Ephrin A5 | 5 | |
| AL136092/L37361 | Ephrin B1 | *X* | |

*(continues)*

TABLE I *continued*

| GenBank Accession No. (gene/cDNA) | Name | Chromosome | Gene ID |
|---|---|---|---|
| L38734 | Ephrin B2 | 13 | |
| U66406/U57001 | Ephrin B3 | 17 | |
| | *Halobacterium* sp. NRC-1 | | |
| AE005073 | HCPA | | hcpA/VNG1637G |
| AE005106 | halocyanin-like, HCPB | | hcpB/VNG2196G |
| AE005022 | HCPC | | hcpC/VNG0795G |
| AE005046 | HCPD | | hcpD/VNG1188G |
| AE005007 | HCPE | | VNG0573C |
| AE005008 | HCPF | | VNG0586C |
| AE005021 | HCPG | | pcy/VNG0786G |
| AE004988 | HCPH | | fbr/VNG0249G |

Both fossil records (Knoll, 1992) and sequence data (Wang *et al.*, 1999; Heckman *et al.*, 2001) indicate that the last common ancestor of plants and animals—some sort of aerobic unicellular protist that had already internalized the mitochondrial endosymbiont, an oxygenrespiring bacteria—existed at least 1.8 billion years ago. Plants diverged into a separate lineage when that unicellular eukaryote incorporated a second prokaryotic endosymbiont, the cyanobacterial chloroplast (Fig. 2) (Margulis, 1996; Margulis *et al.*, 2000). Since then, the formerly free-living mitochondria and chloroplasts started to reduce their own genomes by extensive transfer of genes from the organelles into the nucleus of the host cell. For instance, only 46 genes from a total of 3168 protein-coding genes in the cyanobacteria *Synechocystis* sp. PCC 6803 genome have been retained in contemporary chloroplasts (Race *et al.*, 1999; Rujan and Martin, 2001). It is believed that this gene reallocation process was favorable because it helped prevent deleterious mutations, which accumulate more rapidly in asexual genomes of organelles than they do in the sexual nuclear genomes. For example, the gene for the cupredoxin plastocyanin has been transferred from chloroplast to nucleus, in the process acquiring an N-terminal extension, called a transit peptide, that directs the precursor protein back into the chloroplast to serve its original

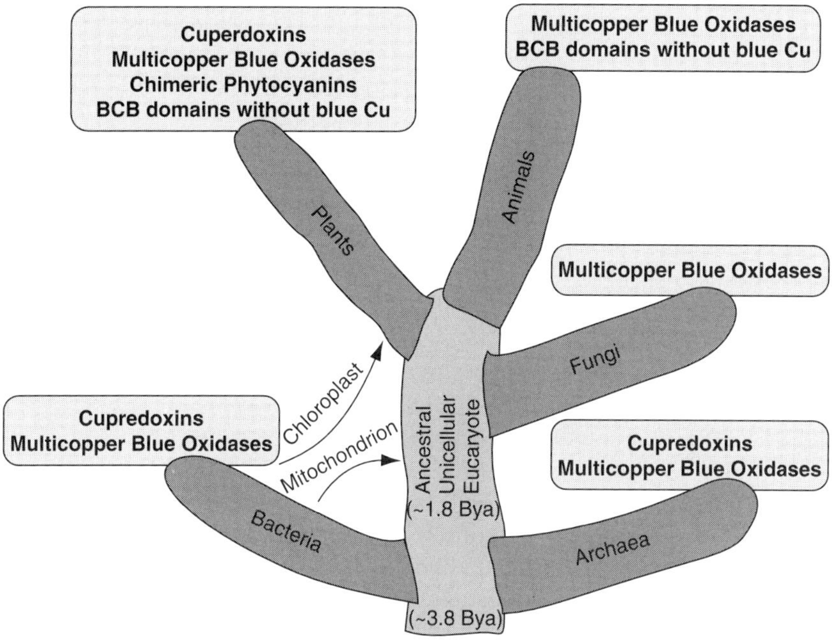


FIG. 2. The "Tree of Life" showing distribution of BCB domain-containing proteins in Archaea, Bacteria, and Eukaryota as well as in three major kingdoms of Eukaryota: animals, plants, and fungi. Arrows indicate the mitochondrial and chloroplast endosymbiotic events. Bya, billion years ago.

bacterial function. Subunit II of cytochrome *c* oxidase (COXII) is a second example of an organelle-localized BCB domain protein. Its gene, however, remained within the organelle and is one of the 6–15 proteins still encoded by mitochondrial DNA. Curiously, a copy of an almost complete genome of mitochondria including the fragment that encodes COXII has been found recently in human chromosome 17 as well as in apes (GenBank Accession No. AF227907).

The rest of the known sequences of all four classes of BCB domain-containing proteins feature signal peptides in their precursors, indicating that they are translocated across the bacterial cytoplasmic membrane or are translated on the endoplasmic reticulum-bound ribosomes and sent to the secretory pathway in eukaryotes. Thus they are located in the bacterial periplasm, secreted into the extracellular milieu, or anchored to the cell surface.

## III. Folding Topology of the BCB Domains and Spectroscopic and Structural Properties of the Blue Copper Sites

The BCB domain-containing proteins evolved by gene duplication and fusion with other structural modules, resulting in diverse subcellular localization and the ability to carry out complex reactions that often differ from that of the ancestral protein. They are also modified by extensive amino acid substitutions and insertions. Remarkably, throughout these complex evolutionary processes they maintained a uniform folding topology, which can be described as an eight-standed Greek key β-barrel, organized by two β-sheets, even though they often display insignificant (less than 10%) sequence identity. Adman and co-workers have performed extensive structural similarity analyses of BCB domains of cupredoxins and those of nitrite reductases and multicopper oxidases, which led to the specification of six different structural classes (see Murphy *et al.*, 1997a). The first class consists of pseudoazurin, plastocyanin, and amicyanin. The CuA-binding domain of subunit II of cytochrome *c* oxidase and the related domain of quinol oxidase, which has lost its copper-binding site, are grouped into the second class. The C-terminal CuA domain of nitrous oxide reductase should also be included in this structural class (Brown *et al.*, 2000b). Azurins form the third class. The cupredoxin auracyanin B, whose structure was recently reported, folds into a molecule quite similar to azurin; hence it should be considered another member of the azurin structural class (Bond *et al.*, 2001). The first domains of nitrite reductase and ascorbate oxidase and domains 1, 3, and 5 of ceruloplasmin form the fourth class. The fifth class contains domains 2, 4, and 6 of ceruloplasmin, domain 2 of nitrite reductase, and domains 2 and 3 of ascorbate oxidase and cupredoxin rusticyanin. The plant phytocyanins represent the sixth structural class.

The metal-binding site is located at the "northern" or top end of the barrel where the copper ion is coordinated with two imidazolate nitrogens, Nσ1, from two His residues and a thiolate from a Cys residue. These ligands are found in all naturally occuring spectroscopically and structurally characterized BCB domains with a blue copper site. They occupy equatorial coordination positions and form a nearly trigonal plane with a varying degree of copper displacement, which is dictated by the strength of the interaction with the fourth ligand occupying the axial coordination position. In most cases it is a weakly coordinated thioether of a Met or a strongly coordinated carbonyl oxygen from a Gln. The latter is found primarily in stellacyanins and as a rare case also in a plantacyanin from tobacco (GenBank Accession No. AF172853; McClure *et al.*, 2000). In azurins, a backbone carbonyl oxygen weakly interacts at the second axial position. In many instances, however, the protein does not provide any liganding residue at the axial position. In these cases the amino acid sequence alignments indicate that they carry a noncoordinating Leu, Phe, Ile, or Val residue at the position of Met and Gln in other BCB domains. The importance of the axial ligand in the fine-tuning of the redox potential is discussed below.

Solvent is usually excluded from the blue copper site, which is buried ~6 Å inside the protein, having only the His ligand from the copper-binding loop exposed to the surface. The phytocyanins, stellacyanin and plantacyanin (cucumber basic protein), are exceptions, in which both His ligands are solvent exposed and the copper ion is only 3 Å beneath the protein surface. This situation makes the copper center in this family of blue copper proteins more accessible to low-molecular-weight solutes (see Section V).

In the amino acid sequence three of the four ligands, His, Cys, and the axial ligand, are located at the C-terminus and in the protein they are positioned in a loop known as a copper-binding loop connecting two C-terminal polypeptide strands (so-called loop 7–8). The fourth ligand, His, is approximately 40 amino acids upstream of that triad and is located on one of the β-strands, inside the rigid β-sandwich of the protein matrix. Sequence characteristics of loop 7–8, such as amino acid composition and length, determined by the number of amino acids separating liganding residues, are important measures of the fine phylogenetic relationships between BCB domain proteins housing a blue copper site. This is especially true for cupredoxins. The shortest loop structure is found in amicyanins and has the following sequence pattern: Cys-X(2)-His-X(2)-Met. In plastocyanins, pseudoazurins, and halocyanins, it is Cys-X(2)-His-X(4)-Met. In azurins it is Cys-X(4)-His-X(3)-Met, while in all phytocyanins, multicopper oxidases, and factor VIII and in cupredoxins rusticyanin and sulfocyanins, it has the following pattern: Cys-X(4)-His-X(4)-[Met, Gln, Phe, Val, Leu, Ile]. The loop sequence is unusually long in nitrite reductases, Cys-X(7–9)-His-X(4)-Met. The loop also governs a fine network of hydrogen bonding that fixes ligands in their proper positions and determines the electronic structure of the copper chromophore. Not surprisingly, the substitution of the entire loop 7–8 in a blue copper protein with a loop sequence derived from a different family of blue copper proteins can generate a new blue copper site on the scaffolding of the host β-barrel (Buning *et al.*, 2000). However, none of the proteins produces a site with spectroscopic properties matching those of the donor protein, which indicates that the robust structure of the β-sandwich may contribute significantly to the fine organization of its blue copper site. It has been also shown that a CuA on the matrix of a cupredoxin can be successfully generated by such a loop substitution strategy (Dennison *et al.*, 1995; Hay *et al.*, 1996).

A series of mutagenesis studies revealed that a blue copper site can also be engineered by different ligand combinations on the scaffolding of a host BCB domain and as a unique case in the zinc-binding site of the Cu,Zn superoxide dismutase (Lu *et al.*, 1993). From these combinations only Cys ligand was indispensable (Canters and Gilardi, 1993; Faham *et al.*, 1997). However, these mutants usually do not produce a stable blue copper site under physiologically relevant conditions and were extremely pH sensitive or were susceptible to the oxidation of the thiolate ligand (Den Blaauwen and Canters, 1993; Germanas *et al.*, 1993; Hammann *et al.*, 1997; Karlsson *et al.*, 1997; Messerschmidt *et al.*, 1998; Murphy *et al.*, 1993; Pascher *et al.*, 1993). Thus it appears that throughout evolution, nature has chosen those three equatorial ligands, N(His), N(His), and S(Cys), for the formation of stable blue copper-binding sites.

Intriguingly, the blue copper sites, especially those with a carbonyl oxygen at the axial coordination position, display high affinity for $Zn^{2+}$ ions. Mutants in which the Met is replaced by Gln or Glu preferentially bind $Zn^{2+}$ when expressed in heterologous systems, e.g., *Escherichia coli*. Examples include azurin, amicyanin, nitrite reductase, and possibly also plastocyanin (Diederix *et al.*, 2000; Hibino *et al.*, 1995; Murphy *et al.*, 1995; Nar *et al.*, 1992a; Romero *et al.*, 1993). In the case of azurin it has been shown that both wild-type and the Met—Gln mutant have the same affinity for both $Zn^{2+}$ and $Cu^{2+}$ (Romero *et al.*, 1993). In addition, EXAFS studies showed that some preparations of blue copper proteins purified from their natural sources also contain small fractions of Zn derivatives (DeBeer George, personal communication).

The coordination geometry of the naturally occurring blue copper sites seems to be dependent on the nature and the availability of the ligand at the axial position. Crystallographic data are available for proteins with all currently known naturally occurring ligand combinations (Fig. 3). Thus, the three-coordinate blue copper sites with NNS donors, characterized in ceruloplasmin and laccase, are trigonal planar (Zaitseva *et al.*, 1996; Ducros *et al.*, 1998). The five-coordinate NNSSO donor site in azurins with two weak, S(Met) and O(backbone carbonyl), axial interactions forms a trigonal bipyramidal geometry (Norris *et al.*, 1983). The four-coordinate NNSO donor site in stellacyanins with a strong O(Gln) axial ligand (Hart *et al.*, 1996), and the remaining cases, organized by NNSS donors with a weak S(Met) axial ligand, are trigonal pyramidal/distorted tetrahedral (Guss and Freeman, 1983; Guss *et al.*, 1988; Petratos *et al.*, 1988; Durley *et al.*, 1993).

Importantly, such coordination geometries within the polypeptide matrix can effectively accommodate a copper ion in both its Cu(II) and its Cu(I) states. This fact is in marked contrast to that of inorganic copper complexes, where Cu(II) displays preferentially tetragonal coordination,

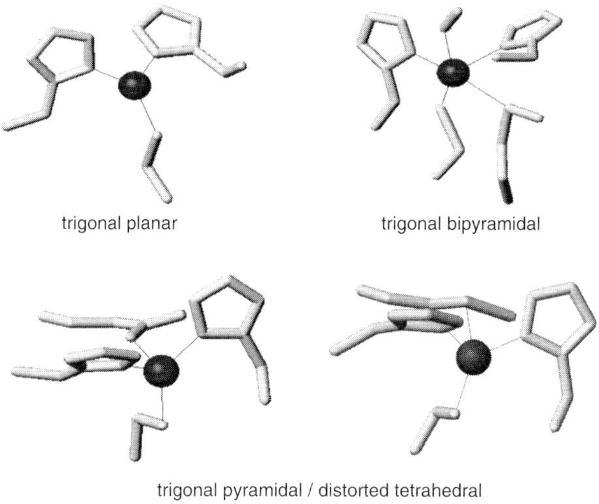


FIG. 3. Geometries of the type 1 copper sites of various blue copper proteins. The trigonal planar geometry is the type 1 site of laccase from *Coprinus cinereus* (PDB Code 1A65). The trigonal bipyramidal geometry shown is the copper site of azurin from *Pseudomonas aeruginosa* (PDB Code 1AZU). The trigonal pyramidal/distorted tetrahedral sites are of the stellacyanin from *Cucumis sativus* (PDB Code 1JER), NNSO site, and of the plastocyanin from *Populus nigra* (PDB Code 1PLC) NNSS site.

whereas Cu(I) strongly prefers a tetrahedral or trigonal planar coordination. The only example of a trigonal planar Cu(II)-utilizing NNS ligand set has been recently reported (Holland and Tolman, 1999). This unique feature of the biological blue copper sites led in the 1960s to the formulation of "entatic" or "rack" concepts (Malmström, 1964; Vallee and Williams, 1968) (see also Gray *et al.*, 2000; Malmström, 1994; Williams, 1995). Both concepts state that the unusual metal-binding site is not dictated by the presence of the metal but is already predetermined by the energetic constraints of the polypeptide, which force a geometry that effectively accommodates both oxidation states. Indeed, a series of elegant crystallographic studies by the Freeman group in the late 1970s and early 1980s proved that there are no significant changes, exceeding the limits of uncertainty of protein crystallography, in the overall protein structures in cupredoxin plastocyanin whether the copper site is reduced, oxidized, or in the apo form (Colman *et al.*, 1978; Guss and Freeman, 1983; Garrett *et al.*, 1984; Guss *et al.*, 1986). Subsequently, a similar situation has also been documented for other cupredoxins (Petratos *et al.*, 1988, 1995; Nar *et al.*, 1992b; Durley *et al.*, 1993; Shepard *et al.*, 1993; Vakoufari *et al.*, 1994; Dodd *et al.*, 2000). More importantly, the ligand configuration remained unchanged even when copper was substituted with other metals. The only observed change was for the surface-exposed His ligand in the reduced protein, which, at low, nonphysiological pH, protonates and flips away from copper. This ligand-flipping effect has been documented only for amicyanins, plastocyanins, and pseudoazurins (Guss *et al.*, 1986; Lommen and Canters, 1990; Lommen *et al.*, 1991; Vakoufari *et al.*, 1994; Zhu *et al.*, 1998). It is important to note that the apoproteins are very vulnerable to proteolysis and rapidly degrade *in vivo*, pointing to the importance of the copper ion in stabilization of the rigid β-barrel protein structure (Li and Merchant, 1995).

The constrained nature of the copper center in BCB domains reduces its reorganization energy, which is considered an important feature for their function in long-range electron transfer processes. They are capable of tunneling electrons, usually over 10- to 12-Å distances, intramolecularly within the same protein (in the case of multicopper oxidases and nitrite reductases) or intermolecularly between a donor and an acceptor protein (in the case of cupredoxins) in a thermodynamically favorable environment.

Three important thermodynamic and spectroscopic properties of biological blue copper sites distinguish them from inorganic copper complexes.

1. The first feature are their unusually high redox potentials compared to that of the inorganic Cu(II)/Cu(I) redox couple (+150 mV). In cupredoxins potentials range between +180 and +680 mV, while the highest value of +1000 mV has been estimated for the three-coordinate site in ceruloplasmin (Machonkin *et al.*, 1998). The redox potential is one of the important properties allowing one to predict the site and the mode of action of a particular blue copper protein, thus coupling its function and the thermodynamics of its copper site. It has been established by a series of mutagenesis studies that the nature of the axial ligand is one of the main components in the fine-tuning of the redox potential: a hydrophobic residue increases the redox potential, while a hydrophilic residue has the opposite effect (Pascher *et al.*, 1993; Hall *et al.*, 1999). However, the axial ligand should not be considered as the sole contributor. Other factors, such as solvent accessibility and the topology of the extensive hydrogen bonding network that organizes the copper site, also have a dramatic effect (Hoitink and Canters, 1992; Libeu *et al.*, 1997; Dong *et al.*, 1999). In addition, it is conceivable that the dipole environment of the protein in the vicinity of copper may contribute to the tuning. It is also important to note that the redox potential of a particular cupredoxin decreases substantially, by 70–100 mV (demonstrated for amicyanin and rusticyanin) (Giudici-Orticoni *et al.*, 1999; Zhu *et al.*, 1998), when it is

bound to its redox partner compared to that of complex free protein. This important message indicates that one would have to expect that the thermodynamic properties of the blue copper sites in their *in vivo* natural environments could be quite different from those determined for *in vitro* preparations.

2. The second property is the strong electronic absorption band centered at around 600 nm with an $\varepsilon \cong 4000\ M^{-1}cm^{-1}$ that is responsible for the intense blue color. It has been assigned to a thiolate sulfur to copper charge transfer transition ($S_{Cys}$-Cu CT) (Solomon *et al.*, 1976; Solomon and Lowery, 1993). Blue copper sites also exhibit a most unusual ground-state EPR spectrum in which the hyperfine coupling in the $g_z/g_{\parallel}$ region is reduced by a factor of 3 compared to that of Cu(II) inorganic complexes (Malmström and Vänngård, 1960; Holm *et al.*, 1996). All of these peculiar spectroscopic properties are attributed to the thiolate bonding to copper, which is very strong and covalent in nature (Penfield *et al.*, 1985; Holm *et al.*, 1996). The beautiful blue coloration was the main criterion in the discovery of these proteins, which in some cases dates from almost a century ago. An intense blue-colored band on a chromatographic column would easily attract researchers' attention. In fact, laccase was probably one of the first enzymes to be isolated. It was described by French biochemist Bertrand in the mid 1890s and 40 years later, Keilin and Mann (1940) showed that it was a copper enzyme.

Another interesting spectroscopic feature is the large shift of the ligand field d–d transition to the near-infrared region, 750 nm, which is shifted even further to 820 nm in Gln harboring blue copper sites, such as stellacyanins. The blue copper sites have been further classified based on their spectroscopic properties as "perturbed" and "classic" (LaCroix *et al.*, 1996, 1998). The perturbed sites are characterized by a shorter and therefore stronger axial ligand bonding than that of classic sites. In addition, perturbed sites display a third intense band in the optical absorption spectrum centered at 450 nm, which is not well developed in classic sites and has been also assigned to $S_{Cys}$-Cu CT. Their EPR signals are rhombic with a well-resolved hyperfine structure in the $g_x$ region ($g_z > g_y > g_x$), while classic sites have an axial signal ($g_z/g_{\parallel} > g_y = g_x/g_{\perp}$). It has been noted that the ratio of the intensities of two $S_{Cys}$-Cu CT transitions, $\varepsilon_{450}/\varepsilon_{600}$, positively correlates with the degree of copper displacement from the equatorial plane (Han *et al.*, 1993; Lu *et al.*, 1993). Those with a high ratio display a rhombic EPR signal and a lower $\nu_{Cu}$ stretching frequency in the resonance Raman spectra, correlating with a weaker Cu–S bond and a stronger axial interaction.

3. The third property is the high level of stability of the thiolate ligand, which is unusual for a transition metal-thiolate coordination. All organometallic complexes featuring such bonding are air sensitive and suscep-

tible to the oxygenation of sulfur ligands to form sulfoxides or sulfones (Grapperhaus and Darensbourg, 1998). In contrast, the blue copper sites protect their thiolates from such unwanted side reactions. It is believed that such stability is determined by the reduced overall negative charge on the thiolate due to hydrogen-bonding interactions involved directly with sulfur ligand and adjacent residues.

The copper site architecture is stabilized by an extensive network of H bonding that controls its peculiar electronic properties. One H bond, between the backbone amide proton of the residue immediately adjacent to the upstream His ligand (Asn in most cupredoxins and can also be Pro, Thr, Ser, Asp, Ala, or Gly in some BCB domains) and the S$\gamma$ atom of the Cys ligand, is considered an important signature of the blue copper sites. In addition, this residue is involved in two hydrogen bondings with the residue adjacent to the Cys ligand. These H-bonding interactions are common to almost all blue copper sites and apparently play an important role in fixing the thiolate ligand into its proper position.

## IV. Cupredoxins

The members of this family of blue copper proteins are composed of a single BCB domain and function as electron shuttle proteins in a variety of energy conversion systems operating in bacterial periplasm and chloroplast photosynthetic membranes (Adman, 1985, 1991). Plant genomes and some bacterial genomes contain at least two different genes for cupredoxins. The phytocyanin plantacyanin could be also attributed to this family, although for phytocyanins a radically different function(s), other than long-range electron transfer, has been proposed (see Section V). Cupredoxins are abundant in archaea, mostly in extreme haloalkaliphiles. For instance, at least seven different genes encoding cupredoxins can be identified in the genome of an archaeon *Halobacterium* sp. NRC-1 (Ng *et al.*, 2000). One of them, HCPG, displays sequence identity to plastocyanins (33%), while the others are distantly related to known cupredoxin sequences (see Fig. 4). To date, no cupredoxins have been found in vertebrates, nematodes, insects, or fungi.

With a few exceptions, cupredoxins are freely diffusible proteins. They accept and donate a single electron to their redox partners during which process the protein-bound copper oscillates between Cu(II) and Cu(I). The cupredoxin and its redox partners form a transient complex that will dissociate upon a successful electron transfer act. Therefore, the protein–protein interactions between a diffusible cupredoxin and its redox partner may not be as specific as one might expect. Indeed, the binding

```
Nhal                       MKDISRRRFVLGTGATVAAATLAGCNGNGNGNGNGNGEPDTPEGRADQFLTDNDALMYDGDITDETGQDEVVVVTGAGNNG-------
hcpB                           MTRLDDTALSRRGVLRAAAGTATAVAAGTAATGAAAAQAYDGWFSGDAKGGAVGNYDGTTVDRTGTDEVTIDVGVDANGG------
hcpD                                 MTDSDSAVTRRRVLQGSAGAGAAAAGIGGFAAGGAAQSASIDDWLGDVSNYDG-VRDETGSSEVTVDVGVDANGG-------
hcpC                                             MLKSTGAAVTATLLAGCSGGGDTGDDESDSESGEEATGNEAG-GGEEASQSGGNTIAVAPDGG-------
hcpH                           DHDAYEAFESSFETYASGLESGTGDVSVVRRASLRAQFAVAGDASAAPVGDAASSGEG-DSQPELSGGPNVVSGVPDDADHVIDMN
hcpF                         MKRRDFLKIATGSAGGAAAVGAVGARSSSSSSAVHQQAPGNATGNGTAMDNSTAGNETAAGGDSESGDGGGATTKTVKVGPN----N
hcpE   MGGSSVPPMISVLERRTCGRALSVAASRGRVANGQTKVFSSRDSLAARMNSETDVGVSRRSVLRGGVAAAAGAAATGTAAAAEGEGGDGGGEKVVKVGPNG-------
hcpG                                                                      MHRRAFLAGGTTLSVGVLAGCIGPSVSKSDYDIGMVSNGFVPQEP------
SynPC                                                                    MKVLASFARRLSLFAVAAVLCVGSFFLSAAPASAQTVAIKMGADNG------
ND2                                                                         YPTSGYPDVDRWLSPVHDYDG----TVVTDTTTVTVGDTSSG-------
ND1                                                     CTDRTTTGTTNQQPTTTHTETQPSEMSLHEWVTSANTPESWTDHRGDATVSIAVGGHDLP-------

                                    Cu                                                            Cu Cu   Cu
Nhal   -FAFDPAAIRVDVGTTVTWEWTGDG--GAHNVVSEPESDFEFES-----------------------DRVDEEGFTFEQTFD-DEGVALYVCTPHRAQGMYGAVIVE
hcpB   TYGFGPAAVRVDPGTTVSFEWVSN----THNILIDAQPDGADWGGH---------------------DPIENTGFTHTHTFD-TEGVYTYYCNPHLTMGMKGAIVVGA
hcpD   AYGFGPAAVRVDPGTTVTFEWVSN----THNVAVESQPDSADWGGV---------------------EEINNEGYTDSHTFE-TAGVYTYFCSPHRAMGMKGAVVVGD
hcpC   -FKFEPSELTVSVGDTVTWEF----EESSHNVSAWPDMHDKISIPDGASGFGTMDKDGNQ-------FAVVDAGETLEHTFE-TAGEYTYVCMPHVASGMIGTVVVEE
hcpH   AVAYDPAELTVSAGDTVAWTHADG---EAHSVTAYEDDIPADATYWASGGFDSQDAAKTGWAN---GKGAVTEGESYVHTFE-TTGEHEYYCIPHKNLGMEGTIVVEE
hcpF   QNVFDPAEVYVSPGDTVKWVW---EGSTGHNVHATSVPDEAGWDAQ---------------------TEIVSAPHEYSHTFEGPTGEYSYVCDPHKSLGMKGKVIVNE
hcpE   SNVFKPAEMYVKPGTKVRFVW----ESGGHNVHATEVPGDADWGVS---------------------TDIAGPPKEYTHTFDGPTGEYNYVCTPHASLGMKGTIIVTD
hcpG   VEGEDMATHQLTAGDTVTWANTGSR---NHTVTAYGSGLPEGASYFASGGFDTEPDARRAWQDNVEGGGNISPGATYEVTLT-VPGEYYYFCIPHEGAGMRGKVVVDD
SyPC   MLAFEPSTIEIQAGDTVQWVNNKLA---PHNVVVEGQPELSHK------------------------DLAFSPGETFEATFS-EPGTYTYYCEPHRGAGMVGKIVVQ
hND2   STQFDPLVLKVPEGTTVQWKWAGEESGGAHNVAFKSAAIGIQ-------------------------ELRSEPGVHVTHTFN-STGTYRYFCSPHSELSGRGAIIVE
hND1   -YAFQPALTRVAPGTTVTWEWTGYG--GAHNVIAVDSAFDSG-------------------------DSTATEDHTFTHTFT-EIGTHRYVSEPDRESGMKGIIAVEE
```

FIG. 4. Multiple sequence alignment of the precursors of seven cupredoxin sequences (hcpB through hcpG) and two BCB domains of the hcpA gene product (ND1 and ND2) identified in the genome of an archaeon, *Halobacterium* sp. NRC-1, with that of halocyanin (Nhal) from *Natronobacterium pharaonis* and plastocyanin from a cyanobacterium, *Synechocystis* sp. PCC 6803.

constants of such complexes are usually very low, in the millimolar range. They can oxidize or reduce *in vitro* different redox macromolecules, often structurally distinct from those of their natural redox partners, as long as they display a favorable redox potential. Further, under some nutritional stress conditions, most organisms can efficiently substitute cupredoxins with other redox proteins. It can be either another cupredoxin or a cytochrome. We provide numerous examples of such substitutions below.

All of this suggests that the theory of the optimization of the electron tunneling pathways in biological macromolecules (Beratan *et al.*, 1987) is more likely applicable for intramolecular electron transfer processes that occur in multicopper proteins with enzymatic activity, such as multicopper oxidases and nitrite reductase where the type 1 sites and the catalytic copper sites are fixed in their stationary positions within the rigid structure of the same polypeptide. This theory has been recently reexamined and, instead of a specific path, a "tunneling tube" concept has been introduced, which now considers multiple pathways as forming a "tube" (Regan and Onuchic, 1999). In addition, a dynamic coupling of a tunneling electron and vibrational motions of the protein matrix has been recently proposed (Daizadeh *et al.*, 1997).

### *A. Plastocyanin*

Plastocyanin is the most studied cupredoxin with respect to its structure and function. It is synthesized in the cytosol as a 160- to 170-amino-acid precursor polypeptide, consisting of a 60- to 70-residue transit peptide followed by 97- to 99-amino-acid mature protein (Rother *et al.*, 1986; Smeekens *et al.*, 1985). The transit peptide has a bipartite structure containing all the information necessary to translocate the precursor plastocyanin across the chloroplast envelope and thylakoid membrane to its final destination in the thylakoid lumen (Smeekens *et al.*, 1986). The *Arabidopsis* genome has two different plastocyanin-encoding genes, both of which are localized on chromosome 1. The At1g20340/F14010.6 gene is in the top arm, while T23E18.3 is in the bottom arm. They display almost 80% amino acid sequence identity and it has been shown that both have the same suborganellar localization—the thylakoid lumen (Kieselbach *et al.*, 2000). Two separate plastocyanin sequences in a single organism have been identified in other plant species as well, e.g., poplar and tobacco (Dimitrov *et al.*, 1993) (see also GenBank Accession Nos. Z50185 and Z50186). Although both gene products have not been functionally characterized simultaneously in the same plant species, it is highly likely that they carry out the same function. Plastocyanin is a key component of the photosynthetic electron transfer chain where it accepts an electron from the membrane-bound cyt *f* of the cyt $b_6/f$ complex and donates it to the photosystem I complex housing the photo-oxidized reaction center $P700^+$. Interestingly, the cDNA corresponding to the At1g20340/F14010.6 gene was first identified during a screen for *Arabidopsis* cDNAs capable of restoring recombination proficiency and DNA damage resistance in *E. coli* (Pang *et al.*, 1993).

Plastocyanin is one of the most abundant copper proteins in plant photosynthetic tissues. It has been estimated to be present at a stoichiometry of $8 \times 10^6$ molecules per cell in the green algae *Chlamidomonas reinhardtii* (Moseley *et al.*, 2000). It is also believed that this freely diffusible electron carrier exists in the thylakoid lumen as a pool of both oxidized and reduced proteins. Under copper-deficient conditions ($<9 \times 10^6$ Cu ions per cell) (Moseley *et al.*, 2000), some green algae and cyanobacteria activate the gene encoding a structurally distinct (predominantly α-helical) protein, cyt *c6*, which is capable of fully substituting plastocyanin function (Kunert *et al.*, 1976; Wood, 1978; Merchant and Bogorad, 1986). Interestingly, the activation of the cyt *c6* gene is not accompanied by a complete repression of the plastocyanin gene. The translated plastocyanin precursor successfully translocates the mature apoprotein into the lumen where it degrades rapidly (Li and Merchant, 1995). These important observations indicate that the metal ion is a key *in vivo* stabilizing factor for this particular metalloprotein. It should be noted that this switch is relatively unusual—plants produce only plastocyanin, while some green algae and cyanobacteria utilize only cyt *c6*.

The high-resolution crystal structures for oxidized and reduced forms, as well as NMR solution structures, are available for both plastocyanin and cyt *c6* from numerous sources (Guss and Freeman, 1983; Guss *et al.*, 1986; Kerfeld *et al.*, 1995; Frazao *et al.*, 1995; Schnackenberg *et al.*, 1999). There is largely no structural difference between the reduced and the oxidized states of these proteins. Furthermore, their surface potentials display no changes that would explain a possible recognition mechanism distinguishing the reduced or oxidized proteins for their corresponding redox partners. In both proteins, two possible docking sites for the redox partners, and accordingly two electron transfer entry sites, have been proposed. In plastocyanins, one is the surface-exposed His ligand located in a hydrophobic surface environment called the "hydrophobic patch." The second site is through a conserved tyrosine residue linking to the Cys ligand. This tyrosine is surrounded by conserved acidic residues referred to as the "acidic patch." The hydrophobic patch has been identified in virtually all cupredoxins, whereas the acidic patch appears to be specific to plastocyanins. A series of mutagenesis and cross-linking experiments, as well as experiments involving chemically modified proteins, revealed that the interaction between plastocyanin and its electron donor cytochrome *f* is highly electrostatic, pointing to the conserved Tyr of the acidic patch as the electron entry site (Kannt *et al.*, 1996). However, NMR analysis of the molecular dynamics in the complex of spinach plastocyanin and the soluble domain of turnip cytochrome *f* revealed that the complex has a single orientation that includes both hydrophobic and acidic patches (Fig. 5) (Ejdebäck *et al.*, 2000; Ubbink *et al.*, 1998). The surface-exposed His has been identified as an electron entry site for a pathway that couples the Cu and heme Fe at a distance of ~11 Å. It has been proposed that the first interaction between soluble oxidized

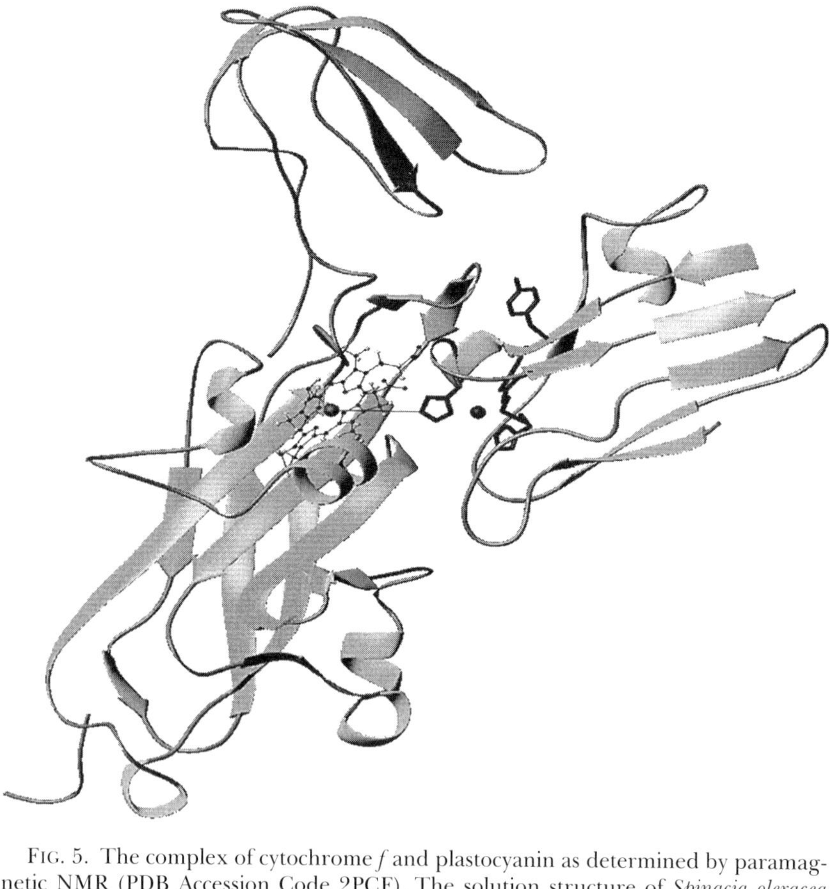

FIG. 5. The complex of cytochrome *f* and plastocyanin as determined by paramagnetic NMR (PDB Accession Code 2PCF). The solution structure of *Spinacia oleracea* plastocyanin was determined by NMR, while cytochrome *f* was modeled from the previous crystal structure of the soluble domain of *Brassica rapa* cytochrome *f* (PDB Code 1CTM), with only the contacts between the two proteins determined by NMR. The distance shown is between the heme Fe atom in cytochrome *f* and the εN of the His-87 copper ligand in plastocyanin.

plastocyanin and membrane-bound cytochrome *f* is mostly electrostatic without any fixed orientation; the orientation becomes fixed when the hydrophobic interactions are involved (Ubbink *et al.*, 1998).

In reduced plastocyanin at pH 4.0, the imidazole of the surface-exposed His rotates 180° around the Cβ–Cγ bond and flips away from copper. Such a structural rearrangement makes the copper coordination trigonal planar with Sγ(Cys), Nσ(His), and Sσ(Met) as ligands and moves the copper toward the Met by shortening the Cu–Sσ(Met) bond from 2.90 to 2.51 Å. This geometry stabilizes the copper in its reduced state and the protein becomes redox inactive. It has been proposed to occur also *in vivo* and is thought to serve as a cellular structural switch in case the electron transfer needs to be turned off. A similar effect has been also observed in pseudoazurin and amicyanin, albeit at relatively high pH (6.7) in the case of amicyanin. On the other hand, it should be noted that there is no evidence that the local pH in the vicinity of plastocyanin in the thylakoid lumen could decrease to the values that would allow His ligand protonation.

The docking of plastocyanin and cytochrome *c6* to their electron acceptor, the photosytem, I complex, has been identified to occur through the PsaF subunit of the complex (Hippler *et al.*, 1998). Surprisingly, no putative binding site for a known redox cofactor could be identified in the sequence of the PsaF polypeptide.

### *B. Amicyanin and Azurin*

Amicyanin is found in methylotrophic bacteria, which utilize methylated amines as their only energy source. Its expression is induced by methylamine, and inactivation of the amicyanin gene in *Paracoccus denitrificans* results in complete loss of its ability to grow on methylamine (Van Spanning *et al.*, 1990). These facts strongly suggest that amicyanin is a key component of the methylamine-driven electron transfer chain. The conversion of methylamine into formaldehyde is catalyzed by the enzyme methylamine dehydrogenase (MADH), which is a tetramer consisting of two small (15 kDa) and two large (46 kDa) subunits. Similar to amicyanin, MADH is also substrate inducible. In the bacterial chromosome, the gene encoding its small subunit is located immediately upstream of the amicyanin gene, indicating that they are cooperatively regulated (Chistoserdov *et al.*, 1994). Each small subunit houses a unique catalytic site, tryptophan tryptophylquinone (TTQ), which is formed by an intricate posttranslational modification involving two tryptophan side chains. One tryptophan is oxidized to an orthoquinone and covalently cross-linked to the indole ring of the second tryptophan (McIntire *et al.*, 1991). Amicyanin serves as the direct electron acceptor for MADH and transfers it to a c-type cytochrome, which subsequently donates the electron to the $aa_3$ cytochrome oxidase, a member of the family of heme–copper oxidases. Crystal structures of all four components are currently available, which makes it the only known electron transfer chain that is fully characterized structurally (Chen *et al.*, 1994; Iwata *et al.*, 1995).

The structures of the binary complex of amicyanin and MADH, as well as the ternary complex including cytochrome c551, have been determined (Chen *et al.*, 1992, 1994). Amicyanin is in contact with both the large and the small subunits of MADH. Similar to plastocyanin, the interaction of amicyanin with its electron donor occurs via the hydrophobic surface centered around the surface-exposed His ligand. The TTQ is oriented in such a way that its indole moiety is pointed toward the region of the small subunit that is in contact with amicyanin and its surface-exposed edge is approximately 10 Å away from the copper (Fig. 6). The interaction between amicyanin and cytc551 in the structure of the ternary complex possibly does not correspond to the optimal geometry for an efficient electron transfer and needs further assignment.

FIG. 6. Electron transfer complex between methylamine dehydrogenase and amicyanin from *Paracoccus dentrificans* (PDB Accession Code 2MTA). The distance shown is between εN of the redox cofactor tryptophan tryptophylquinone of methylamine dehydrogenase and the εN of the His-95 ligand of amicyanin.

It has been predicted that the His ligand flip cannot occur in amicyanin complexed with MADH because of steric hindrance due to close van der Waals contacts with residues of the neighboring small subunit of the MADH molecule (Zhu *et al.*, 1998).

Azurin is the most prominent member of the cupredoxin family. It was used for more than a decade as an excellent model protein for engineering and redesigning blue copper sites (Canters and Gilardi, 1993). Its amino acid sequence was one of only a few dozen protein sequences available in the mid 1960s (Dayhoff *et al.*, 1965) and in 1987 its gene was one of the first cupredoxins to be cloned (Canters, 1987) (plastocyanin gene sequences became available a year earlier). Moreover, it was one of the first cupredoxins to be structurally characterized (Adman *et al.*, 1978). Azurin was also the first cupredoxin for which efficient procedures were established allowing for production of a mutant cupredoxin in large quantities (Karlsson *et al.*, 1989). The information gained from these pioneering studies indeed has had a great impact on our understanding of many peculiar aspects of blue copper sites with respect to their mechanistic properties and electronic structure. Azurin has also been extensively studied as a useful model for characterizing long-range electron transfer processes within a polypeptide matrix (Farver *et al.*, 1993; Wuttke and Gray, 1993).

Surprisingly, its biological redox partners remain largely unknown. It has been implicated in anaerobic nitrite respiration and it has been shown that azurin can donate electrons to nitrite reductase, a function that is proposed to be carried out by another cupredoxin, pseudoazurin (see Section IV, E). On the other hand, azurin is not an inducible protein and denitrifying bacteria express azurin constitutively under aerobic conditions.

### *C. Rusticyanin*

Rusticyanin is found in *Thiobacillus ferrooxidans*, an acidophilic, chemolithotrophic sulfur bacterium utilizing $Fe^{2+}$ and reduced sulfur compounds as the sole energy source (Rawlings, 2001). These cells do not produce rusticyanin when grown only on a reduced sulfur source. Similar to other substrate-inducible bacterial cupredoxins, the rusticyanin gene is transcriptionally activated when soluble iron is introduced. It has been estimated to constitute almost 5% of the total cell protein when *T. ferrooxidans* grew autotrophically on an iron source. The reduced sulfur compounds are eventually oxidized to sulfuric acid, which acidifies the medium close to pH 2.0, though cells maintain their internal pH near neutral values. Rusticyanin itself does not carry out $Fe^{2+}$ oxidation and its redox potential, + 680 mV, is the highest among the currently characterized cupredoxins. Other iron-oxidizing bacteria, e.g., *Leptospirillum ferrooxidans*, produce a cytochrome *a* that substitutes for rusticyanin functionally (Takai *et al.*, 2001).

Little is known about the redox partners of rusticyanin, although a diheme cytochrome, cyt *c4*, has been implicated because of its ability to form a complex with rusticyanin. The complex formation between these two proteins at pH 4.8 induces a dramatic decrease, by almost 100 mV, in the redox potential of rusticyanin, while the potentials of both hemes in the cytochrome remain unchanged. Interestingly, complex formation is also accompanied by changes in the electronic absorption spectrum of rusticyanin that are reminiscent of those observed for the uncomplexed protein at high pH (above pH 7.0) (Giudici-Orticoni *et al.*, 1999).

### *D. Auracyanin, Halocyanin, and Sulfocyanin*

These cupredoxins are predicted to be cell surface proteins attached via a lipid anchor covalently bound to the N-terminus of the protein (McManus *et al.*, 1992; Scharf and Engelhard, 1993). Halocyanin was the first cupredoxin purified from an archaeon, haloalkaliphilic *Natronobacterium pharaonis*, which grows in high pH (around 10–11) and extreme salinity (30%) environments (Scharf and Engelhard, 1993; Mattar *et al.*, 1994). The occurrence of a blue copper protein, sulfocyanin, in another archaeon, *Sulfolobus acidocaladarius*, was first predicted from its gene sequence (Castresana *et al.*, 1995), and recently it has been purified as a recombinant protein displaying spectroscopic properties typical for a blue copper protein (Komorowski and Schäfer, 2001). Three different auracyanins, labeled A, B1, and B2, have been characterized from *Chloroflexus aurantiacus*, a gliding thermophilic photosynthetic bacterium (Trost *et al.*, 1988; McManus *et al.*, 1992). This bacterium is only distantly related to other photosynthetic organisms and is believed to have acquired its photosynthetic capabilities by lateral gene transfer rather than by evolution from an ancestral eubacterial photosynthetic cell. Two forms of auracyanin B, B1 and B2, exhibit 38% amino acid sequence identity with the auracyanin A form and are derived from the same gene product by differing degrees of N-terminal proteolytic processing (McManus *et al.*, 1992; Van Driessche *et al.*, 1999). Auracyanins B1 and B2 are glycosylated while auracyanin A is not. A and B forms also display distinct spectroscopic properties. Auracyanins B1 and B2 exhibit an axial EPR signal while auracyanin A has a rhombic EPR. Accordingly, the electronic absorption spectrum of the A-form features a second $S_{Cys}$-Cu CT transition band, at 450 nm, in addition to the main 600-nm band.

The sequences of auracyanin B, halocyanin, and sulfocyanin deduced from their gene sequences reveal an unusually long N-terminal extension featuring a hydrophobic domain similar to signal peptides found in other bacterial cupredoxins that directs the protein into the periplasm. This signal peptide is followed by a segment, that in the case of halocyanin, has an Asn-Gly doublet occurring consecutively seven times. In sulfocyanin, this segment is rich in Ser residues while in auracyanin B it is rich in Pro and Ala residues. In the DNA sequence of auracyanin B two AUG translational initiation sites, both of which follow a potential ribosome-binding site, can be identified (GenBank Accession No. U78046). These two AUG sites can generate either a 95- or a 60-amino-acid-long extension. However, only the nucleotide sequence flanking the second AUG displays characteristics of a strong translational initiation site, which requires a purine, usually A, at position −3 and a G at position +4 with respect to the first nucleotide of AUG. Therefore, the auracyanin B form with a 60-amino-acid-long N-terminal extension is probably predominant.

The Ser- and Pro/Ala-rich segments in these proteins appear to serve as a signal for attachment of lipid anchors. Such posttranslational modifications have also been suggested for halocyanin and for an outer membrane lipoprotein from *Neisseria gonorrhoeae* with a high degree of sequence identity to azurins (Gotschlich and Seiff, 1987). The *N. gonorrhoeae* lipoprotein features a 60-amino-acid-long N-terminal extension with a hydrophobic domain and a Pro/Ala-rich segment similar to that found in the precursor sequence of auracyanin B.

It is believed that auracyanin A also undergoes a similar posttranslational modification, albeit with a different mechanism since it has been reported that its precursor does not have an N-terminal Pro/Ala segment even though it does contain a signal peptide (Bond *et al.*, 2001). The N-terminus of the mature protein was found to be blocked by a group, predicted to be an acetyl-*N*-cysteine-*S*-glycerol found in other lipoproteins (Van Driessche *et al.*, 1999). The precise lipid attachment sites in these proteins have not been determined. The Pro/Ala-rich segment in *N. gonorrhoeae* azurin has been identified as the epitope reactive with the H.8 monoclonal antibodies, which indicates that it is not processed during the maturation of the proteins and possibly serves as a tether to space the protein from the cell surface, thus providing more flexibility for their contact with the redox partners (Kawula *et al.*, 1987).

The redox partners of these proteins have yet to be identified, although it has been shown that auracyanins can donate electrons to the membrane-bound cytochrome *c*-554, which is the direct electron donor for the photooxidized bacterial reaction center $P870^+$ (McManus *et al.*, 1992). However, whether it is their proper *in vivo* function remains uncertain. The sulfocyanin gene is in the same operon with the components of the respiratory electron transfer chain and, since *Su. acidocaladarius* completely lack *c*-type cytochromes, it is implicated as a substrate for the CuA-containing terminal oxidase. Interestingly, the occurrence of this operon in archaea has been interpreted as an indication that aerobic metabolism (respiration) evolved earlier than photosynthesis (Castresana *et al.*, 1995). The photosynthetic reaction centers are found only in bacteria and no chlorophyll-based photosynthesis has yet been detected in archaea.

The X-ray crystal structure of auracyanin B has been recently reported. It revealed a structure quite similar to that of azurin (Bond *et al.*, 2001). The root mean square deviations between the positions of 89 Cα atoms of the auracyanin B and *Alcaligenes denitrificans* azurin have been estimated to be only 0.795 Å.

### *E. Pseudoazurin (Nitrite Reductase)*

(Although nitrite reductases do not belong to the cupredoxin family, they are discussed together with pseudoazurin because of their close functional relationships).

Under limited oxygen conditions some bacteria are capable of utilizing nitrate/nitrite as an energy source. Nitrite reductase and pseudoazurin are components of that respiratory electron transfer chain that sequentially reduces $NO_3^-/NO_2^-$ to molecular nitrogen. Their sequences are highly conserved throughout species and display more than 60% amino acid sequence identity. A copper-containing nitrite reductase has been also reported to occur in the fungus *Fusarium oxysporum* (Kobayashi and Shoun, 1995).

Nitrite reductases are homotrimers in solution and they also crystallize as a triangular homotrimer, with a shape similar to that of the multicopper oxidases (Godden *et al.*, 1991; Grossmann *et al.*, 1993). Each monomer is composed of two BCB domains harboring a blue copper site in domain 1 and a catalytic mononuclear copper site located at the interface of two subunits and coordinated by three His residues, two from domain 1 and the third from domain 2 of the adjacent subunit. In addition, $H_2O$ or $OH^-$ coordinates with the catalytic copper site and is displaced upon substrate binding (Adman *et al.*, 1995; Murphy *et al.*, 1997b). The solvent binding and its replacement with the substrate have been also confirmed by ENDOR studies (Howes *et al.*, 1994). $NO_2^-$ binds to that position in a bidentate fashion through both oxygen atoms. Its reduction occurs using electrons transferred from the reduced blue copper site through a Cis-His pathway similar to that characterized in multicopper blue oxidases. The blue copper site and catalytic copper are separated by a distance of 12.6 Å, which is comparable to those estimated for the two other structurally characterized electron transfer pairs, amicyanin–MADH and plastocyanin–cyt *f*, as well as for the multicopper oxidases and $N_2O$ reductase discussed below. In the ascorbate-reduced form of the enzyme there was no solvent molecule found in the catalytic copper site, while the geometry of the ligand environment at the blue copper site remains unperturbed (Murphy *et al.*, 1997b). The blue copper site is 7 Å beneath the surface of the protein molecule and similar to cupredoxins, it has one of its His ligands exposed to the solvent.

Three proteins have been observed to donate electrons to the blue copper site of nitrite reductases: pseudoazurin, azurin, and *c*-type cytochromes. The biological electron donors for the three proteins are not known. Interestingly, even cytochrome *c* from eukaryotes (horse or yeast) can act as an efficient electron donor and is widely used in many laboratories for activity studies (e.g., see Olesen *et al.*, 1998). However, only pseudoazurin is currently considered to be an *in vivo* reaction partner of nitrite reductase. In this context it is important to note that nitrite anions can efficiently oxidize some cupredoxins (Nersissian *et al.*, 1985), including azurin. Interestingly, the kinetics of azurin reduction by nitrite reductase has been shown to be biphasic, with a fast initial linear phase followed by an extended nonlinear phase (Dodd *et al.*, 1995). Instead of a copper-containing nitrite reductase, many denitrifying microorganisms utilize its functional isologue, a $cd_1$-type diheme cytochrome, which can also accept electrons from azurin, pseudoazurin, and cytochrome *c*.

## V. Phytocyanins

Phytocyanins are plant-specific proteins that constitute a large family of single BCB domain-containing proteins (Nersissian *et al.*, 2001). They share a remarkably high degree of sequence identity and are distinctly different from plastocyanins and from other cupredoxins of bacterial origin. More than 80 full-length sequences of phytocyanin precursors deduced from either genomic DNAs or cDNAs are currently available. Most of them are from a single species, *Ar. thaliana*. At least 42 phytocyanin-related genes in the haploid genome of this plant species can be identified, which makes it one of the largest gene families known in plants (see Fig. 7). Numerous partial sequences closely resembling phytocyanins can be also identified in ESTs from a number of different plant species, including pine, soybean, tomato, rice, tobacco, cotton, maize, ice plant, alfalfa, barley, wheat, and potato. Phytocyanins are further divided into three distinct subfamilies: stellacyanins, plantacyanins, and uclacyanins (Nersissian *et al.*, 1998). Members of fourth subfamily, the early nodulin proteins, are identified by their high level of sequence similarity to the phytocyanins but they appear from their sequences to have lost their blue copper-binding capabilities (Nersissian *et al.*, 2001).

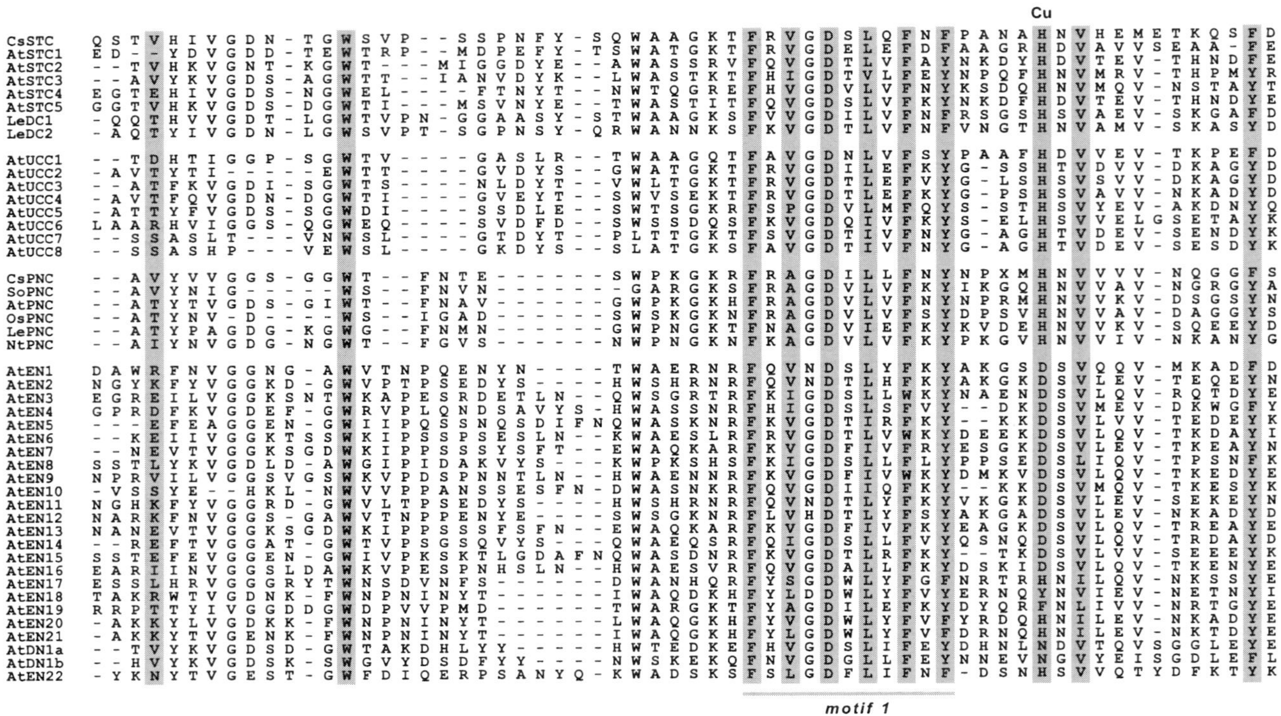


FIG. 7. Alignment of the amino acid sequences of the copper-binding domains of 36 members of the phytocyanin family identified in the haploid genome of *Arabidopsis thaliana* (5 stellacyanins, AtSTC1 through AtSTC5; 8 uclacyanins, AtUCC1 through AtUCC8; 1 plantacyanin, AtPNC; and 22 early nodulins, AtEN1 through AtEN22, as well as two BCB domains of dinodulin 1, AtDN1a and AtDN1b). The sequences of four nodulins (AtEn23-AtEN26) with an N-terminal Gly, Ser, and Trp rich domain, and dinodulin 2 (AtDN2) are omitted from the alignment for simplicity. For the gene names and their identification see Table I. The sequences of phytocyanins characterized by X-ray crystallography (CsSTC, cucumber stellacyanin; CsPNC, cucumber basic protein; SoPNC, spinach plantacyanin), and those with

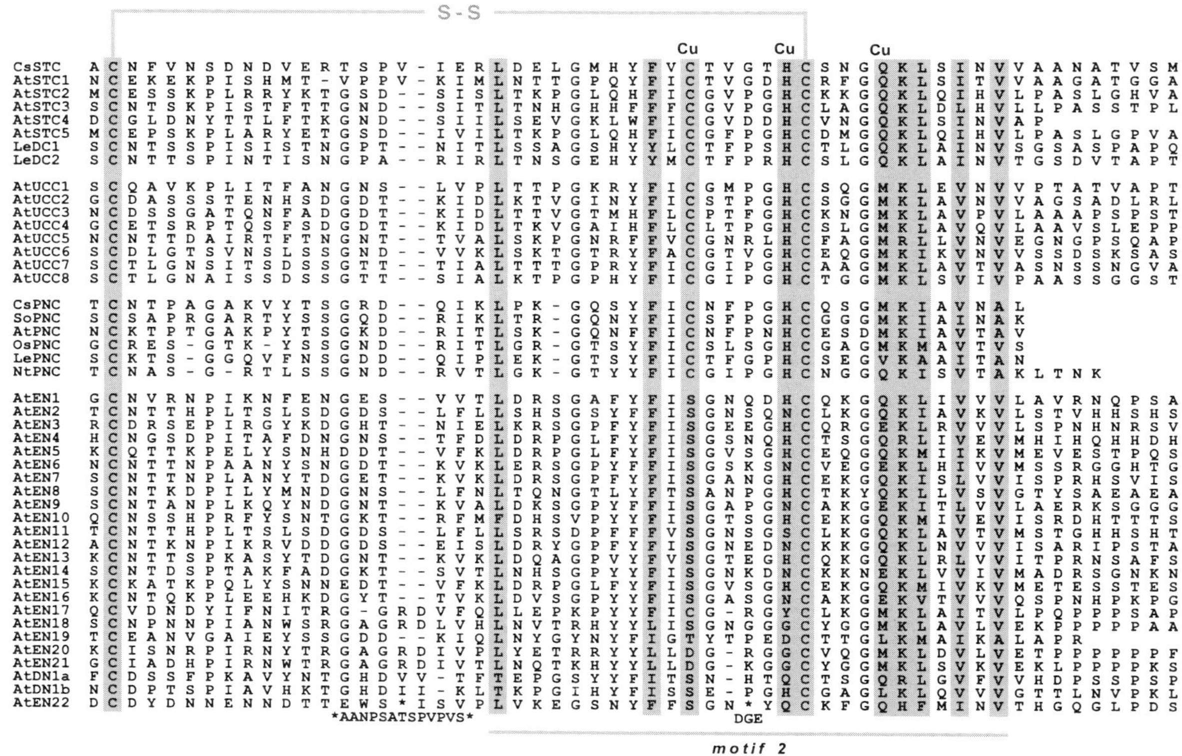

FIG. 7. *continued*

and LeDC2), and the shortest BCB domain protein, rice plantacyanin (OsPNC), are also included in the alignment. Two short amino acid stretches from nodulin AtEN22 sequence are removed for clarity and displayed below the sequence alignment. The positions of these stretches are marked with asterisks. The residues that are identical or display conserved side-chain characteristics in all 46 sequences are highlighted in gray. The amino acids involved in copper binding are marked with Cu. S–S denotes conserved Cys residues involved in disulfide bridge formation. The two conserved sequence motifs are marked with a bar.

Cucumber stellacyanin and spinach plantacyanin are the only phytocyanins for which sequence information is currently available for both the mature proteins, determined by protein sequencing, and the precursor proteins, deduced from the cDNA (Mann *et al.*, 1992, 1996; Nersissian *et al.*, 1996, 1998). With the exception of the plantacyanins, most phytocyanins are chimeric proteins in their predicted or known mature forms. They are composed of two structurally distinct sequence domains, a 100- to 109-amino-acid BCB domain followed by a domain that varies in length between 30 and 220 amino acids, lacks any obvious consensus sequence, and resembles heavily glycosylated arabinogalactan proteins (AGP) (Nothnagel, 1997).

All phytocyanins are processed in the endoplasmic reticulum as evidenced by the presence of signal peptides in their precursor sequences that target proteins to the secretory pathway. In addition, precursors of stellacyanins, uclacyanins, and early nodulins contain 16–25 amino acid hydrophobic peptides at the C-termini with sequence characteristics typical for a glycosylphosphatidylinositol (GPI)-anchoring signal. The occurrence of both a signal peptide and a C-terminal hydrophobic domain in the same precursor is a well-documented indication that the mature protein receives a covalently attached lipid moiety, GPI, which anchors it to the cell surface. Importantly, some stellacyanins have been purified from plant tissues as soluble proteins and, where the amino acid sequences are available, they lack the C-terminal hydrophobic peptides. Since the hydrophobic tail is removed when the GPI anchor is attached, this information suggests that phytocyanins may be freed from GPI anchors, possibly by a phosphatidyl-specific phospholipase, and released into the extracellular matrix. Further, the yield of phytocyanins during the preparative purification from plant tissues is significantly improved by increasing the salt concentrations in the extraction buffers. Such high-salt treatment is especially effective for the strongly basic plantacyanins, suggesting that despite the fact that plantacyanins lack the AGP-like domain and a GPI-anchoring signal, they still may be embedded within the negatively charged pectin network and can be released from it by high concentrations of $Ca^{2+}$ ions. Interestingly, it is thought that pectins serve as recognition molecules that alert plant cells to the presence of foreign organisms, either symbionts or pathogens. Thus, currently available experimental and structural data clearly indicate that phytocyanins are cell surface-attached proteins that may also be found circulating in the extracellular milieu, as diffusible redox macromolecules.

The classification of the four subfamilies, stellacyanins, plantacyanins, uclacyanins, and early nodulins, is based (i) on their spectroscopic features, (ii) on precursor as well as mature protein domain organization,

(iii) on identity or availability of residues involved in blue copper binding, and (iv) on the fact that a representative from each of the subfamilies has been characterized in a single species, *Ar. thaliana*, which excludes interpretation of such sequence diversity among phytocyanins as simply a result of their diverse species of origin. In addition, pairwise sequence comparison shows that these proteins are indeed clustered into four different groups, exhibiting in their single BCB domains a high degree of sequence identity, 50–80%, within the same subfamily, but only 30–40% identity between different subfamilies.

### *A. Stellacyanin*

Stellacyanin was the first phytocyanin to be described in the literature and is one of the most thoroughly studied biochemically and biophysically, although its function is still unknown. It was isolated in 1940 as a blue pigment coproduct during the purification of the multicopper oxidase laccase from the extracellular secretion of the Japanese lacquer tree *Rhus vernicifera* (Keilin and Mann, 1940). The amino acid sequence of this protein revealed that it possesses a Gln residue at the position of the axial ligand for the copper atom (Bergman *et al.*, 1977). Proteins displaying spectroscopic properties similar to those of *R. vernicifera* stellacyanin have been subsequently purified from horseradish, cucumber, zucchini, and spinach (Aikazyan and Nalbandyan, 1979; Marchesini *et al.*, 1979; Paul and Stigbrand, 1970; Sarkissian and Nalbandyan, 1983). Their amino acid sequences also feature a Gln residue at the position of the axial ligand (Mann *et al.*, 1992; Schinina *et al.*, 1996; Van Driessche *et al.*, 1995). The other ligands are two His residues and a Cys residue. Stellacyanins resemble uclacyanin and early-nodulin subfamily proteins in that they are chimeric proteins consisting of a copper-binding domain and an AGP-like domain. They have GPI-anchoring signals in their precursors. Stellacyanins and uclacyanins carry several N-linked glycosylation sites in their copper-binding domains through Asn residues and numerous O-linked glycosylation sites in their AGP-like domains through hydroxyproline, Ser, and Thr residues. There are five different stellacyanin genes in *Arabidopsis*.

### *B. Plantacyanin*

Plantacyanins are strongly basic proteins with an isoelectric point close to 11. First isolated from cucumber in 1974, plantacyanins have since been characterized from various other plant species (Markossian *et al.*, 1974; Aikazyan and Nalbandyan, 1981). They exhibit relatively high sequence identity to stellacyanin and spectroscopic properties similar to those of stellacyanin, although they generally utilize a Met residue as the axial ligand for copper (Murata *et al.*, 1982; Mann *et al.*, 1996; Nersissian *et al.*, 1998). However, there are two exceptions. First, tomato plantacyanin (GenBank Accession No. AF243181) is the only known single BCB domain blue copper protein that has a noncoordinating residue (Val) at the position of the axial ligand, a feature previously seen only in multicopper oxidases, such as laccases and ceruloplasmin. Second, tobacco plantacyanin (GenBank Accession No. AF172853; McClure *et al.*, 2000) has a Gln residue similar to stellacyanins (see Fig. 7).

Plantacyanins and stellacyanins were originally grouped into a separate protein family within the cupredoxins and named phytocyanins (Ryden and Hunt, 1993). This suggestion was based mainly on comparison of the protein sequences of cucumber plantacyanin (also known as cucumber basic protein) and *R. vernicifera* stellacyanin available at that time. Its validity became more convincing as more sequences and crystal structures of these proteins became available. Plantacyanins are single-domain proteins in their mature form and neither an AGP-like domain nor a GPI-anchoring signal is found in their precursors. Although they have been purified from cucumber and spinach as nonglycosylated proteins, there is one potential N-linked glycosylation site in the sequences of soybean (GenBank Accession No. AW185058), pine (GenBank Accession No. BG275625), and chickpea (GenBank Accession No. AJ012693) proteins. Plantacyanin is represented in the *Arabidopsis* genome by a single gene. However, in soybean and rice at least three different ESTs that encode full-length plantacyanin precursor sequences can be identified. One of the rice plantacyanins (GenBank Accession Nos. BE040691 and BE040721) is the smallest BCB domain identified to date. It has only 90 amino acids (see Fig. 7). Spinach plantacyanin is 91 amino acids long (Nersissian *et al.*, 1998).

### *C. Uclacyanin*

In the mature form, these proteins are predicted to consist of a copper-binding domain and an AGP-like domain and to have a GPI anchor. Although the copper-binding site in these proteins is predicted to be arranged from the same residues as in plantacyanins, 2His, Cys, and Met, their spectroscopic properties closely resemble those of the Gln99Met mutant form of cucumber stellacyanin rather than those of the plantacyanins (Nersissian *et al.*, 1998). *Arabidopsis* UCC1 is the only uclacyanin for which the copper-binding abilities and spectroscopic properties have been characterized. There are eight different uclacyanin genes in the *Arabidopsis* genome.

### *D. Early Nodulin*

These phytocyanin-related proteins were first identified as early nodulins in soybean (labeled as GmENOD55 or GmN#315) (De Blank *et al.*, 1993; Kouchi and Hata, 1993) and subsequently also in *Medicago truncatula* (labeled as MtN16 and MtN20) (Greene *et al.*, 1998). The name early nodulin is derived from the name of a plant organ, the root nodule, which is formed on infection of leguminous plants with diazotrophic bacteria. Legumes recognize these bacteria as symbionts and house them in nodule cells, thus entering a mutually beneficial relationship that results in the conversion by the bacterial enzyme nitrogenase of molecular nitrogen into ammonia, which the plant can use. The nodule formation is triggered by host-specific bacterial Nod factors that activate plant genes encoding proteins called nodulins. These proteins are further classified as early or late based on the timing of their expression relative to nodule formation. In their precursors GmN, MtN16, and MtN20 display a domain organization identical to that of stellacyanins and uclacyanins. Accordingly, they possess a signal peptide, a BCB domain followed by an AGP-like domain, and a GPI-anchoring signal. In addition, in their BCB domain, they exhibit a high degree of sequence identity with the analogous domain of phytocyanins. However, despite such extensive similarity, these proteins lack the amino acid residues that are crucial for the formation of a blue copper site in other phytocyanins, although the possibility that they may bind another metal, even copper, with a configuration different from that of a blue copper, should not be entirely ruled out. In its haploid genome, *Arabidopsis* possesses 22 genes encoding such polypeptides even though it is not a legume. In addition, there is a group of four genes (AtEN23–AtEN26) each encoding a BCB domain containing polypeptide, which similarly to early-nodulins do not have a blue copper binding site. However, they display a significantly different precursor domain organization where the BCB domain is fused with an N-terminal 150–180 amino acid domain composed of predominantly Gly, Ser, and Trp residues.

Most phytocyanin genes contain two exons that are separated by an intron. In all cases the intron occurs between the first and second nucleotides of the codon for a nonconserved amino acid, corresponding to Gln-38 in cucumber stellacyanin, which is located within the consensus sequence motif 1 of the copper-binding domain. It is also important to note that all four amino acids involved in copper binding (2His, Cys, and Met or Gln) as well as conserved bridging cysteines are located in the second exon. The genes for MtN16 and MtN20 and all of their *Arabidopsis* homologues are identical to phytocyanins in their exon–intron organization, including the location of the splice site. This fact together with the overall sequence similarity strongly suggests that these proteins have a common evolutionary origin.

Nucleotide sequence analysis of the genes encoding these proteins can provide important clues as to whether early nodulins lacking a blue copper-binding site and phytocyanins that possess such a site are from the same line of descent. The vast majority of the early nodulins have a Ser residue in the position of the copper ligand Cys in uclacyanins, plantacyanins, and stellacyanins. The Cys residue provides its side chain sulfur for the copper coordination and numerous mutagenesis studies have shown that it is the only indispensable ligand and is crucial for the formation of a functional blue copper site. Therefore, conversion of Cys to Ser, or vice versa Ser to Cys, was the most dramatic mutation event that might have occurred during the evolution of BCB domain-containing proteins. It would convert a redox-active copper protein into one that would presumably lack that important metal and therefore would have a radically different function, now having a Ser residue in its putative "active" site. Conversely, it would generate a novel function determined by the presence of the copper atom on the scaffolding of the protein, which once had a Ser residue.

To date, at least 70 full-length or truncated early-nodulin precursor sequences from a number of different plant species can be identified in the GenBank database; these sequences are derived from nuclear DNA or EST/cDNA sequences. Ser is encoded by a set of six different codons, TCG, TCA, TCT, TCC, AGT, and AGC, which are equally represented in the *Arabidopsis* protein coding genes; hence one would expect that all six would be also found for that active site Ser residue in early-nodulin nucleotide sequences. Remarkably, the active site Ser in all 70 sequences is exclusively encoded by AGT or AGC codons, indicating that it is highly likely that they originated from the only two codons of Cys, TGT and TGC, by a single T-A point mutation. Therefore, one should conclude that early nodulins that lack their blue copper-binding sites evolved from corresponding blue copper proteins.

Crystal structures of three phytocyanins are currently available. Two are for plantacyanins, from cucumber (also known as cucumber basic protein) (Guss *et al.*, 1988, 1996) and from spinach (Einsle *et al.*, 2000), and one is for the recombinant BCB domain of cucumber stellacyanin (Hart *et al.*, 1996). The three proteins display folding topology identical to one another, suggesting that phytocyanins fold into a uniform structure, which can be designated as a phytocyanin fold. As a historical note, the crystallization of the cucumber basic protein and its preliminary crystallographic data were reported in 1977, before any structure of a blue copper protein was available (Colman *et al.*, 1977). However, the structure was solved in 1988 only by application of the then newly developed method of multiple wavelength anomalous dispersion phasing (Guss *et al.*, 1988). It was the first structure of a metalloprotein to be solved by this method using the metal atom as the anomalous scatterer.

Three important features of the phytocyanin fold distinguish it from those of other BCB domain folds. First, one of the two β-sheets that form the barrel is less H-bonded and exhibits a severe degree of twist. This feature affects the topology of the barrel, flattening it and making it more open on one side, compared to other BCB domains. Second, there is a disulfide bridge in the phytocyanins that appears to play a crucial role in stabilizing the overall structure, in particular the copper-binding loop, as one of the bridging cysteines directly follows the His ligand located in the copper-binding loop. The third and perhaps the most striking structural feature of the phytocyanins is the extensive exposure of the copper site. The imidazole rings of both His ligands are completely solvent exposed, pointing their copper-distal nitrogen atoms away from the surface of the protein molecule. Such a structural arrangement allows the copper to reside only about 3 Å below the surface of the protein, making it more accessible relative to other BCB domains where the copper is at least 5 Å from the surface with only one His ligand solvent exposed. In stellacyanin the ε-carbonyl oxygen of a Gln is strongly coordinated at a distance of 2.2 Å, while in plantacyanins and most other BCB domains harboring copper the thioether sulfur of a Met residue is weakly coordinated at a distance ranging between 2.6 Å in cucumber plantacyanin to 3.15 Å in *Al. denitrificans* azurin. The structure of cucumber stellacyanin in its reduced form displays a relatively large structural difference in the ligand environment compared to that of its oxidized form. In the reduced structure, the axial site Gln ligand rotates around the Cβ–Cγ bond and moves the ε-oxygen ligand away from the copper by almost 0.5 Å (Stine *et al.*, manuscript in preparation). Such a redox-dependent structural rearrangement has no analogue in any other naturally occurring blue copper protein and appears to be unique for stellacyanins and perhaps also other phytocyanins with Met as an axial ligand. It may significantly increase the inner-sphere reorganization barrier, making their participation in long-range electron transfer processes unlikely. Therefore, it was suggested that phytocyanins are involved in redox reactions with low-molecular-weight compounds rather than with protein electron mediators (Nersissian *et al.*, 1998, 2001). For the same reasons a similar suggestion has been made for nitrosocyanin with a "red" copper site (see Section X).

Stellacyanin is also an important example of a heavily glycosylated protein for which the structure has been determined without its glyco components. It demonstrated that the carbohydrate moieties have virtually no effect on the folding topology of the polypeptide core of this particular glycoprotein. One of the three glycosylation sites in

*R. vernicifera* stellacyanin can be modeled on the structure of its cucumber counterpart close to the vicinity of the copper-binding site. Such an arrangement would apparently create an extensive carbohydrate shield at the copper-binding site of *R. vernicifera* stellacyanin. It would possibly make the entire northern end of the protein surface inaccessible for direct docking to another protein molecule, such as a putative redox partner, through specific protein–protein interaction. On the other hand, it is most probable that the copper would still remain accessible to low-molecular-weight redox compounds. Structural studies clearly indicate that with the solvent-exposed copper site, high reorganization energy, and extensive carbohydrate shield at the active site, it is unlikely that phytocyanins participate in long-range electron transfer processes through an interaction with large-redox macromolecules.

However, the function of phytocyanins is as yet unknown, although a growing body of recent literature data provides intriguing hints that phytocyanins may participate in a wide range of physiological processes that involve cell wall signaling pathways. Examples include lignin formation, oxygen activation, pathogenesis, stress, cell differentiation, organogenesis, and finally cell-to-cell signaling.

Far less is known about the molecular mechanisms of cell-to-cell signaling in plants than in animals. Unlike animal cells, plant cells have cell walls that house surface molecules that mediate cell–cell and wall–nucleus communications initiating organogenesis and other important morphological processes. In some cases these signals extend progressively to greater distances from the initiation site via a transport conduit that requires secretion of specific molecules into the extracellular milieu. Wall surface receptors also allow plants to discriminate their own cells from foreign cells in pollen-style interactions thereby promoting self-compatibility in plants. Apparently, it is the wall-to-wall communication that orchestrates the initial events that occur during the interaction of plants with microbes. This communication may determine whether they are seen as pathogens or symbionts. In other instances, the wall may initiate synthesis and deposition of lignin to armor itself against invading fungal and bacterial pathogens or to repair mechanical wounds.

Putative functions of phytocyanins based on literature data published over the past few years can be grouped into three different categories:

*1. Cell to Cell Signaling, Cell Differentiation, and Organogenesis*

(a) *Nicotiana alata* plantacyanin binds to the self-compatibility factor $S_{C10}$-RNase and therefore may be implicated in pollen rejection (McClure *et al.*, 2000), (b) Analyses of ESTs from immature female sexual organ revealed that *Marchantia polymorpha* plantacyanin is expressed during sexual differentiation (Nagai *et al.*, 1999), (c) The alfalfa plantacyanin gene shows nodule-specific expression before the onset of nitrogen fixation (Jimenez-Zurdo *et al.*, 2000), (d) Several cDNAs encoding early-nodulin subfamily proteins were identified as early nodulins in soybean and *M. truncatula* (De Blank *et al.*, 1993; Greene *et al.*, 1998; Kouchi and Hata, 1993), (e) Three early-nodulin subfamily proteins have been identified as *Nicotiana tabacum* L. embryogenic pollen-abundant phosphoproteins that appear in the cells undergoing the dedifferentiation process from immature pollen grains to embryogenic cells (Kyo *et al.*, 2000), (f) Morning glory cDNA encoding an early-nodulin subfamily protein was isolated during a screen for the genes involved in floral initiation (Yoshizaki *et al.*, 2000).

Interestingly, plant AGPs have been also implicated in cellular processes that are identical to those described above for phytocyanins (Nothnagel, 1997). As we have already mentioned, most phytocyanins in their mature form are predicted to be composed of a BCB domain and a domain with sequence characteristics reminiscent of those of described for AGPs. Thus, phytocyanins are interesting examples of the recently developed "Rosetta stone sequence" concept, which postulates that when two different proteins also occur in parallel as a fused, larger composite protein, it is an indication that they are functionally related and may even physically interact (Eisenberg *et al.*, 2000; Marcotte, 2000).

*2. Lignin Formation*

(a) A cDNA encoding a uclacyanin has been identified as corresponding to the gene specifically related to the lignification processes in pea pods (Drew and Gatehouse, 1994). (b) Phytocyanins are abundant in xylem, highly specialized cells involved in lignin formation and accumulation (Allona *et al.*, 1998; Sterky *et al.*, 1998). (c) A cDNA encoding a stellacyanin has been identified and characterized from differentiating xylem of loblolly pine along with arabinogalactan and proline-rich proteins (Zhang *et al.*, 2000).

*3. Oxygen Activation, Pathogenesis, and Stress*

(a) The *bcb* gene encoding one of the *Arabidopsis* stellacyanins (AtSTC1) is a light-negatively regulated gene (Van Gysel *et al.*, 1993). (b) The same gene is activated by oxidative stress in concert with other oxidative stress-inducible genes such as those that encode superoxide dismutase, peroxidase, and glutathione *S*-transferase (Miller *et al.*, 1999; Richards *et al.*, 1998). (c) Uclacyanin cDNA has been isolated from loblolly pine in a screen for water deficit stress-inducible genes (Chang *et al.*, 1996). (d) Stellacyanin cDNA has been isolated from the pepper cDNA library from hypersensitive response lesions of leaves infected with an avirulent strain of *Xanthomonas campestris* pv. *vesicatoria* by a differential screen against the library from healthy leaves (Jung and Hwang, 2000).

The occurrence of multigene families in eukaryotes is often an indication of tissue-specific and/or developmentally regulated expression. The fact that phytocyanins are represented in *Arabidopsis* as a large gene family hints that they may be involved in vital physiological processes by carrying out related functions in different tissues. The preliminary data extracted from the *Arabidopsis* DNA microarray database at the Stanford DNA Microarray Facility (at the URL http://genome-www4.stanford.edu/MicroArray/SMD/index.html) show that indeed phytocyanins display a preferential tissue expression. Thus, uclacyanin 2 (AtUCC2) is highly expressed in roots, while AtUCC3 and AtUCC8 are expressed in flowers (for the GenBank accession numbers, see Table I). One of the five stellacyanin genes, *bcb* (AtSTC1), shows no expression in roots and is regulated by the machinery that controls circadian rhythms in plants. The most striking finding is that plantacyanin (AtPNC) appears to be strictly flower specific. Its gene is one of the highly expressed genes in flower with a $R/G = 27.4$ against leaves and has virtually no expression elsewhere. The early-nodulin AtEN12 is also flower specific.

These studies are providing a clear message. In summary, stellacyanins and uclacyanins with their redox-active copper seem to be involved in two major processes, defense and lignification, which have long been thought to be orchestrated by redox-active components of the cell wall. Early-nodulin subfamily proteins that do not have a blue copper site, as predicted from their amino acid sequences, participate in cell–cell communications that initiate organogenesis in plants, such as nodulation, floral initiation, and sexual differentiation. Surprisingly, redox-active plantacyanins are also involved in these processes. In addition, plantacyanins are implicated in the mechanisms that control self-compatibility in plants.

### *E. Dicyanin and Dinodulin*

BCB domains with high sequence identity to those of phytocyanins are also found in a recently identified novel class of blue copper proteins in which two such domains are fused together into a single polypeptide. One of them, named dicyanin (GenBank Accession No. AF243180), was identified in tomato and is composed of two stellacyanin-like BCB domains, both of which feature a blue copper-binding site with a Cys, a Gln, and two His ligands. They display 45–60% sequence identity with each other and with stellacyanins. Similar to the known mature stellacyanin sequences, both domains are followed by a 35-amino-acid AGP-like domain. Two genes encoding proteins with a duplicated BCB domain organization has been also identified in the *Arabidopsis* genome. Unlike dicyanin, the BCB domains in this protein lack copper-binding sites and are reminiscent of the early-nodulin subfamily proteins. Therefore, it was designated dinodulin. Both dicyanin and dinodulin precursors harbor an N-terminal signal peptide, and dicyanin in addition harbors a C-terminal GPI-anchoring signal, indicating that these proteins and phytocyanins follow the same posttranslational secretory pathway. Similar sequences can be also identified among ESTs from other plant species.

Tomato dicyanin has been purified as a 26-kDa recombinant protein, which is a monomer in solution and binds two copper atoms per protein molecule with spectroscopic properties (EPR, UV–Vis) quite similar to those of stellacyanins (Nersissian, Hill, Valentine, unpublished data).

Prior to the identification of tomato dicyanin, all known blue copper proteins were organized with either one BCB domain, as in the case of cupredoxins and phytocyanins, or three or six BCB domains, as in the case of multicopper blue oxidases. Nitrite reductases are composed of two domains, but they are trimeric proteins both in solution and in the crystalline state, which forms a ceruloplasmin-like six domain structure. In addition, the BCB domains in nitrite reductases are structurally quite different from one another. Tomato dicyanin is the only known blue copper protein with a duplicated BCB domain organization that is a monomer in solution. Therefore, dicyanin and dinodulin are candidates for that missing intermediary link in the evolution of single BCB domain proteins to multiple BCB domain proteins. One could speculate that such a transformation might have occurred through stepwise gene fusion events, i.e., addition of one BCB domain at a time, which would form two- and three-BCB-domain proteins. The genes encoding six-BCB-domain proteins are apparently formed by triplication of a gene encoding a two-BCB-domain protein.

Intriguingly, our BLAST analysis revealed that a gene encoding a duplicated BCB domain protein can be also identified in the genome of an archaeon, *Halobacterium* sp. NRC-1 (GenBank Accession No. AE005073, gene *hcpA*), which is the richest in cupredoxin sequences among prokaryotes, containing at least seven such sequences. One of the domains in this archaeal polypeptide lacks a blue copper-binding site while the other features the three invariant ligands, two His and a Cys, required for the formation of such a site (see Fig. 4). In addition they display ~30% sequence identity with each other and with the seven other cupredoxin sequences identified in the same genome.

## VI. Ephrins

Ephrin ligands are signaling molecules that, along with their corresponding Eph receptor tyrosine kinases, participate in vital cell–cell signaling pathways that initiate pattern formation and morphogenesis in metazoa (Flanagan and Vanderhaeghen, 1998; Wilkinson, 2001). The recently solved crystal structure of mouse ephrin B2 ligand revealed that it displays a phytocyanin-like folding topology (Toth *et al.*, 2001), thus providing important evidence that proteins similar to phytocyanins may be present in animals as well. Ephrin ligands are subdivided into two classes, A and B. Both are cell surface-attached proteins either via a GPI anchor, which is reminiscent of the situation with phytocyanins, or via a C-terminal short transmembrane segment. In their extracellular domains, which constitute active core sequences, they display 20–30% sequence identity between the members of the A and B classes, and the identity further increases to 40–50% between members of the same class. Therefore, one should expect that both classes of ligands might have a common folding topology, which, based on the structure of mouse ephrin B2, appears to be similar to the phytocyanin fold. In addition, it has been recently shown that ephrin-mediated events become repulsive by cleavage and diffusion of the extracellular domain into the intercellular milieu (Hattori *et al.*, 2000). Again, this has also been documented for phytocyanins. Thus, it is highly likely that plants and animals have utilized the same structurally conserved protein modules for their own evolution from a common ancestral unicellular eukaryote into multicellularity.

## VII. Multicopper Oxidases

In this section we will provide only a brief summary of multicopper blue oxidases and some of the novel members of these family that we identified because of their unique sequence characteristics. For more detailed information we direct the readers to excellent reviews and a book that were published in the past few years (Solomon *et al.*, 1996; Messerschmidt, 1997).

Multicopper blue oxidases are synthesized as a single polypeptide chain, which is composed of three BCB domains in the case of laccases (LC) and ascorbate oxidases (AO) and six such domains in ceruloplasmin (CP) and hephaestin (HP). Structurally they are arranged in a triangular manner. These enzymes, along with heme-copper oxidases (cytochrome *c* oxidases and quinol-oxidases) and a cyanide-resistant alternative oxidase found in mitochondria of plants and fungi, are the only known enzymes capable of catalyzing four-electron reduction of dioxygen to water. In the

1970s, the spectroscopic characteristics of these multicopper proteins led to the specification of three distinct types of copper coordination in biology (Malkin and Malmström, 1970; Malmström, 1982). The type 1 copper is the "blue" copper site, which was discussed above. The multicopper oxidases have at least one type 1 site (three in the case of ceruloplasmin and hephaestin), which is the primary electron acceptor from the substrate. The catalytic site of oxygen reduction is a trinuclear site composed of a type 2 copper and a pair of copper atoms labeled as type 3 copper (Fig. 8). The type 2 copper is characterized with a weaker absorption in the visible region and larger hyperfine interactions in the EPR signal. It is coordinated with two His ligands and a solvent molecule. The type 3 site is a pair of copper ions, each coordinated by three His ligands and a bridging moiety, presumably hydroxide, which is believed to couple them antiferromagnetically, thus rendering them EPR silent. In the catalytic trinuclear site, there is no bridging ligand between type 2 and type 3 coppers. The type 2 copper is usually very labile (Calabrese *et al.*, 1988; Ducros *et al.*, 1998).

The type 1 copper site, which is the substrate oxidation site, is maintained in the C-terminal BCB domain (domain 3 in AO and LC and

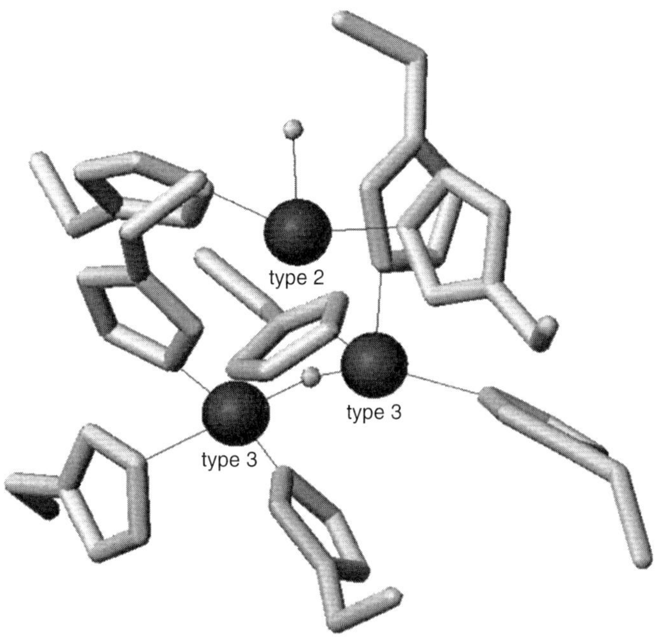


FIG. 8. The trinuclear catalytic copper site of human ceruloplasmin (PDB Accession Code 1KCW). The two small spheres are oxygen atoms from water.

domain 6 in CP and HP) while the trinuclear catalytic site is located in the interface of this domain and the N-terminal BCB domain 1. Each flanking domain provides four histidines for type 2 and type 3 copper coordination; the histidines are arranged in the amino acid sequence in a conserved pattern of four His-X-His motifs. One motif, His-Cys-His, contains the Cys ligand of the type 1 copper and two His ligands, one for each of the type 3 coppers. These ligands couple the type 1 copper and trinuclear catalytic center at a distance of 10–12 Å for an efficient electron transfer through a pathway known as "Cis-His." The His-Cys-His-X-X-X-His-X-X-X-X-Met(Leu, Ile, Phe in most laccases) motif, localized close to the C-terminus of the last BCB domain, is an important sequence signature for multicopper oxidases.

### *A. Laccases, Ascorbate Oxidases, and Pectinesterases*

Laccases are found in bacteria, fungi, plants, and insects. They catalyze oxidation of a variety of phenolic and inorganic substances with $K_m$ values ranging between 1 and 10 mM (Xu, 1996, 1997). Because of such broad substrate specificity, a substrate-binding pocket could not be identified in the crystal structure of the enzyme from the fungus *Coprinus cinereus* (Ducros *et al.*, 1998, 2001). They are involved in sporogenesis in bacteria (Donovan *et al.*, 1987), while in insects they function in sclerotization of cuticles (Sugumaran *et al.*, 1992). Some bacterial LCs have been identified as bilirubin oxidases (Koikeda *et al.*, 1993) and phenoxazinone synthases (Freeman *et al.*, 1993) and are also identified because of their link to a copper-resistance phenotype of a particular microorganism (Lee *et al.*, 1994; Mellano and Cooksey, 1988). In addition, in some isolated cases bacterial LCs carry out a $Mn^{2+}$ oxidizing activity (see below). There are at least 34 bacterial LC sequences currently available in the GenBank database, which includes the only known multicopper oxidase from an archaeon, hyperthermophilic *Pyrobaculum aerophilum* (GenBank Accession No. AE009845, gene PAE1888).

In plants, laccases are implicated in wound responses and lignin biosynthesis (Bao *et al.*, 1993; Dean and Eriksson, 1994; Driouich *et al.*, 1992; LaFayette *et al.*, 1999; O'Malley *et al.*, 1993; Richardson *et al.*, 2000; Sterjiades *et al.*, 1992, 1996). Contrary to plant LC function, the fungi LCs are apparently involved in the degradation of the same lignin (Eggert *et al.*, 1996). Because of their ability to degrade toxic phenolic compounds and lignin, fungal LCs have great industrial importance.

Laccases are usually monomers and are considered to be the simplest blue copper oxidases. A fungal genome may express multiple LC isoforms that differ by their substrate specificity, pH optimum, and redox potentials (Germann *et al.*, 1988; Wahleithner *et al.*, 1996; Xu, 1996; Yaver and

Golightly, 1996). They are usually induced by their corresponding substrates, suggesting a tight gene regulation. In most cases, LCs are extracellular glycoproteins. Immunolocalization studies of a laccase from sycamore maple revealed that the protein is localized only in the lignifying cell wall of xylem and in the epidermal cells (Driouich *et al.*, 1992). It has been also found only among cell wall-associated proteins and it can be solubilized from the differentiating xylem by 1 M $CaCl_2$, which hints of a possible interaction with pectins (Bao *et al.*, 1992).

In the *Arabidopsis* genome, LCs are represented as a large gene family composed of at least 18 different genes (see Table I), which, based on their sequence characteristics, can be separated into two groups. The first group consists of 16 laccases, which are 50–70% identical to one another (see Fig. 9, AtLC1 through AtLC16). The second group is composed of two sequences (see Table I; AtLC17 and AtLC18) that share 80% amino acid sequence identity and are closely related to a bacterial LC (35% identity) localized in the outer spore coat and known to be involved in brown pigmentation during sporogenesis in *Bacillus subtilis* (Donovan *et al.*, 1987), but are only 10–12% identical to the LCs of the first group. Apparently those 2 LCs have a bacterial origin and are a result of a lateral gene transfer from a bacterial genome into the plant genome.

Ascorbate oxidases are glycoproteins that are found mainly in plants. They are homodimers, catalyzing the oxidation of the only known substrate, ascorbate, into semidehydroascorbate radical, which subsequently disproportionates into dehydroascorbate and ascorbate (Messerschmidt and Huber, 1990; Messerschmidt *et al.*, 1989). However, the role of this activity and whether it is physiologically relevant remain largely unknown. The deglycosylated enzyme displays enzymatic activity virtually identical to that of the glycosylated form, suggesting that carbohydrate moieties have no effect on folding of the polypeptide and are not involved in substrate binding (D'Andrea *et al.*, 1993).

Immunohistochemical localization studies on zucchini ascorbate oxidase revealed a wide range of tissue distribution (Chichiricco *et al.*, 1989; Hayashi and Morohashi, 1993; Lin and Varner, 1991; Ohkawa *et al.*, 1989). AO is found in stems, leaves, flowers, fruits, and seeds at both early and late developmental stages and in both differentiating and undifferentiating cells. At the cellular level, the enzyme is cell wall associated and present at high concentrations in epidermal tissues, especially in cucurbit species.

*Arabidopsis thaliana* has three different genes encoding ascorbate oxidases. They display 50–70% sequence identity with one another and only 20–25% identity with the proteins of the LC family. They all have a Met as the axial ligand for blue copper, whereas most plant laccases have Leu or Ile (see Fig. 9). A multicopper blue oxidase has also been characterized from fungus *Acremonium* sp. HI-25 and identified as AO because of its

```
PE1   MKQKHLLLLGFLLAYCFSSVFVINAEDPYLFFTWTVTYGTRSPLGVPQQVILINGQFPGPPIEGVTNNNIVVNVINKLDEP-----FLITWNGIKQRKMSWQDGVL-GTNCPIQPKSSWTYHFQLKDQIGTYAYFASTSMHRASGAFGALNVNQRSVIF-VPY
Pe2   MKQTNLLVCKLFIGALFWLGSVLVNAEDPYMFYTWTVTYGTRSPLGVPQQVILINGQFPGPAIEAVTNNNIVVNLINKLDEP-----FLITWNGVKQRRTSWQDGVL-GTNCPIQPNSNWTYQFQLKDQIGTYTYFASTSLHRASGAFGALNINQRSVIT-TPY
PE3   MASRKTTSLLNHLLLLGALTLLSSLVIVKGESPYKFYTWTVTYGIISPLGVPQQVILINGQFPGPKLEVVTNDNIILNLINKLDQP-----FLLTWNGIKQRKNSWQDGVL-GTNCPIQPNSNFTYKFQTKDQIGTFNYFPSTAFHKAAGGFGAINVYARPGIP-IPY
PE4   MAVGRSGGTILLFCLSFFAAVTAESPYRFFDWNVTYGDIYPLGVRQQGILINGQFPGPDIHSVTNDNLIINVHNSLDEP-----FLISWNGVQNRRNSYVDGMY-GTTCPIPPRSNYTYILQVKDQIGSFYYFPSLAFHKAAGGFGGIRILSRPGIP-VPF
PE5   MAGSASFAAALFIGLSLLFAVTAEDPYRFFEWNITYGDIYPLGVRQQGILINGAFPGPDIHSVTNDNLIINVYNSLDEP-----FLLSWNGIQQRRNSFVDGVY-GTTCPIPPGKNYTYILQMKDQIGSFYYFPSLGFHKAAGGFGGIRILSRPRIP-VPF
PE6   MEVKSVNTTAMILGLFFLISFVAAEDPYKFFEWHVTYGNISPLKVAQQGILINGKFPGPDIAAVTNDNLIINVFNHLDEP-----FLISWSGIRNWRNSYQDGVY-GTTCPIPPGKNYTYALQVKDQIGSFYYFPSLGFHKAAGGFGAIRISSRPRIP-VPF
PE7   MKVKSMNTRAMITTLLFLISLAFAEDPYRFFEWHVTYGNISPLGVAQQGILINGKFPGPDIISITNDNLIINVFNHLDEP-----FLLSWNGIRNWKNSFQDGVY-GTMCPIPPGKNYTYALQVKDQIGSFYYFPSLGFHKAAGGFGGIRISSRALIP-VPF
PE8   MRGSCKVSIVLLLVLINGVLGDNPYRFFTWKITYGDIYPLGVKQQGILINGQFPGPHIDAITNDNIIISVFNYLKEP-----FLISWNGVQQRKNSWQDGVV-GTTCPIPPGKNFTYVIQVKDQIGSFYYFPSLAFHKAAGAFGAIRVWSRPRIP-VPF
PE9   MEWWLNGGVWMMMMTTTIISFVKAEDTLFYNWRVTYGKIALDTLPRRGILINGQFPGPEIRSLTNDNLVINVQNDLDDP-----FLLSWNGVHMRKNSYQDGVY-GTNCPIPPGKNYTYDFQVKDQVGSYFYFPSLAVQKAAGGYGSLRIYSLPRIP-VPF
PE10  MCWWLNGAVWTMMMMTISIISFVQADDPYRFFDWRVTYGNISPLGIPQRGILINGQYPGPDIYSVTNDNLIINVHNDLDEP-----FLLSWNGVQLRKNSYQDGVY-GTTCPIPPGKNYTYAIQVKDQIGSFFYFPSLAVHKAAGGFGGFRILSRPRIP-VPF
PE11  MKGGVKLLAVCLCVATATVMMVQAEDPYFHHVWNVTYGTASPLGVPQQVILINGQFPGPNINSTSNNNVIVNVFNNLDEP-----FLITWAGIQHRKNCWQDGTA-GTMCPIPPGQNFTYHFQPKDQIGSYFYYPTTAMHRAAGGFGGLRVNSRLLIP-VPY
PE12  MRGVKLLAACLYLAAAATVVVRAEDPYFHHVWNVTYGTVSPLGVPQQVILINGQFPGPNVNSTSNNNVIINVFNNLDEP-----FLLTWNGIQHRKNCWQDGTP-GTMCPIMPGTNYTYHFQPKDQIGSYFYYPSTAMHRSAGGFGGLRVNSRLLIP-VPY
PE13  MEGRLLTVLVCLVSTVAIVNAGDPYFFHTWNVTYGTASPLGVPQKVILINGQFPGPNLNSTSNNNVVINVFNHLDEP-----FLLTWSGIQHRKNCWQDGVA-GTSCPIPAGQNFTYHFQPKDQIGSYFYYPTTSLHRFAGGFGGLRVNSRLLIP-VPY
PE14  MQGGRLLTVLVCLASTVALVSAGDPYFYYTWNVTYGTAAPLGIPQQVILINGQFPGPNLNSTSNNNVVINVFNNLDEP-----FLLTWSGLQHRKNSWQDGVT-GTSCPIPAGTNFTYHFQPKDQIGSYFYYPSTALHRFAGGFGGLRVNSRLLIP-VPY
PE15  MAATDFFFAFVFSFALIFGFSFAGDPYVSYDFTLSYITASPLGVPQQVIAVNGKFPGPVINATTNYNVHVNVLNHLDEP-----LLLTWPGVQMRRNSWQDGVL-GTNCPIPPNWNFTYDFQLKDQIGSYFYSPSLNFQRASGGFGALIINNRDLVP-IPF
PE16  MAATCSLLASFLLCFALLSAVSFAADPFVSYDFRVSYLTASPLGVPQQVIAVNGQFPGPLLNATTNYNVVVNVFNHLDEP-----LLLTWPGIQMRRNSWQDGVL-GTNCPIPPRWNFTYQFQVKDQIGSFFYSPSLNFQRASGGFGPIVINNRDIIP-IPF
PE17  MDLFKILLLVFFVNISFCFAADPYSFYNFEVSYITASPLGVPQQVIAINGKFPGPTINVTTNENLVVNVRNKLDEG-----LLLHWNGIQQRRVSWQDGVL-GTNCPIPPKWNWTYEFQVKDQIGSFFYFPSLHFQRASGGFGSFVVNPRAIIP-VPF
PE18  MRCFPPPLWCTSLVVFLSVTGALAADPYVFFDWTVSYLSASPLGTRQQVIGINGQFPGPILNVTTNWNVVMNVKNNLDEP-----LLLTWNGIQHRKNSWQDGVL-GTNCPIPSGWNWTYEFQVKDQIGSFFYFPSTNFQRASGGYGGIIVNNRAIIP-VPF
LC1   MGFLFLFCYLLAFLGYSPVDAAVKKYQFDVQVKNISRICNAKPIVTVNGMFPGPTVYAREGDRVIINVTNHVQYN-----MSIHWHGLKQYRNGWADGPAYITQCPIQTGQSYLYDFNVTGQRGTLWWHAHILWLR-ATVYGAIVILPAPGKP-YPF
LC2   MVWFLFLVSFFSVFPAPSESMVRHYKFNVVMKNVTRLCSSKPTVTVNGRYPGPTIYAREDDTLLIKVVNHVKYN-----VSIHWHGVRQVRTGWADGPAYITQCPIQPGQVYTYNYTLTGQRGTLWWHAHILWLR-ATVYGALVILPKRGVP-YPF
LC3   MVFPIRILVLFALLAFPACVHGAIRKYTFNVVTKQVTRICSTKQIVTVNGKFPGPTIYANEDDTILVNVVNNVKYN-----VSIHWHGIRQLRTGWADGPAYITQCPIKPGHSYVYNFTVTGQRGTLWWHAHVLWLR-ATVHGAIVILPKLGLP-YPF
LC4   MALQLLLAVFSCVLLLPQPAFGITRHYTLEIKMQNVTRLCHTKSLVSVNGQFPGPKLIAREGDQVLIKVVNQVPNN-----ISLHWHGIRQLRSGWADGPAYITQCPIQTGQSYVYNYTIVGQRGTLWYHAHISWLR-STVYGPLIILPKRGVP-YPF
LC5   MVTWVLNYLLVAFLFAISYNIDAASAGITRHYQFDIQLKNITRLCKTKTIVTVNGKFPGPRVTAREGDNLQIKVVNHVSNN-----ISIHWHGIRQLRSGWADGPSYVTQCPIRMGQSYVYNFTVTGQRGTLWWHAHIQWMR-ATVYGPLIILPKLHQP-YPF
LC6   MENLGFLIISTFLLLFTTLLPYSSASTTRRFHFNVEWKKVTRLCHTKQLLTVNGQYPGPTVAVHEGDIVEIKVTNRIAHN-----TTIHWHGLRQYRTGWADGPAYITQCPIRSKQSYTYRFKVEDQRGTLLWHAHHSWQR-ASVYGAFIIYPR--QP-YPF
LC7   MTTVHTFSIILFFCSLFSASLIIAKVQHHDFVIQETPVKRLCKTRNAITVNGMFPGPTLEVNNGDTLEVKVHNRARYN-----ITIHWHGVRQIRTGWADGPEFVTQCPIRPGKSYTYRFTIQGQEGTLWWHAHSSWLR-ATVYGALIIHPTPGSS-FPF
LC8   MDVTKSLLCFISFVAFLLFSSVAEANKAHHHEFIIQATKVKRLCETHNSITVNGMFPGPMLVVNNGDTLVVKVINRARYN-----ITIHWHGVRQMRTGWADGPEFVTQCPIRPGSSYTYRFTIQGQEGTLWWHAHSSWLR-ATVYGSLLVFPPAGSS-YPF
LC9   MESFRRFSLLSFIALLAYFAFLASAEHHVHQFVITPTPVKRLCRTHQSITVNGQYPGPTLVVRNGDSLAITVINRARYN-----ISIHWHGIRQLRNPWADGPEYITQCPIRPGQTYTYRFKIEDQEGTLWWHAHSRWLR-ATVYGALIIYPRLGSP-YPF
LC10  MEQLRPFFLLLAIFVASLVNAEVHFHEFVIQETPVKRLCRVHNSITVNGQFPGPTLEVRNGDSLVITAINKARYN-----ISLHWHGIRQMRNPWADGPEYITQCPIQPGGSYTYRFTMEDQEGTLWWHAHSRWLR-ATVYGALIIRPPLSSPHYPF
LC11  MSHSFFNLFLISLFLYNNCIAHHYTFTVREVPYTKLCSTKAILTVNSQFPGPIIKVHKGDTIYVNVQNRASEN-----ITMHWHGVEQPRNPWSDGPEYITQCPIRPGSDFLYKVIFSIEDTTVWWHAHSSWTR-ATVHGLIFVYPRPPQI-LPF
LC12  MEFKLNIPNTIIKTLQTIVFFLFVLLAFQIAEAEIHHHTFKIKSKAYTRLCNTNKILTVNGEFPGPTLKAYRGDKLIVNVINNANYN-----ITLHWHGARQIRNPWSDGPEYVTQCPIRPGESYVYRIDLKVEEGTIWWHAHSQWAR-ATVHGAFIVYPKRGSS-YPF
LC13  MTSSAVPSLFRLSFLLFTLQVMNIGRIGAATRFYQFKVQTIRLTRLCQTNEIVTVNKKFPGPAISAQEDDRIVIKVINMTPYN-----TTIHWHGIKQKRSCWYDGPSYITQCPIQSGQSFTYNFKVAQQKGTFLWHAHFSWLR-ATVYGPLIVYPKASVP-YPF
LC14  MPRLHHYLSNCAFLVLLLFSSIASAAVVEHVLHIQDVVVKPLCKEQIIPAANGSLPGPTINVREGDTLVVNVINNSTYN-----VTIHWHGVFQLKSVWMDGANMITQCPIQPGYNFTYQFDITGQEGTLLWHAHVVNLR-ATLHGALVIRPRSGRP-YPF
LC15  MPRVHHSLSNCAFLVLLLFSSIASAAIVEHVLHVKDVVVTPLCKEQMIPIVNGSLPGPTINVREGDTLVVHVINKSTYN-----VTIHWHGVFQLKSVWMDGANMITQCPIQPSNNFTYQFDITGQEGTLLWHAHVVNLR-ATIHGALIIRPRSGRP-YPF
LC16  MEGVRVPIACALILLAISSITSASIVEHTFNVQNLTVSRLCKRQVITVVNGSLPGPTIRVKEGDSLVIHVLNHSPHN-----ITIHWHGIFHKLTVWADGPSMITQCPIQPGQRYAYRFNITGQEGTLWWHAHASFLR-ATVYGALVIRPKSGHS-YPF
FLC   MFKNLLSFALLAISVANAQIVNSVDTMTLTNANVSPDGFTRAGILVNG-VHGPLIRGGKNDNFELNVVNDLDNPTMLRPTSIHWHGLFQRGTNWADGADGVNQCPISPGHAFLYKFTPAGHAGTFWYHSHFGTQYCDGLRGPMVIYDDNDPHAALY
AO1   MMRPKRSSDTVHVFNLMVLCFIALFFSSVLCQGKIRRFKWEVKYEFKSPDCFEKLVITINGKFPGPTIKAQQGDTIVVELKNSFMTEN----VAVHWHGIRQIGTPWFDGVEGVTQCPILPGEVFIYQFVV-DRPGTYMYHSHYGMQRESGLIGMIQVSPPATEP----
AO2   MAVIVWWLLTVVVVAFHSASAAVVESTWEVEYKYWWPDCKEGIVMAINGQFPGPTIDAVAGDTVIIHVVNKLSTEG----VVIHWHGIRQKGTPWADGAAGVTQCPINPGETFTYKFIV-DKAGTHFYHGHYGMQRSSGLYGMLIVRSPK-------
AO3   MGVWWIVLVVAVLTHTASAAVREYHWEVEYKYWSPDCKEGAVMTVNGEFPGPTIKAFAGDTIVVNLTNKLTTEG----LVIHWHGIRQFGSPWADGAAGVTQCAINPGETFTYNFTV-EKPGTHFYHGHYGMQRSAGLYGSLIVDVAKGKS----
ZAO   SQIRHYKWEVEYMFWAPNCNENIVMGINGQFPGPTIRANAGDSVVVELTNKLHTEG----VVIHWHGILQRGTPWADGTASISQCAINPGETFFYNFTV-DNPGTFFYHGHLGMQRSAGLYGSLIVDPPQGKK----
                                                                               2 3                                                        3 3
```

FIG. 9. (*continues*)

```
PE1   P--KPDADFTLLVSDWYK-MGHKELQRRLDSSRALPP-PDGLLINGASKG---------------------------LVFTGQHGKIYRFRISNVGISTSINFRIQGHMMTLVEVEGSHTLQEVYESLDIHVGQSVTVLVTLKA--PVKDYFIVASTRFTKP------
PE2   P--TPDGDFTLLVSDWFSNMTHKDLRKSLDAGSALPL-PDALLINGVS---------------------------KGLIFTGQQGKTYKFRVSNVGIATSINFRIQNHTMSLIEVEGAHTLQESYESLDVHVGQSMTVLVTLKA--SVRDYFIVASTRFTKP------
PE3   P--LPTADFTLLVGDWFK-TNHKTLQQRLDSGGVLPF-PDGMLINGQT----------------------------QSTFSGDQGKTYMLRISNVGLSSTFNFRIQGHTMKVVEVEGSHVIQTDYDSLDIHVGQSLAVLVTLNQ--SPKDYYIVASTRFIRS------
PE4   A--DPAGDYTVLIGDWYK-FNHTDLKSRLDRGRKLPS-PDGILINGRS---------------------------NGATLNVEQGKTYRLRISNVGLQDSLNFRIQNHRMKLVEVEGTHTLQTMFSSLDVHVGQSYSVLITADQ--SPRDYYVVVSSRFTDK------
PE5   P--DPAGDTTVLIGDWYK-ANHTDLRAQLDNGKKLPL-PDGILINGRS---------------------------SGATLNVEQGKTYRFRISNVGLQDSLNFRIQDHKMKVVEVEGTHTLQTTFSSLDVHVGQSYSVLVTADQ--TPRDYYVVVSSRFTSN------
PE6   P--APAGDYTVLIGDWYK-TNHKDLRAQLDNGGKLPF-PDGILINGRG---------------------------SGATLNIEPGKTYRLRISNVGLQNSLNFRIQNHKMKLVEVEGTHTIQTPFSSLDVHVGQSYSVLITADQ--PAKDYYIVVSSRFTSK------
PE7   P--TPADDYTLLVGDWYK-TNHKDLKAQLDNGGKLPL-PDGILINGRS---------------------------SGATLNIEPGKTYRLRISNVGLQNSLNFRIQNHTMKLVEVEGRYTIQNLFSSLDVHVGQSYSVLITADQ--PAKDYYVVVSSRFTSK------
PE8   S--PPDGDFWLLAGDWYK-TNHYVLRRLLEAGRNLPN-PDGVLINGRG--------------------------WGGNTFTVQPGKTYRFRISNVGVATSLNFRIQGHTMKLVEVEGSHTVQNIYTSLDIHLGQSYSVLVTANQ--APQDYYIVISSRFTRK------
PE9   P--EPAEDFTFLVNDWYR-RNHTTLKKILDGGRKLPLMPDGVMINGQG-------------------------VSTVYSITVDKGKTYRFRVSNVGLQTSLNLEILGHQLKLIEVEGTHTVQTMYTSLDIHVGQTYSFLVTMDQ--PPQNYSIVVSTRFINA------
PE10  P--EPAGDFTFLIGDWFK-HDHKVLKAILDRGHKLPL-PQGKSIILSH--------------------------HLFFVVEQLSGKTYRFRISNVGLQHTLNFRIQGHQMKLVEVEGTHTVQSMYTSLDIHVGQSYSVLVTMDQ--PDQDYDIVVSTKFVAK------
PE11  A--DPEDDYTILINDWYT-KSHTQLKKFLDSGRTIGR-PDGILINGKS---------------------GKTDGSDKPLFTLKPGKTYRVRICNVGLKASLNFRIQNHKMKLVEMEGSHVLQNDYDSLDVHVGQCFGVIVTADQ--EPKDYYMIASTRFLKK------
PE12  A--DPEDDYTVLIGDWYT-KSHTQLKKFLDSGRTLGR-PDGILINGKS---------------------GKGDGSDAPLFTLKPGKTYRVRICNVGLKTSLNFRIQNHKLKLVEMEGSHVLQNDYDSLDVHVGQCYGTILTANQ--EAKDYYMVASSRFLKS------
PE13  A--DPEDDYTVLLGDWYT-AGHTALKNFLDSGRTLGL-PNGVLINGKS---------------------GKVGGKNEPLFTMKPGKTYKYRLCNVGFKSTLNFRIQNHKMKLVEMEGSHVIQNDYDSLDVHVGQCFSVLVTANQ--AAKDYYMVASTRFLKK------
PE14  A--DPEDDRTILINDWYA-KSHTALKNFLDSGRTLGS-PDGVLINGKS---------------------GKLGGNNAPLFTMKPGKTYKYRICNVGFKSTLNFRIQGHKMKLVEMEGSHVLQNDYDSLDVHVGQCFAVLVTADQ--VAKNYYMVASTRFLKK------
PE15  T--EPDGEIIFIIGDWYT-QNHTALRRILDSGKELGM-PDGVLINGKG--------------PFKYNSSVP-DGIEHETVNVDPGKTYRIRVHNVGISTSLNFRIQNHKLLLIETEGRYTSQMNFTDFDVHVGQSYSFLVTMDQN-ATSDYYIVASARFVNE---TVW
PE16  P--QPDGELIFIIGDWYT-QDHKALRRALDSGKELGM-PDGVLINGKG--------------PYKYNSSVP-DGIDYLTFHVEPGKTYRIRVHNVGISTSLNFRIQNHSLLLVETEGHYTSQANFTDFDVHVGQSYSFLVTMDQD-ATSDYYIVASARFVNE---TVW
PE17  S--TPDGDITVTIGDWYI-RNHTALRKALDDGKDLGM-PDGVLINGKG--------------PYRYNDTLVADGIDFETITVHPGKTYRLRVSNVGISTSLNFRIQGHNLVLAESEGSYTVQQNYTSLDIHVGQSYSFLVTMDQN-ASSDYYIVASARVVNE---TIW
PE18  A--LPDGDVTLFISDWYT-KSHKVLRKDVESKNGLRP-PDGIVINGFG--------------------------PFASNGTYLLGRTYRFRVHNSGIATSLNFRIQNHNLLLVETEGSYTIQQNYTNMDIHVGQSFSFLVTMDQS-GSNDYYIVASPRFATS------
LC1   P--QPYQESNIILGEWWNK-DVETAVNQANQLGAPPPMSDAHTINGKPGP-------------LFPCSEKHTF-----VIEAEAGKTYLLRIINAALNDELFFGIAGHNMTVVEIDAVYTKPFTTKAILLGPGQTTNVLVKTD--RSPNRYFMAASPFMD-APVSVD-
LC2   P--KPDNEKVIVLGEWWKS-DTENIINEALKSGLAPNVSDSHMINGHPGP-------------VRNCPSQG-Y-----KLSVENGKTYLLRLVNAALNEELFFKVAGHIFTVVEVDAVYVKPFKTDTVLIAPGQTTNVLLTAS--KSAGKYLVTASPFMD-APIAVD-
LC3   P--KPHREEVIILGEWWKS-DTETVVNEALKSGLAPNVSDAHVINGHPGF-------------VPNCPSQ----------AVESGKTYMLRLINAALNEELFFKIAGHRFTVVEVDAVYVKPFNTDTILIAPGQTTTALVSAA--RPSGQYLIAAAPFQDSAVVAVD-
LC4   A--KPHKEVPMIFGEWFNA-DTEAIIRQATQTGGGPNVSDAYTINGLPGP-------------LYNCSAKDTF-----RLRVKPGKTYLLRLINAALNDELFFSIANHTVTVVEADAIYVKPFETETILIAPGQTTNVLLKTKSSYPSASFFMTARPYVT-GQGTFD-
LC5   P--KPYKQVPILFGEWFNA-DPQAVVQQALQTGAGPNASDAHTFNGLPGP-------------LYNCSTKDTY-----KLMVKPGKTYLLRLINAALNDELFFTIANHTLTVVEADACYVKPFQTNIVLLGPGQTTNVLLKTKPIYPNATFYMLARPYFT-GQGTID-
LC6   SGSHIQSEIPIILGEWWND-DVDNVEKAMMKTGAGAKVSDAYTLNGLPGP-------------LYPCSTKDTF-----TATVDAGKTYILRIINAALNNELFVAVANHTLTVVEVDAVYTKPVHTKAIMIAPGQTTTLLLRADQ-LSGGEFLIAATPYVT-SVFPFN-
LC7   P--KPDRQTALMLGEWWNA-NPVDVINQATRTGAAPNISDAYTINGQPGD-------------LYNCSTKETV-----VVPINSGETSLLRVINAALNQPLFFTVANHKLTVVGADASYLKPFTTKVLMLGPGQTTDVLLTADQ-PPK-RYYIAARAYQSAQNAPFD-
LC8   T--KPHRNVPLLLGEWWDA-NPVDVLRESIRTGGAPNNSDAYTINGQPGD-------------LYKCSSQDTT-----VVPINVGETILLRVINSALNQPLFFTVANHKLTVVGADASYLKPFTTNVIVLGPGQTTDVLITGDQ-PPN-RYYMAARAYQSAQNAPFG-
LC9   S--MPKRDIPILLGEWWDR-NPMDVLKQAQFTGAAANVSDAYTINGQPGD-------------LYRCSRAGTI-----RFPIFPGETVQLRVINAGMNQELFFSVANHQFTVVETDSAYTKPFTTNVIMIGPGQTTNVLLTANQ-RPG-RYYMAARAYNSA-NAPFD-
LC10  PV-IPKREITLLLGEWWDR-NPMDVLNLAQFTGAAPNISDAFTINGQPGD-------------LYRCSSQETL-----RFLVGSGEIVLLRVINSALNQELFFGVANHKLTVVAADASYTKPFSTNVIMLGPGQTTDVLLTADQ-PPA-HYYMAAHAYNSA-NAAFD-
LC11  P--KADHEVPIILGEWWKR-DVREVVEEFVRTGGAPNVSDALTINGHPGF-------------LYPCSKSDTF-----HLTVEKGKTYRIRMVNAAMNLPLFFAIANHSLTVVSADGHYIKPIKATYITISPGETLDMLLHADQ-DPERTYYMAARAYQSGN-IDFN-
LC12  P--KPHREIPLILGEWWKKENIMHIPGKANKTGGEPAISDSYTINGQPGY-------------LYPCSKPETF-----KITVVRGRRYLLRIINAVMDEELFFAIANHTLTVVAKDGFYLKHFKSDYLMITPGQSMDVLLHANQ-RPN-HYFVAARAYSSAFGAGFD-
LC13  K--KPFNEHTILLGEYWLK-NVVELEQHVLESGGPPPPADAFTINGQPGP-------------NYNCSSKDVY-----EIQIVPRKIYLLRLINAGINMETFFTIANHRLTIVEVDGEYTKPYTTERVMLVPGQTMNILVTADQ--TVGRYSMAMGPYESAKNVKFQ-
Lc14  P--KPYKEVPIVFQQWWDT-DVRLLQLR------PAPVSDAYLINGLAGD-------------SYPCSENRMF-----NLKVVQGKTYLLRIVNAALNTHLFFKIANHNVTVVAVDAVYSTPYLTDVMILTPGQTVDALLTADQAIG--KYYMATLPYISAIGIPT-P
LC15  P--KPYKEVPLIFQQWWDT-DVRLLELR------PAPVSDAYLINGLAGD-------------SYPCSKNRMF-----NLKVVQGKTYLLRIINAALNTHLFFKIANHNVTVVAVDAVYTTPYLTDVMILTPGQTIDAILTADQPIG--TYYMAIIPYFSAIGVPASP
LC16  P--KPHKEVPILFGEWWNT-DVVALEEAAIATGVPPNNSDAYTINGRPGN-------------LYPCSKDRMF-----SLNVVKGKRYLLRIINAAMNIQLFFKIANHRLTVVAADAVYTAPYVTDVIVIAPGQTIDALLFADQSVDT-SYYMAAHPYASAPAVPF--
FLC   D--EDDENTIITLADWYHI--PAPSIQG-------AAQPDATLINGKGRY--------------VGGPAAELS-----IVNVEQGKKYRMRLISLSCDPNWQFSIDGHELTIIEVDGQLTEPHTVDRLQIFTGQRYSFVLDANQ--PVDNYWIRAQPNKGRNGLAGTF
Ao1   EPFTYDYDRNFLLTDWYHKSMSEKATGLASIPFKWVGEPQSLMIQGRGRFNCSNNLTTPPSLVSGVCNVSNAD-CSRFILTVIPGKTYRLRIGSLTALSALSFQIEGHNLTVVEADGHYVEPFTVKNLFVYSGETYSVLLKADQ-NPRRNYWITSSIVSRPATTPPPA
AO2   ERLIYDGEFNLLLSDWWHQSIHAQELALSSRPMRWIGEPQSLLINGRGQFNCSQAAYFNKGGEKDVCTFKENDQCAPQTLRVEPNRVYRLRIASTTALASLNLAVQGHQLVVVEADGNYVAPFTVNDIDVYSGETYSVLLKTNA-LPSKKYWISVGVRGREPKTPPQA
AO3   ERLRYDGEFNLLLSDWWHEAIPSQELGLSSKPMRWIGEAQSILINGRGQFNCSLAAQFSNNTSLPMCTFKEGDQCAPQILHVEPNKTYRIRLSSTTALASLNLAVQGHKLVVVEADGNYITPFTTDDIDIYSGESYSVLLTTDQ-DPSQNYYISVGVRGRKPNTTTQA
ZAO   EPFHYDGEINLLLSDWWHQSIHKQEVGLSSKPIRWIGEPQTILLNGRGQFDCSIAAKYDSN--LEPCKLKGSESCAPYIFHVSPKKTYRIRIASTTALAALNFAIGNHQLLVVEADGNYVQPFYTSDIDIYSGESYSVLITTDQ-NPSENYWVSVGTRARHPNTPPPG
```

FIG. 9. (*continues*)

```
PE1   -ILTTTGILSYQGSK----------IRPSHPLP----IGPTYHIHWSMKQARTIRLNLTANAARPNPQGSFHYGTIPINRTFVLANGRA--------MINGKLRYTVNRVSYVNPATPLKLADWFNIPG--VFNFKTIMNIP--TPG-PS--------------ILGT
PE2   -VLTTTASLRYQGSK----------NAAYGPLP----IGPTYHIHWSMKQARTIRMNLTANAARPNPQGSFHYGTIPINRTLVLANAAT--------LIYGKLRYTVNRISYINPTTPLKLADWYNISG--VFDFKTIISTP--TTG-PA--------------HIGT
PE3   -KLSVMGLLRYSNSR----------VPASGDPP----ALPPGELVWSMRQARTFRWNLTANAARPNPQGSFHYGMISPTKTFVFSNSAP--------LINGKQRYAVNGVSYVKSETPLKLADHFGISG--VFSTNAIQSVP--SNS-PP--------------TVAT
PE4   -IITTTGVLRYSGSS----------TPASGPIP----GGPTIQVDWSLNQARAIRTNLTASGPRPNPQGSYHYGLIPLIRTIVFGSSAG--------QINGKQRYGVNSVSFVPADTPLKLADFFKISG--VYKINSISDKP--TYG-GL--------------YLDT
PE5   -VLTTTGIFRYSNSA----------GGVSGPIP----GGPTIQIDWSLNQARAIRTNLSASGPRPNPQGSYHYGMINTTRTIRLASSAG--------QVDGKQRYAVNSVSFKPADTPLKIADYFKIDG--VYRSGSIQYQP--TGG-GI--------------YLDT
PE6   -ILITAGVLHYSNSA----------GPVSGPIP----EAP-IQLRWSFDQARAIKTNLAASGPRPNPQGTYHYGKIKVTRTIKLASSAG--------NINGKQRYAVNSASFYPTDTPLKLADYFKIAG--VYNPGSIPDQP--THG-AI--------------YPVT
PE7   -ILTTTGVLHYSNSV----------APVSGPIP----DGP-IKLSWSFNQARAIRTNLTASGPRPNPQGSYRYGVINITRTIRLANNLG--------HIEGKQRYAVNSASFYPADTPLKLVDYFKIDG--VYKPGSISDQP--TNG-AI--------------FPTT
PE8   -VLTTTSILHYSNSR----------KGVSGPVP----NGPTLDIASSLYQARTIRRNLTASGPRPNPQGSYHYGLIKPGRTIILANSAP--------WINGKQRYAVNGASFVAPDTPLKLADYFKIPG--VFNLGSIPTSP--SGGNGG--------------YLQS
PE9   -EVVIRATLHYSNSK----------GHKIITAR----RPDPDDVEWSIKQAQSIRTNLTASGPRTNPQGSYHYGKMKISRTLILESSAA--------LVKRKQRYAINGVSFVPSDTPLKLADHFKIKD--VFKVGTIPDKP--RRGGGI--------------RLDT
PE10  -KLLVSSTIHYSNSR----------HSHSSSANSVHVQQPADELDWSIKQARSIRTNLTASGPRPNPQGSYHYGRIKISRTLILESSAA--------LVKRKQRYAINGVSFVPGDTPLKLADYFKIKG--VFKMGSIPDKP--RRGRGM--------------RMET
PE11  -PLTTTGLLRYEGGK----------GPASSQLP----AAP-VGWAWSLNQYRSFRWNLTASAARPNPQGSYHYGKINITRTIKLVNTQG--------KVDGKLRYALSGVSHTDPETPLKLAEYFGVADK-VFKYDTISDNPNPDQIKNI--------------KIEP
PE12  -VITTTGLLRYEGGK----------GPASSQLP----PGP-VGWAWSLNQFRSFRWNLTASAARPNPQGSYHYGKINITRTIKLVNTQG--------KVDGKLRYALNGVSHTDPETPLKLAEYFGVADK-VFKYDSITDNPTPEQIKSI--------------KIVP
PE13  -ELSTVGVIRYEGSN----------VQASTELP----KAP-VGWAWSLNQFRSFRWNLTSNAARPNPQGSYHYGKINITRSIKLVNSKS--------VVDGKVRFGFNGVSHVDTETPLKLAEYFQMSEK-VFKYNVIKDEP-AAKITAL--------------TVQP
PE14  -EVSTVGVMSYEGSN----------VQASSDIP----KAP-VGWAWSLNQFRSFRWNLTASAARPNPQGSYHYGKINITRTIKLANTKN--------LVNGKVRFGFNGVSHVDTETPLKLAEYFGMSEK-VFKYNVIKDEP-AAKITTL--------------TVEP
PE15  QRVTGVGILHYSNSK----------GPASGPLPV--SATDVNHPWSAMNQPRAIKQNTSASGARPNPQGSFHYGQINITRTYILRSLPPT-------KINGKLRATLNGISFVNPSTPMRLADDHKVKG--DYMLD-FPDRP--LDEKLP--------------RLSS
PE16  QRVTGVAILHYSNSK----------GPVSGPLPV--PKTDVSSPWSAMSQPKTIRQNTSASGARPNPQGSFHYGQINITNTYILRSLPPT-------IINGALRATLNGISFVNPSTPVRLADRNKVKG--AYKLD-FPDRP--FNRPL---------------RLDR
PE17  RRVTGVGILKYTNSK----------GKAKGQLPP--GPQDEFDKTFSMNQARSIRWNVSASGARPNPQGSFKYGSINVTDVYVLRNMPPV-------TISGKRRTTLNGISFKNPSTPIRLADKLKVKD--VYKLD-FPKRP--LTGPA---------------KVAT
PE18  IKASGVAVLRYSNSQ----------GPASGPLPD--PPIELDT-FFSMNQARSLRLNLSSGAARPNPQGSFKYGQITVTDVYVIVNRPPE-------MIEGRLRATLNGISYLPPATPLKLAQQYNISG--VYKLD-FPKRP--MNRHP---------------RVDT
Lc1   -NKTVTAILQYKG-------------VPNTVLPILPKLPLPNDTSFALDYNGKLKSLNTPNFPALVPLKVDRRLFYTIGLGINACPT------CVNG---TNLAASINNITFIMPKT-ALLKAHYSNIS-GVFRTDFPDRPPKAFNYTGVPLT------ANLGTSTGT
LC2   -NVTATATVHYSG-------------TLSSSPTILTLPPPQNATSIANNFTNSLRSLNSKKYPALVPTTIDHHLFFTVGLGLNACPT------CKAGN-GSRVVASINNVTFIMPKT-ALLPAHYFNTS-GVFTTDFPKNPPHVFNYSGGSVT-------NMATETGT
LC3   -NRTATATVHYSG-------------TLSATPTKTTSPPPQNATSVANTFVNSLRSLNSKTYPANVPITVDHDLLFTVGLGINRCHS------CKAGN-FSRVVAAINNITFKMPKT-ALLQAHYFNLT-GIYTTDFPAKPRRVFDFTGKPPS-------NLATMKAT
LC4   -NSTVAGILEYEPPKQTKGAH--SRTSIKNLQLFKPILPALNDTNFATKFSNKLRSLNSKNFPANVPLNVDRKFFFTVGLGTNPCNHKN-NQTCQGPTNTTMFAASISNISFTMPTK-ALLQSHYSGQSHGVYSPKFPWSPIVPFNYTGTPPN-------NTMVSNGT
LC5   -NTTVAGILQYQ--------H--HTKSSKNLSIIKPSLPPINSTSYAANFTKMFRSLASSTFPANVPKVVDKQYFFAIGLGTNPCPK---NQTCQGPTNTTKFAASINNVSFILPNKTSLLQSYFVGKSKNVFMTDFPTAPIIPFNYTGTPPN-------NTMVSRGT
LC6   -NSTTVGFIRYTGKTKPENSVNTRRRRRLTAMSTVVALPNMLDTKFATKFSDSIKSLGSAKYPCKVPTKIDKRVITTISLNLQDCPLN---QTCDGYAG-KRFFASMNNISFVRPPI-SILESYYKKQSKGVFSLDFPEKPPNRFDFTGVDPVS-----ENMNTEFGT
LC7   -NTTTTAILQYK-------------KTTTTSKPIMPVLPAFNDTNTVTSFSRKFKSL----RNVVVPKTIDDNLFFTIGLGLDNCPKKFPKSRCQGLNG-TRFTASMNNVSFVLPSNFSLLQAHSNGIP-GVFTTDFPSKPPVKFDYTGNNISR-----ALFQPVKGT
LC8   -NTTTTAILQYKSAPCCGVGGGSGTKKGNSFKPIMPILPAYNDTNTVTRFSQSFRSL----RRAEVPTEIDENLFVTIGLGLNNCPKNFRSRRCQGPNG-TRFTASMNNVSFALPSNYSLLQAHHHGIP-GVFTTDFPAKPPVKFDYTGNNISR-----SLYQPDRGT
LC9   -NTTTTAILQYVNAPTRRG------RGRGQIAPVFPVLPGFNDTATATAFTNRLRYW----KRAPVPQQVDENLFFTVGLGLINCANPN-SPRCQGPNG-TRFAASMNNMSFVLPRSNSVMQAYYQGTP-GIFTTDFPPVPPVQFDYTG-NVSR-----GLWQPIKGT
LC10  -NTTTTAILKYKDASCVTL------QAKSQARAIPAQLPGFNDTATAAAFTAQMKSP----SKVKVPLEIDENLFFTVGLGLFNCPTPN-TQRCQGPNG-TRFTASINNVSFVFPKQNSIMQAYYQGTPTGVFTTDFPPTPPVTFDYTG-NVSR-----GLWQPTRGT
LC11  -NSTTIGILSYTSSCKAKT---------SSFSGYYPTLPFYNDTSAAFGFFTKIKCL----FSGQVPVQISRRIITTVSINLRMCPQNS----CEGPNG-SRLAASMNNISFVTPSHVDILKAYYYHIK-GVYGTRFPEFPPLIFNFTAENQPL-----FLETPRLAT
LC12  -KTTTTAILQYKGDTLNR------------IKPILPYLPPYNRTEASTRFTNQFRSQ----RPVNVPVKINTRLLYAISVNLMNCSDDR--P-CTGPFG-KRFSSSINNISFVNPS-VDILRAYYRHIG-GVFQEDFPRNPPTKFNYTGENLPF--------PTRFGT
LC13  -NTSAIANFQYIG-------------ALPNNVTVPAKLPIFNDNIAVKTVMDGLRSLN---A-VDVPRNIDAHLFITIGLNVNKCNSENPNNKCQGPRK-GRLAASMNNISFIEPKV-SILEAYYKQLE-GYFTLDFPTTPEKAYDFVNGAPND---IANDTQAANGT
LC14  DIKPTRGLIVYQG----------ATSSSSP---AEPLMPVPNDMSTAHRFTSNITSLVGGPHWTPVPRHVDEKMFITMGLGLDPCPAG---TKCIGPLG-QRYAGSLNNRTFMIPERISMQEAYFYNIS-GIYTDDFPNQPPLKFDYTKFEQR-TNNDMKMMFPERKT
LC15  DTKPTRGLIVYEG----------ATSSSSP---TKPWMPPANDIPTAHRFSSNITSLVGGPHWTPVPRHVDEKMFITMGLGLDPCPSN---AKCVGPLD-QRLAGSLNNRTFMIPERISMQEAYFYNIT-GVYTDDFPDQPPLKFDFTKFEQHPTNSDMEMMFPERKT
LC16  PNTTTRGVIHYGG----------ASKTGRSKPVLMPKLPSFFDTLTAYRFYSNLTALVNGPHWVPVPRYVDEEMLVTIGLGLEACADN---TTC--P----KFSASMSNHSFVLPKKLSILEAVFHDVK-GIFTADFPDQPPVKFDYTNPNVT--QTNPGLLFTQKST
FLC   ANGVNSAILRYAG--------------AANADPTTSANPNPAQLNEA-DLHALIDPAAP---GIPTPGAADVNLRFQLGFSGG------------------RFT--INGTAYESPSVPTLLQIMSGAQS-------ANDLLPAG------------------------
AC1   TAVLNYYPNHPRRRP--------------PTSESSNIVPEWNDTRSRLAQSLAIKARRG--FIHALPENSDKVIVLLNTQNEVNGYR--------------R--WSVNNVSYHHPKT--PYLIALKQNLTNAFDWRFTAPENYDSRNYDIFAKPLN-----ANATTSD
AC2   LTVINYVDATESR----------------PSHPPP-VTPIWNDTDRSKSFSKKIFAAKG--YP-KPPEKSHDQLILLNTQNLYEDYT--------------K--WSINNVSLSVPVT--PYLGSIRYGLKSAYDLK-SPAKKLIMDNYDIMKPPPN-----PNTTKGS
AC3   LTILNYVTAPASKL---------------PSSPPP-VTPRWDDFERSKNFSKKIFSAMG--SP-SPPKKYRKRLILLNTQNLIDGYT--------------K--WAINNVSLVTPAT--PYLGSVKYNLKLGFNRK-SPPRSYRMD-YDIMNPPPF-----PNTTTGN
ZAO   LTLLNYLPNSVSKL---------------PTSPPP-QTPAWDDFDRSKNFTYRITAAMG--SP-KPPVKFNRRIFLLNTQNVINGYV--------------K--WAINDVSLALPPT--PYLGAMKYNLLHAFDQN-PPPEVFPED-YDIDTPPTN-----EKTRIGN
```

FIG. 9. (*continues*)

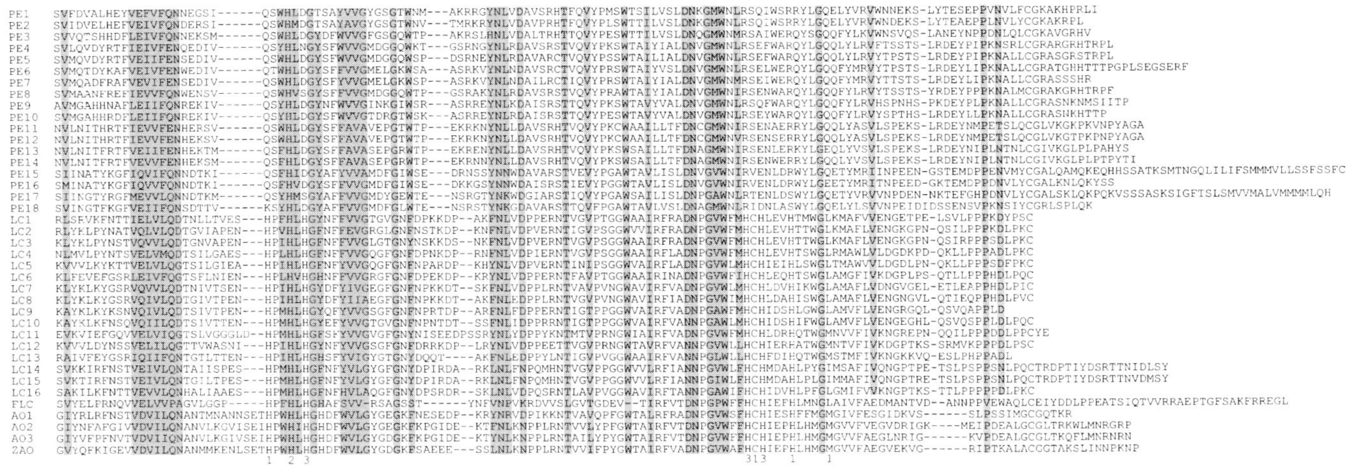

FIG. 9. Amino acid sequence alignment of the precursors of 3 ascorbate oxidases (AtAO1 through AtAO3), 16 laccases (AtLC1 through AtLC16), and 18 "pectinesterasess" (AtPE1 through AtPE18) identified in the haploid genome of *Arabidopsis thaliana*. The sequence of the precursor of laccase from a fungus *Coprinus cinereus* (FLC) and mature ascorbate oxidase from zucchini (ZAO), which have been crystallographically characterized, are also included in the alignment. The sequences of two *Arabidopsis* laccases, AtLC17 and AtLC18, which display sequence identity with the bacterial laccase from *Bacillus subtilis* but not with other *Arabidopsis* laccases, are not included. The sequence of the pectinesterase AtPE19 needs further corrections, therefore it has also been omitted. The alignment was performed using CLUSTAL W and was visually inspected and edited to ensure a largely optimized alignment. The amino acids that are identical or display conserved side-chain characteristics in 30 or more sequences are highlighted in gray. The numbers 1, 2, and 3 mark the amino acid residues that are involved in type 1, 2, and 3 copper coordination, respectively.

ability to oxidize ascorbate with no laccase-type activity (Hirose *et al.*, 1994).

In addition, the *Arabidopsis* genome contains a large gene family (at least 19 different genes), members of which are labeled in the GenBank annotations as pectinesterases (PE) and display a high level of sequence identity (25–30%) to AO and LC. However, like early nodulins, dinodulins, and ephrins, they lack the amino acids involved in any of the three types of copper binding (see Fig. 9). The occurrence of such a large protein family indicates that PEs may play an important role(s) in plant physiology, which remains to be determined. Similar sequences have been also identified in tobacco, labeled NTP303 (Weterings *et al.*, 1992), and in rape, labeled Bp10 (Albani *et al.*, 1992), where they display a pollen-specific expression pattern.

### *B. Ceruloplamin and Hephaestin*

Ceruloplasmin is organized by internally triplicated sequence modules, A1, A2, and A3, displaying high sequence identity with one another (Dwulet and Putnam, 1981; Koschinsky *et al.*, 1986; Takahashi *et al.*, 1983, 1984). Each of the A modules is composed of two BCB domains. There are three type 1 copper sites in BCB domains 2, 4, and 6 and a single trinuclear catalytic copper site, which, similar to that found in laccases and ascorbate oxidases, is localized at the interface of the two proximal domains, 1 and 6 (Zaitseva *et al.*, 1996). The type 1 copper in domain 6 is the most likely site for substrate oxidation because its Cys ligand forms the Cis-His pathway to the trinuclear cluster, which is commonly seen in other multicopper oxidases. The type 1 copper in domain 2 has a three-coordinate ligand arrangement, with a nonligating Leu residue positioned at the axial coordination site, a situation reminiscent of that seen in plant and fungal laccases. It is permanently reduced in the resting enzyme and has a potential of ~1000 mV (Machonkin *et al.*, 1998). Therefore, it is not implicated in the oxidase activity since such a high potential is beyond the range of being physiologically relevant. The significance of the type 1 copper in domain 4 is not known. Ceruloplasmin-bound copper accounts for almost 95% of the copper found in human plasma.

Ceruloplasmin is a multifunctional enzyme capable of oxidizing phenols and aromatic amines (Musci *et al.*, 1999). It can also efficiently oxidize Fe(II) to Fe(III), which is currently considered its main *in vivo* biological function. The ferroxidase activity of this enzyme was first reported in 1960 (Curzon and O'Reilly, 1960) and it was later suggested that such activity is important for loading iron into the transferrin, since it binds only Fe(III) (Osaki, 1966). Recent studies on ceruloplasmin knockout mice demonstrated that they indeed exhibit a severe impairment of iron efflux from hepatocytes, thus supporting its involvement in iron metabolism (Harris *et al.*, 1999).

Ceruloplasmin is synthesized in the liver as a precursor with a signal peptide directing it to the endoplasmic reticulum. It incorporates Cu provided by the Wilsons disease-associated Cu-ATPase, ATP7B, and is secreted into the bloodstream. The copper-deficient ceruloplasmin in Wilson's disease patients has been found to rapidly degrade. A GPI-anchored form of ceruloplasmin has been recently identified as the product of an alternately spliced ceruloplasmin transcript that generates a hydrophobic C-terminal GPI-anchoring signal, similar to those found in phytocyanins (Patel and David, 1997). The GPI-anchored form of ceruloplasmin is preferentially expressed in brain.

An autosomal recessive disorder of iron metabolism, known as aceruloplasminemia, has been linked to the mutations in the gene encoding ceruloplasmin, which further supports its connection with iron homeostasis (Yoshida *et al.*, 1995). The disease is manifested by a marked iron deposition in a number of organs including brain, where it results in neurodegeneration. In one case, the mutation generates an early termination codon at residue 991. The translated ceruloplasmin thus would lack the three ligands involved in the type 1 copper site of domain 6 as well as the two His ligands of the trinuclear catalytic site. Such a truncation would certainly generate an inactive enzyme.

A second member of the ceruloplasmin family multicopper oxidases with six BCB domains was recently identified as the causative agent of sex-linked anemia (sla) in mice (Vulpe *et al.*, 1993). It was named hephaestin and shown to be expressed mostly in the small intestine and the colon, where it is presumably involved in gastrointestinal iron uptake. Hephaestin displays a high level of sequence identity to ceruloplasmin and differs from it only by an additional C-terminal transmembrane domain, which anchors the protein to the cell membrane. A 582-nucleotide in-frame deletion in the mRNA for hephaestin *sla* mice has been identified compared to normal animals. The mice with such a mutation are unable to release iron from enterocytes (intestinal epithelial cells) into the circulation, which results in severe anemia. The GPI-anchored form of ceruloplasmin could potentially also mediate similar cellular iron efflux in the central nervous system. There is a transferrin-independent iron uptake system that requires Fe(III) to be reduced to Fe(II) at the cell surface for uptake to occur (DeSilva *et al.*, 1996). Ceruloplasmin would oxidize Fe and prevent its uptake by this mechanism. Briefly, the role of ceruloplasmin is most likely to prevent excessive intracellular iron accumulation by tightly controlling iron efflux and inhibiting its uptake.

In the yeast *Sa. cerevisiae* the functional homologue of ceruloplasmin is Fet3. It is a multicopper oxidase that displays ferroxidase activity similar to that of ceruloplasmin (Dancis *et al.*, 1994; De Silva *et al.*, 1995), but it is much closer to laccases with respect to its domain organization and sequence characteristics than to ceruloplasmin or hephaestin. Fet3 is an extracellular membrane-bound protein where it operates together with the iron transporter Ftr1 as a high-affinity iron uptake system. While Fet3 is implicated in high-affinity iron uptake in the yeast *Sa. cerevisiae*, ceruloplasmin appears to promote an opposite iron conduit, i.e., efflux from the cells.

Three different genes encoding putative multicopper blue oxidases have been recently characterized from different microorganisms that are capable of oxidizing $Mn^{2+}$ to $Mn^{3+}$. Two, the *mofA* gene product from *Leptothrix discophora* (Corstjens *et al.*, 1997) and the *cumA* gene product from *Pseudomonas putida* GB-1 (Brouwers *et al.*, 1999), are similar to laccases in their three-BCB-domain arrangement. All amino acids involved in the coordination of the three different types of copper are conserved in these proteins. The third gene, *mnxG* from *Bacillus* sp. SG-1, encodes a polypeptide that displays an internally triplicated sequence organization similar to that found in ceruloplasmin and hephaestin (Van Waasbergen *et al.*, 1996). In this protein the three sequence modules, A1, A2, and A3, are approximately 400 amino acids long and can be aligned with a 30% amino acid sequence identity. However, unlike in ceruloplasmin and hephaestin, the conserved HCHXXXHXXXXM motif in this protein is located in the second BCB domain of the A1 module, while no obvious copper-binding sites can be identified in the BCB domains of A2 and A3. The A1 module of bacterial "ceruloplasmin" and the A3 modules of human ceruloplasmin and hephaestin display only 10–13% identity. The occurrence of a ceruloplasmin-like sequence in a prokaryote may have very important evolutionary implications and needs further investigation.

The catalytic activity of these proteins is likely to be $Mn^{2+}$ oxidation since inactivation of the corresponding genes, *mofA*, *cumA*, and *mnxG*, resulted in the complete loss of the ability of their respective hosts to oxidize $Mn^{2+}$. Additionally, in all three cases the $Mn^{2+}$ oxidizing activity was inhibited by metal chelating agents and was restored by addition of $Cu^{2+}$. Such an unusual catalytic activity for a multicopper oxidase has been also recently demonstrated for a laccase from the fungus *Trametes versicolor* (Höfer and Schlosser, 1999).

## VIII. Coagulation Factors V and VIII

Factor VIII (F8) and Factor V (F5) are key regulatory elements of the blood coagulation cascade of sequential proteolytic activation of serine proteinases (Fig. 10A). Unlike many other components of the cascade neither F8 nor F5 displays proteolytic activity or any known enzymatic activity of its own (Antonarakis, 1995; Kaufman, 1992). They are proteolytically converted into their activated forms, F8a and F5a, by other coagulation enzymes such as Factor Xa and thrombin. F8a acts as a cofactor in the reaction of proteolytic activation of Factor X by Factor IXa, and in the presence of $Ca^{2+}$ and phospholipids it increases the $V_{max}$ of this reaction by 10,000-fold. F5a is a cofactor for Factor Xa in the transformation of prothrombin into thrombin, which converts fibrinogen to fibrin, the terminal component of the cascade. Although Factor Xa itself is capable of cleaving prothrombin to thrombin, the presence of F5a increases its thrombin-generating proteolytic activity by 270,000-fold. The mechanisms of these two processes are poorly understood.

The deficiency of F8a activity in blood causes an X-chromosome-linked hereditary bleeding disorder, hemophilia A, which affects primarily males with an incidence of 1 in 5000. The molecular origin of F8 deficiency has been well characterized (Kemball-Cook *et al.*, 1998). More than 600 unique mutations, of which nearly 400 are point mutations, have been identified in the F8 gene of hemophilia A patients. Similarly, F5a deficiency also leads to a severe bleeding disorder. The F5 gene is localized on chromosome 1 and one of the mutations, converting residue R506 into Q, is considered to be a genetic risk factor for thrombosis (Dahlbäck, 1997; Dahlbäck *et al.*, 1993; Egan *et al.*, 1997). It was estimated that nearly 4–6% of Dutch and Swedish population carry this mutant allele.

The nucleotide sequence of the F8 gene and of its corresponding cDNA was reported in 1984 and was the first sequence for a BCB domain-containing protein to be determined (Gitschier *et al.*, 1984; Toole *et al.*, 1984; Vehar *et al.*, 1984; Wood *et al.*, 1984). It encodes a 2351-amino-acid-long polypeptide with a complex domain organization (see Fig. 1). The first 19 amino acids constitute the signal peptide, which is followed by a triplicated 330-residue-long A domain, a large ~900-residue-long B domain, and a duplicated 150-amino-acid-long C domain, which are arranged in the following pattern: A1-A2-B-A3-C1-C2. A similar domain organization is also found in the precursor of F5.

### *A. Posttranslational Processing and Activation*

Following cleavage of the leader peptide and nonspecific proteolysis in domain B, F8 is secreted and circulates in blood as a two-chain (heavy chain and light chain) molecule bound to its biological carrier in the serum, von Willebrand factor (vWF), with a ratio of 1:50. While the light chain is a homogeneous 80-kDa fragment composed of domains A3, C1, and C2, the heavy chain is extremely heterogeneous in size (90 to

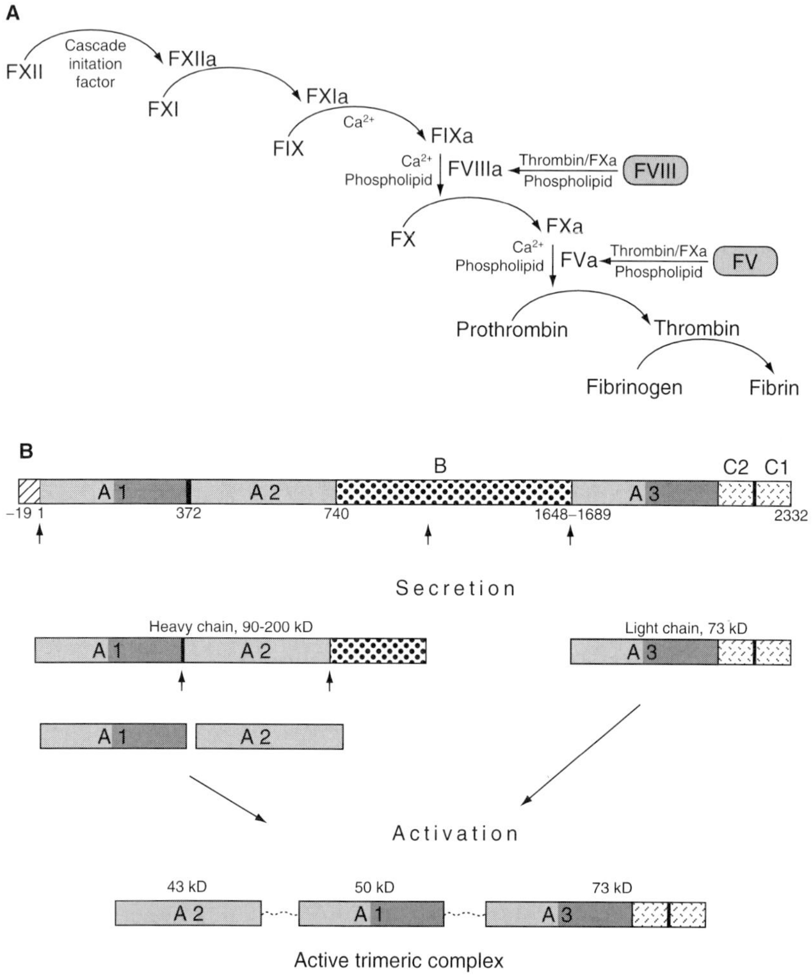


FIG. 10. Schematic presentation of blood coagulation cascade (A) and the activation mechanism of Factor VIII (B).

200 kDa) and is composed of domains A1 and A2 and possibly various sizes of cleaved domain B (Fulcher *et al.*, 1983). Neither chain shows coagulation activity when separated (Kaufman, 1992) (Fig. 10B).

On cleavage by thrombin at two sites within the heavy chain (Arg-372 and Arg-740), and one site in the light chain (Arg-1689), F8 dissociates from the vWF and forms an active heterotrimer, F8a, composed of a 50-kDa domain A1, a 43-kDa domain A2, and a 73-kDa light chain

(A3-C1-C2) (see Fig. 10B). All three components of this complex are necessary for activity (Eaton *et al.*, 1986). This heterotrimer is stabilized and held together by $Ca^{2+}$ ions, although it is highly labile and pH sensitive. At pH 8.0, the 43-kDa (domain A2) subunit dissociates from the complex, resulting in its inactivation. The 50-kDa (domain A1) and 72-kDa (domains A3-C1-C2) subunits form a relatively stable, albeit inactive, complex. The purified 43-kDa subunit is capable of fully restoring the cofactor activity of F8 upon its addition to inactive heterodimer missing domain A2 (Fay *et al.*, 1991). In addition, simultaneous expression of two separate DNA fragments encoding domain A2 and a construct of A1-A3-C1-C2, respectively, resulted in an active heterotrimer. This indicates that various domains of F8 do not require the presence of other domains for their correct folding; i.e., they can fold independently and undergo purification as autonomous polypeptide units.

### *B. Properties of Different Domains*

Domain B is heavily glycosylated, has no known function, and is proteolytically released during the activation of both F8 and F5. Furthermore, its sequence is not conserved between F8 from different organisms or between F8 and F5. In addition, it displays no homology with any of the currently available sequences in GenBank.

Domains C1 and C2 are homologous to each other and to the different chains of discoidins (carbohydrate-binding proteins from amebae *Dictyostelium discoideum*) and mouse milk fat globular membrane protein and they also display sequence identity between F8 and F5.

Domains A1, A2, and A3 in F8 are homologous to one another. In addition, they display a remarkably high level of sequence identity (almost 40%) to the similar domains of F5, ceruloplasmin, and hephaestin. Amino acid sequence alignment of the A domains revealed that, of the three sets of amino acid ligands involved in three type 1 copper-binding sites in ceruloplasmin and hephaestin, only those in BCB domains 2 and 6 are conserved in F8 (see Table II). No such sites are present in the sequences of F5. The type 2 copper-binding His residues (H99 and H1957 in human F8 and H85 and H1815 in human F5) are conserved and match the similar residues involved in type 2 copper coordination in ceruloplasmin and hephaestin. However, of the six His ligands of the pair of type 3 coppers, only one is conserved in both F8 and F5. Thus, they apparently lack the trinuclear catalytic copper site of multicopper oxidases.

A number of different expression systems have been developed that allow production of F8a in quantities sufficient for the purposes of both biochemical studies and clinical application for the treatment of

TABLE II
*The Known or Predicted Copper Ligands in All Currently Available Sequences of Hephaestin (Human, hHP; Mice, mHP), Ceruloplasmin (Human, hCP; Mice, mCP; rat, rCP; Sheep, sCP), Factor V (Human, hFV; Mice, mFV; Pig, pFV, Bovine, bFV), and Factor VIII (Human, hFVIII; Mice, mFVIII; Pig, pFVIII, Dog, dFVIII)*[a]

| | | Type 1 | | | | Type 3 | |
|---|---|---|---|---|---|---|---|
| | | Dom-2 | Dom-4 | Dom-6 | Type 2 | Cu1 | Cu2 |
| hHP | | His-281 | His-633 | His-977 | His-103 | His-105 | His-165 |
| | | Cys-324 | Cys-676 | Cys-1023 | His-980 | His-163 | His-982 |
| | | His-329 | His-681 | His-1028 | | His-1024 | His-1022 |
| | | Met-334 | Met-686 | Met-1033 | | | |
| mHP | | His-281 | His-632 | His-976 | His-103 | His-105 | His-165 |
| | | Cys-324 | Cys-675 | Cys-1022 | His-979 | His-163 | His-981 |
| | | His-329 | His-680 | His-1027 | | His-1023 | His-1021 |
| | | Met-334 | Met-685 | Met-1032 | | | |
| hCP | | His-276 | His-637 | His-975 | His-101 | His-103 | His-163 |
| | | Cys-319 | Cys-680 | Cys-1021 | His-978 | His-161 | His-980 |
| | | His-324 | His-685 | His-1026 | | His-1022 | His-1020 |
| | (ttg) | Leu-329 | Met-690 | Met-1031 | | | |
| mCP | | His-275 | His-632 | His-971 | His-101 | His-103 | His-162 |
| | | Cys-318 | Cys-675 | Cys-1017 | His-974 | His-160 | His-976 |
| | | His-323 | His-680 | His-1022 | | His-1018 | His-1016 |
| | (ttg) | Leu-328 | Met-685 | Met-1027 | | | |
| rCP | | His-275 | His-631 | His-969 | His-101 | His-103 | His-162 |
| | | Cys-318 | Cys-674 | Cys-1015 | His-972 | His-160 | His-974 |
| | | His-323 | His-679 | His-1020 | | His-1016 | His-1014 |
| | (ttg) | Leu-328 | Met-684 | Met-1025 | | | |
| sCP | | His-276 | His-631 | His-958 | His-101 | His-103 | His-163 |
| | | Cys-319 | Cys-674 | Cys-1004 | His-961 | His-161 | His-963 |
| | | His-324 | His-679 | His-1009 | | His-1005 | His-1003 |
| | (ttg) | Leu-329 | Met-684 | Met-1014 | | | |
| hFVIII | | His-267 | Leu-649 | His-1954 | His-99 | Val-101 | His-161 |
| | | Cys-310 | Cys-692 | Cys-2000 | His-1957 | Leu-159 | Ser-1959 |
| | | His-315 | Phe-697 | His-2005 | | Leu-2001 | Glu-1999 |
| | | Met-320 | Met-702 | Met-2010 | | | |
| mFVIII | | His-268 | Leu-649 | Gln-1922 | His-100 | Val-102 | His-162 |
| | | Cys-311 | Cys-692 | Cys-1968 | His-1925 | Met-160 | Ser-1927 |
| | | His-316 | Phe-697 | His-1973 | | Leu-1969 | Glu-1967 |
| | | Met-321 | Met-702 | Met-1978 | | | |

(*continues*)

TABLE II *continued*

| | Type 1 | | | | Type 3 | |
|---|---|---|---|---|---|---|
| | Dom-2 | Dom-4 | Dom-6 | Type 2 | Cu1 | Cu2 |
| pFVIII | His-268 | Leu-649 | His-1736 | His-100 | Val-102 | His-162 |
| | Cys-311 | Cys-692 | Cys-1782 | His-1739 | Leu-160 | Ser-1741 |
| | His-316 | Leu-697 | His-1787 | | Leu-1783 | Glu-1781 |
| | Met-321 | Met-702 | Met-1792 | | | |
| dFVIII | His-262 | Leu-643 | His-1946 | His-100 | Val-102 | His-162 |
| | Cys-305 | Cys-686 | Cys-1992 | His-1949 | Phe-160 | Ser-1951 |
| | His-310 | Leu-691 | His-1997 | | Leu-1993 | Glu-1991 |
| | Met-315 | Met-696 | Met-2002 | | | |
| hFV | Phe-239 | Leu-594 | His-1812 | His-85 | Gln-87 | His-147 |
| | Ser-282 | Ser-637 | Thr-1858 | His-1815 | Tyr-145 | His-1817 |
| | His-287 | Pro-642 | Asn-1863 | | Glu-1859 | Asp-1857 |
| | Met-292 | Leu-647 | Met-1868 | | | |
| mFV | Phe-238 | Leu-592 | His-1771 | His-84 | Gln-86 | Tyr-146 |
| | Ser-281 | Thr-635 | Thr-1817 | His-1774 | Tyr-144 | His-1776 |
| | His-286 | Pro-640 | Asn-1822 | | Glu-1818 | Asp-1816 |
| | Met-291 | Leu-645 | Met-1827 | | | |
| pigFV | Phe-239 | Leu-593 | His-1846 | His-85 | Gln-87 | Tyr-147 |
| | Ser-282 | Thr-636 | Thr-1892 | His-1849 | Tyr-145 | His-1851 |
| | His-287 | Pro-641 | Asn-1897 | | Glu-1893 | Asp-1891 |
| | Met-292 | Leu-646 | Met-1902 | | | |
| bFV | Phe-239 | Leu-598 | His-1799 | His-85 | Gln-87 | Tyr-147 |
| | Ser-282 | Thr-641 | Thr-1845 | His-1802 | Tyr-145 | His-1804 |
| | His-287 | Pro-646 | Ile-1850 | | Glu-1846 | Asp-1844 |
| | Met-292 | Leu-651 | Met-1855 | | | |

[a]Residues with nonliganding side-chain characteristics that match the amino acid sequences of FV and FVIII with known copper ligands in ceruloplasmin are also presented. Codons for the conserved Leu residue at the position of the axial ligand in the blue copper site of BCB domain 2 of CPs are in parentheses. Residue numbers are for the mature proteins.

hemophilia A patients. Surprisingly, neither the protein purified directly from plasma nor that produced in heterologous expression systems has been shown to possess any spectroscopic features characteristic of a type 1 copper. There are several possible explanations for this lack of type 1 copper. It may have been lost during the purification of the protein, since the purification procedures described in the literature always include several buffers containing at least 3 mM sodium azide and reducing agents such as dithiothreitol and 2-mercaptoethanol. In some cases, they also include low concentrations of non-ionic detergents such as Tween 20 and 80. The purification procedures usually are lengthy and long exposure of the protein to these agents would certainly release copper from the protein. In addition, in many expression systems, the recombinant protein is secreted into the medium, which in some cases can increase the volume of the starting material up to tens of liters. Therefore, immunoaffinity chromatography on monoclonal antibodies against heavy and light chains of F8 needed to be performed. The elution of the protein bound to its antibody-affinity column requires extremely harsh conditions, such as application of 2–3 M NaSCN or 2 M KI, which may have contributed to the loss of copper.

Only one copper ion has been found in the preparations of both F8a and F5a. It was identified as a type 2 copper (Bihoreau *et al.*, 1994; Mann *et al.*, 1984; Tagliavacca *et al.*, 1997). It is believed that the single copper ion is not involved in any redox reaction and instead it plays a structural role by stabilizing the association of domain A1 with domain A3 in the active trimeric complex. This is a very unusual role for a $d^9$ redox-active transition metal in biology. Mutant F8 in which the type 2 copper ligand His-1957 was replaced with Ala displayed secretion, active complex assembly, and activity similar to that of wild-type protein, while a mutant in which the second ligand for the type 2 copper, His-99, was replaced with Ala was partially defective for secretion and had low levels of active complex formation and activity (Tagliavacca *et al.*, 1997).

Remarkably, the mutation of the type 1 copper site Cys-310 to Ser in the A1 domain resulted in an inactive enzyme that was partially defective for secretion from the cell (Tagliavacca *et al.*, 1997). The occurrence of intact type 1 copper sites in F8 is also supported by the fact that their putative cysteine ligand residues were shown not to be involved in disulfide bridge formation (McMullen *et al.*, 1995). Thus the F8 has five half-cysteines in each A domain. Analysis of the disulfide bond pattern in F8 showed that four half-cysteines are involved in the formation of two disulfide bonds, while the fifth remains free. The free Cys residues have been identified as Cys-310 in domain A1, Cys-692 in domain A2, and Cys-2000 in domain A3. All these Cys residues match analogous residues involved in type 1 copper binding in ceruloplasmin. However, as mentioned above, only two of the domains, A1 and A3, contain the complete set of ligands for the type 1 copper.

The purification of homogeneous F8 and F5 from plasma is extremely difficult because of (1) its very low concentration in plasma, 0.2 μg/ml; (2) its susceptibility to nonspecific proteolysis, which occurs both *in vitro* and *in vivo*; (3) the relatively unstable nature of its activated form, e.g., dissociation of the A2 domain from the heterotrimeric active complex; and (4) the fact that a large amount of the protein circulates as a complex with vWF, an oligomeric 16,000-kDa glycoprotein. All of these factors have rendered X-ray crystallographic studies extremely difficult. Therefore only homology-derived structure modeling studies based on structures of nitrite reductase and ceruloplasmin are currently available for F8 and F5 (Pan *et al.*, 1995; Pemberton *et al.*, 1997; Villoutreix and Dahlbäck, 1998). These models indicate that as is seen in ceruloplasmin, the A domains in F8 and F5 also consist of two BCB domains. A putative type 2 copper site between BCB domains 1 and 6 can be successfully built in the model. Remarkably, no discussion was provided regarding the occurrence of blue copper sites in the structural models of F8.

## IX. BCB Domains with a Binuclear CuA Site

BCB domains that house a binuclear copper-binding site, known as CuA, are found in virtually all aerobic prokaryotes, both Archaea and Bacteria. They are also encoded by the mitochondrial genomes of all Eukaryotes and serve as the subunit II of cytochrome *c* oxidase, the terminal oxidase of the mitochondrial respiratory electron transfer chain. In addition, many bacteria utilize a homologous to the cytochrome *c* oxidase enzyme, quinol oxidase, which uses quinols instead of Cyt c as electron donor and in contrast to the cytochrome *c* oxidases, its subunit II lacks a CuA binding site. The CuA site is best described as an association of two blue copper sites bridged by two Cys sulfur ligands, one from each site. The other coordination positions are filled by a nitrogen of a His at each copper, a thioether of a Met, and a backbone carbonyl oxygen, which in proteins from different sources is derived from a Glu, Gln, or Trp residue (Fig. 11A). It accepts electrons from cytochrome *c* and passes them to the heme centers bound to subunit I. When the site is in the oxidized state, the single unpaired electron is completely delocalized between the two copper atoms, resulting in a mixed-valence $Cu^{1.5+}–Cu^{1.5+}$ binuclear center (reviewed in Beinert, 1997).

A similar domain is also found in nitrous oxide ($N_2O$) reductase, the terminal reductase of the electron transfer chain in many denitrifying bacteria utilizing $NO_3^-/NO_2^-$ as energy source. $N_2O$ reductase catalyzes the two-electron reduction of $N_2O$ to molecular nitrogen. Throughout different species $N_2O$ reductases display conserved sequence characteristics and are composed of an ~500-amino-acid N-terminal catalytic domain, referred to as the CuZ domain, which is followed by an ~100-amino-acid BCB domain harboring the binuclear CuA site. The recently reported crystal structure of the protein from *Pseudomonas nautica* revealed that the CuZ is an unusual polynuclear catalytic copper site—a tetracopper cluster coordinated by seven conserved His residues and

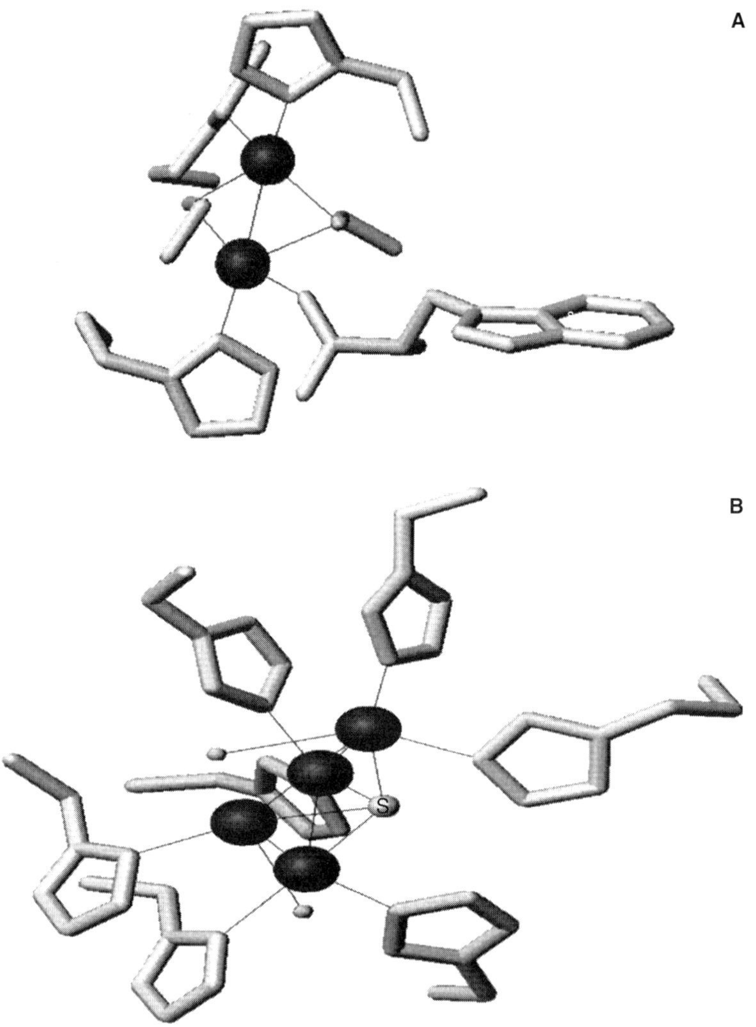


FIG. 11. The binuclear CuA (A) and tetranuclear CuZ (B) copper-binding sites of nitrous oxide reductase from *Pseduomonas nautica* (PDB Accession Code 1QN1). The sulfur atom in the tetranuclear copper site is marked with an S.

three hydroxide ions (Fig. 11B) (Brown *et al.*, 2000b). The cluster has the shape of a distorted tetrahedron around a bridging ligand, which was originally modeled in the structure as an $OH^-$ where two Cu atoms were bound to two copper ions and two Cu atoms were bound to three copper ions (Brown *et al.*, 2000b). However, Raman spectroscopic investigations of isotopically labeled enzyme have conclusively demonstrated that the CuZ center has an acid-labile inorganic sulfur ligand (Alvarez *et al.*, 2001; Rasmussen *et al.*, 2000), and in agreement with the spectroscopic data, a higher resolution structure of the same enzyme revealed that indeed the bridging moiety is a sulfur atom (Brown *et al.*, 2000a). It makes CuZ the first copper–sulfur cluster identified in biological macromolecules.

In the crystal structure $N_2O$ reductase is organized as a dimer in which the CuA center of one monomer is in close proximity (10.2 Å) to the CuZ tetranuclear catalytic center of the second monomer.

## X. Nitrosocyanin

Another recently reported crystal structure is of a mononuclear copper protein that has a classic BCB domain folding topology and an unusual copper-binding site labeled as red (Lieberman *et al.*, 2001). The protein, nitrosocyanin, was isolated from the chemoautotrophic nitrifying bacterium *Nitrosomonas europaea*, which utilizes energy driven from oxidation of ammonia to nitrite. It is homotrimer of a 112-residue BCB domain both in solution and in crystalline form. The spectroscopic properties and ligand organization of the red copper site are quite distinct from those of blue copper sites (Hooper and Arciero, 1999). Two components of the $S_{Cys}$-Cu CT transition are blue-shifted with a dominant band at 390 nm ($\varepsilon = 5500\ M^{-1}cm^{-1}$) and a less intense band at 498 nm ($\varepsilon = 1500\ M^{-1}cm^{-1}$). In addition, there is a d–d transition band at 726 nm ($\varepsilon = 600\ M^{-1}cm^{-1}$). The EPR spectrum has been reported to be typical for a normal tetragonal copper (Lieberman *et al.*, 2001). Its redox potential is substantially lower than that of the inorganic Cu(II)/Cu(I) redox couple and was estimated to be only +85 mV. The copper center is five-coordinate with a square pyramidal geometry (Fig. 12). A C-terminal copper-binding loop features the triad of the liganding residues composed of two His and a Cys in a sequence spaced similar to that of the copper-binding loops of plastocyanin, pseudoazurin, and halocyanin. At the position of the axial ligand Met in cupredoxins, however, a His residue is found in nitrosocyanin. Additionally, the upstream His ligand is replaced with a Glu, which coordinates in a monodentate fashion, providing its carboxylate oxygen for copper coordination. Surprisingly, the fifth ligand has been identified to be a solvent molecule, a highly unusual feature for any long-range electron transfer protein. This is despite the fact that the red copper site is entirely buried inside the protein molecule due to an extended β-hairpin, not a part of the β-barrel, that efficiently covers the copper site of the neighboring monomer. With the β-hairpin, the BCB domain in nitrosocyanin is reminiscent

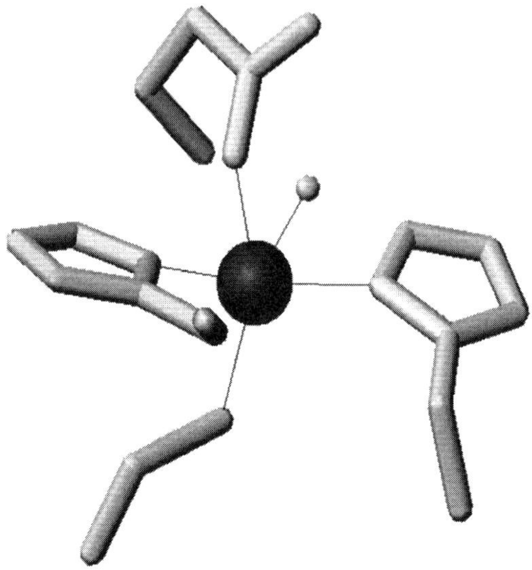

FIG. 12. The "red copper" site of nitrosocyanin from the bacteria *Nitrosomonas europaea* (PDB Accession Code 1IBY).

of those of similar domains in multicopper oxidases rather than reminiscent of cupredoxins. The hydrogen-bonding interactions involved the sulfur ligand, and donor atoms from residues adjacent to other ligands appear to be conserved in both blue and red copper sites. In the reduced protein, no ligand-flipping effect was observed for the surface-exposed His (located in the same position as in copper-binding loops of blue copper sites), which is a part of a hydrophobic patch similar to that of cupredoxins. However, on reduction it moves away from the copper by 0.4 Å, but still maintains its bonding interactions. Simultaneously, the solvent molecule leaves the copper coordination sphere, which becomes four-coordinate. Such redox-dependent ligand rearrangements would substantially increase the reorganization energy of nitrosocyanin compared to that of most blue copper sites. Therefore, it is unlikely that nitrosocyanin participates in long-range electron transfer processes and possibly it would carry out some unspecified sort of redox catalysis. Intriguingly, such properties are also suggested for the phytocyanin family proteins (see Section V).

## ACKNOWLEDGMENTS

We thank Drs. Edith Gralla and Hans Freeman for critically reading the manuscript and for invaluable discussions, and we thank Soshanna Zittin for helpful comments. We apologize to our colleagues whose work was not referenced because of space limitations. This work was supported by NIH GM28222.

## References

Adman, E. T. (1985). *In* "Metalloproteins," (P. Harrison, ed.), pp. 1–42. McMillan, London.

Adman, E. T. (1991). *Adv. Protein Chem.* **42**, 145–197.

Adman, E. T., Godden, J. W., and Turley, S. (1995). *J. Biol. Chem.* **270**, 27458–27474.

Adman, E. T., Stenkamp, R. E., Sieker, L. C., and Jenson, L. H. (1978). *J. Mol. Biol.* **123**, 35–47.

Aikazyan, V. T., and Nalbandyan, R. M. (1981). *Biochim. Biophys. Acta* **667**, 421–432.

Aikazyan, V. T., and Nalbandyan, R. M. (1979). *FEBS Lett.* **104**, 127–130.

Albani, D., Sardana, R., Robert, L. S., Altosaar, I., Arnison, P. G., and Fabijanski, S. F. (1992). *Plant J.* **2**, 331–342.

Allona, I., Quinn, M., Shoop, E., Swope, K., St. Cyr, S., Carlis, J., Riedl, J., Retzel, E., Campbell, M. M., Sederoff, R., and Whetten, R. W. (1998). *Proc. Natl. Acad. Sci. USA* **95**, 9693–9698.

Alvarez, M. L., Ai, J. Y., Zumft, W., Sanders-Loehr, J., and Dooley, D. M. (2001). *J. Am. Chem. Soc.* **123**, 576–587.

Antonarakis, S. E. (1995). *Thromb. Haemost.* **74**, 322–328.

The *Arabidopsis* Genome Initiative (2000). *Nature* **408,** 796–815.

Bao, W. L., O'Malley, D. M., Whetten, R., and Sederoff, R. R. (1993). *Science* **260**, 672–674.

Bao, W. L., O'Malley, D. M., and Sederoff, R. R. (1992). *Proc. Natl. Acad. Sci. USA* **89**, 6604–6608.

Beinert, H. (1997). *Eur. J. Biochem.* **245**, 521–532.

Beratan, D. N., Onuchic, J. N., and Hopfield, J. J. (1987). *J. Chem. Phys.* **86**, 4488–4498.

Bergman, C., Gandvick, E.-K., Nyman, P. O., and Strid, L. (1977). *Biochem. Biophys. Res. Commun.* **77**, 1052–1059.

Bihoreau, N., Pin, S., De Kersabiec, A. M., Vidot, F., and Fontaine-Aupart, M. P. (1994). *Eur. J. Biochem.* **222**, 41–48.

Bond, C. S., Blankenship, R. E., Freeman, H. C., Guss, J. M., Maher, M. J., Selvaraj, F. M., Wilce, M. C. J., and Willingham, K. M. (2001). *J. Mol. Biol.* **306**, 47–67.

Brouwers, G. J., de Vrind J. P. M., Corstjens, P. L. A. M., Cornelis, P., Baysse, C., and de Vrind-de Jong, E. W. (1999). *Appl. Environ. Microbiol.* **65**, 1762–1768.

Brown, K., Djinovic-Carugo, K., Haltia, T., Cabrito, I., Saraste, M., Moura, J. J. G., Moura, I., Tegoni, M., and Cambillau, C. (2000a). *J. Biol. Chem.* **275**, 41133–41136.

Brown, K., Tegoni, M., Prudencio, M., Pereira, A. S., Besson, S., Moura, J. J., Moura, I., and Cambillau, C. (2000b). *Nat. Struct. Biol.* **7**, 191–195.

Buning, C., Canters, G. W., Comba, P., Dennison, C., Jeuken, L., Melter, M., and Sanders-Loehr, J. (2000). *J. Am. Chem. Soc.* **122**, 204–211.

Calabrese, L., Carbonaro, M., and Musci, G. (1988). *J. Biol. Chem.* **263**, 6480–6483.

Canters, G. W. (1987). *FEBS Lett.* **212**, 168–172.

Canters, G. W., and Gilardi, G. (1993). *FEBS Lett.* **325**, 39–48.

Castresana, J., Lübben, M., and Saraste, M. (1995). *J. Mol. Biol.* **250**, 202–210.

Chang, S. J., Puryear, J. D., Dias, M. A. D. L., Funkhouser, E. A., Newton, R. J., and Cairney, J. (1996). *Phys. Plantar.* **97**, 139–148.

Chen, L., Durley, R., Poliks, B. J., Hamada, K., Chen, Z., Mathews, F. S., Davidson, V. L., Satow, Y., Huizinga, E., Vellieux, F. M. D., and Hol, W. G. J. (1992). *Biochemistry* **31**, 4959–4964.

Chen, L., Durley, R. C. E., Mathews, F. S., and Davidson, V. L. (1994). *Science* **264**, 86–90.

Chichiricco, G., Ceru, M. P., D'Alessandro, A., Oratore, A., and Avigliano, L. (1989). *Plant Sci.* **64**, 61–66.

Chistoserdov, A. Y., Chistoserdova, L. V., McIntire, W. S., and Lidstrom, M. E. (1994). *J. Bacteriol.* **176**, 4052–4065.

Colman, P. M., Freeman, H. C., Guss, J. M., Murata, M., Norris, V. A., Ramshaw, J. A. M., and Venkatappa, M. P. (1978). *Nature* **272**, 319–324.

Colman, P. M., Freeman, H. C., Guss, J. M., Murata, M., Norris, V. A., Ramshaw, J. A. M., Venkatappa, M. P., and Vickery, L. E. (1977). *J. Mol. Biol.* **112**, 649–650.

Corstjens, P. L. A. M., de Vrind, J. P. M., Goosen, T., and de Vrind-de Jong, E. W. (1997). *Geomicrobiol. J.* **14**, 91–108.

Curzon, G., and O'Reilly, S. (1960). *Biochem. Biophys. Res. Commun.* **2**, 284–286.

Dahlbäck, B. (1997). *Semin. Hemat.* **34**, 217–234.

Dahlbäck, B., Carlsson, M., and Svensson, P. J. (1993). *Proc. Natl. Acad. Sci. USA* **90**, 1004–1008.

Daizadeh, I., Medvedev, E. S., and Stuchebrukhov, A. A. (1997). *Proc. Natl. Acad. Sci. USA* **94**, 3703–3708.

Dancis, A., Yuan, D. S., Haile, D., Askwith, C., Eide, D., Moehle, C., Kaplan, J., and Klausner, R. D. (1994). *Cell* **76**, 393–402.

D'Andrea, G., Salucci, M. L., Pitari, G., and Avigliano, L. (1993). *Glycobiology* **3**, 563–565.

Dayhoff, M. O., Eck, R. V., Chang, M. A., and Sochard, M. R. (1965). "Atlas of Protein Sequence and Structure". National Biomedical Research Foundation, Silver Spring, MD.

Dean, J. F. D., and Eriksson, K-E. L. (1994). *Holzforschung* **48**, 21–33.

De Blank, C., Mylona, P., Yang, W. C., Katinakis, P., Bisseling, T., and Franssen, H. (1993). *Plant Mol. Biol.* **22**, 1167–1171.

Den Blaauwen, T., and Canters, G. W. (1993). *J. Am. Chem. Soc.* **115**, 1121–1129.

Dennison, C., Vijgenboom, E., de Vries, S., van der Oost, J., and Canters, G. W. (1995). *FEBS Lett.* **365**, 92–94.

De Silva, D. M., Askwith, C. C., Eide, D., and Kaplan, J. (1995). *J. Biol. Chem.* **270**, 1098–1101.

De Silva D. M., Askwith, C. C., and Kaplan, J. (1996). *Physiol. Rev.* **76**, 31–47.

Diederix, R. E. M., Canters, G. W., and Dennison, C. (2000). *Biochemistry* **39**, 9551–9560.

Dimitrov, M. I., Donchev, A. A., and Egorov, T. A. (1993). *Biochim. Biophys. Acta* **1203**, 184–190.

Dodd, F. E., Abraham, Z. H. L., Eady, R. R., and Hasnain, S. S. (2000). *Acta Crystallogr. Sect. D* **56**, 690–696.

Dodd, F. E., Hasnain, S. S., Hunter, W. N., Abraham, Z. H. L., Debenham, M., Kanzler, H., Eldridge, M., Eady, R. R., Ambler, R. P., and Smith, B. E. (1995). *Biochemistry* **34**, 10180–10186.

Dong, S. L., Ybe, J. A., Hecht, M. H., and Spiro, T. G. (1999). *Biochemistry* **38**, 3379–3385.

Donovan, W., Zheng, L., Sandman, K., and Losick, R. (1987). *J. Mol. Biol.* **196**, 1–10.

Drew, J. E., and Gatehouse, J. A. (1994). *J. Exp. Bot.* **45**, 1873–1884.

Driouich, A., Laine, A. C., Vian, B., and Faye, L. (1992). *Plant J.* **2**, 13–24.

Ducros, V., Brzozowski, A. M., Wilson, K. S., Brown, S. H., Østergaard, P., Schneider, P., Yaver, D. S., Pedersen, A. H., and Davies, G. J. (1998). *Nat. Struct. Biol.* **5**, 310–316.

Ducros, V., Brzozowski, A. M., Wilson, K. S., østergaard, P., Schneider, P., Svendson, A., and Davies, G. J. (2001). *Acta Crystallogr. Sect. D* **57**.

Durley, R., Chen, L., Lim, L. W., Mathews, F. S., and Davidson, V. L. (1993). *Protein Sci.* **2**, 739–752.

Dwulet, F. E., and Putnam, F. W. (1981). *Proc. Natl. Acad. Sci. USA* **78**, 2805–2809.

Eaton, D., Rodriguez, H., and Vehar, G. A. (1986). *Biochemistry* **25**, 505–512.

Egan, J. O., Kalafatis, M., and Mann, K. G. (1997). *Protein Sci.* **6**, 2016–2027.

Eggert, C., Temp, U., Dean, J. F. D., and Eriksson, K-E. L. (1996). *FEBS Lett.* **391**, 144–148.
Einsle, O., Mehrabian, Z., Nalbandyan, R., and Messerschmidt, A. (2000). *J. Biol. Inorg. Chem.* **5**, 666–672.
Eisenberg, D., Marcotte, E. M., Xenarios, I., and Yeates, T. O. (2000). *Nature* **405**, 823–826.
Ejdebäck, M., Bergkvist, A., Karlsson, B. G., and Ubbink, M. (2000). *Biochemistry* **39**, 5022–5027.
Faham, S., Mizoguchi, T., Adman, E. T., Gray, H. B., Richards, J. H., and Rees, D. C. (1997). *J. Biol. Inorg. Chem.* **2**, 464–469.
Farver, O., Skov, L. K., Pascher, T., Karlsson, B. G., Nordling, M., Lundberg, L. G., Vänngård, T., and Pecht, I. (1993). *Biochemistry* **32**, 7317–7322.
Fay, P. J., Haidaris, P. J., and Smudzin, T. M. (1991). *J. Biol. Chem.* **266**, 8957–8962.
Flanagan J. G., and Vanderhaeghen, P. (1998). *Annu. Rev. Neurosci.* **21**, 309–345.
Frazao, C., Soares, C. M., Carrondo, M. A., Pohl, E., Dauter, Z., Wilson, K. S., Hervas, M., Navarro, J. A., De La Rosa, M. A., and Sheldrick, G. M. (1995). *Structure* **3**, 1159–1169.
Freeman, J. C., Nayar, P. G., Begley, T. P., and Villafranca, J. J. (1993). *Biochemistry* **32**, 4826–4830.
Fulcher, C. A., Roberts, J. R., and Zimmerman, T. S. (1983). *Blood* **61**, 807–811.
Garrett, T. P., Clingeleffer, D. J., Guss, J. M., Rogers, S. J., and Freeman, H. C. (1984). *J. Biol. Chem.* **259**, 2822–2825.
Germanas, J. P., Di Bilio, A. J., Gray, H. B., and Richards, J. H. (1993). *Biochemistry* **32**, 7698–7702.
Germann, U. A., Muller, G., Hunziker, P. E., and Lerch, K. (1988). *J. Biol. Chem.* **263**, 885–896.
Gitschier, J., Wood, W. I., Goralka, T. M., Wion, K. L., Chen, E. Y., Eaton, D. H., Vehar, G. A., Capon, D. J., and Lawn, R. M. (1984). *Nature* **312**, 326–330.
Giudici-Orticoni, M. T., Guerlesquin, F., Bruschi, M., and Nitschke, W. (1999). *J. Biol. Chem.* **274**, 30365–30369.
Godden, J. W., Turley, S., Teller, D. C., Adman, E. T., Liu, M. Y., Payne, W. J., and Le Gall, J. (1991). *Science* **253**, 438–442.
Gotschlich, E. C., and Seiff, M. E. (1987). *FEMS Microbiol. Lett.* **43**, 253–255.
Grapperhaus, C. A., and Darensbourg, M. Y. (1998). *Acc. Chem. Res.* **31**, 451–459.
Gray, H. B., Malmström, B. G., and Williams, R. J. P. (2000). *J. Biol. Inorg. Chem.* **5**, 551–559.
Greene, E. A., Erard, M., Dedieu, A., and Barker, D. G. (1998). *Plant Mol. Biol.* **36**, 775–783.
Grossmann, J. G., Abraham, Z. H. L., Adman, E. T., Neu, M., Eady, R. R., Smith, B. E., and Hasnain, S. S. (1993). *Biochemistry* **32**, 7360–7366.
Guss, J. M., and Freeman, H. C. (1983). *J. Mol. Biol.* **169**, 521–563.
Guss, J. M., Harrowell, P. R., Murata, M., Norris, V. A., and Freeman, H. C. (1986). *J. Mol. Biol.* **192**, 361–387.
Guss, J. M., Merritt, E. A., Phizackerley, R. P., and Freeman, H. C. (1996). *J. Mol. Biol.* **262**, 686–705.
Guss, J. M., Merritt, E. A., Phizackerley, R. P., Hedman, B., Murata, M., Hodgson, K. O., and Freeman, H. C. (1988). *Science* **241**, 806–811.
Hall, J. F., Kanbi, L. D., Strange, R. W., and Hasnain, S. S. (1999). *Biochemistry* **38**, 12675–12680.
Hammann, C., van Pouderoyen, G., Nar, H., Rüth, F-X. G., Messerschmidt, A., Huber, R., den Blaauwen, T., and Canters, G. W. (1997). *J. Mol. Biol.* **266**, 357–366.
Han, J., Loehr, T. M., Lu, Y., Valentine, J. S., Averill, B. A., and Sanders-Loehr, J. (1993). *J. Am. Chem. Soc.* **115**, 4256–4263.
Harris, Z. L., Durley, A. P., Man, T. K., and Gitlin, J. D. (1999). *Proc. Natl. Acad. Sci. USA* **98**, 10812–10817.

Hart, P. J., Nersissian, A. M., Herrmann, R. G., Nalbandyan, R. M., Valentine, J. S., and Eisenberg, D. (1996). *Protein Sci.* **5**, 2175–2183.
Hattori, M., Osterfield, M., and Flanagan, J. G. (2000). *Science* **289**, 1360–1365.
Hay, M., Richards, J. H., and Lu, Y. (1996). *Proc. Natl. Acad. Sci. USA* **93**, 461–464.
Hayashi, R., and Morohashi, Y. (1993). *Plant Physiol.* **102**, 1237–1241.
Heckman, D. S., Geiser, D. M., Eidell, B. R., Stauffer, R. L., Kardos, N. L., and Hedges, S. B. (2001). *Science* **293**, 1129–1133.
Hibino, T., Lee, B. H., Takabe, T., and Takabe, T. (1995). *J. Biochem.* **117**, 101–106.
Hippler, M., Drepper, F., Haehnel, W., and Rochaix, J. D. (1998). *Proc. Natl. Acad. Sci. USA* **95**, 7339–7344.
Hirose, J., Sakurai, T., Imamura, K., Watanabe, H., Iwamoto, H., Hiromi, K., Itoh, H., Shin, T., and Murao, S. (1994). *J. Biochem.* **115**, 811–813.
Höfer, C., and Schlosser, D. (1999). *FEBS Lett.* **451**, 186–190.
Hoitink, C. W. G., and Canters, G. W. (1992). *J. Biol. Chem.* **267**, 13836–13842.
Holland, P. L., and Tolman, W. B. (1999). *J. Am. Chem. Soc.* **121**, 7270–7271.
Holm, R. H., Kennepohl, P., and Solomon, E. I. (1996). *Chem. Rev.* **96**, 2239–2314.
Hooper A. B., and Arciero, D. M. (1999). *J. Inorg. Biochem.* **74**, 166.
Howes, B. D., Abraham, Z. H. L., Lowe, D. J., Bruser, T., Eady, R. R., and Smith, B. E. (1994). *Biochemistry* **33**, 3171–3177.
Iwata, S., Ostermeier, C., Ludwig, B., and Michel, H. (1995). *Nature* **376**, 660–669.
Jimenez-Zurdo, J. I., Frugier, F., Crespi, M. D., and Kondorosi, A. (2000). *MPMI* **13**, 96–106.
Jung, H. W., and Hwang, B. K. (2000). *MPMI* **13**, 136–142.
Karlsson, B. G., Pascher, T., Nordling, M., Arvidsson, R. H., and Lundberg, L. G. (1989). *FEBS Lett.* **246**, 211–217.
Karlsson, B. G., Tsai, L. C., Nar, H., Sanders-Loehr, J., Bonander, N., Langer, V., and Sjölin, L. (1997). *Biochemistry* **36**, 4089–4095.
Kannt, A., Young, S., and Bendall, D. S. (1996). *Biochim. Biophys. Acta* **1277**, 115–126.
Kaufman, R. J. (1992). *Annu. Rev. Med.* **43**, 325–339.
Kawula, T. H., Spinola, S. M., Klapper, D. G., and Cannon, J. G. (1987). *Mol. Microbiol.* **1**, 179–185.
Keilin, D., and Mann, T. (1940). *Nature* **145**, 304.
Kemball-Cook, G., Tuddenham, E. G. D., and Wacey, A. I. (1998). *Nucleic Acids Res.* **26**, 216–219.
Kerfeld, C. A., Anwar, H. P., Interrante, R., Merchant, S., and Yeates, T. O. (1995). *J. Mol. Biol.* **250**, 627–647.
Kieselbach, T., Bystedt, M., Hynds, P., Robinson, C., and Schroder, W. P. (2000). *FEBS Lett.* **480**, 271–276.
Knoll, A. (1992). *Science* **256**, 622–627.
Kobayashi, M., and Shoun, H. (1995). *J. Biol. Chem.* **270**, 4146–4151.
Koikeda, S., Ando, K., Kaji, H., Inoue, T., Murao, S., Takeuchi, K., and Samejima, T. (1993). *J. Biol. Chem.* **268**, 18801–18809.
Komorowski, L., and Schäfer, G. (2001). *FEBS Lett.* **487**, 351–355.
Koschinsky, M. L., Funk, W. D., van Oost, B. A., and MacGillivray, R. T. (1986). *Proc. Natl. Acad. Sci. USA* **83**, 5086–5090.
Kouchi, H., and Hata, S. (1993). *Mol. Gen. Genet.* **238**, 106–119.
Kunert, K. J., Bohme, H., and Boger, P. (1976). *Biochim. Biophys. Acta* **449**, 541–553.
Kyo, M., Miyatake, H., Mamezuka, K., and Amagata, K. (2000). *Plant Cell Physiol.* **41**, 129–137.
LaCroix, L. B., Randall, D. W., Nersissian, A. M., Hoitink, C. W. G., Canters, G. W., Valentine, J. S., and Solomon, E. I. (1998). *J. Am. Chem. Soc.* **120**, 9621–9631.
LaCroix, L. B., Shadle, S. E., Wang, Y. N., Averill, B. A., Hedman, B., Hodgson, K. O., and Solomon, E. I. (1996). *J. Am. Chem. Soc.* **118**, 7755–7768.

LaFayette, P. R., Eriksson, K-E. L., and Dean, J. F. D. (1999). *Plant Mol. Biol.* **40**, 23–35.
Lee, Y. A., Hendson, M., Panopoulos, N. J., and Schroth, M. N. (1994). *J. Bacteriol.* **176**, 173–188.
Li, H. H., and Merchant, S. (1995). *J. Biol. Chem.* **270**, 23504–23510.
Libeu, C. A. P., Kukimoto, M., Nishiyama, M., Horinouchi, S., and Adman, E. T. (1997). *Biochemistry* **36**, 13160–13179.
Lieberman, R. L., Arciero, D. M., Hooper, A. B., and Rosenzweig, A. C. (2001). *Biochemistry* **40**, 5674–5681.
Lin, L. S., and Varner, J. E. (1991). *Plant Physiol.* **96**, 159–165.
Lommen, A., and Canters, G. W. (1990). *J. Biol. Chem.* **265**, 2768–2774.
Lommen, A., Pandya, K. I., Koningsberger, D. C., and Canters, G. W. (1991). *Biochim. Biophys. Acta* **1076**, 439–447.
Lu, Y., LaCroix, L. B., Lowery, M. D., Solomon, E. I., Bender, C. J., Peisach, J., Roe, J. A., Gralla, E. B., and Valentine, J. S. (1993). *J. Am. Chem. Soc.* **115**, 5907–5918.
Machonkin, T. E., Zhang, H. H., Hedman, B., Hodgson, K. O., and Solomon, E. I. (1998). *Biochemistry* **37**, 9570–9578.
Malkin, R., and Malmström, B. G. (1970). *Adv. Enzymol. Related Areas Mol. Biol.* **33,** 177–244.
Malmström, B. G. (1964). *In* "Oxidases and Related Redox Systems," King, T. E., Mason, H. S., and Morrison, M. (Eds.), pp. 207–216. Wiley, New York.
Malmström, B. G. (1982). *Annu. Rev. Biochem.* **51,** 21–59.
Malmström, B. G. (1994). *Eur. J. Biochem.* **223,** 711–718.
Malmström, B. G., and Vänngård, T. (1960). *J. Mol. Biol.* **2,** 118–129.
Mann, K., Eckerskorn, C., Mehrabian, Z., and Nalbandyan, R. M. (1996). *Biochem. Mol. Biol. Int.* **40,** 881–887.
Mann, K., Schäfer, W., Thoenes, U., Messerschmidt, A., Mehrabian, Z., and Nalbandyan, R. (1992). *FEBS Lett.* **314,** 220–223.
Mann, K. G., Lawler, C. M., Vehar, G. A., and Church, W. R. (1984). *J. Biol. Chem.* **259,** 12949–12951.
Marchesini, A., Minelli, M., Merkle, H., and Kroneck, P. M. (1979). *Eur. J. Biochem.* **101,** 77–84.
Marcotte, E. M. (2000). *Curr. Opin. Struct. Biol.* **10,** 359–365.
Margulis, L. (1996). *Proc. Natl. Acad. Sci. USA* **93,** 1071–1076.
Margulis, L., Dolan, M. F., and Guerrero, R. (2000). *Proc. Natl. Acad. Sci. USA* **97,** 6954–6959.
Markossian, K. A., Aikazyan, V. T., and Nalbandyan, R. M. (1974). *Biochim. Biophys. Acta* **359,** 47–54.
Mattar, S., Scharf, B., Kent S. B. H., Rodewald, K., Oesterhelt, D., and Engelhard, M. (1994). *J. Biol. Chem.* **269,** 14939–14945.
McClure, B. A., Cruz-Garcia, F., Beecher, B., and Sulaman, W. (2000). *Ann. Bot.* **85** (Suppl. A), 113–123.
McIntire, W. S., Wemmer, D. E., Chistoserdov, A., and Lidstrom, M. E. (1991). *Science* **252,** 817–824.
McManus, J. D., Brune, D. C., Han, J., Sanders-Loehr, J., Meyer, T. E., Cusanovich, M. A., Tollin, G., and Blankenship, R. E. (1992). *J. Biol. Chem.* **267,** 6531–6540.
McMullen, B. A., Fujikawa, K., Davie, E. W., Hedner, U., and Ezban, M. (1995). *Protein Sci.* **4,** 740–746.
Mellano, M. A., and Cooksey, D. A. (1988). *J. Bacteriol.* **170,** 2879–2883.
Merchant, S., and Bogorad, L. (1986). *Mol. Cell. Biol.* **6,** 462–469.
Messerschmidt A. (Ed.) (1997). "Multicopper Oxidases." World Scientific, Singapore.
Messerschmidt, A., and Huber, R. (1990). *Eur. J. Biochem.* **187,** 341–352.
Messerschmidt, A., Prade, L., Kroes, S. J., Sanders-Loehr, J., Huber, R., and Canters, G. W. (1998). *Proc. Natl. Acad. Sci. USA* **95,** 3443–3448.

Messerschmidt, A., Rossi, A., Ladenstein, R., Huber, R., Bolognesi, M., Gatti, G., Marchesini, A., Petruzzelli, R., and Finazzi-Agro, A. (1989). *J. Mol. Biol.* **206,** 513–529.

Miller, J. D., Arteca, R. N., and Pell, E. J. (1999). *Plant Physiol.* **120,** 1015–1023.

Moseley, J., Quinn, J., Eriksson, M., and Merchant, S. (2000). *EMBO J.* **19,** 2139–2151.

Murata, M., Begg, G. S., Lambrou, F., Leslie, B., Simpson, R. J., Freeman, H. C., and Morgan, F. J. (1982). *Proc. Natl. Acad. Sci. USA* **79,** 6434–6437.

Murphy, L. M., Strange, R. W., Karlsson, B. G., Lundberg, L. G., Pascher, T., Reinhammar, B., and Hasnain, S. S. (1993). *Biochemistry* **32,** 1965–1975.

Murphy, M. E. P., Lindley, P. F., and Adman, E. T. (1997a). *Protein Sci.* **6,** 761–770.

Murphy, M. E. P., Turley, S., and Adman, E. T. (1997b). *J. Biol. Chem.* **272,** 28455–28460.

Murphy, M. E. P., Turley, S., Kukimoto, M., Nishiyama, M., Horinouchi, S., Sasaki, H., Tanokura, M., and Adman, E. T. (1995). *Biochemistry* **34,** 12107–12117.

Musci, G., Bellenchi, G. C., and Calabrese, L. (1999). *Eur. J. Biochem.* **265,** 589–597.

Nagai, J.-I., Yamato, K. T., Sakaida, M., Yoda, H., Fukuzawa, H., and Ohyama, K. (1999). *DNA Res.* **6,** 1–11.

Nar, H., Huber, R., Messerschmidt, A., Filippou, A. C., Barth, M., Jaquinod, M., van de Kamp, M., and Canters, G. W. (1992a). *Eur. J. Biochem.* **205,** 1123–1129.

Nar, H., Messerschmidt, A., Huber, R., van de Kamp, M., and Canters, G. W. (1992b). *FEBS Lett.* **306,** 119–124.

Nersissian, A. M., Babayan, M. A., Sarkissian, L. K., Sarukhanian, E. G., and Nalbandyan, R. M. (1985). *Biochim. Biophys. Acta* **830,** 195–205.

Nersissian, A. M., Hart, P. J., and Valentine J. S. (2001). *In* "Handbook of Metalloproteins," Messerschmidt, A., Huber, R., Wieghardt, K., and Poulos, T. (Eds.), Vol.II, pp. 1219–1234, Wiley, New York.

Nersissian, A. M., Immoos, C., Hill, M. G., Hart, P. J., Williams, G., Herrmann, R. G., and Valentine, J. S. (1998). *Protein Sci.* **7,** 1915–1929.

Nersissian, A. M., Mehrabian, Z. B., Nalbandyan, R. M., Hart, P. J., Fraczkiewicz, G., Czernuszewicz, R. S., Bender, C. J., Peisach, J., Herrmann, R. G., and Valentine, J. S. (1996). *Protein Sci.* **5,** 2184–2192.

Ng, W. V., Kennedy, S. P., Mahairas, G. G., Berquist, B., Pan, M., Shukla, H. D., Lasky, S. R., Baliga, N. S., Thorsson, V., Sbrogna, J., Swartzell, S., Weir, D., Hall, J., Dahl, T. A., Welti, R., Goo, Y. A., Leithauser, B., Keller, K., Cruz, R., Danson, M. J., Hough, D. W., Maddocks, D. G., Jablonski, P. E., Krebs, M. P., Angevine, C. M., Dale, H., Isenbarger, T. A., Peck, R. F., Pohlschroder, M., Spudich, J. L., Jung, K. H., Alam, M., Freitas, T., Hou, S. B., Daniels, C. J., Dennis, P. P., Omer, A. D., Ebhardt, H., Lowe, T. M., Liang, R., Riley, M., Hood, L., and DasSarma, S. (2000). *Proc. Natl. Acad. Sci. USA* **97,** 12176–12181.

Norris, G. E., Anderson, B. F., and Baker, E. N. (1983). *J. Mol. Biol.* **165,** 501–521.

Nothnagel, E. (1997). *Int. Rev. Cytol.* **174,** 195–291.

Ohkawa, J., Okada, N., Shinmyo, A., and Takano, M. (1989). *Proc. Natl. Acad. Sci. USA* **86,** 1239–1243.

Olesen, K., Veselov, A., Zhao, Y. W., Wang, Y. S., Danner, B., Scholes, C. P., and Shapleigh, J. P. (1998). *Biochemistry* **37,** 6086–6094.

O'Malley, D. M., Whetten, R., Bao, W. L., Chen, C. L., and Sederoff, R. R. (1993). *Plant J.* **4,** 751–757.

Osaki, S. (1966). *J. Biol. Chem.* **241,** 5053–5059.

Pan, Y., Defay, T., Gitschier, J., and Cohen, F. E. (1995). *Nat. Struct. Biol.* **2,** 740–744.

Pang, Q., Hays, J. B., and Rajagopal, I. (1993). *Nucleic Acids Res.* **21,** 1647–1653.

Pascher, T., Karlsson, B. G., Nordling, M., Malmström, B. G., and Vänngård, T. (1993). *Eur. J. Biochem.* **212,** 289–296.

Patel, B. N., and David, S. (1997). *J. Biol. Chem.* **272,** 20185–20190.

Paul, K. G., and Stigbrand, T. (1970). *Biochim. Biophys. Acta* **221,** 255–263.

Pemberton, S., Lindley, P., Zaitsev, V., Card, G., Tuddenham, E. G. D., and Kemball-Cook, G. (1997). *Blood* **89,** 2413–2421.

Penfield, K. W., Gewirth, A. A., and Solomon, E. I. (1985). *J. Am. Chem. Soc.* **107,** 4519–4529.

Petratos, K., Dauter, Z., and Wilson, K. S. (1988). *Acta Crystallogr. Sect. B* **44,** 628–636.

Petratos, K., Papadovasilaki, M., and Dauter, Z. (1995). *FEBS Lett.* **368,** 432–434.

Race, H. L., Herrmann, R. G., and Martin, W. (1999). *Trends Genet.* **15,** 364–370.

Rasmussen, T., Berks, B. C., Sanders-Loehr, J., Dooley, D. M., Zumft, W. G., and Thomson, A. J. (2000). *Biochemistry* **39,** 12753–12756.

Rawlings, D. E. (2001). *Hydrometallurgy* **59,** 187–201.

Regan, J. J., and Onuchic, J. N. (1999). *Adv. Chem. Phys.* **107.**

Richards, K. D., Schott, E. J., Sharma, Y. K., Davis, K. R., and Gardner, R. C. (1998). *Plant Physiol.* **116,** 409–418.

Richardson, A., Duncan, J., and McDougall, G. J. (2000). *Tree Physiol.* **20,** 1039–1047.

Romero, A., Hoitink, C. W., Nar, H., Huber, R., Messerschmidt, A., and Canters, G. W. (1993). *J. Mol. Biol.* **229,** 1007–1021.

Rother, C., Jansen, T., Tyagi, A., Tittgen, J., and Herrmann, R. G. (1986). *Curr. Genet.* **11,** 171–176.

Rujan, T., and Martin, W. (2001). *Trends Genet.* **17,** 113–120.

Ryden, L. G., and Hunt, L. T. (1993). *J. Mol. Evol.* **36,** 41–66.

Sarkissian, L. K., and Nalbandyan, R. M. (1983). *Biosci. Rep.* **3,** 915–920.

Scharf, B., and Engelhard, M. (1993). *Biochemistry* **32,** 12894–12900.

Schinina, M. E., Maritano, S., Barra, D., Mondovi, B., and Marchesini, A. (1996). *Biochim. Biophys. Acta* **1297,** 28–32.

Schnackenberg, J., Than, M. E., Mann, K., Wiegand, G., Huber, R., and Reuter, W. (1999). *J. Mol. Biol.* **290,** 1019–1030.

Shepard, W. E. B., Kingston, R. L., Anderson, B. F., and Baker, E. N. (1993). *Acta Crystallogr. Sect. D* **49,** 331–343.

Smeekens, S., Bauerle, C., Hageman, J., Keegstra, K., and Weisbeek, P. (1986). *Cell* **46,** 365–375.

Smeekens, S., Degroot, M., van Binsbergen, J., and Weisbeek, P. (1985). *Nature* **317,** 456–458.

Solomon, E. I., Hare, J. W., and Gray, H. B. (1976). *Proc. Natl. Acad. Sci. USA* **73,** 1389–1393.

Solomon, E. I., and Lowery, M. D. (1993). *Science* **259,** 1575–1581.

Solomon, E. I., Sundaram, U. M., and Machonkin, T. E. (1996). *Chem. Rev.* **96,** 2563–2605.

Sterjiades, R., Dean, J. F. D., and Eriksson, K-E. L. (1992). *Plant Physiol.* **99,** 1162–1168.

Sterjiades, R., Ranocha, P., Boudet, A. M., and Goffner, D. (1996). *Anal. Biochem.* **242,** 158–161.

Sterky, F., Regan, S., Karlsson, J., Hertzberg, M., Rohde, A., Holmberg, A., Amini, B., Bhalerao, R., Larsson, M., Villarroel, R., Van Montagu, M., Sandberg, G., Olsson, O., Teeri, T. T., Boerjan, W., Gustafsson, P., Uhlen, M., Sundberg, B., and Lundeberg, J. (1998). *Proc. Natl. Acad. Sci. USA* **95,** 13330–13335.

Stine, J., Nersissian, A. M., Valentine, J. S., and Hart, P. J. Manuscript in preparation.

Sugumaran, M., Giglio, L. B., Kundzicz, H., Saul, S., and Semensi, V. (1992). *Arch. Insect. Biochem. Physiol.* **19,** 271–283.

Tagliavacca, L., Moon, N., Dunham, W. R., and Kaufman, R. J. (1997). *J. Biol. Chem.* **272,** 27428–27434.

Takahashi, N., Bauman, R. A., Ortel, T. L., Dwulet, F. E., Wang, C. C., and Putnam, F. W. (1983). *Proc. Natl. Acad. Sci. USA* **80,** 115–119.

Takahashi, N., Ortel, T. L., and Putnam, F. W. (1984). *Proc. Natl. Acad. Sci. USA* **81,** 390–394.

Takai, M., Kamimura, K., and Sugio, T. (2001). *Eur. J. Biochem* **268,** 1653–1658.

Toole, J. J., Knopf, J. L., Wozney, J. M., Sultzman, L. A., Buecker, J. L., Pittman, D. D., Kaufman, R. J., Brown, E., Shoemaker, C., Orr, E. C., Amphlett, G. W., Foster, W. B., Coe, M. L., Knutson G. J., Fass, D. N., and Hewick, R. M. (1984). *Nature* **312,** 342–347.

Toth, J., Gelinas, A. D., Bethoney, K. A., Bard, J., Stafford, W. F., III, Cutforth, T., and Harrison, C. J. (2001). "Abstracts of 15th West Coast Protein Crystallography Workshop," Pacific Grove, CA, March 25–28.

Trost, J. T., McManus, J. D., Freeman, J. C., Ramakrishna, B. L., and Blankenship, R. E. (1988). *Biochemistry* **27,** 7858–7863.

Ubbink, M., Ejdebäck, M., Karlsson, B. G., and Bendall, D. S. (1998). *Structure* **6,** 323–335.

Vakoufari, E., Wilson, K. S., and Petratos, K. (1994). *FEBS Lett.* **347,** 203–206.

Vallee, B. L., and Williams, R. J. P. (1968). *Proc. Natl. Acad. Sci. USA* **59,** 498–505.

Van Driessche, G., Dennison, C., Sykes, A. G., and van Beeumen, J. (1995). *Protein Sci.* **4,** 209–227.

Van Driessche, G., Hu, W., van de Werken, G., Selvaraj, F., McManus, J. D., Blankenship, R. E., and van Beeumen, J. J. (1999). *Protein Sci.* **8,** 947–957.

Van Gysel, A., Van Montagu, M., and Inze, D. (1993). *Gene* **136,** 79–85.

Van Spanning, R. J., Wansell, C. W., Reijnders, W. N., Oltmann, L. F., and Stouthamer, A. H. (1990). *FEBS Lett.* **275,** 217–220.

Van Waasbergen, L. G., Hildebrand, M., and Tebo, B. M. (1996). *J. Bacteriol.* **178,** 3517–3530.

Vehar, G. A., Keyt, B., Eaton, D., Rodriguez, H., O'Brien, D. P., Rotblat, F., Oppermann, H., Keck, R., Wood, W. I., Harkins, R. N., Tuddenham, E. G. D., Lawn, R. M., and Capon, D. J. (1984). *Nature* **312,** 337–342.

Villoutreix, B. O., and Dahlbäck, B. (1998). *Protein Sci.* **7,** 1317–1325.

Vision, T. J., Brown, D. G., and Tanksley, S. D. (2000). *Science* **290,** 2114–2117.

Vulpe, C., Levinson, B., Whitney, S., Packman, S., and Gitschier, J. (1993). *Nat. Genet.* **3,** 7–13.

Wahleithner, J. A., Xu, F., Brown, K. M., Brown, S. H., Golightly, E. J., Halkier, T., Kauppinen, S., Pederson, A., and Schneider, P. (1996). *Curr. Genet.* **29,** 395–403.

Wang, D. Y. C., Kumar, S., and Hedges, S. B. (1999). *Proc. R. Soc. London Ser. B Biol. Sci.* **266,** 163–171.

Weterings, K., Wim, R. L., van Aarssen, R., Kortstee, A., Spijkers, J., van Herpen, M., Schrauwen, J., and Wullems, G. (1992). *Plant Mol. Biol.* **18,** 1101–1111.

Wilkinson, D. G. (2001). *Nat. Rev. Neurosci.* **2,** 155–164.

Williams, R. J. P. (1995). *Eur. J. Biochem.* **234,** 363–381.

Wood, P. M. (1978). *Eur. J. Biochem.* **87,** 9–19.

Wood, W. I., Capon, D. J., Simonsen, C. C., Eaton, D. L., Gitschier, J., Keyt, B., Seeburg, P. H., Smith, D. H., Hollingshead, P., Wion, K. L., Delwart, E., Tuddenham, E. G. D., Vehar, G. A., and Lawn, R. M. (1984). *Nature* **312,** 330–337.

Wuttke, D. S., and Gray, H. B. (1993). *Curr. Opin. Struct. Biol.* **3,** 555–563.

Xu, F. (1996). *Biochemistry* **35,** 7608–7614.

Xu, F. (1997). *J. Biol. Chem.* **272,** 924–928.

Yaver, D. S., and Golightly, E. J. (1996). *Gene* **181,** 95–102.

Yoshida, K., Furihata, K., Takeda, S., Nakamura, A., Yamamoto, K., Morita, H., Hiyamuta, S., Ikeda, S., Shimizu, N., and Yanagisawa, N. (1995). *Nat. Genet.* **9,** 267–272.

Yoshizaki, M., Furumoto, T., Hata, S., Shinozaki, M., and Izui, K. (2000). *Biochem. Biophyys. Res. Commun.* **268,** 466–470.

Zaitseva, I., Zaitsev, V., Card, G., Moshkov, K., Bax, B., Ralph, A., and Lindley, P. (1996). *J. Biol. Inorg. Chem.* **1,** 15–23.

Zhang, Y., Sederrof, R. R., and Allona, I. (2000). *Tree Physiol.* **20,** 457–466.

Zhu, Z. Y., Cunane, L. M., Chen, Z. W., Durley, R. C. E., Mathews, F. S., and Davidson, V. L. (1998). *Biochemistry* **37,** 17128–17136.

# CYTOCHROME *c* OXIDASE

SHINYA YOSHIKAWA

Department of Life Science, Himeji Institute of Technology, and CREST, Japan Science and Technology Corporation (JST), Kamigohri Akoh, Hyogo 678–1297, Japan



## I. Introduction

Cytochrome *c* oxidase has four redox-active metal sites, hemes a and $a_3$, $Cu_A$ and $Cu_B$ (Ferguson-Miller and Babcock, 1995, Malmström, 1990). Both prosthetic groups of hemes a and $a_3$ sites are heme A, which is a derivative of protoheme in which the vinyl group at position 2 of the






0065-3233/02 $35.00

porphyrin periphery is replaced with a hydroxyl farnesyl-ethyl group and the methyl group at position 8 is replaced with a formyl group (Fig. 1) (Caughey *et al.*, 1975). The electronic effects of the formyl group at position 8 are the origin of the very bright blue-green color when heme A is in the ferrous low-spin state. When the enzyme is fully reduced, heme A is found in a six-coordinate state in the heme a site and in a five-coordinate state in the heme $a_3$ site. Heme $a_3$ is the site for binding of $O_2$ and respiratory inhibitors such as CO, cyanide ($CN^-$), and azide ($N_3^-$). $Cu_B$ is located close to heme $a_3$ and thus appears to play an important role in the reduction of $O_2$ bound at heme $a_3$ (Caughey *et al.*, 1976). Thus, the site including two metals is called the dioxygen reduction site. The direct electron-accepting site from cytochrome *c* is $Cu_A$, a dinuclear copper site (Tsukihara *et al.*, 1995, Iwata *et al.*, 1995). However, it is well established that this site is reversibly reduced by one electron equivalent (Kronech *et al.*, 1988, 1990). Electrons are transferred from $Cu_A$ to heme $a_3$ via heme a, which is fixed by two histidine imidazole groups (Hill, 1994).

This enzyme is embedded in the mitochondrial inner membrane in eukaryotic cells and in the cytoplasmic membrane in prokaryotic cells (Ferguson-Miller and Babcock 1995, Malmström, 1990). Cytochrome *c* oxidase reduces dioxygen ($O_2$) to water with electrons from cytochrome *c*

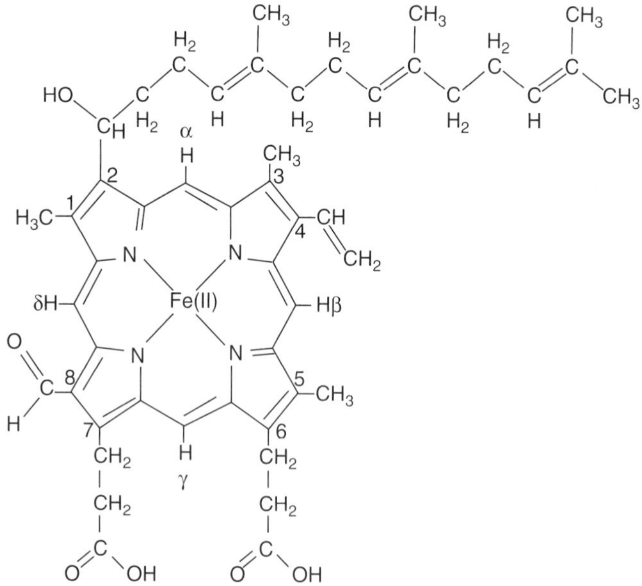

FIG. 1. Structure of heme A. This heme is characterized by a hydroxyfarneslyethyl group at position 2 and a formyl group at position 8.

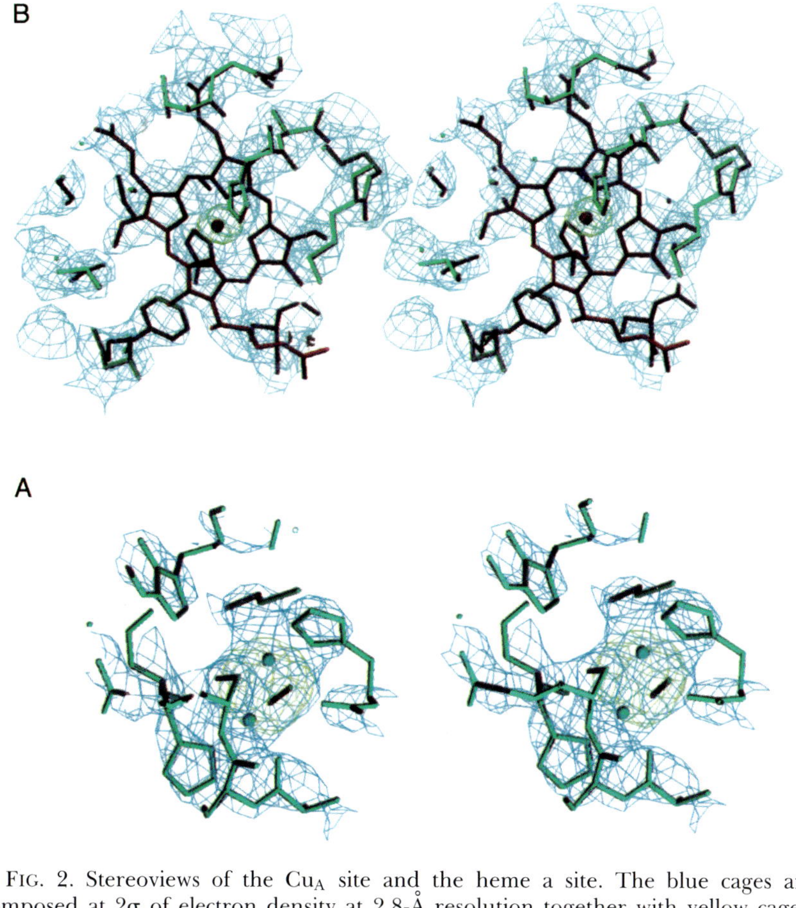


FIG. 2. Stereoviews of the $Cu_A$ site and the heme a site. The blue cages are composed at $2\sigma$ of electron density at 2.8-Å resolution together with yellow cages, drawn at $10\sigma$ of the native anomalous difference Fourier at 4.5-Å resolution. (A) $Cu_A$ site: the two blue balls show the positions of two copper atoms. (B) Heme a site: the dark structure represents the porphyrin plane with a dark ball showing the position of iron.

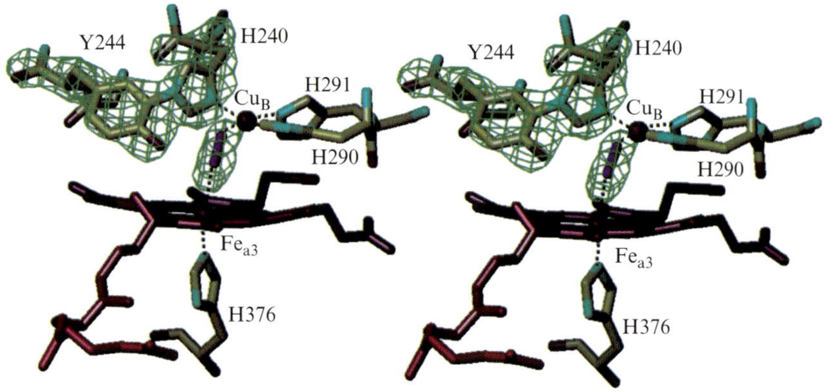


FIG. 3. X-ray structure of the $O_2$ reduction site of bovine heart cytochrome *c* oxidase in the fully oxidized state at 2.3-Å resolution. The cages show the $|F_o - F_c$ difference Fourier map of the oxidized form calculated omitting His-240, Tyr-244, and any ligand between $Fe_{a3}$ and $Cu_B$ from the $F_c$ calculation. The cages are drawn at the $7\sigma$ level ($1\sigma = 0.00456\,e^-/\text{Å}^3$).

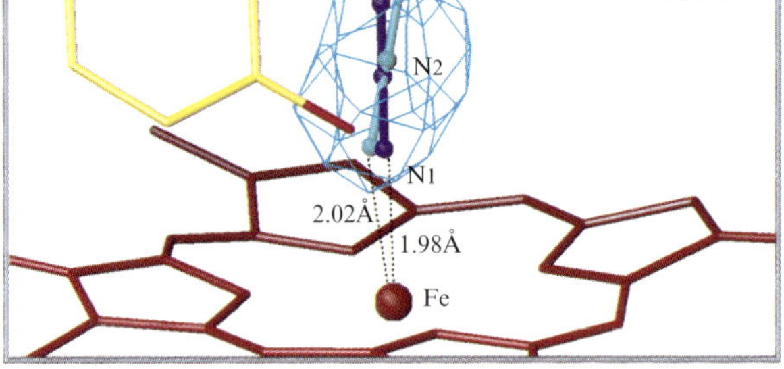


FIG. 5a. X-ray structures of bound respiratory inhibitors. (A) Two possible azide-binding modes to the $O_2$ reduction site. A difference electron density map for the azide bound at the $O_2$ reduction site obtained from the crystals at 2.9-Å resolution. The electron density cage is drawn at the $4\sigma$ level.

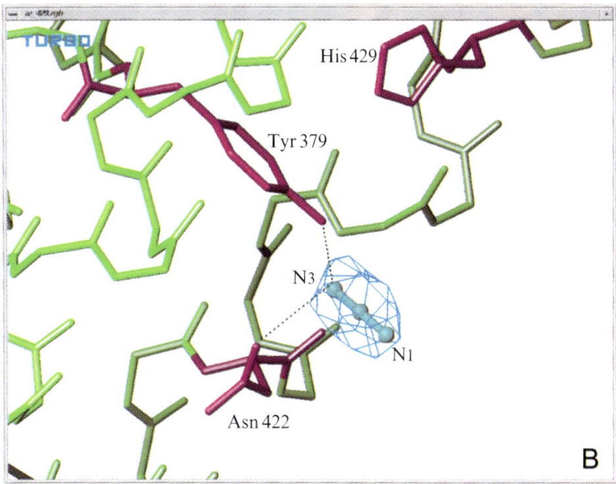


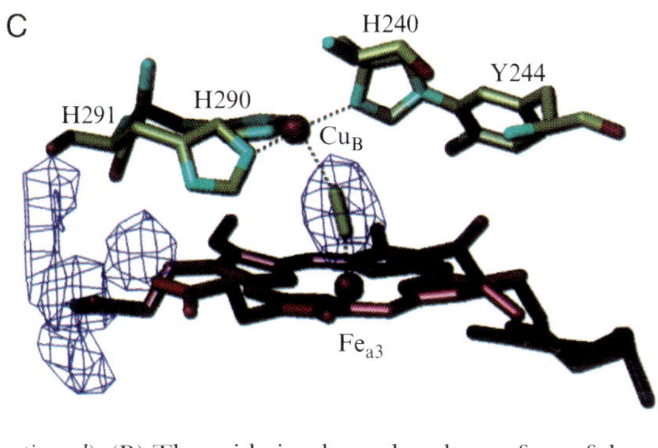


FIG. 5b, c (*continued*) (B) The azide ion bound at the surface of the transmembrane region. The azide is terminally hydrogen-bonded with Tyr-379 and Asn-422. A difference density map was obtained at the same resolution as in (A). (C) A difference electron density map for CO bound at $Fe_{a3}$ at the $3\sigma$ level was obtained from the crystals of the CO derivative at 2.8-Å resolution.

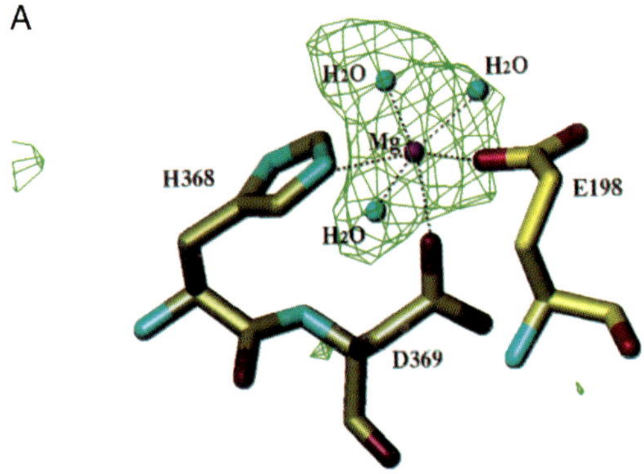


FIG. 6a. X-ray structures of redox-inactive metal sites. (A) A difference electron density map for $Mg^{2+}$ and its water ligands at the $4\sigma$ level ($1\sigma = 0.0436\,e^-/\text{Å}^3$) obtained from crystals of the fully oxidized form at 2.3-Å resolution. The two amino acids H368 and D369 belong to subunit I and amino acid E198 belongs to subunit II.

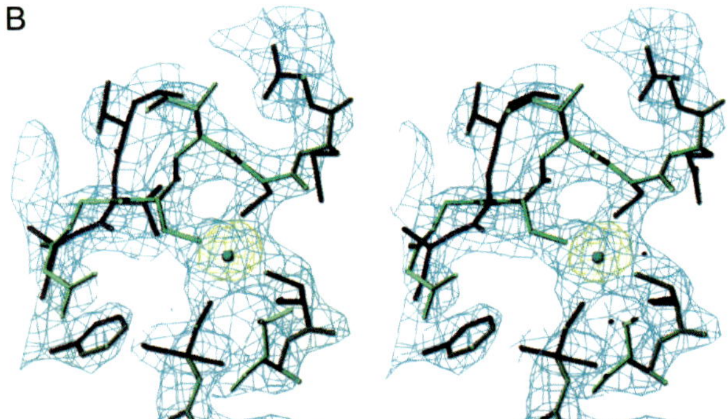


FIG. 6b. (B) Stereo pairs of the Zn site fitted in blue cages composed at $2\sigma$ of electron density at 2.8-Å resolution together with yellow cages, drawn at $10\sigma$, of the native anomalous difference Fourier at 4.5-Å resolution.

(on the intermembrane side of mitochondria or the periplasmic side of prokaryotic cells). Protons are supplied from the inside of the membrane (the matrix space of mitochondria or the cytosolic space of prokaryotic cells). Thus, the $O_2$ reduction in this enzyme is an electrogenic four-electron reduction process that produces an electrochemical potential or a proton motive force across the membrane. This electrogenic process is coupled with proton translocation across the membrane (Wikström, 1977). It is widely accepted that four proton equivalents are transferred (pumped) across the membrane coupled with complete reduction of each equivalent of $O_2$ to two equivalent of water (Babcock and Wikström, 1992; Babcock *et al.*, 1995; Malmström, 1990). The proton motive force created by $O_2$ reduction and proton pumping is utilized for ATP formation by ATP synthase. Thus, cytochrome *c* oxidase is a key enzyme in cell respiration. In fact, it has been estimated that 90% of biological oxygen consumption is carried out by this enzyme (Malmström, 1973).

Warburg (1924) was responsible for the identification of this enzyme as a hemoprotein via the photochemical action spectrum for CO-inhibited yeast cells about 70 years ago. Later, Keilin and Hartree (1939) discovered two types of *a*-type cytochromes, *a* and $a_3$, based on the difference in the inhibitor-binding ability and the autoxidizability, and they identified cytochrome $a_3$ as the enzyme, the CO compound for which Warburg had shown in his photochemical action spectrum. Furthermore they showed that only the cytochrome *a* portion oxidized cytochrome *c* and reduced cytochrome $a_3$. Thus, they named the two cytochromes (*a* and $a_3$) cytochrome *c* oxidase (Keilin and Hartree 1938a). Because of the extremely strong absorption of hemes in the enzyme, copper sites were discovered about 22 years after the discovery of the two *a*-type cytochromes in the isolated enzyme (Takemori, 1960; Griffths and Wharton, 1961). Another important finding regarding the function of this enzyme is the proton-pumping ability shown conclusively by Wikström, in 1977. The intriguing $O_2$ reduction and proton-pumping functions of this enzyme and the physiological importance of this enzyme have made cytochrome *c* oxidase one of the most extensively and actively studied enzymes in the history of biochemistry. Determinations of atomic-resolution X-ray structures of cytochrome *c* oxidase from bovine heart and bacteria in 1995 (Tsukihara *et al.*, 1995, Iwata *et al.*, 1995) have greatly stimulated this research field. In this chapter, the contributions of X-ray structural information for understanding the mechanism of this enzyme will be discussed, as will the importance of careful work in the determination of chemical structures, which is often tedious and laborious.

The catalytic sites of most enzymes are not exposed to the bulk water phase. Thus, substrates fixed at a catalytic site are found in an extremely anisotropic environment, which is impossible to attain in homogeneous chemical reaction systems in solution. Thus, X-ray structures of the catalytic site including the prosthetic group are indispensable for elucidation of the reaction mechanism of any enzyme. In fact, the level of understanding of an enzyme reaction mechanism is usually determined by the level of the resolution of the change in the three-dimensional structure of the catalytic site during the course of the enzyme reaction. Thus, for elucidation of the mechanism of an enzyme, the first step, which is very often the most difficult, is purification of the enzyme from the cell in its intact state for crystallization. Establishment of the purification procedure for an enzyme is also critical for determination of the chemical composition of the enzyme and for collection of the reproducible data required to study its catalytic function. The cytochrome *c* oxidases, particularly those contained in mammalian cells, are extremely large multicomponent membrane proteins. Much effort over many years was required for the establishment of the purification method.

## II. Purification and Crystallization

### *A. Purification of Bovine Heart Cytochrome c Oxidase*

The first isolation of cytochrome *c* oxidase was carried out from bovine heart muscle by Yakushiji and Okunuki as early as 1941. They used a natural detergent, sodium cholate, to solubilize the enzyme from the mitochondrial inner membrane, followed by ammonium sulfate fractionations in the presence of the same detergent to remove contaminant proteins in the solubilized extract. However, the spectroscopically pure isolated enzyme preparation showed essentially no enzyme activity. Later, in 1959, Yonetani, who was a graduate student of Okunuki, discovered that various synthetic non-ionic detergents with much lower critical micelle concentrations than that of cholate displaced sodium cholate effectively to restore the enzyme activity. This work showed that the cytochrome *c* oxidase preparation first isolated in 1941 was enzymatically active but inhibited by sodium cholate. Many isolation procedures have been reported after Yonetani's report. The number of methods for purification of bovine heart cytochrome *c* oxidase may have been as large as the number of research groups studying this enzyme (Caughey *et al.*, 1976). Unfortunately, the spectral and enzymatic properties of the enzyme isolated via the various methods were not completely consistent with one another, suggesting that most of the preparations did not represent the native state of the enzyme. In fact, the efficiency of the detergent treatment for solubilization of a membrane protein is species dependent and very sensitive to various factors including pH, temperature, and concentrations of the detergent and protein. Thus, establishment of a highly reproducible purification method giving a homogeneous preparation of any membrane protein is usually very difficult. Furthermore, no criteria for evaluation of the qualities of these preparation were available before crystallization conditions of the enzyme were established in 1995 (Tsukihara *et al.*, 1995; Iwata *et al.*, 1995). Crystallization is the best method for removing denatured protein molecules, and thus the molecular conformation of a preparation purified via crystallization is likely to represent the native conformation. It should be noted that sodium cholate is still one of the most effective detergents thus far examined for solubilization of bovine heart enzyme from the mitochondrial inner membrane without denaturation. Because of its high critical micelle concentration, cholate in the solubilized enzyme fraction is quite effectively replaced with many synthetic non-ionic detergents. The best crystals of bovine heart cytochrome *c* oxidase thus far are obtained from the enzyme preparation isolated by the procedure essentially identical to the 60-year-old Yakushiji–Okunuki method (Yakushiji and Okunuki, 1941).

The above purification method for crystalline bovine heart cytochrome *c* oxidase is as follows [more details are given in Yoshikawa *et al.* (1977)]. Fresh bovine heart muscle, free of fat and connective tissue, taken from one beef heart (about 850 g), is minced and suspended in 20 mM sodium phosphate, pH 7.4, and homogenized mechanically. The homogenate is centrifuged to remove cell debris, leaving the mitochondrial inner membrane fraction in the supernatant. The mitochondrial membrane fraction is collected by centrifugation at the same speed as for the first centrifugation after the pH of the supernatant is lowered with acetic acid. The precipitate is suspended in 0.1 M sodium phosphate buffer, pH 7.4. This is essentially the same procedure as that of Keilin and Hartree (1938b), which eliminates a step in which minced muscle is washed with a large volume of cooled water. In fact, a large part of the mitochondrial inner membrane fraction is lost in the washing step, giving a very low yield of the final preparation: about 10% or lower (unfortunately, once a procedure is published, it is rarely improved upon). The mitochondrial inner membrane fraction is then treated with 3.3% sodium cholate in the presence of ammonium sulfate at 33% saturation for extraction of cytochrome *c* oxidase. The extract containing various other membrane proteins embedded in the mitochondrial inner membrane in addition to cytochrome *c* oxidase is obtained as the supernatant by high-speed centrifugation. These contaminant proteins are effectively removed by repeated ammonium sulfate fractionations in the presence of sodium cholate, followed by ammonium sulfate fractionation in the presence of a non-ionic detergent such as alkyl polyoxyethylene or sugar alkyl-type detergents. The final preparation dissolved in sodium phosphate buffer, pH 7.4, containing a non-ionic detergent is concentrated with an Amicon diaflow apparatus for crystallization. Crystals of the enzyme appear without amorphous precipitate. Thus, the crystals can easily be collected by centrifugation. Optimal crystals are obtained from the enzyme preparation stabilized with decylmaltoside (Tsukihara *et al.*, 1995). About 0.6 g of crystalline sample is obtained from about 850 g of beef heart muscle in 4 working days. This method, which does not require column chromatography, provides a high yield in a short working time.

### *B. Homogeneity of the Fully Oxidized $O_2$ Reduction Site*

It is well known that the $O_2$ reduction site of bovine heart cytochrome *c* oxidase in the fully oxidized state exhibits variable reactivity to cyanide and ferrocytochrome *c*, which is dependent on the method of purification (Moody, 1996). Some preparations react with cyanide extremely slowly at an almost immeasurable rate and are known as the "slow" form. Other preparations, which react at a half-life of about 30 s, are known as the "fast" form (Brandt *et al.*, 1989). Electronic absorption spectra of the slow- and fast-form preparations exhibit Soret bands at 418 and 424 nm, respectively. The two forms often coexist in a single preparation (Baker *et al.*, 1987). Both forms exhibit an identical visible–Soret spectrum in the fully reduced state. The slow-form preparation can be converted to the fast form by dithionite reduction followed by reoxidation with $O_2$. The fast form thus obtained returns to the slow form spontaneously at a rate much slower than the enzymatic turnover rate. Thus, the slow form is unlikely to be involved in the enzymatic turnover (Antonini *et al.*, 1977). It should be noted that no clear experimental evidence has been reported for direct involvement of the fast form in the enzyme turnover, although its direct involvement has been widely accepted. The third species of the fully oxidized $O_2$ reduction site, which appears in the partially reduced enzyme, reacts with cyanide $10^3$–$10^4$ times more rapidly than the fast form (Jones *et al.*, 1984). In the absence of a reducing system, no interconversion is detectable between the slow and the fast forms (Brandt *et al.*, 1989). Thus, the heterogeneity is expected to inhibit the crystallization of this enzyme. In fact, the enzyme preparations providing crystals showing X-ray diffraction at atomic resolution are the fast form preparation.

### *C. Crystallization of Membrane Proteins*

The most critical parameter for crystallization of any class of protein, membrane bound or not, is homogeneity of its three-dimensional molecular structure. According to Anfinsen (1973), the three-dimensional structure of a protein in the physiologically active state is the thermodynamically most stable structure. Thus, any protein is likely to be in an essentially unique conformation in the cell and ready to be crystallized if it is isolated from the cell without denaturation. Any integral protein has unique properties in solution, including isoelectric point, solubility, sedimentation behavior, and affinities to various ligands. On the other hand, denatured protein molecules are unlikely to be in a unique conformation to provide a homogeneous solution property since they are not in the most thermodynamically stable state. Thus, proteins in the denatured state are usually difficult to isolate from the cell extract. Improvement of purification procedures for minimizing denaturing effects typically increases the yield of the preparation. Modification of the three-dimensional structure of a protein influences crystallization conditions very sensitively. For example, our purification procedure for bovine heart cytochrome *c* oxidase provides perfect reproducibility in terms of the yield and the spectral and enzymatic properties. However, the quality of the crystals greatly depends on the batch of the purified preparation, and the X-ray diffraction resolution may vary from 1.8 to 2.8 Å.

Recently, various equipment and reagents for crystallization have become commercially available. However, if the purified protein is not homogeneous and intact, the modern equipment and reagents provide no advantage. The importance of improving purification methods for protein crystallization is often not recognized.

Most of the membrane proteins embedded in biological membranes have fairly large hydrophilic membrane surface moieties. Thus, membrane proteins are stabilized effectively by exposing the molecular surface of the transmembrane portion to a hydrophobic environment and the extramembrane regions to a hydrophilic environment. In other words, membrane proteins are unstable in both organic and aqueous media. Thus, the best way to isolate membrane proteins from biological membranes is to solubilize the proteins with detergents that can provide local hydrophobic environments that stabilize the hydrophobic surface of the protein (Garavito and Picot, 1990). In fact, most crystals of membrane proteins that diffract X rays up to atomic resolution are solubilized with detergents (Landau and Rosenbusch, 1996). During the solubilization process, membrane phospholipids interacting with the hydrophobic protein surface are exchanged with detergent molecules. Thus, differences in the phospholipids and detergents could influence the stability of the isolated membrane proteins. In fact, the structure of detergents strongly affects crystallization conditions (Shinzawa-Itoh *et al.*, 1995). For example, a difference in two methylene units ($-CH_2-$) in the alkyl chain of an alkyl maltoside was critical for crystallization of bovine heart cytochrome *c* oxidase, although stability of the isolated preparation was scarcely influenced by the structural difference (S. Yoshikawa, unpublished results). Recently many synthetic detergents have become commercially available. However, the specificity of the detergent structure is likely to be extremely high and greatly dependent on the protein species. Thus, the design and synthesis of detergents for each membrane protein are desirable for the crystallization process.

It should be noted that crystallization is the best method for isolation of large multicomponent proteins such as cytochrome *c* oxidase. Any isolation method other than crystallization typically results in denaturation of the protein. For example, in cation-exchange column chromatography, the distribution of the negatively charged groups on the surface of the matrix of the column is not precisely matched to the distribution of positive charges on the protein surface. Furthermore, unfavorable interactions between the hydrophobic surface of the protein and the matrix of the chromatographic column are likely to occur. These electrostatic and hydrophobic interactions may cause destabilization of the integrity of the three-dimensional structure. In small-protein purification, a small amount of damage in the three-dimensional structure could decrease the solubility in aqueous solution significantly. Thus, the denatured enzyme molecules are easily removed. However, large proteins with many subunits not very tightly bound near the molecular surface are likely to suffer denaturation or deletion of small polypeptides by such purification procedures without a significant decrease in solubility. Such slightly denatured proteins surely inhibit crystallization and they are very difficult to remove. Thus, crystallization is crucial for determination of the composition of large multicomponent proteins as described in the next section.

## III. Composition of Bovine Heart Cytochrome *c* Oxidase

In order to elucidate the reaction mechanism of cytochrome *c* oxidase, the complete structure of the enzyme must be determined. The first step in this process is the complete determination of its composition. X-ray crystallographic analysis at high resolution was required in addition to chemical analysis for crystalline enzyme preparation.

### *A. Metal Composition*

The involvement of copper ions as the prosthetic group of this enzyme was recognized as a result of chemical and spectroscopic (especially copper EPR) analyses of the purified preparation during the years

1958–1961 (Okunuki *et al.*, 1958; Takemori, 1960; Griffiths and Wharton, 1961; Yonetani, 1961). Several reports on iron and copper analysis have been published (Okunuki *et al.*, 1958; Takemori, 1960; Griffiths and Wharton, 1961; Yonetani, 1961; Yoshikawa *et al.*, 1977, 1988; Steffens *et al.*, 1987; Mochizuki *et al.*, 1999). Cytochrome *c* oxidase functions as a copper chelator (Yuan *et al.*, 1998). Thus, the earlier papers reported fairly high values for the Cu/Fe ratio, such as 4.5 (Okunuki *et al.*, 1958) and 1.7 (Yonetani, 1961). However, such adventitious copper is readily removed by treatment with EDTA. On the other hand, bovine heart cytochrome *c* oxidase preparations purified by any method that does not include a crystallization step contain significant amounts of contaminant iron, which is strongly bound to the protein. The contaminant iron ions can be removed only with crystallization, as shown in Table I, indicating that the contaminating metals were bound to contaminating proteins (Mochizuki *et al.*, 1999). The iron content of the twice-crystallized preparation is typically identical to that of the preparation crystallized three times. Thus, no contaminant iron is likely to be contained in the twice- and thrice-crystallized samples. The contents of other metals do not vary with repeated crystallization. Any constituent not removed with repeated crystallization is typically part of the intact protein. Thus, Table 1 exhibits the metals that are intrinsic components of bovine heart cytochrome *c* oxidase. It has been proposed that zinc and magnesium are intrinsic components of bovine heart cytochrome *c* oxidase based on the finding that both metals are detectable consistently in the purified preparation without crystallization (Einarsdottir and Caughey, 1984, 1985).

TABLE I

*Effect of Crystallization on the Metal Content of Bovine Heart Cytochrome c Oxidase*[a]

| Preparation | Atoms / enzyme molecule | | | | $\varepsilon^{red}_{604-630\,nm}$ ($mM^{-1}cm^{-1}$) |
|---|---|---|---|---|---|
| | Fe | Cu | Mg | Zn | |
| Before crystallization[b] | 2.90 | 2.90 | 1.04 | 0.97 | 32.2 |
| Once-crystallized[b] | 2.21 | 3.04 | 1.07 | 1.01 | 42.2 |
| Twice-crystallized[b] | 2.06 | 3.00 | 1.06 | 1.00 | 45.6 |
| Thrice-crystallized[b] | 2.00 | 2.95 | 1.04 | 1.04 | 46.6 |
| Twice-crystallized, averaged[c] | 2.00 (0.07) | 2.95 (0.12) | 1.02 (0.08) | 1.04 (0.04) | 46.6 (1.16) |

[a]All of the values were determined assuming that the thrice-crystallized sample contains two iron atoms per molecule of enzyme.

[b]Enzyme stabilized with $CH_3(CH_2)_{11}(OCH_2CH_2)_8OH$ was used.

[c]Averaged values with (SD) for three and four enzyme preparations stabilized with $CH_3(CH_2)_{11}(OCH_2CH_2)_8OH$ and $CH_3(CH_2)_{12}(OCH_2CH_2)_{23}OH$, respectively.

The physiological roles of these metals have not been identified. Thus, the possibility exists that these metals are copurified contaminants. Table I indicates the amounts of metals insensitive to repeated crystallization; this provides strong evidence that they are intrinsic components. This is the case for the amount of copper (3 Cu/enzyme), which is consistent with the X-ray structure as described below.

An X-ray structure of the fully oxidized enzyme at 2.3-Å resolution shows a sodium site near the intermembrane side as described below (Yoshikawa *et al.*, 1998). Prior to publication of the X-ray structure, a sodium site had not been considered. This result shows that X-ray structural analysis is indispensable for determination of the metal content of proteins.

### *B. Subunit Composition*

Kadenbach *et al.* (1983) proposed that bovine heart cytochrome *c* oxidase contains 13 different protein subunits. The only experimental result supporting this proposal was the finding that 13 different subunits are reproducibly detectable in SDS–PAGE. The physiological roles of these subunits (other than the two subunits containing the redox-active metal sites) are not yet known. As a result, the proposal was not well accepted until the X-ray structure of bovine heart cytochrome *c* oxidase at 2.8-Å resolution was published, and it was found that these subunits bind specifically to the two subunits containing the catalytic sites (Tsukihara *et al.*, 1995). The specific assembly is the only experimental result showing that these subunits without the transition metals are intrinsic components of this enzyme.

### *C. Phospholipids*

It has long been proposed that cytochrome *c* oxidase contains various phospholipids as intrinsic constituents and that two or three cardiolipins are essential for its catalytic activity (Robinson *et al.*, 1980; Robinson, 1982; Sedlák and Robinson, 1999). However, the composition and structures of phospholipids in the enzyme are still matters of controversy. No experimental evidence has been obtained to discount the proposal that all phospholipids suggested are not intrinsic components of the enzyme. Thus, a careful phospholipid analysis of the crystalline enzyme and an X-ray structure determination are indispensable for the determination of the phospholipid composition.

An X-ray structure of bovine heart cytochrome *c* oxidase at 2.8-Å resolution showed eight phospholipid molecules, including five phosphatidylethanolamines and three phosphatidylglycerols (Tsukihara *et al.*,

1996). The fatty acid tails are inserted into the transmembrane region of the protein and the head groups are placed on either one of the levels of interfaces between the transmembrane region and the two hydrophilic moieties on both sides of the protein. The specific binding of these phospholipids to the protein subunits strongly suggests that these phosphlipids are not contaminants but intrinsic constituents similar to cofactors of this enzyme. Some of the fatty acid tails near the transmembrane region are unusually short and appear to be mobile. No cardiolipin is detectable in the X-ray structure at this resolution. However, some of the phosphatidylglycerols near the molecular surface may actually be exposed parts of cardiolipins, which are, for the most part, invisible in the X-ray structure due to their mobility. Thus, determination of the chemical structures of phospholipids by methods other than X-ray structural analysis, such as mass spectrometric analysis, is useful for acquiring a complete picture of the phospholipids. In addition, identification of unsaturated C-C bonds in the fatty acid tails by mass spectral analysis is also desirable since resolution of electron density high enough for identification of the unsaturated bonds is not easy to attain by X-ray methods.

## IV. X-Ray Structures of Cytochrome *c* Oxidase

X-ray structures of 2.8-Å resolution of bovine heart cytochrome *c* oxidase with the metals in the fully oxidized state were reported in 1995 (Tsukihara *et al.*, 1995). The X-ray structure of cytochrome *c* oxidase from *Paracoccus denitrificans* in the fully oxidized azide-bound state at 2.8-Å resolution was also published in the same week (Iwata *et al.*, 1995). The structure and location of the metal sites of the two enzymes are astonishingly similar at that resolution. Later, the resolution of the bovine enzyme structure was improved to 2.3Å (Yoshikawa *et al.*, 1998). However, resolution of the *Paracoccus* enzyme has been improved to 2.7-Å resolution (Ostermeier *et al.*, 1997). Recently another bacterial ba3-type oxidase at 2.3-Å resolution (Soulimane *et al.*, 2000) and *Escherichia coli* quinol oxidase at 3.5-Å resolution were reported (Abramson *et al.*, 2000). X-ray structures of the protein and its redox-active metal sites are discussed in terms of the bovine enzyme below.

### *A. X-Ray Structures of Redox-Active Metal Sites*

At 2.8-Å resolution, the electron density of $Cu_A$ is spherical (Fig. 2A, see color insert) (Tsukihara *et al.*, 1995). However, the three-dimensional organization of the amino acids surrounding the spherical electron density strongly suggests that the $Cu_A$ site is a dinuclear copper center. The two copper atoms are bridged by two cysteine thiols and provide a square planar arrangement. Each copper ion has two other amino acid-derived ligands in a tetrahedral arrangement with the copper atom placed at the geometrical center of the tetrahedral. The two tetrahedrals are symmetrically opposed to each other. However, at 2.3-Å resolution, significant distortion from the very symmetrical arrangement observed at 2.8-Å resolution is detectable. The observation of the dinuclear structure is fully consistent with the multifrequency EPR data that suggest a dinuclear copper center with one electron delocalized between the two copper atoms. This structure is also consistent with the redox titration data indicating that $Cu_A$ is a single electron-accepting site.

The dinuclear copper center is in the extramembrane moiety of subunit II, while the other three redox-active metal sites are located in subunit I, the largest subunit with 12 transmembrane helices. It should be noted that the location of $Cu_A$, which is the direct electron acceptor from cytochrome c, conclusively shows that the molecular surface is on the intermembrane space side. Thus the extramembrane moiety on the other side is exposed to the matrix space of mitochondria. One of the hemes is in a six-coordinated state with two axial imidazole ligands (Fig. 2B, see color insert), while the other has only one imidazole ligand and a copper is placed very near the heme (Fig. 3, see color insert) (Yoshikawa *et al.*, 1998). Thus, the former is assigned to heme a and the latter to heme $a_3$. The copper should be $Cu_B$. The hydroxyl farnesylethyl group of heme a is essentially in a fully extended conformation while that of heme $a_3$ is in a U-shaped conformation. Both heme planes are arranged vertically to the membrane surfaces, which are estimated by the locations of the transmembrane helices. All four propionate groups are directed toward the intermembrane side. It should be noted that perhaps no method other than X-ray crystallography is able to determine the orientation of the heme plane and the conformation of the hydroxyl farnesylethyl group at position 2 of the heme periphery.

The role of the long alkyl side chain at position 2 of heme a has been an intriguing question since the determination of the structure of heme A about 26 years ago by the careful and extensive work of Caughey *et al.* (1975). Based on the structure they determined, they proposed that the alkyl side chain provided a pathway for electron transfer that could be precisely controlled by the conformation of the chain (Fig. 4A) (Caughey *et al.*, 1975). On the other hand, Woodruff *et al.* (1991) proposed that coordinations of the C–C double bond in the terminal isoprenoide unit to both metals could provide a strategy for proton pumping of this enzyme (Fig. 4B). Both proposals are intriguing, and determination of the conformation of the side chain has been an important subject in this field. Unfortunately the X-ray structure does not show any of the suggested

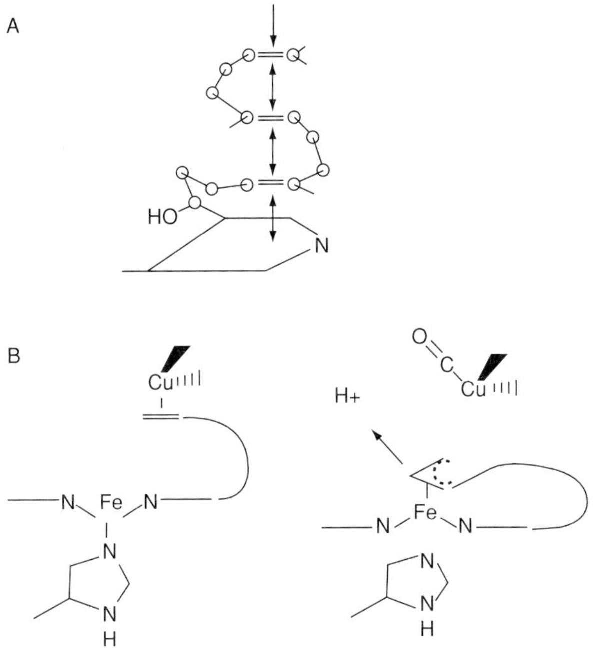


FIG. 4. Proposed functions of the hydroxyfarnesylethyl group of heme A. (A) A possible electron transfer pathway formed by overlapping of π-electron orbitals in the alkyl chain with that of the pyrrole. (B) A side-on coordination of the terminal double bond to $Cu_B$ and a coordination of a deprotonated form of the double bond to $Fe_{a_3}$. A proposed proton-pumping mechanism including the redox-coupled change in the two coordination states.

structures in either of the heme a groups in this enzyme. However, the unique structure of the side chains, which no other constituent in this enzyme contains, must have some important structural or functional role yet to be elucidated.

The second copper center near the heme $a_3$ iron, which is readily assignable as $Cu_B$, has three histidine imidazole groups as the ligands (Fig. 3). The copper atom in this site is located approximately at the center of the triangle formed by three imidazole nitrogen atoms coordinated to the copper atom. The position of the copper atom is about 1.0 Å from the heme normal for heme $a_3$. The $Cu_B$–$Fe_{a_3}$ distance is 4.7 Å. In the fully oxidized state, a peroxide is bridged between the two metal atoms (Fig. 3). The $|F_o - F_c|$ difference Fourier map of the oxidized form at 2.3-Å resolution indicates that the $Cu_B$–O distance is 2.16 Å and the $Fe_{a_3}$–O distance is 2.52 Å (Yoshikawa *et al.*, 1998) The latter is significantly longer than a normal metal coordination bond. Thus it is possible that a hydroperoxo ligand bound at $Cu_B$(H–O–O–$Cu_B$) is coordinated to $Fe_{a_3}$. The weaker coordination to $Fe_{a_3}$ is consistent with the high spin state of $Fe_{a_3}$ (Van Gelder and Beinert, 1969; Tweedle *et al.*, 1978; Moss *et al.*, 1978). One of the unexpected findings in X-ray structural analysis of bovine heart cytochrome *c* oxidase is a covalent link between a nitrogen atom of one of the imidazoles of histidines bound to $Cu_B$ (His-240) and an *ortho* carbon of the phenol group of a tyrosine (Tyr-244) as shown in Fig. 3 (Yoshikawa *et al.*, 1998). The cross link was detectable only in the fully oxidized and reduced states at 2.3- and 2.35-Å resolution, respectively, though the electron density maps at 2.8- to 2.9-Å resolution for the two amino acid side chains are consistent with a direct C–N single bond (Ostermeier *et al.*, 1997). With this bond, Tyr-244 is fixed near the $O_2$ reduction site. The significance of this structure will be discussed below.

In the fully reduced state, ligands are not detectable between the two metals (Yoshikawa *et al.*, 1998). By complete reduction of the fully oxidized form to the fully reduced form, the $Fe_{a_3}$–$Cu_B$ distance is increased from 4.9 to 5.2 Å by movement of $Cu_B$ with $Fe_{a_3}$ fixed. No significant structural change was detectable near the dioxygen reduction site on the redox change except for the movement of $Cu_B$ and the deletion of the peroxide between the two metals. In the fully reduced state $Cu_B^+$ is in a trigonal planar coordination, which is very stable. Azide binding to the fully oxidized form induces a migration of $Cu_B$, which causes an increase in the $Fe_{a_3}$–$Cu_B$ distance to 5.3 Å (Yoshikawa *et al.*, 1998; Fei *et al.*, 2000). The three histidine imidazoles bound to $Cu_B$ are clearly seen in this form, in contrast to the structure of the azide-bound fully oxidized form of bacterial cytochrome *c* oxidase where one of the histidine imidazoles bound to $Cu_B$ is missing in the electron density map (Iwata *et al.*, 1995). Bovine heart cytochrome *c* oxidase exhibits three histidine imidazoles clearly in the electron density maps of all the forms thus far obtained including the fully oxidized, fully reduced, fully oxidized azide-bound, and fully reduced CO-bound forms (Tsukihara *et al.*, 1995; Yoshikawa *et al.*, 1998; Fei *et al.*, 2000). Three histidine imidazoles are also found in the fully oxidized and reduced forms of the bacterial enzyme (Michel and Harrenga, 1999). Except for the azide-bound $O_2$ reduction site, X-ray structures of the redox-active metal sites in bovine and bacterial enzymes are essentially identical.

An X-ray structure of the azide-bound fully oxidized form at 2.9-Å resolution shows two possible coordination states of azide to the $O_2$ reduction site, namely, a bridging of the two metals (with $Fe_{a_3}^{3+}$–N and $Cu_B^{2+}$–N distances of 2.02 and 1.87 Å, respectively) and a terminal binding of azide to $Fe_{a_3}^{3+}$ with a significant interaction with $Cu_B^{2+}$ ($Fe_{a_3}^{3+}$–N distance of 1.98 Å and $Cu_B^{2+}$–N distance of 2.44 Å) (Fig. 5A, see color insert) (Fei *et al.*, 2000). However, infrared results provide strong evidence for terminal binding. The X-ray structure shows an azide bound to a nonmetal site in the transmembrane region (Fig. 5B, see color insert). The azide forms two hydrogen bonds with Tyr-379 and Asn-422 on the molecular surface in the middle of the transmembrane region, which is covered with detergents. Thus, azide in the protonated form ($HN_3$) is likely to arrive at the binding site from the medium. His-429 near the azide-binding site probably serves as a proton-accepting site for $HN_3$ (Fei *et al.*, 2000). This structure is consistent with infrared results, as discussed below.

The X-ray structure of the CO-bound fully reduced form was also determined at 2.8-Å resolution (Fig. 5C, see color insert) (Yoshikawa *et al.*, 1998). The structure of the $O_2$ reduction site shows that CO coordinates to $Fe_{a_3}$ in a bent end-on fashion. The distances of $Fe_{a_3}$–C and $Cu_B$–O are 1.90 and 2.47 Å, respectively. The $Cu_B$–O distance, 2.47 Å, indicates very weak interactions between the two atoms. This structure is consistent with the infrared results as described below. The stability of the cuprous copper compound in the trigonal planar coordination provides the weak interaction.

### *B. Redox-Inactive Metal Sites*

The magnesium site has been identified at 2.8-Å resolution, despite its atomic size. In fact, at 2.8-Å resolution the electron density of the magnesium atom is too small to identify. However, a metal site found at 2.8-Å resolution was identified as the magnesium site since the metal was coordinated by Glu-198 in subunit II and Asp-369 and His-368 in subunit I, which had fortunately been identified as the magnesium ligand by the mutagenesis investigations of Ferguson-Miller's group (see Hosler *et al.*, 1995; Espe *et al.*, 1995). At 2.3-Å resolution, the magnesium site was determined to be in a tetragonal bipyramidal coordination, including three water molecules in addition to the three amino acids at 2.3-Å resolution (Yoshikawa *et al.*, 1998) as shown in Fig. 6A (see color insert). This metal site is bridged between subunits I and II via the three amino acids as described above. The second redox-inactive metal site, Zn, was found in a tetragonal coordination state with four cysteine sulfhydryl groups within subunit Vb and located on the matrix side of the enzyme surface (Fig. 6B, see color insert) (Tsukihara *et al.*, 1995) These two metal sites had been detected via a metal analysis (Einansdottir and Caughey, 1984 and 1985). The physiological roles of these metals are still unknown. Thus, the X-ray structures are the only experimental results that show that these metals are intrinsic constituents of this enzyme.

The sodium site is in a slightly distorted trigonal bipyramidal coordination geometry with five ligands, including a carboxyl group of Glu-40, three peptide carbonyl groups of Glu-40, Gly-45, and Ser-441, and one water molecule (Fig. 6C, see color insert) (Yoshikawa *et al.*, 1998). The corresponding site of the *Paracoccus* enzyme is replaced with a calcium ion (Ostermeier *et al.*, 1997).

### *C. Subunit Conformations and Assembly*

Bovine heart cytochrome *c* oxidase is in a dimer state in the asymmetric unit of the crystal as shown in Fig. 7 (see color insert) (Tsukihara *et al.*, 1996). Thirteen different subunits were identified in each monomer in the X-ray structure of the fully oxidized enzyme at 2.8-Å resolution. The top view from the intermembrane side indicates a fairly strong interaction between the two monomers. The middle portion of the side view is readily identified as the transmembrane region by the large cluster of α-helices. This part was composed mainly of 28 α-helices as had been predicted by the amino acid sequences. The Cα backbone traces show that most of the α-helices are not arranged strictly perpendicularly to the membrane surfaces, in contrast to the prediction by the amino acid sequences. Thus, most of α-helices in the X-ray structure are longer than those predicted by the amino acid sequences. The three largest subunits, subunits I, II, and III, form a core portion and the other 10 nuclear-encoded subunits surround the core as shown in Figs. 7C and 7D. In the X-ray structure at 2.8-Å resolution, 3560 of 3606 amino acid residues were identified in the asymmetric unit composed of a dimer. Only 23 of 1803 amino acid residues per monomer were not detectable in the electron density map. Most of the undetectable residues are in the N- and C-terminals, which are exposed to the bulk water phase.

Subunit I, which includes two hemes and $Cu_B$, is the largest subunit composed mainly of the 12 transmembrane α-helices. On the other hand, subunit II has a large extramembrane moiety on the intermembrane side, which is in a β-barrel conformation holding the $Cu_A$ site 7 Å from the nearest surface atom. The extramembrane moiety is anchored above subunit I by two transmembrane α-helices. Subunits I and II are the only subunits the physiological roles of which are known. The third mitochondrially encoded subunit, subunit III, does not have a large extramembrane moiety. A large V-shaped cleft on the intermembrane side is formed between the two helix bundles, including two and five helices, respectively. The cleft is filled with three phospholipids, two phosphatidylethanolamines, and one phosphatidylglycerol. The opening is covered with the extramembrane moiety of a nuclear-encoded subunit, subunit VIa. Subunit III contacts subunit I in the transmembrane region, without any direct contact with subunit II. This assembly of the three mitochondrially encoded subunits conflicts with the proposal that the physiological role of subunit III is to stabilize the two functional subunits, subunits I and II.

Each of 7 of 10 nuclear-encoded subunits has a single transmembrane helix. The largest nuclear-encoded subunit, subunit IV, resembles a dumbbell with two extramembrane moieties. Subunit VIc also has two extramembrane moieties on both sides, one of which is in an extended conformation protruding to the matrix space (Fig. 7D). Subunits VIIa, VIIb, VIIc, and VIII, each with an extramembrane moiety on one end, look like screws (Fig. 7). Subunit VIa also has only one extramembrane moiety on the intermembrane side. However, 10 residues of the N-terminal are in an extended conformation and make contact with two α-helices of subunit I of the other monomer in the transmembrane region. Thus, this moiety provides a U-shaped turn near the N-terminus. The bridging contact contributes to the stability of the dimer state.

A nuclear-encoded subunit, subunit VIb, without a transmembrane α-helix placed near the noncrystallographic twofold symmetry axis in the asymmetric unit, is in contact with the corresponding subunit in the other monomer (Fig. 7D). On the other hand, subunit Vb, which is another nuclear-encoded subunit without a transmembrane helix and attached to subunits I and III on the matrix side, also provides monomer–monomer contact by an interaction with the same subunit in the other monomer. This zinc-containing subunit has a zinc finger motif in the amino acid sequence between Cys-60 and Cys-85, although the conformation does not suggest the physiological function of the zinc finger.

Adjacent to subunit Vb, another nuclear-encoded subunit, subunit Va without a transmembrane helix, is placed under subunit I in contact with the extramembrane moieties of subunits IV and VIc (Fig. 7D). This subunit contains five α-helices, each with four to five turns forming a right-handed superhelix. At 2.8-Å resolution, no significant contact between subunits Va and I was detectable. However, the contacts with subunit IV and VIc appear to be insufficient for stable assembly of the subunit in the enzyme. The possibility exists that water molecules are located between subunits I and Va.

Subunits II, VIa, and VIb are arranged on the top of subunit I and form a concave surface including a surface atom near the $Cu_A$ site (Fig. 7D). The concave surface is large enough to bind two cytochrome *c* molecules. However, a theoretical prediction of cytochrome *c* binding to the enzyme based on the X-ray structures of both proteins does not seems very straightforward, particularly for the weakly bound site. Crystallization of the enzyme–cytochrome *c* complex is desirable.

As shown in Fig. 7, the transmembrane helices are not always in parallel but interact with their adjacent helices with intersecting angles of 0°, 20°, or 50°. It has been established that these intersecting angles provide the most stable helix–helix contact (Cothia *et al.*, 1977). Thus, one of the possible physiological roles of the nuclear-encoded subunits is stabilization of the three-dimensional structure of the core subunits. In fact, the highest resolution of an X-ray structure of *Paracoccus* cytochrome *c* oxidase, not including the nuclear-encoded subunits (2.7 Å) (Ostermeier *et al.*, 1997), is definitely lower than that of bovine heart enzyme (2.3 Å) (Yoshikawa *et al.*, 1998) at present.

It should be noted that the complete amino acid sequences of the 13 subunits of the enzyme, carefully determined mainly by Buse *et al.* (1986) before crystallographic analysis, greatly contributed to the successful determination of the conformation and assembly of these subunits.

## V. Functions of the Redox-Active Metal Sites in This Enzyme

### *A. Spectral Properties of the Redox-Active Metal Sites*

Keilin and Hartree (1938a) discovered that cytochrome *c* oxidase contains two types of hemes, a and $a_3$, as described above. After isolation of this enzyme by Okunuki *et al.* (1958), many unsuccessful attempts were made to separate the two components. The failure is now understandable since, as described above, the two hemes are located on a single subunit. In order to elucidate the functions of these hemes, determination of their absorption spectra is required. The first extensive work for the spectral separation was reported by Yonetani in 1960, followed by Vanneste in 1966. A respiratory inhibitor, cyanide, reacts with ferric heme $a_3$ strongly and stabilizes the ferric state so that even excess amounts of dithionite are not able to reduce it. On the other hand, heme a is ligand-insensitive and is easily reduced with dithionite or other reducing systems even in the presence of cyanide. Therefore, by addition of slightly excess amounts of dithionite to cytochrome *c* oxidase treated with cyanide, only the heme a moiety can be reduced, leaving the cyanide-bound heme $a_3$ in the ferric state. A difference spectrum of the dithionite-reduced enzyme in the presence of a sufficient amount of cyanide versus the fully oxidized enzyme treated with the same amount of cyanide provides the redox difference spectrum of heme a. By subtracting the redox difference spectrum of heme a from the difference spectrum of the fully reduced enzyme versus that of the fully oxidized enzyme, both in the absence of cyanide, the redox difference spectrum of heme $a_3$ can be determined, since the redox difference spectrum in the absence of cyanide is composed of the redox difference spectra of hemes a and $a_3$. However, it should be noted that these procedures are possible only if the spectrum of one component is not influenced by any change in the other component. As described above, the two hemes are arranged quite close to one another. The structure strongly suggests some interactions between them, consistent with a spectroscopic investigation (Blair *et al.*, 1982). Thus, it is desirable to determine the spectra of the two hemes independently without making any assumption. Vanneste (1966) presented the absolute spectra of hemes a and $a_3$ in both oxidation states using the photochemical action spectrum of the CO-bound form of this enzyme. However, no procedure for evaluation of the extinction coefficient of CO–bound heme $a_3$ by the photochemical action spectrum is given in the paper. The evaluation is impossible nonempirically.

Nevertheless, the redox difference spectra of the two hemes are consistent with those predicted from the coordination (or spin state) of the hemes. That is, a strong and sharp hemochrome-type peak is observed at 605 nm for the difference spectrum of heme a, while heme $a_3$ exhibits a weak absorption in the visible region. Redox difference spectra of the two hemes are quite similar in the Soret region, showing intense positive bands at 445 nm and negative bands at 410–425 nm. Many kinetic results have been analyzed with the extinction coefficients given by Vanneste (1966).

The absorption spectrum of $Cu_A$ in the visible–Soret region is much weaker than the spectra of the two hemes. However, it is well established that $Cu_A$ in the oxidized state contributes to the broad absorption band near 830 nm significantly (Wharton and Tzagoloff, 1964; Beinert *et al.*, 1980). However, the contribution of heme a to the 830-nm band should not be ignored (Caughey *et al.*, 1976).

EPR spectra of $Cu_A$ in the fully oxidized and cyanide-bound states were assigned to the cupric type 1 copper (Van Gelder and Beinert, 1969). This was confirmed by low-temperature magnetic circular dichroism (MCD) spectroscopy (Greenwood *et al.*, 1983). However, a recent multifrequency EPR measurement of $Cu_A$ in the oxidized state suggests that $Cu_A$ is a dinuclear cupric copper center with an unpaired electron delocalized between two cupric copper atoms ($Cu^{1.5+} \cdots Cu^{1.5+}$) (Kroneck *et al.*, 1990). On the other hand, the absence of EPR signals for $Fe^{3+}_{a_3}$ and $Cu^{2+}_B$ has been interpreted as antiferromagnetic coupling between the two metals (Van Gelder and Beinert, 1969). The antiferromagnetic coupling is supported by magnetic susceptibility measurements (Tweedle *et al.*, 1978; Moss *et al.*, 1978) and by low-temperature MCD measurements (Thomson *et al.*, 1982). Similarly, the cyanide-bound form of the fully oxidized enzyme has no EPR signal assignable to $Fe^{3+}_{a_3}$ and $Cu^{2+}_B$ (Van Gelder and Beinert, 1969). A low-temperature MCD spectrum of the fully oxidized cyanide-bound form shows a peak at 1946 nm, suggesting ferromagnetic coupling between low-spin ferric heme $a_3$ and a $Cu_B^{2+}$ bridged by cyanide (Thomson *et al.*, 1981).

A spectral characterization of $Cu_B$ has not been reported except for an EPR signal detectable in the presence of a reducing system and $O_2$ where presumably $Cu_B$ is in the cupric state and $Fe_{a_3}$ is in the ferric state (Reinhammer *et al.*, 1980).

### *B. Redox Properties of Cytochrome c Oxidase*

The number of electron equivalents required for complete reduction of this enzyme in the fully oxidized state is one of the most fundamental properties of the enzyme to be carefully determined. However, the determination for cytochrome *c* oxidase is not easy in contrast to those for small hydrophilic metalloproteins, since complete anaerobiosis for an aqueous solution of a detergent-solubilized large membrane protein is extremely difficult to attain. Furthermore, the stability of the enzyme is also very critical for the time-consuming process of complete removal of $O_2$ from the enzyme. Considering these factors, few reports are reliable.

Cytochrome *c* oxidase has perhaps the strongest reactivity to $O_2$ among biological macromolecules. Thus a trace amount of residual $O_2$ in the system would perturb accurate redox titrations. On the other hand, $O_2$, which is very hydrophobic, is difficult to remove from hydrophobic material, the detergent-solubilized cytochrome *c* oxidase. For attainment of complete anaerobiosis for the membrane protein solution, repeated washings of the solution with oxygen-free $N_2$ are required. The procedure takes several hours at room temperature. The important points for the titration apparatus are as follows. The anaerobic titration system should be constructed with metal tubing and glassware. Rubber or plastic stoppers placed on the port of a Thumberg-type cuvette for inserting the syringe needle for introducing titrants to the main chamber should be covered with $N_2$-saturated water. Gas-tight syringes with Teflon pistons are tight enough for this titration. An example of the redox titration system is shown in Fig. 8 (Mochizuki *et al.*, 1999). Dithionite is a stable and effective reductant under anaerobic conditions. An aqueous solution of dithionite under strictly anaerobic conditions can be kept in at least 7 days at ambient temperature without any detectable change in concentration. The bovine heart cytochrome *c* oxidase, stabilized with a dodecyl polyoxyethyrene monoether, $CH_3(CH_2)_{11}O(CH_2(H_2O)_8H$, is soluble in aqueous solution without additional detergent because of the low critical micelle concentration. Complete anaerobiosis for the enzyme preparation at 7.5 μM is attainable with three cycles of evacuation and equilibration of the system with the purified $N_2$ gas by the system given in Fig. 8. For a highly concentrated enzyme solution, such as that required for infrared measurements, a different system has been reported in which $O_2$ is removed by prolonged exposure of the viscous solution to wet and $O_2$-free $N_2$ gas without evacuation of the viscous solution (Yoshikawa *et al.*, 1995).

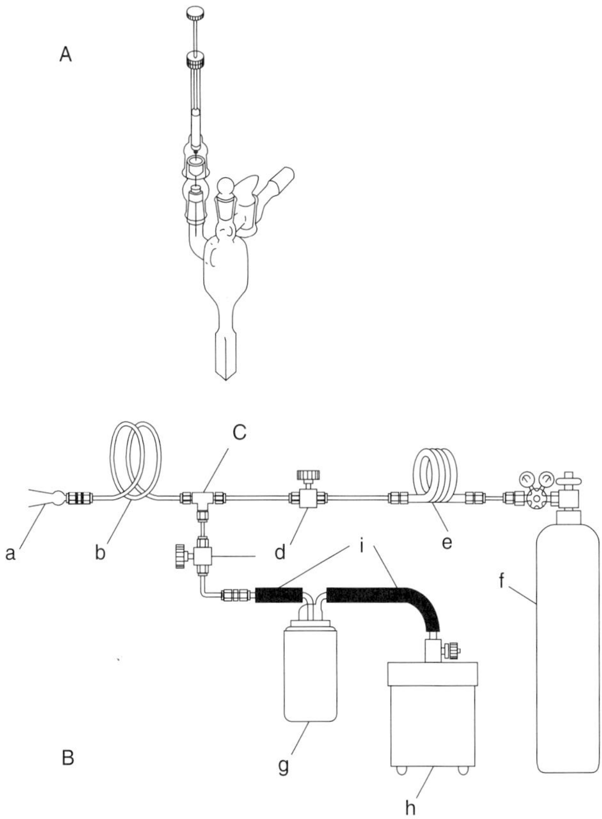


FIG. 8. The anaerobic titration system including the Thumberg-type cuvette for the solution including the enzyme and detergent at low concentration. (A) A sketch of the Thumberg-type cuvette with the titrant syringe. (B) The vacum line system for the Thumberg-type cuvette. (a) A female joint for placing the Thumberg-type cuvette; (b) spiral stainless steel tubing; (c) a T-shaped stainless steel tube; (d) metal two-way stopcocks; (e) a tube containing catalyst for complete removal of trace amounts of $O_2$ in $N_2$; (f) a nitrogen tank for ultrapure nitrogen gas; (g) a water trap; (h) a vacuum pump; and (i) vacuum rubber tubing.

Typical results obtained from a reductive titration of bovine heart cytochrome *c* oxidase are shown in Fig. 9 (Mochizuki *et al.*, 1999). Two important aspects should be noted. Six electron equivalents are required for complete reduction of the fully oxidized enzyme as prepared. The spectral changes during the reductive titration at 830, 605, and 445 nm are due mainly to the reduction of $Cu_A$, heme a, and hemes a and $a_3$, respectively. The insets in Fig. 9 show that the three titration curves are essentially identical to one another. The slopes of these titration curves are shallow in the sections corresponding to the initial one to two electron equivalents compared to the rest of the curves. This shallow slope is not due to incomplete anaerobiosis since both slopes are shallower under incomplete anaerobiosis, and increasing the number of the evacuation–$N_2$-saturation cycle up to 10 times did not affect the titration curve after the three cycles given in Fig. 9. The titration results are consistant with the X-ray structure of the $O_2$ reduction site of the fully oxidized form as given in Fig. 3. In the initial lag phase, the peroxide bridging $Fe_{a_3}$ and $Cu_B$ is reduced to give an oxidized form that requires four electron equivalents for the complete reduction.

Several findings in the above results are not consistent with earlier reports (Yoshikawa *et al.*, 1995; Van Gelder, 1966; Tiesjema *et al.*, 1973; Schroedl and Hartzell, 1977; Babcock *et al.*, 1978; Blair *et al.*, 1986; Steffens *et al.*, 1993). It has been widely accepted that four electron equivalents are sufficient for complete reduction of the fully oxidized enzyme as prepared. However, most of the previous titrations were performed in the presence of electron transfer mediators. In the presence of electron transfer mediators, such as phenazine methosulfate (PMS) under anaerobic conditions, the bovine heart enzyme purified with crystallization also showed a four-electron reduction without the initial lag phase as observed in Fig. 9. A catalytic amount of PMS induced a small spectral change corresponding to the initial lag phase. These results suggest that electron transfer mediators in other titration experiments also induce autoreductions to provide the enzyme form that receives four electrons for the complete reduction.

None of the redox titration results in previous papers showed the completely synchronized titration curves in the three wavelength regions as given in Fig. 9. This inconsistency is most likely caused by the quality of the purified preparation. The fast-form preparation was used for the experiments published in 1999 (Fig. 9), whereas other published results were likely to be obtained from the slow form or mixtures of slow and fast forms. The reduction rates of hemes a and $a_3$ with dithionite are identical in the fast form, but reduction of heme $a_3$ is much slower than that of heme a in the slow form (Moody, 1996).

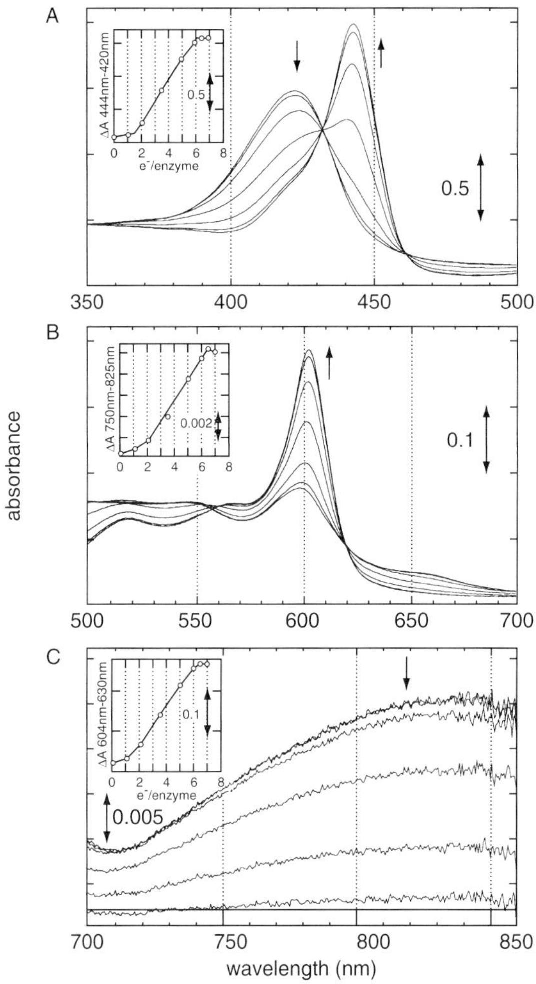


FIG. 9. A reductive titration of the crystalline bovine heart cytochrome *c* oxidase with dithionite. Absolute spectra for each oxidation state are shown for the Soret (A) and visible (B) regions. The difference spectra against the spectrum in the fully reduced state are given for the near-infrared region (C). The insets show titration curves against the electron equivalent per enzyme. The reaction mixture contained 7.5 μM bovine heart cytochrome *c* oxidase in 0.1 M sodium phosphate buffer, pH 7.4. The enzyme preparation was stabilized with a synthetic non-ionic detergent, $CH_3(CH_2)_{11}(OCH_2CH_2)_8OH$. The light path was 1 cm.

### *C. Ligand-Binding Studies*

The heme iron $Fe_{a_3}$ in the $O_2$ reduction site is very reactive not only with $O_2$ but also various transition-metal ligands or respiratory inhibitors, such as CO, $CN^-$, and $N_3^-$ (Malmström, 1990). Thus, these respiratory inhibitors have been used for probing the functions of the catalytic site since the discovery of this enzyme by Warburg (1924). The recent development in X-ray structural studies on the enzyme are expected to stimulate additional ligand-binding studies to elucidate the function of the $O_2$ reduction site. In addition to the extensive studies on the effects of these inhibitors on enzyme activity and absorption spectra of the enzyme in various oxidation states since the pioneering works by Warburg (1924) and Keilin and Hartree (1938a), the introduction of the infrared technique to this enzyme system by Caughey *et al.* (1976) should not be ignored.

#### *1. CO*

CO reacts with $Fe_{a_3}^{2+}$ to yield a characteristic absorption maximum near 590 nm. The CO-bound form is photosensitive. Thus, the flash-photolysis technique has been involved in most of the kinetic investigations of the reaction of cytochrome *c* oxidase with $O_2$, as described below.

It is widely accepted that reduction of both $Fe_{a_3}$ and $Cu_B$ is required for CO binding to the enzyme (Wilson and Nelson, 1982; Wever *et al.*, 1977, Wilson and Miyata, 1977). In other words, CO bound at $Fe_{a_3}^{2+}$ increases the redox potential of $Cu_B$. On the other hand, the two-electron-reduced enzyme CO derivative (the mixed-valence CO form) can be prepared by incubation of the enzyme with CO under anaerobic conditions, without addition of external reductant (Tzagoloff and Wharton, 1965; Babcock *et al.*, 1978). CO may act as an electron donor to the two metals by an unknown mechanism. Interestingly, under limited partial pressure of CO, CO-induced reduction was negligibly slow (Yoshikawa and Caughey, 1982). These electron equivalents may originate from some amino acid side chains. Perhaps this CO-induced change in the redox potential of $Cu_B$ is the first experimental evidence showing the interaction between $Fe_{a_3}$ and $Cu_B$.

The infrared band of the C–O stretch vibration of CO bound at $Fe^{2+}$ or $Cu^+$ appears in the region between 1900 and 2100 $cm^{-1}$ in a narrow window in the very intense infrared background due to water and proteins (Maxwell and Caughey, 1978). The position and half-band width of the stretch band are greatly influenced by the coordination structure and environment, such as species of the metal and the trans-ligand, the structures of the heme periphery, the geometry of the ligand binding to the metal ion, and the interactions between the bound CO and amino acid side chains located on the distal side. Thus, CO is definitely the most sensitive infrared probe for hemoproteins. Furthermore, the C–O stretch band is an effective tool for quantitative analysis of heme iron content (Yoshikawa *et al.*, 1977). In fact, the quantitative determination of the ratio of heme $a_3$ to heme a using the infrared technique had been the most reliable method until the X-ray structure of the enzyme appeared (Yoshikawa *et al.*, 1977).

The CO stretch band of the fully reduced CO form of bovine heart cytochrome *c* oxidase is observed at 1963 $cm^{-1}$. The band with the half-band width of 4 $cm^{-1}$ was much narrower than those of other hemoproteins such as hemoglobins, myoglobins, peroxidases, and cytochrome P450 (Yoshikawa *et al.*, 1977; Yoshikawa and Caughey, 1982). However, the position and integrated intensity of the band strongly suggest a bent end-on type coordination, essentially the same as those of other hemoproteins, which is consistent with the X-ray structure given above (Yoshikawa *et al.*, 1977; Yoshikawa and Caughey, 1982). The narrowness of the band suggests that the bound CO is tightly fixed at the binding site. The CO molecules bound to hemoglobins and myoglobins are also fixed by hydrogen bonds provided by the distal histidines. However, the $Cu_B$–O distance (2.5 Å) in the X-ray structure suggests that the interaction between $Cu_B$ and the bound ligand is significantly weaker than the interaction provided by a hydrogen bond. Thus, CuB does not contribute significantly to the stabilization of the CO binding.

The C–O stretch band at 1963 $cm^{-1}$ in the fully reduced state is shifted slightly but significantly to 1965 $cm^{-1}$ on oxidation of both $Cu_A^{1+}$ and $Fe_a^{2+}$ (Yoshikawa and Caughey, 1982). The result suggests interactions between the $O_2$ reduction site and the other redox-active metal sites, indicating that the independence of the absorption spectra of hemes a and $a_3$ is unlikely. The spectral independence is the key assumption for determination of the redox difference spectra of hemes a and $a_3$ as described above.

The binding of CO to $Cu_B^{1+}$ was discovered by low-temperature Fourier transform infrared (FTIR) spectroscopy at 2062 $cm^{-1}$ (Alben *et al.*, 1981), which was similar to the C–O stretch band of hemocyanin–CO complexes (Fager and Alben, 1972). The band was obtained by flash photolysis of the CO-bound enzyme at liquid nitrogen temperature. The 2062-$cm^{-1}$ band was stable below 140 K. Subsequently, transient binding of CO as well as $O_2$ to $Cu_B^{2+}$ was also observed at ambient temperature by visible–Soret rapid reaction measurements (Woodruff *et al.*, 1991; Blackmore *et al.*, 1991). The lifetime of $Cu_B$–CO at ambient temperature was determined to be about 1.5 μs by transient infrared spectrometry (Woodruff *et al.*, 1991). However, astonishingly, the lifetime in $D_2O$ was estimated to be about 1.5 ms, $10^3$ times longer than in $H_2O$ (Iwase *et al.*, 1999). The physiological relevance of this ligand reactivity of $Cu_B$ is under debate.

*2. $CN^-$*

Cyanide is a potent respiratory inhibitor that is much more reactive with $Fe^{3+}$ than with $Fe^{2+}$. Thus, this reagent has compensated for CO, which lacks reactivity to $Fe^{3+}$, in studies of the function of the $O_2$ reduction site, since the historical work of Keilin and Hartree (1938b) for identification of hemes a and $a_3$. However, cyanide is reactive also to ferrous iron although much more weakly than to ferric iron. Among

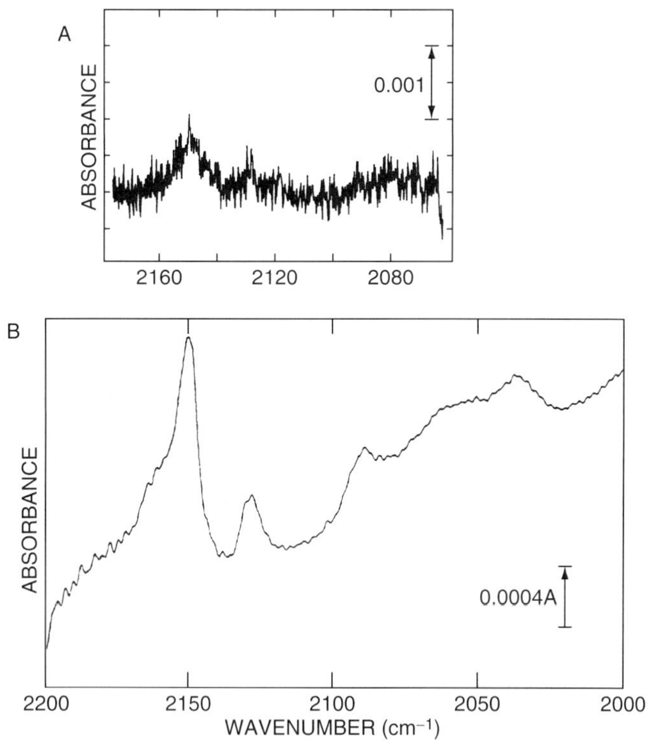


FIG. 10. Improvement of the infrared spectrophotometer. Infrared spectra of fully oxidized bovine heart cytochrome *c* oxidase cyanide derivatives measured with (A) a dispersive infrared spectrophotometer (Perkin–Elmer Model 180) and (B) a FTIR spectrometer equipped with a mercury/cadmium/tellurium detector (Perkin–Elmer Model 1800). Concentrations of the enzyme (0.7 mM) and cyanide (19.4 M) were identical in both measurements.

the many hemoproteins, only horseradish peroxidase (Yoshikawa *et al.*, 1985) and cytochrome *c* oxidase in the reduced state form stable cyanide derivatives (Yoshikawa and Caughey, 1990). As observed for the CO derivative, $Fe_{a_3}^{2+}$–$CN^-$ exhibits an α-band near 590 nm (Caughey *et al.*, 1976). Astonishingly a recent report indicates cyanide reactivity with a bacterial $ba_3$-type cytochrome *c* oxidase only in the reduced state and not in the oxidized state (Soulimane *et al.*, 2000). X-ray structural determinations of the cyanide derivatives will provide many important insights with regard to the function of the $O_2$ reduction site.

Cyanide is a very weak acid and has a p*K* near 10. The mechanisms of HCN and $CN^-$ reacting with the metal site are completely different from each one another. Determination of the reactive form of cyanide is not trivial for understanding the function of the $O_2$ reduction site. The report in 1948 by Stannard and Horecker should not be ignored. In the report, the inhibitory effect of cyanide on cytochrome *c* oxidase activity was carefully examined at various pH values and it was concluded that HCN was the reactive form.

Cyanide is also an effective infrared probe (Yoshikawa *et al.*, 1985). A drawback of this reagent as an infrared probe is its infrared intensity, which is much weaker than that of CO. However, as given in Fig. 10, the recent development in the infrared technique has solved this problem with the introduction of a mercury/cadmium/tellurium (MCT) detector (Fig. 10) (Yoshikawa *et al.*, 1995). The C–N stretch vibrational band is sensitive to many factors, such as the oxidation state and species of the coordinating metal, the structures of porphyrin ring substituents, and the ligand trans to the cyanide and protein structure (Yoshikawa *et al.*, 1985). This technique can be quite effectively applied for determination of the protonation state of the cyanide bound at a metal site. Possible binding modes of cyanide to a ferric iron are shown by Structures (I), (II), and (III). Infrared spectroscopy is the best method for identification of these

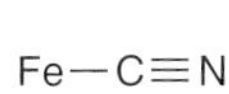


( I )

( II )

( III )

modes of coordination. Isotopic shifts of the C–N stretch band of $CN^-$ in aqueous solution by labeling $C^{12}$ and $N^{14}$ with $^{13}C$ and $^{15}N$, respectively, are very near those expected from simple reduced mass calculations. That is, stretch vibrations of $^{12}C-^{14}N$, $^{13}C-^{14}N$, and $^{12}C-^{15}N$ for $CN^-$ were observed at 2079, 2037, and 2048 $cm^{-1}$, respectively. The amounts of the isotopic shifts by $^{13}C$ and $^{15}N$ are only 1.5 and 1.2 $cm^{-1}$ smaller than the calculated values, respectively. On the other hand, for HCN, the C–N stretch vibration peaks appeared at 2093, 2063, and 2060 $cm^{-1}$ for $^{12}C-^{14}N$, $^{13}C-^{14}N$, and $^{12}C-^{15}N$, respectively. The isotopic shift for $^{13}C$ was 10.8 $cm^{-1}$ smaller than calculated, while the shift for $^{14}N$ was only 0.5 $cm^{-1}$ greater than calculated. This unusual isotopic shift for $^{13}C$ substitution is caused by the C–H vibration that perturbs the vibration of the $^{13}C$ atom. These isotopic shift effects for various hemin cyanides, cyanide complexes of methemoglobins, metmyoglobins, and peroxidases were essentially the same as expected from the simple reduced mass calculations (Yoshikawa *et al.*, 1985). These results provide the cyanide-binding mode shown in Structure (I). It is well known that the Fe–C chemical bond is much weaker than the H–C bond. Thus, Fe in the coordination of Structure (I) does not perturb significantly the vibrational movement of the carbon atom. Another method for identification of the two binding modes as shown in Structures (I) and (II) is to examine the $H_2O/D_2O$ exchange effect on the C–N stretch band. The C–N stretch band of cyanide in Structure (I) would not be affected by the introduction of $D_2O$. However the C–N stretch band of DCN in solution is at a position 106 $cm^{-1}$ lower than that of HCN. Thus, a large $D_2O$ exchange shift is expected for Structure (II). The third coordination structure [Structure (III)] provides a far lower wavenumber than those of the other two structures, since the triple-bond character is likely to be weakened by the side-on binding. All cyanide C–N stretch bands observed in various oxidation states show the isotopic and $H_2O/D_2O$ exchange effects expected for Structure (I). That is, this coordination indicates that $CN^-$ is bound at $Fe_{a_3}$, in contrast to the conclusion of Stannard and Horecker (1948). These results indicate that HCN approaches $Fe_{a_3}^{3+}$ to form $Fe_{a_3}^{3+}-CN^-$, leaving a proton. In other words, the two results strongly suggest that a charge-neutralizing group is placed near the $O_2$ reduction site.

The absorption spectral change in the fully oxidized state induced by cyanide is characterized by a red shift of the Soret band peak position to 428 nm, regardless of the original position (418–424 nm) (Yoshikawa and Caughey, 1990). $Fe_{a_3}^{2+}-CN^-$ showing absorption near 590 nm is autoxidizable. Cyanide strongly stabilizes the ferric state of $Fe_{a_3}$ so that excess amounts of dithionite cannot reduce $Fe_{a_3}^{3+}-CN^-$ of the enzyme at the concentrations appropriate for visible–Soret measurements at neutral pH (about 5–10 μM). However, $CN^-$ does not stabilize $Cu_B^{2+}$. Three electron equivalents are required for complete reduction of the metal sites (except for $Fe_{a_3}^{3+}$) in the presence of $CN^-$ (Yoshikawa *et al.*, 1995).

As shown in Fig. 10, $Fe_{a_3}^{3+}$–$CN^-$ in the fully oxidized state has a C–N stretch band at 2151.5 $cm^{-1}$. However, this band splits into bands at 2131.4 and 2091.0 $cm^{-1}$ on reduction of the metal sites except for the $Fe_{a_3}^{3+}$–$CN^-$ site. The transition between the 2151.5-$cm^{-1}$ band and the pair of 2131.4 and 2091.0-$cm^{-1}$ bands is linearly dependent on the electron equivalents added to the system. At intermediate oxidation states between the fully oxidized state and the three-electron-reduced state, approximately one equivalent of cyanide is required to saturate the cyanide-binding site. These results indicate that reduction of $Cu_B^{2+}$ induces the split and shift of the 2151-$cm^{-1}$ band. This is the clearest evidence for the interaction between $Cu_B$ and the ligand bound at $Fe_{a_3}$. The two bands observed in the partially reduced state are likely to be induced by the two types of ligand environment. For example, one of the histidine imidazoles coordinated to $Cu_B^{1+}$ could be partly deprotonated and create a significantly different polar environment in the vicinity of the bound cyanide. In the $Cu_B^{2+}$ state, no such equilibrium is present (Yoshikawa *et al.*, 1995). X-ray structures of these cyanide derivatives in various oxidation states and pH at high resolution would contribute to an improvement of understanding the mechanism of the cyanide inhibition.

*3. $N_3^-$*

Azide is another classical respiratory inhibitor. The reactive species for the enzyme inhibition is $HN_3$ (Stannard and Horecker, 1948). On the other hand, azide bound to cytochrome *c* oxidase shows several azide asymmetric stretch bands depending on the oxidation state. All these bands observed below 2051 $cm^{-1}$ are located significantly lower than the band of $HN_3$ at 2147 $cm^{-1}$ (Yoshikawa and Caughey, 1992, McCoy and Caughey, 1970). Thus, azide is also bound to the enzyme in the negatively charged form, indicating that the charge-neutralizing group must be placed near the azide-binding sites.

Azide very weakly perturbs the visible–Soret spectrum of the fully oxidized enzyme (Yoshikawa and Caughey, 1992, Li and Palmer, 1993). Azide has no reactivity with $Fe_{a_3}^{2+}$. Thus, in the presence of azide, heme a is preferentially reduced. The peak position of the hemochrome spectrum of heme a shifts from 601 nm in the absence of azide to 597 nm in the presence of azide at liquid $N_2$ temperature. It was proposed that the spectral shift was due to azide binding to heme a (Gilmour *et al.*, 1967). However, the spectral shift of heme a is probably induced by azide binding at heme $a_3$.

In the fully oxidized state, two azide asymmetric stretch bands were detected at 2051 and $2041\,cm^{-1}$. A terminally labeled azide, $^{15}N-^{14}N-^{14}N$, gave three bands at 2039, 2035, and $2024\,cm^{-1}$. The terminally labeled azide ion free in aqueous solution gives a single band while the azide bound at a metal site usually shows a split in the band position depending on which end of the azide ion is bound to the metal. The magnitude of the shifts suggests that the latter two bands at 2035 and $2024\,cm^{-1}$ shift from the 2041-$cm^{-1}$ band, and the 2039-$cm^{-1}$ band is from the 2051-$cm^{-1}$ band without splitting (Yoshikawa and Caughey, 1992). The results are consistent with the X-ray structure of the azide-bound oxidized enzyme as described above (Fei *et al.*, 2000). In the partially reduced state, clear azide bands were observed at 2016 and $2004\,cm^{-1}$. Each band splits into two bands, indicating that azide is binding to $Fe_{a_3}^{3+}$. The multiple bands again suggest multiple ligand environments as in the case of cyanide. Cyanide displaces all azide bands including the 2051-$cm^{-1}$ band in the fully oxidized state, which is assigned to that of azide bound on the transmembrane surface. This result suggests a strong interaction between $Fe_{a3}$ and the azide-binding site on the transmembrane surface.

*4. Other Ligands*

NO is another strong infrared-active ligand to heme. Special caution for investigation of this reactive gaseous reagent is required in experiments with metalloproteins since it reacts with the transition metals in the oxidized states to induce chemical modifications of the ligand. Careful infrared studies on fully reduced cytochrome *c* oxidase–NO derivatives show NO stretch bands at 1610 and $1670\,cm^{-1}$. The former is due to $Fe_{a_3}^{2+}$–NO since myoglobin–NO and hemoglobin–NO show bands at 1612 and at $1617\,cm^{-1}$. Thus, the 1670-$cm^{-1}$ band is assigned to $Cu_B^{1+}$–NO (Zhao *et al.*, 1994). It should be noted that this is the first reliable evidence for the binding of two ligands at the $O_2$ reduction site.

Two formyl reagents, hydroxylamine and hydrazine, react with the fully reduced bovine heart cytochrome *c* oxidase to induce spectral shifts in the visible–Soret region, suggesting transitions from ferrous high spin to low spin states. The spectral changes are strongly pH-dependent, indicating that the deprotonated forms (i.e., $NH_2OH$ and $NH_2NH_2$) are the reactive species (Kubota and Yoshikawa, 1993). No report of these formyl reagents binding to another hemoprotein has been published.

These ligand-binding studies including cyanide, azide, and the formyl reagents show that only uncharged reagents are accessible to the $O_2$ reduction site. The results are consistent with possible $O_2$ pathways in the hydrophobic environments as observed in the X-ray structure

(Tsukihara *et al.*, 1996). The charge-neutralizing group for cyanide and azide near the $O_2$ reduction site could be a site for donating acidic protons to $O_2$ or the $O_2$ adducts bound at the $Fe_{a3}$–$Cu_B$ site.

Among the four redox-active metal sites, the physiological roles of $Cu_A$ and $Fe_{a3}$ are quite obvious. The former receives electrons from cytochrome *c*, while the latter binds $O_2$. However, the role of $Cu_B$ is still essentially unknown. These ligand-binding studies, especially the infrared studies, indicate various interactions or influences of $Cu_B$ on the bound ligands.

### *D. Steady State Kinetics*

The electron transfer from cytochrome *c* to $O_2$ catalyzed by cytochrome *c* oxidase was studied with initial steady state kinetics, following the absorbance decrease at 550 nm due to the oxidation of ferrocytochrome *c* in the presence of catalytic amounts of cytochrome *c* oxidase (Minnart, 1961; Errede *et al.*, 1976; Ferguson-Miller *et al.*, 1976). Oxidation of cytochrome *c* oxidase is a first-order reaction with respect to ferrocytochrome *c* concentration. Thus initial velocity can be determined quite accurately from the first-order rate constant multiplied by the initial concentration of ferrocytochrome *c*. The initial velocity depends on the substrate (ferrocytochrome *c*) concentration following the Michaelis–Menten equation (Minnart, 1961). Furthermore, a second catalytic site was found by careful examination of the enzyme reaction at low substrate concentration (Ferguson-Miller *et al.*, 1976). The $K_m$ value was about two orders of magnitude smaller than that of the enzyme reaction previously found. The multiphasic enzyme kinetic behavior could be interpreted by a single catalytic site model (Speck *et al.*, 1984). However, this model also requires two cytochrome *c* sites, catalytic and noncatalytic.

The cytochrome *c* oxidase reaction under initial steady state conditions is as follows:

$$4 \text{ ferrocytochrome } c + O_2 + 4H^+ \rightarrow 4 \text{ ferricytochrome } c + 2H_2O.$$

Under typical experimental conditions, the enzyme system is saturated with $O_2$ and $H^+$. Thus this enzyme system includes four substrates and four products. However, the initial steady state kinetics of this enzyme system obeys a simple Michaelis–Menten equation (a rectangular hyperbolic relation) for each kinetic phase of the two phases at low and high ferrocytochrome *c* concentrations as described above. This result indicates that the four ferrocytochromes *c* react with the enzyme in a ping-pong fashion in each substrate concentration range. That is, each ferroferrocytochrome *c* reacts with the enzyme after the previous cytochrome *c* in the oxidized state is released from the enzyme. Cytochrome *c* does not form any ternary complex. It should be noted that complete steady state kinetic analysis including product inhibition analysis has not yet been completed.

The X-ray structure of bovine heart cytochrome *c* oxidase at 2.8-Å resolution (Tsukihara *et al.*, 1996) exhibits a concave surface on the molecular surface on the intermembrane side, large enough to accommodate two cytochrome *c* molecules as described above. On the other hand, a stable 1:1 cytochrome *c* oxidase cytochrome *c* complex, in low ionic strength only, has long been known. The complex can be isolated even by exclusion chromatography. This result strongly suggests that the two cytochrome *c*-binding sites on the concave surface have largely different affinities for cytochrome *c*, consistent with the X-ray structure of the concave surface, which is unlikely to provide two cytochrome *c*-binding sites with similar affinities. Furthermore the high-affinity phase in steady state kinetic results disappears at high ionic strength (Ferguson-Miller *et al.*, 1976). This strongly suggests that cytochrome *c* at the high-affinity site induces the high-affinity kinetic phase. The extremely stable complex of the enzyme with cytochrome *c* seems consistent with the single catalytic site model in which the strongly bound cytochrome *c* is not catalytic.

### *E. Internal Electron Transfer Kinetics*

Internal electron transfer was studied mainly with the flow-flash method developed by Gibson and Greenwood (1963) almost 40 years ago. The reaction of the reduced form of cytochrome *c* oxidase with $O_2$ is too fast to follow with a conventional stopped apparatus with a dead time of about 1–3 ms. The reaction of fully reduced enzyme with $O_2$ finishes within about 0.1 ms at 25°C. In the method, a CO-saturated enzyme solution in one syringe is mixed with air-saturated buffer in a second syringe. Then, the mixed enzyme solution is introduced into a cell within several milliseconds for photolysis by a Xenon lamp flash for initiation of the oxidation of fully reduced enzyme with $O_2$. The rapid mixing of the CO-bound enzyme with $O_2$ is crucial since the CO-bound form of this enzyme is fairly rapidly oxidized by $O_2$, due to the extremely low $K_m$ for $O_2$. Since the historical work by Gibson and Greenwood, many investigative results for the absorbance changes at various wavelengths have been published (Hill *et al.*, 1986). Absorption spectra of transition metal compounds are extremely sensitive to the oxidation state of the metal ions, although essentially no information for the structure of the ligand at heme $a_3$ is included in the absorption spectrum of heme $a_3$. After extensive work on this reaction, including many improvements of the flow-flash apparatus, the internal electron transfer pathway was established as (cytochrome c)–($Cu_A$)–(heme a)–(heme $a_3$–$Cu_B$) (Hill, 1994) in 1994.

The features of the X-ray structure of bovine heart cytochrome *c* oxidase are consistent with the electron transfer pathway determined by kinetic investigation. The location of the four redox-active metal sites is given in Fig. 11 (see color insert) (Tsukihara *et al.*, 1996). One of the amino acids coordinated to $Cu_A$, His-204 of subunit II, is hydrogen-bonded to a peptide carbonyl group of Arg-438 of subunit I. The amide group of Arg-439 is hydrogen-bonded to a propionate group of heme a. This hydrogen bond network provides an effective electron transfer pathway between $Cu_A$ and heme a. The double-bond character of the peptide bond would facilitate a facile electron transfer. Heme $a_3$ is located very close to heme a. One of the histidine ligands of heme a, His-378, is only one amino acid residue away from His-376, the fifth ligand of heme $a_3$ in the amino acid sequence. The structure suggests an effective "through-bond" electron transfer pathway between the two hemes. The peripheral groups of the two hemes are located close to one another at a distance of 3.5 Å, which is close enough for effective "through-space" electron transfers. As shown in Fig.11, a system including hydrogen and coordination bonds and a magnesium site connects one of the propionates of heme $a_3$ with $Cu_A$ directly. A hydrogen bond network including His-204 of subunit II and Arg-438 of subunit I also connects $Cu_A$ with the other propionate of heme $a_3$. Both networks seem to facilitate efficient electron transfer. However, as stated above, no direct electron transfer from $Cu_A$ to heme $a_3$ has been detected. The accuracy of the flow-flash experiments suggests that the direct electron transfer from $Cu_A$ to heme $a_3$ at the rate slower than 20% of the rate for the indirect electron transfer via heme a would not have been detected. Theoretical calculations for the X-ray structure (Fig. 11) showed that the electron transfer from $Cu_A$ to heme a is much faster than the electron transfer through the other two pathways, which confirms the experimental results (Ragan *et al.*, 1998). Then, what are the physiological roles of these structures, which are unlikely to be effective electron transfer pathways? The possibility that these structures play structural roles in the stabilization of the conformation and subunit assembly of this enzyme should not be ignored. The hydrogen bond networks located between subunits I and II, given in Fig. 11, are surrounded by many aromatic amino acid side chains. Thus, hydrogen bonds in a hydrophobic environment could be very effective for facilitating the specific and stable assembly of the two subunits. Another question yet to be answered is the physiological role of heme a. $O_2$ is reduced at the $Fe_{a_3}$–$Cu_B$ site. This brings up the question of why electrons must be transferred from $Cu_A$ to heme $a_3$ via heme a, and not directly. A significantly large amount of chemical energy is required for synthesis of heme a. What is the physiological purpose of placing heme a for this enzyme? Further discussion of this point is presented below.

### *F. $O_2$ Reduction Mechanism*

#### *1. Chemistry of $O_2$ Reduction*

As given in Fig. 12, a one-electron reduction of molecular oxygen, $O_2$, in the ground state (or the triplet state) is energetically strongly disfavored, while a two-electron reduction of $O_2$ to the peroxide level is energetically very favored (Caughey *et al.*, 1976). These chemical properties of $O_2$ contribute greatly to the stability of oxygenated hemoglobins and myoglobins. In these globins, $O_2$ binds to protoheme in the ferrous state. The ferrous iron ($Fe^{2+}$) readily releases one electron equivalent, but not two. However, the ligated $O_2$ does not receive a single electron equivalent readily. Furthermore, the globin heme sites are isolated from other heme sites. Thus, there is no facile way for the bound $O_2$ to receive two electron equivalents. This is the main reason for the stability of the oxygenated globins. In fact, isolated protoheme in solution is readily oxidized with $O_2$, since a second electron is available to the oxygenated hemes from another ferrous heme in solution, as suggested by autoxidation of a cobalt compound (Mori and Weil, 1967), as shown in Fig. 13. In fact, with the addition of small-molecule reducing agents such as hydroquinone, which are readily accessible to the $O_2$-binding site, these oxygenated globins are "oxidized" to metglobins. Therefore, the presence of $Cu_B$, which had been known long before the X-ray structure was determined, seems critical for the facile reduction of $O_2$ at the $O_2$ reduction site of cytochrome *c* oxidase, and a dioxygen reduction mechanism as given in Fig. 14 was proposed by Caughey's group (see Yoshikawa *et al.*, 1977). In this scheme, the two cobalt ions in Fig. 13 are replaced with $Fe^{2+}$ and $Cu^{+}$. This scheme was widely accepted when it was proposed. However, an important mechanistic implication of this mechanism has

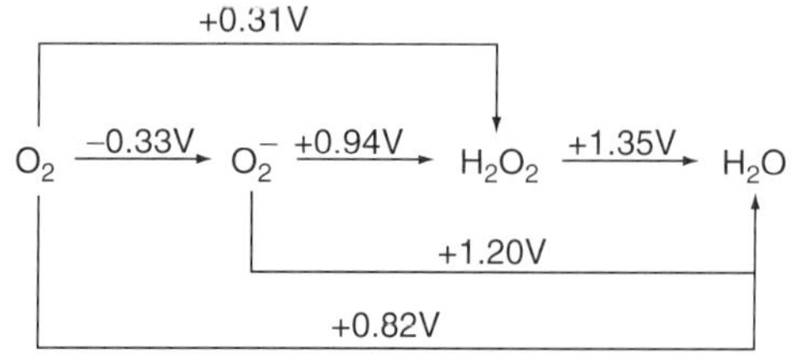


FIG. 12. Standard oxidation–reduction potentials for the steps involved in the conversion of $O_2$ to water at 25°C and pH 7.

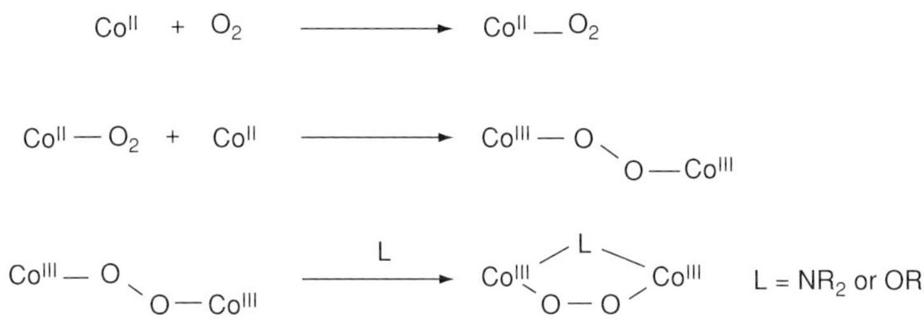


FIG. 13. Formation of a μ-peroxo-bridged complex from a nitrogen ligand complex of cobalt(II) and $O_2$. The μ-peroxo-bridged complex is stabilized in the presence of a second bridging ligand, L, such as amino or hydroxo.

not been well accepted. That is, the oxidation of the $O_2$-bound form to the μ-peroxo form must be extremely rapid, since this reaction step is rate-limited by the electron transfer from $Cu_B^{1+}$ to an oxygen atom on the distal side. An electron transfer at a distance of 2–3 Å could be as fast as on the order of a picosecond. On the other hand, formation of the $O_2$-bound form is rate-limited by the $O_2$ transfer in the protein from the molecular surface to the $O_2$ reduction site, which is likely to be determined by molecular motions of the protein. Thus formation of the $O_2$-bound form is probably much slower than degradation of the $O_2$-bound form. In other words, detection of the $O_2$-bound form during the course of the $O_2$ reduction by this enzyme is very difficult. Thus, assignment of an intermediate species only by the absorption spectral property and the order of appearance during the $O_2$ reduction process is impossible.

### 2. *Spectroscopic Identification of Intermediate Species*

The first trial for identification of the $O_2$-bound form (the oxygenated form, hereafter) was performed by Chance *et al.* (1975) using pigeon heart mitochondria. They trapped the initial intermediate species at low temperature (−100°C), which had a 590-nm peak quite similar to that of

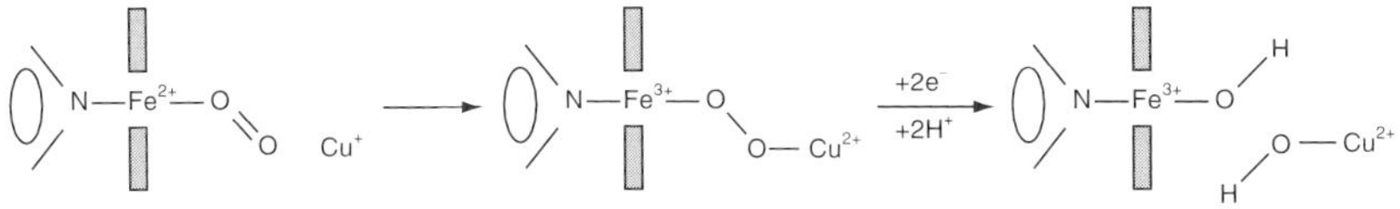


FIG. 14. Bridging peroxide mechanism. Fe and Cu represent $Fe_{a_3}$ and $Cu_B$ in the $O_2$ reduction site, respectively. N represents the nitrogen of the fifth ligand of heme $a_3$. The shaded rectangles represent a side view of the porphyrin plane.

the CO-bound reduced heme $a_3$ but was not sensitive to their photolysis flash. Furthermore, at the low temperature, rebinding of the photolyzed CO was negligibly slow compared with the initial intermediate formation. From these results they concluded that this initial intermediate was the oxygenated form of cytochrome *c* oxidase. They named it compound A, which is still used in this field. However, these two experimental results, the absorption spectrum and the photoinsensitivity, do not provide direct evidence of the $Fe^{2+}$–$O_2$ structure.

Direct evidence for the oxygenated form was obtained by resonance Raman investigations by three different research groups (Varotsis *et al.*, 1990; Ogura *et al.*, 1990; Han *et al.*, 1990). Resonance Raman spectroscopy provides selective enhancement of various vibrational modes of a chromophore that has significant absorbance at the wavelength of the excitation laser light. All three research groups tried to detect the oxygenated form during the course of the $O_2$ reduction process at ambient temperature, initiating the reaction by photolysing the CO compound in the presence of $O_2$. They identified a resonance Raman band at 571 $cm^{-1}$ as the initial intermediate species. The half-life of the bands was about 0.5 ms at 4°C. The band position is consistent with the Fe–$O_2$ vibrational band identified for hemoglobins and myoglobins. In order to confirm the assignment, isotopic shifts for this band were examined as shown in Fig. 15 (Ogura *et al.*, 1993). The isotopic shift obtained by exchanging $^{16}O_2$ with $^{18}O_2$ is as expected from the reduced mass calculation. If $O_2$ binds to $Fe_{a_3}$ in a symmetrical side-on fashion, the half-labeled $O_2$, $^{16}O^{18}O$, provides only one band just halfway between the $^{18}O_2$ and the $^{16}O_2$ bands. If the $O_2$ binding occurs in a terminal fashion, $^{16}O^{18}O$ shows two bands due to Fe$-^{16}O^{18}O$ and Fe$-^{18}O^{16}O$ between the two bands of the two symmetrical $O_2$ molecules ($^{18}O_2$ and $^{16}O_2$). Each mode of the iron-proximal oxygen vibration is influenced by the oxygen atom at the distal side depending on the angle of the O–O bond to the Fe–O bond. Furthermore, if only a single oxygen atom is on $Fe_{a_3}$, $^{18}O^{16}O$ oxygen provides two bands at the same positions as either those of $^{18}O_2$ or $^{16}O_2$. The Fe–$O_2$ band is too weak to detect in the absolute spectra. It is detectable only in the isotopic difference spectra as given in Fig. 15. Spectrum *d* shows that the $^{16}O^{18}O$ spectrum is not the average of the $^{16}O_2$ and $^{18}O_2$ spectra and that the asymmetrically labelled $O_2$ provides two bands that slightly deviate from both the $^{16}O_2$ and the $^{18}O_2$ bands. Thus, completely unexpectedly, the initial intermediate is the $Fe^{2+}$–$O_2$ species in which $O_2$ binds to Fe in a bent end-on fashion very similar to that of hemoglobins and myoglobins. The half-life of this species was about 0.5 ms at 4°C, which is astonishingly long. The next intermediate had a band at 804 $cm^{-1}$, followed by the third intermediate with a band at 785 $cm^{-1}$. Isotopic shift analysis for these two intermediate species using $^{18}O_2$ and $^{18}O^{16}O$

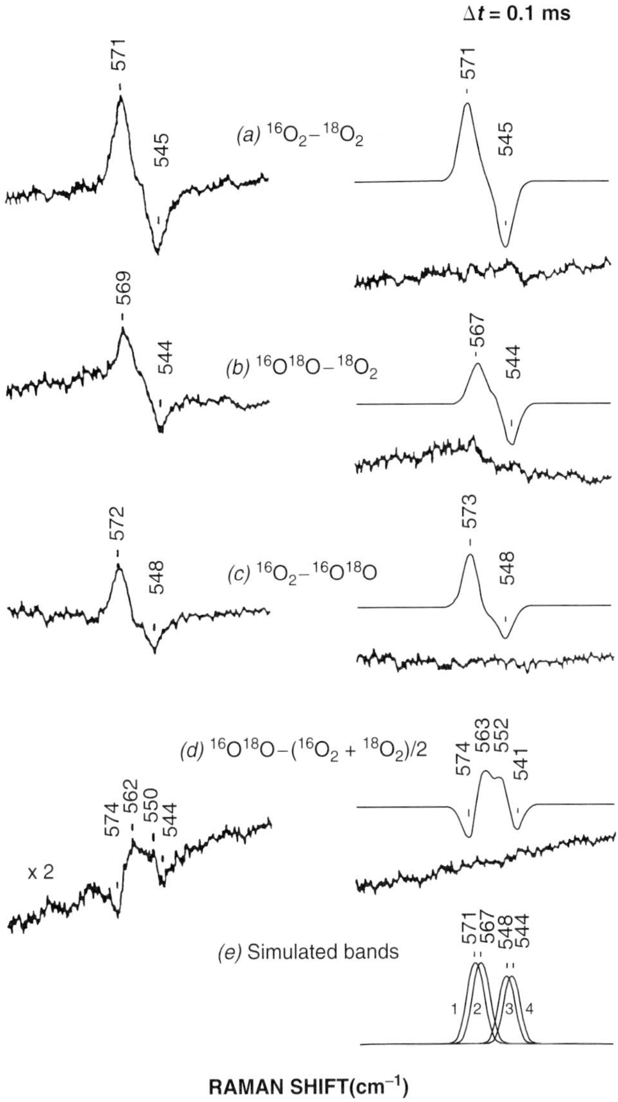


FIG. 15. Resonance Raman spectra of the $Fe^{2+}$–$O_2$ stretching frequency region of bovine heart cytochrome *c* oxidase 0.1 ms after initiation of the reaction of the fully reduced enzyme with $O_2$. Spectra on the left- and right-hand sides are the observed spectra and the calculated spectra with the differences of the observed versus calculated spectra, respectively. Spectrum (*d*) is obtained using the following calculation: [Spectrum (*b*)–Spectrum (*c*)]/2. (*e*) Simulated bands for $Fe-{}^{16}O_2$ (1), $Fe-{}^{16}O^{18}O$ (2), $Fe-{}^{18}O^{16}O$ (3), and $Fe-{}^{18}O_2$ (4). The peak intensity ratio is 6:6:5:5. All bands have the Gaussian band shape with a half-maximal band width of 12.9 $cm^{-1}$.

showed that both bands were due to species with a single oxygen atom, that is, Fe=O species (Ogura *et al.*, 1993).

These two species are also detectable during the course of the reaction of the enzyme with $H_2O_2$. The visible spectral change in the heme $a_3$ site in the $H_2O_2$-enzyme reaction is much easier to determine than the reaction of the fully reduced enzyme with $O_2$, since in the former reaction no absorbance change in heme a is involved. The visible spectral change shows that the 804-$cm^{-1}$ species has a peak at 607 nm and the 785-$cm^{-1}$ species has a peak at 580 nm. These species correspond to the P- and F-forms, respectively (Proshlyakov *et al.*, 1996). It is quite well accepted that the P-form is the intermediate next to the oxygenated form and that the F-form follows the P-form. Originally the P- and F-forms denoted peroxy and ferryl, respectively. Furthermore, one-electron reduction is required for the transition from P to F. However, resonance Raman results indicate that both P- and F-forms have an oxide on the heme $a_3$ iron. Thus, Kitagawa's group has proposed a perferryl oxo form ($Fe^{V}=O$) for the P-form and a ferryl oxo for the F-form (see Kitagawa and Ogura, 1997). Following the F-form, a band assignable to $Fe^{3+}$–OH appeared at 450 $cm^{-1}$ (Ogura *et al.*, 1993; Proshlyakov *et al.*, 1996). Another unexpected finding of the above resonance Raman results is the absence of a peroxide intermediate during the course of $O_2$ reduction.

As described above, resonance Raman spectroscopy is an extremely powerful tool for the identification of intermediate species during the course of $O_2$ reduction. However, the method provides no information on the state of $Cu_B$. Also, close correlations between absorption spectral changes for hemes and $Cu_A$ and the resonance Raman spectral changes during the $O_2$ reduction, which would provide various insights into the reaction mechanism of this enzyme, have not been obtained.

### *3. Mechanism of $O_2$ Reduction*

As described above, the X-ray structure of the fully reduced bovine heart cytochrome *c* oxidase shows that $Cu_B$ is in a trigonal planar cuprous copper structure. Trigonal planar cuprous copper compounds are usually very stable. Thus, $Cu_B$ is likely to be a poor ligand acceptor as well as a poor electron donor. This structure seems essential to provide the unusually stable oxygenated form in the enzyme reaction. Tyr-244, which is fixed with His-240 by covalent linkage, is placed close enough to the dioxygen reduction site for formation of a hydrogen bond with $O_2$ bound at $Fe_{a_3}^{2+}$. On the other hand, Tyr-244 is placed on one end of a hydrogen bond network from the molecular surface on the matrix space (Tsukihara *et al.*, 1996). Thus Tyr-244 could take up protons quite readily from the matrix side for making water molecules. The cross-link between His-240 and Tyr-244 provides a facile electron transfer pathway from $Cu_B$ to Tyr-244. As stated above, the two-electron reduction of $O_2$ is thermodynamically favorable. Thus, once Tyr-244 is hydrogen-bonded to $O_2$ bound at $Fe_{a_3}^{2+}$, reduction of the bound $O_2$ with two electrons, one from $Fe_{a_3}^{2+}$ and the other from the hydroxyl group of Tyr-244, would proceed quite readily. However, the hydrogen bond formation could be controlled by the conformation of Tyr-244 to provide the fairly slow rate of $O_2$ reduction as shown in the resonance Raman Results. The two-electron reduction of the bound $O_2$ results in a ferric hydroperoxo complex of heme $a_3$ ($Fe_{a_3}^{3+}$–OOH) and a phenolic radical of Tyr-244. The facile electron transfer from $Cu_B^{1+}$ via the cross-link between His-240 and Tyr-244 and the proton transfer from the matrix surface to Tyr-244 would regenerate the OH group of Tyr-244. At this point, the OH of Tyr-244, which is acidic, is placed very near the ferric heme-hydroperoxo. Thus, the OH group would provide a substantially strong acidic environment for the hydroperoxo. The acidic environment stimulates the O–O bond break. The controlled rate of conformational change of Tyr-244 for formation of the hydrogen bond to $O_2$ bound at $Fe_{a_3}^{2+}$ may provide a fairly slow hydroperoxo formation compared with the O–O bond cleavage. Thus, the peroxo species is undetectable during the course of the $O_2$ reduction.

It should be noted that when the O–O bond is broken, the $O_2$ molecule bound at $Fe_{a_3}^{2+}$ receives four electron equivalents and yields a water molecule and an oxide on $Fe_{a_3}^{5+}$, at least, formally. The stable oxygenated form and the unstable peroxide intermediate provide the four-electron reduction of $O_2$ at once. As is well known, if an $O_2$ molecule receives four electrons one at a time, three active oxygen species, superoxide, peroxide, and hydroxyl radical, will be produced during the $O_2$ reduction. Cytochrome *c* oxidase must reduce $O_2$ totally without releasing these species, which are extremely toxic to the cell. The four-electron reduction of this enzyme may be the strategy of this enzyme for safe $O_2$ reduction (without damaging cells).

## VI. Proton Transfer Mechanism

### *A. Proton Transfer Pathways within Protein Molecules*

The interior of globular proteins is filled mostly with hydrophobic and aromatic amino acids. Thus specific structures are required for proton transfer inside proteins. It is well accepted that protons are readily transferred via hydrogen bonds. As shown in Fig. 16A, effective proton transfers are possible only in a hydrogen bond between two functional groups, both of which can be either a hydrogen donor or an acceptor. Many waters found in the interior of bovine heart cytochrome *c* oxidase are involved in possible proton transfer pathways. Proton transfers through a fixed

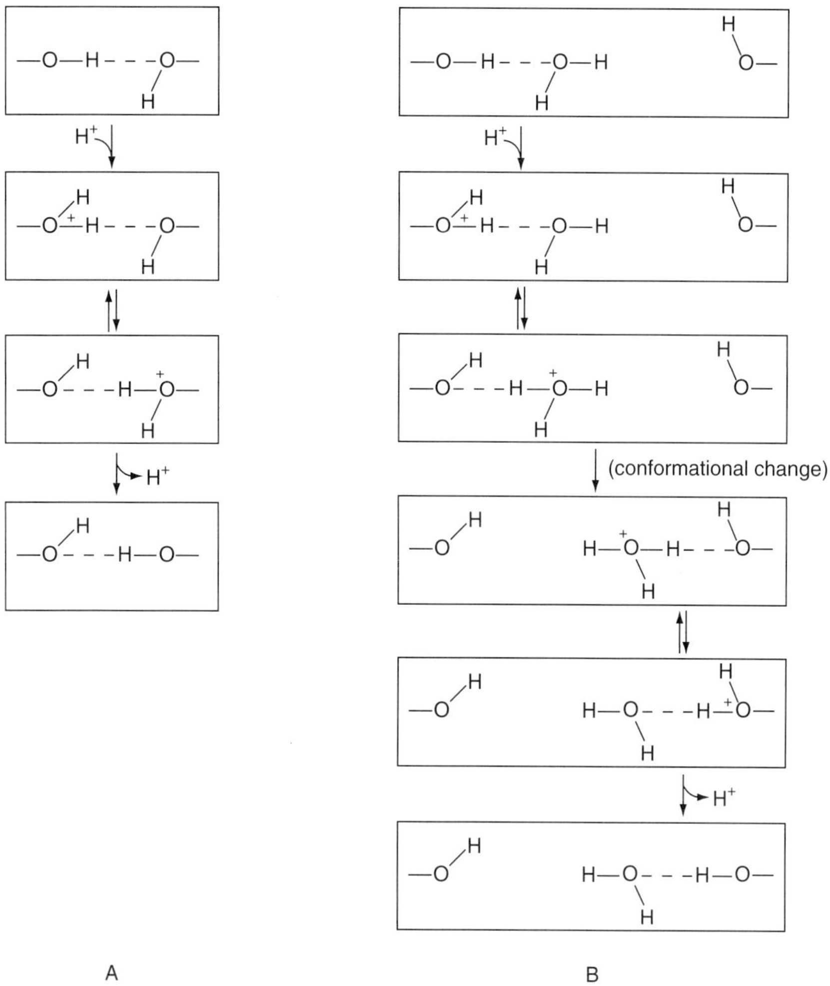


FIG. 16. Possible proton transfers through hydrogen bonds. (A) A possible proton transfer pathway through a hydrogen bond between the two groups, both of which can be either a hydrogen donor or an acceptor. (B) A possible irreversible proton transfer with migration of a hydronium ion.

hydrogen bond are essentially reversible. Thus a hydrogen bond network composed of many hydrogen bonds in tandem provides a simple proton channel for passive proton transport. Proton deliveries inside the enzyme molecules either for making water or for proton pumping must be tightly controlled. Irreversibility could be facilitated effectively by a conformational change as shown in Fig. 16B. In this hydrogen bond system, a water molecule is hydrogen-bonded to a hydroxyl group on the left-hand side in one of the conformations, whereas in the other conformation, the central water migrates to the hydroxyl group on the right side. Various conformational changes could induce the migration of the water molecule. For example, a conformational change near the left hydroxyl group could induce a decrease in the dielectric constant of the environment of the left hydroxyl group relative to that of the right hydroxyl group when the water molecule is protonated. Then, the hydronium ion migrates to the right hydroxyl group to form a new hydrogen bond. In this conformation, no reverse proton transfer to the left hydroxyl group is possible. The conformational change could be driven either by an electron transfer reaction or by the $O_2$ reduction process at the $Fe_{a_3}$–$Cu_B$ site. Thus, a hydrogen bond network with such disconnecting points is likely to drive well-controlled unidirectional proton transfers. In the X-ray structure of bovine heart cytochrome *c* oxidase, large cavities without any clear electron density peak inside, but large enough for trapping more than one water molecule, are detectable, suggesting that mobile waters are located inside the cavities. These cavities provide very effective proton transfer pathways between two protonatable groups on the walls of the cavities. One such hydrogen bond network found in the X-ray structure of bovine heart cytochrome *c* oxidase at 2.8-Å resolution is given schematically in Fig. 17 (Tsukihara *et al.*, 1996). Dark balls denote the positions of water molecules fixed inside the protein molecule. Hydrogen bonds are shown by dotted lines. The oval is a cavity as stated above. The dotted double-headed arrow shows a disconnecting point in which movement of a protonatable amino acid side chain (Lys-319) is required for transferring protons. This structure indicates that Lys-319 controls the proton supply to the $O_2$ reduction site up to Tyr-244. The importance of this residue has also been confirmed by mutagenesis of the corresponding amino acid side chain in bacterial enzymes (Konstantinov *et al.*, 1997).

### *B. Two Types of Proton Transfer for Reduction of $O_2$ to $H_2O$*

Two proton transfer pathways, one for pumping protons and the other for transferring protons to make water molecules at the $O_2$ reduction site, were proposed based on the X-ray structure of the enzyme from soil

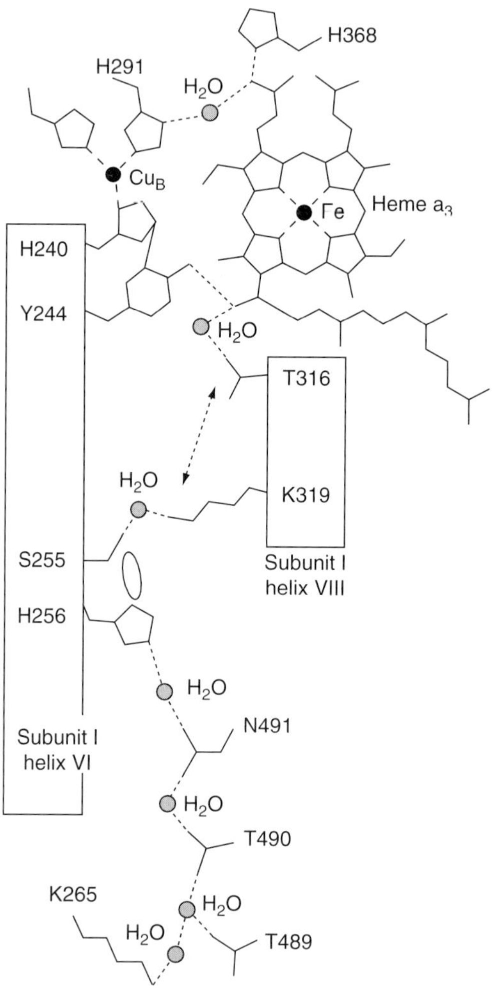


FIG. 17. A hydrogen bond network connecting the matrix space with the $O_2$ reduction site. The amino acid residues on the bottom end are exposed to the matrix space. Dotted lines denote hydrogen bonds. The oval represents space without an electron density peak. The dotted double-headed arrow shows a space where K319 moves toward T316 to form a new hydrogen bond after breaking the hydrogen bond with $H_2O$. This movement provides a unidirectional and irreversible proton transfer.

bacterium, *P. denitrificans* (Iwata *et al.*, 1995). Extensive mutagenesis research has been reported for amino acids in these pathways (Gennis, 1998). By exchanging amino acids in the proposed pathways with non-hydrogen-bond-forming amino acids, the enzyme activity for the turnover reaction is essentially eliminated. These mutated amino acids are placed far from the dioxygen reduction site. Thus, these mutational results are the most conclusive evidence for proton/electron coupling during $O_2$ reduction.

Although the effects of these various mutations on steady state kinetics are essentially identical (the mutations kill enzyme activity completely), two types of mutagenesis effects were observed for the single turnover reaction. Mutations in the amino acids in one of the channels, the K-channel, inhibited seriously the electron input phase while no significant influence was detectable in the electron output or the $O_2$ reduction process (Ädelroth *et al.*, 1998; Vygodina *et al.*, 1998; Konslantinov *et al.*, 1997). This channel, including K319, corresponds to that given in Fig. 17. Another important finding of the mutagenesis studies is that the mutant enzyme, in reaction with $H_2O_2$, is as active as the wild-type enzyme (Zaslavsky and Gennis, 1998). These results strongly suggest that the K-channel is required for electron transfer of the initial two electrons for reduction of $O_2$ to the peroxide level (P-form). On the other hand, mutations in the other channel, which is called the D-channel because of the presence of a glutamate residue, D91, at one end of the channel near the $O_2$ reduction site, inhibit the electron input to the P-form. The mutant enzymes in the fully oxidized state receive electrons to form the fully reduced form as quickly as the wild-type enzyme does (Ädelroth *et al.*, 1998; Konstantinov *et al.*, 1997). On the other hand, it is widely accepted, as described above, that two electrons in the $O_2$ reduction site are indispensable for binding CO and possibly $O_2$ to the site (Wilson and Nelson, 1982; Wever *et al.*, 1977; Wilson and Miyata, 1977). Considering this result, the above two mutant classes suggest two types of proton transfer processes, each coupled to electron transfers either to the fully oxidized $O_2$ reduction site or to the site in the P-state, respectively. The former process creates the two-electron-reduced state to bind $O_2$, and the two electron equivalents are transferred to the bound $O_2$ to provide the P-state, which is electronically equivalent to $Fe^{3+}$–OOH $\cdots Cu^{2+}$. As described above, the resonance Raman results show that in the P-state, the O–O bond has been cleaved by another two electron equivalents possibly from $Fe^{3+}$ to give $Fe^{5+}=O$ and $OH^-$. The resultant P-state ($Fe_{a_3}^{5+}=O$) receives two electron equivalents and one or two protons in the latter process to give $Fe^{3+}$–$OH^-$ or $Fe^{3+}$–$OH_2$. These mutagenesis results indicate that proton transfer coupled to the initial two-electron reduction process occurs via the K-channel and proton transfer to the subsequent two-electron reduction occurs via the D-channel. It should be noted that both K- and D-channels are required for proton transfer for making water molecules. These results are inconsistent with the proposal of Iwata *et al.* (1995) in which proton transfers for making water and for pumping protons are assigned to K- and D-channels, respectively. Why does this enzyme require two channels for transferring protons for making water? One of the possible interpretations is that the two channels are placed for transferring two or three protons almost at once to the $O_2$ reduction site.

Wikström and colleagues proposed that proton pumping is coupled to the transitions from the P-state to the F-state and from the F-state to the fully oxidized state (see Wikström, 1989; Wikström and Morgan, 1992; Verkhovsky *et al.*, 1999) As described above, proton transfers through the D-channel are coupled to these two transitions. Thus, many workers in this field believe that the D-channel transfers protons both for making water and for pumping. However, it should be noted that the experimental results showing proton pumping coupled to the transitions from P to F and F to O do not provide any information about the proton-pumping site. In fact, part of the proton pumps seemed to be coupled with reduction of heme a (Verkhovsky *et al.*, 1999). Furthermore, protons for pumping must be separated from those for making water molecules, otherwise protons to be pumped would be used to make water molecules. A third possible proton channel has been proposed, which is likely to be the channel for transferring protons for pumping, as described below (Yoshikawa *et al.*, 1998).

### *C. Proposed Mechanisms of Proton Pumping by Cytochrome c Oxidase*

It is worthwhile to review proposals for the proton-pumping mechanism of cytochrome *c* oxidase. The first proton-pumping hypothesis drawn with chemical formulas was given by Chan's group (see Gelles *et al.*, 1986), in which protons are pumped by changes in coordination of the CuA site depending on its oxidation state. Unfortunately this proposal did not last very long, since, shortly after its publication, a proton-pumping function was detected in a quinol oxidase from *E. coli*, which lacks the $Cu_A$ site. Thus, they proposed another redox-linked ligand-exchange mechanism for proton pumping at the $Cu_B$ site (Larsen *et al.*, 1992). Later, another mechanism was proposed, which involves a proton transfer driven by a redox-coupled conformational change in the CuB site (Wikström *et al.*, 1994). In this proposal, one of the three imidazoles coordinated to CuB is released from the CuB on reduction. The released imidazole groups could receive two protons at most for pumping. This proposal (the histidine cycle mechanism) is widely accepted since the possible proton-pumping site is completely conserved. In contrast to the proton-pumping mechanism at $Cu_A$, the mechanism includes proton pumping at the $O_2$ reduction site. Furthermore, in the X-ray structure of the *Paracoccus* enzyme in the fully oxidized azide-bound form, one of the three histidine imidazoles coordinated to $Cu_B$ is missing. This suggests that the histidine imidazole is mobile (Iwata *et al.*, 1995). The X-ray structure seems to support the above mechanism. However, the X-ray structures of the bacterial enzyme in both the fully oxidized and the fully reduced states at higher resolution, free from azide, show three histidine imidazoles on $Cu_B$ (Michel and Harrenga, 1999). Thus, these X-ray structures do not support the above proposal, although some transient conformational changes are possible for the imidazole.

Another point in the proposal is that the proton-pumping site is in the dioxygen reduction site, where water molecules are produced by reduction of $O_2$. During $O_2$ reduction, acidic protons are expected to play a critical role. In other words, protons to be pumped at $Cu_B$ would be readily captured by $O_2$ reduction intermediates on $Fe_{a_3}$. X-ray structures of both bacterial and bovine heart cytochrome *c* oxidases in all ligand-binding and oxidation states obtained thus far do not show any structure to prevent contact of protons at $Cu_B$ with the $Fe_{a_3}$ site. One might say that proton pumping and $O_2$ reduction, both at the $O_2$ reduction site, could be differentiated temporally. However, this mechanism includes a redox-coupled coodination structural change in $Cu_B$. The redox reaction in $Cu_B$ cannot be independent of $O_2$ reduction at $Fe_{a_3}$. Thus, the time-sharing mechanism seems impossible.

Rich (1995) has proposed that the proton pump in cytochrome *c* oxidase is driven mainly by electrostatic interactions (or repulsion) between protons in a "proton trap" and protons transferred from the matrix side to the $O_2$ reduction site for neutralizing oxides ($O^{2-}$) are produced by $O_2$ reduction. In this mechanism, a structural change for gating proton transfer from the matrix side to the proton trap is required for a complete cycle of redox-coupled proton pumping. However, no such structural change has been detected.

Papa *et al.* (1998; see also Capitanio *et al.*, 2000) proposed involvement of heme a in proton pumping by cytochrome *c* oxidase from careful examination of the redox potential of the metal sites at various pH. They also have emphasized the importance of the redox-coupled change in p*K* of the protonatable group (redox Bohr effect) in the mechanism of the proton pump of this enzyme. Perhaps, the simplest example of the redox Bohr effect is the p*K* change in heme propionates depending on the oxidation state of the heme iron. In the simplest case, essentially no change in conformation is required for changing the p*K* of the protonatable groups. However, for a detectable change in p*K* of a protonatable group inside a protein, a significant conformational change is required for changing the polarity of the environment of the protonatable group.

Furthermore, changes in the accessibility of the proton-pumping site must be effectively switched depending on the oxidation state of the redox-active metal center. To date, this aspect has not been discussed in detail.

### *D. Fourier Transform Infrared Spectroscopic Examination of Redox-Coupled Conformational Change in the Protein Moiety of Cytochrome c Oxidase*

Infrared spectroscopy is one of the most powerful tools for functional studies of hemoproteins reactive to external ligands with infared absorptions in the triple bond region (1900–2200 $cm^{-1}$) where the background level due to absorptions of proteins and water molecules is quite low as described above. However, recent improvement in the sensitivity and stability of the FTIR apparatus with an MCT detector has enabled infrared spectroscopic examination of the protein moiety also. In fact, one of the most sensitive methods for monitoring the dissociation of a COOH group is infrared spectroscopy.

In the past few years, at least three groups started to report changes in the environments of the COOH group in bacterial and bovine heart enzymes (Hellwig *et al.*, 1996; Lübben and Gerwert, 1996; Puustinen *et al.*, 1997). The C=O stretch vibration of a COOH group is greatly influenced by deprotonation, since the C=O group has a significant amount of single-bond character due to resonance stabilization between C=O and $C–O^-$ groups. Furthermore, the vibration is sensitive to the environment of the group. Another advantage is that the anti-symmetric stretch vibrational mode of the C=O group of COOH in glutamate and aspartate appears outside of the strong background due to the amide and water bands. The deprotonated form shows a symmetrical vibrational band near 1580 $cm^{-1}$. However, these bands are too weak to be detected in a conventional difference spectrum method in which two separate solutions containing strictly identical concentrations of the protein are used for the sample and reference cells. Thus, all three groups use an enzyme solution in a single cell placed in FTIR apparatus. After measurement of the enzyme sample in the cell, the oxidized enzyme is reduced with an electrode installed inside the cell or with photoreduction of the enzyme by laser irradiation. Woodruff *et al.*, (1991) photolysed the CO-bound enzyme to investigate the change in the protein region in FTIR coupled with photodissociation of the CO-complexed enzyme. Most of these measurements were performed for the bacterial mutant enzyme for identification of the observed infrared bounds.

The most noteworthy finding in FTIR of the protein region may be the redox-coupled band shift of glutamate of *Paracoccus* enzyme correspondingto Glu-244 in bovine, which is one of the amino acids in the D-channel. Both of the bands were eliminated by mutating the amino acid. Thus, this change in the FTIR spectrum is probably due to the change in the polarity of the environment of Glu-244. A similar shift in the infrared band was found during the CO dissociation. These findings suggest the successful contribution of this method to understanding the reaction mechanism of this enzyme.

Heme propionates are likely to have a critical role in proton pumping. Thus, these bands were identified by isotope labeling of the carbon atoms of the carboxyl groups using a mutant enzyme deficient in the synthesis of $\delta$-amino levulinate (Behr *et al.*, 1998). The band positions are significantly lower than those of aspartate and glutamate, perhaps due to the strong hydrogen bond provided by a fixed water molecule between the two propionates in each heme. Furthermore, the effects of mutations of amino acids hydrogen-bonded to these four propionate groups were selectively examined for identification of each propionate (Behr *et al.*, 2000). However, the X-ray structure of these propionates and of the amino acids hydrogen-bonded to them suggests that secondary effects due to the mutation of a single amino acid may influence the vibration of the propionate to which the amino acid does not form any hydrogen bond.

A positive infrared band in the oxidized-minus-reduced difference spectrum of bovine heart cytochrome *c* oxidase was observed at $1737\,cm^{-1}$ in $H_2O$ (Hellwig *et al.*, 1999). This band is not detectable in *Paracoccus* cytochrome *c* oxidase or in *E. coli* quinol oxidase. Furthermore, the difference spectrum exhibits a negative band at $1580\,cm^{-1}$. This result indicates that one of the carboxyl groups of glutamate or aspartate is dissociated on reduction of this enzyme. Perhaps this is the first experimental evidence that the COOH of Asp-51 dissociates to $COO^-$ on reduction of the enzyme. However, the authors give a different assignment assuming that Asp-51 is not involved in the proton-pumping function.

### *E. Redox-Coupled Conformational Change in Bovine Heart Cytochrome c Oxidase*

As stated above, for proton transfers inside cytochrome *c* oxidase, redox-coupled conformational changes are required. As shown in Fig. 18 (see color insert for A), a fairly large conformational change, including even a movement of the peptide backbone in a loop region between helices I and II of subunit I of bovine heart cytochrome *c* oxidase, was detected by comparison of the fully oxidized form at 2.3-Å resolution with the fully reduced form at 2.35-Å resolution (Yoshikawa *et al.*, 1998). No significant conformational change is detectable in other moieties of the protein. Among the movements of the several amino acids in the loop, only the change in the conformation of Asp-51 involves a clear change in accessibility. In the fully oxidized state, the carboxyl group of Asp-51 is completely buried inside the protein, so that bulk water molecules have no access to the group. However, on reduction, this group moves toward the molecular surface on the intermembrane side and becomes exposed to water molecules in the bulk water phase. In the fully oxidized state, Asp-51 is hydrogen-bonded to a peptide N–H group of Ser-441 and the hydroxyl group of the same residue. The structure indicates a fairly hydrophilic environment. Thus, the p*K* decrease in the conformation change from the inside to the outside may not be very large. However, the $COO^-$ group fixed at the position by a hydrogen bond would function as a very strong base for the peptide N–H group. Thus, this $COO^-$ will extract very effectively the proton of the imidic acid as discussed below. In the fully oxidized state, Asp-51 is connected with the matrix phase by a hydrogen bond network and water pathway as shown in Fig. 18B. The water pathway was determined by a "random walk" program, which produces a random walk of a probe starting from an arbitrary point giving the atomic size of the probe. By setting the size of the probes smaller than the size estimated by the van der Waals radius of each atom, the molecular motion of the protein can be taken into account. The reliability of the program has been confirmed by applying the program for determination of $O_2$ pathways in hemoglobins that have been well established experimentally. The water path given in Fig. 18B is a simplified scheme. The large water cavity in Fig. 18B is actually separated into three smaller cavities connected by water paths. In any case, water in the bulk water phase on the matrix side is accessible to Arg-38 located on the upper end of the cavity in Fig. 18B. From Arg-38, a hydrogen bond network starts and ends at Asp-51 in the fully oxidized state. Thus, Asp-51 can take up protons from the matrix side when it is hydrogen-bonded to the Ser-441 peptide N–H group in the fully oxidized state. On reduction, the COOH group of Asp-51 is released from the hydrogen bond network and migrates toward the molecular surface where the group is exposed to the water phase on the intermembrane side. In this conformation, the COOH group will be dissociated to the $COO^-$ form. Thus, the conformational change in the amino acid residue strongly suggests that this is the proton-pumping site.

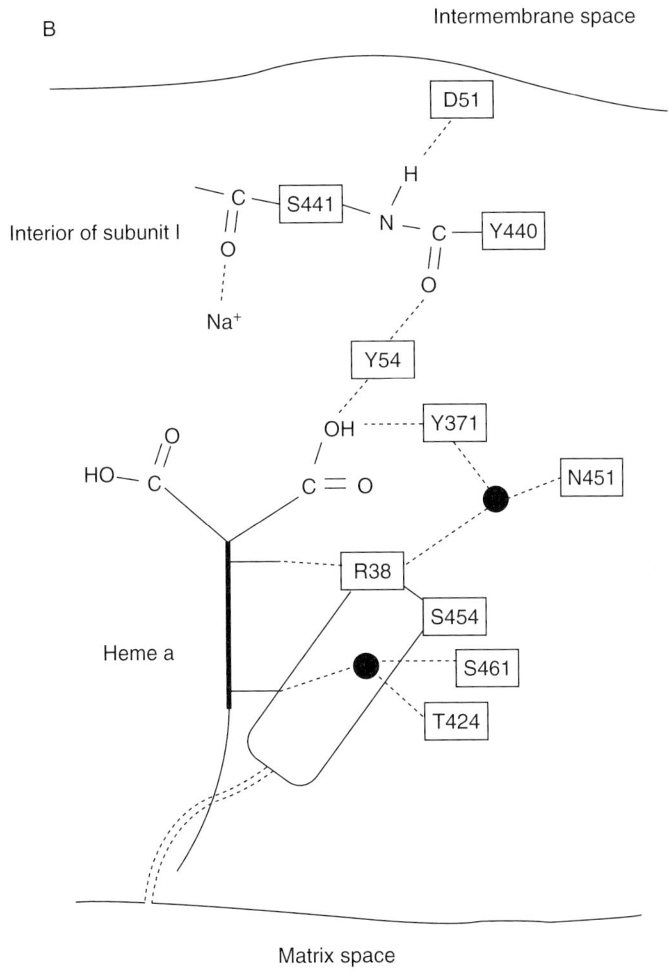


FIG. 18. Redox-coupled conformational change in a loop between helices I and II of subunit I. A stereoview (A, see color insert) and a schematic representation of the hydrogen bond network connecting Asp-51 with the matrix space (B). (A) The molecular surface on the intermembrane side is shown by small dots. Maroon and green sticks represent the structures in the fully oxidized and reduced states. (B) Dotted lines show hydrogen bonds. The rectangle represents a cavity near heme a. The two dotted lines connecting the matrix surface and the cavity represent the water path. The dark balls show the positions of the fixed water molecules.

### *F. Undirectional Proton Transfer through the Peptide Bond within the Hydrogen Bond Network*

It is well known that N–H protons in peptide bonds are quickly exchanged with deuterons. The H–D exchange does not proceed directly through–CO–$N^+$HD–but through an imidic acid–C(OD)=$N^+$H–since the CO group is much more basic than the –NH– group (Perrin, 1989; Perrin *et al.*, 1984). Thus, once the imidic acid is formed, Asp-51 will receive the acidic proton readily to leave the enol form of peptide –C(OH)=N–. The enol form is much less stable than the keto form (–CO=NH–). Thus, the enol form will be readily tautomerized back to the keto form. In this tautomeric transition, a conformational change is required for the proton transfer. However, the conformational change could be very small. The difference in stability between the two tautomers seems sufficient for the energy source of the conformational change. In addition to these two factors involved in facilitating the irreversible proton transfer through the peptide bonds, the irreversible electron transfer from heme a to heme $a_3$ contributes to the undirectional proton transfer. As described above, oxidation of heme iron decreases the p*K* value of the two propionates significantly. This could provide the strong driving force required to protonate the peptide carbonyl group of Tyr-440, which is connected by the short hydrogen bond networks shown in Fig. 18B. The positive charge of Arg-38 is likely to prevent movement of the protons released from heme a propionate in the direction of the water channel. The free energy produced at the $O_2$ reduction site is utilized at least partly for the irreversible protonation event. This structure indicates the physiological role of heme a. On the other hand, heme $a_3$, when it binds $O_2$, oxidizes heme a irreversibly. When heme a is reduced by $Cu_A$, the p*K* value of the propionate increases. However, the propionate cannot take up protons from the peptide group, since the peptide has been tautomerized back to the keto form. Thus, the propionate will be protonated by protons transferred from Arg-38. The Arg-38 will be readily protonated by bulk waters or hydronium ions transferred from the matrix space through the water channel. This proton-pumping pathway (known as the H-channel) does not include the $O_2$ reduction site. Thus, protons to be pumped remain distinct from the protons used for making water molecules.

### *G. Comparison of the H-Channel of Bovine Heart Cytochrome c Oxidase with Those of Another Species*

The key residue of the H-channel of bovine heart cytochrome *c* oxidase, Asp-51, is conserved only in the animal kingdom. The enzyme of plants and bacteria does not have this residue. On the other hand, the amino acids holding the three redox-active metals (with the exception of $Cu_A$) are completely conserved in all biological species. Due to the lack of conservation of Asp-51, its proposed role as the proton-pumping site remains controversial. However, several amino acids, and even water, are capable of reversible proton binding. Thus, Asp-51 can be replaced with many protonatable amino acids. On the other hand, $O_2$ reduction by the enzyme is a very complex process as described above. Thus, no better system for $O_2$ reduction than the heme–copper system has been discovered. This is the reason that D51 is not conserved. However, many amino acids such as Arg-38 in the H-channel are highly conserved. In fact, the X-ray structure of the *Paracoccus* enzyme shows an H-channel very similar to that in bovine heart enzyme (Ostermeier *et al.*, 1997). In the *Paracoccus* H-channel, Asp-51 is replaced with a cavity capable of holding water molecules that could transfer protons (Yoshikawa *et al.*, unpublished results). This finding strongly suggests that all of the heme–copper terminal oxidases have an H-channel coupled to the chemical events at the $O_2$ reduction site.

Recently, mutagenesis of the amino acids involved in the H-channel has been reported (Pfitzner *et al.*, 1998; Lee *et al.*, 2000). However, no appreciable effect for many of these mutations has been obtained with regard to enzyme activity and proton-pumping function. Mutagenesis of the amino acids involved in the water pathway would not have any significant effect on the function of this enzyme, since most of the mutated amino acids are likely to widen the water pathway. On the other hand, mutation of Arg-38 with methionine essentially abolishes the enzyme activity (Kannt *et al.*, 1999).

Mutation of an amino acid influences the environment more or less. Thus, positive evidence of mutagenesis is reliable. For example, mutation of Lys-317 in the K-channel abolishes enzyme activity. Thus, Lys-317 is indispensable for the enzyme activity of this enzyme. However, if no significant effect is detected by mutation of an amino acid, it is not easy to conclude that the amino acid is not directly involved in the function of the enzyme, since secondary effects may compensate for the mutation.

X-ray structures of bovine heart enzyme are available at 2.3-Å resolution in the fully oxidized state and at 2.35-Å resolution in the fully reduced state. However, spectroscopic studies remain a requirement for improvement of our understanding of the function of metal sites.

## Acknowledgments

I sincerely acknowledge Dr. W. S. Caughey, Dr. T. Tsukihara, Dr. K. Shinzawa-Itoh, Dr. T. Ogura, and Dr. T. Kitagawa, who stimulated my enthusiasm for studying cytochrome *c* oxidase during a long-term collaboration with them.

## REFERENCES

Abramson, J., Riistama, S., Larsson, G., Jasaitis, A., Svenson-EK, M., Laakkonen, L., Puustinen, A., Iwata, S., and Wikström, M. (2000). *Nat. Struct. Biol.* **7**, 910–917.

Ädelroth, P., Gennis, R., and Brzezinski, P. (1998). *Biochemistry* **37**, 2470–2476.

Alben, J. O., Moh, P. P., Fiamingo, F. G., and Altschuld, R. A. (1981). *Proc. Natl. Acad. Sci. USA* **78**, 234–237.

Anfinsen, C. B. (1973). *Science* **181**, 223 230.

Antonini, E., Brunori, M., Colosimo, A., Greenwood, C., and Wilson, M. T. (1977). *Proc. Natl. Acad. Sci. USA* **74**, 3128–3132.

Babcock, G. T., Vickery, L. E., and Palmer, G. (1978). *J. Biol. Chem.* **253**, 2400–2411.

Babcock, G. T., and Wikström, M. (1992). *Nature* **356**, 301–309.

Baker, G. M., Noguchi, M., and Palmer, G. (1987). *J. Biol. Chem.* **262**, 595–604.

Behr, J., Hellwig, P., Mäntele, W., and Michel, H. (1998). *Biochemistry* **37**, 7400–7406.

Behr, J., Michel, H., Mäntele, W., and Hellwig, P. (2000). *Biochemistry* **39**, 1356–1363.

Beinert, H., Shaw, R. W., Hamsen, R. E., and Hartzell, C. R. (1980). *Biochim. Biophys. Acta* **591**, 458–470.

Blackmore, R. S., Greenwood, C., and Gibson, Q. H. (1991). *J. Biol. Chem.* **266**, 19245–19249.

Blair, D. F., Bocian, D. F., Babcock, G. T., and Chan, S. I. (1982). *Biochemistry* **21**, 6928–6935.

Blair, D. F., Ellis, W. R., Jr., Wang, H., Gray, H. B., and Chan, S. I. (1986). *J. Biol. Chem.* **261**, 11524–11537.

Brandt, U., Schägger, H., and Von Jagow, G. (1989). *Eur. J. Biochem.* **182**, 705–711.

Buse, G., Steffens, G. J., Steffens, G. C. M., Meinecke, L., and Hensel, S. (1986). *In* "Advanced Methods in Protein Microsequence Analysis" (Wittman-Liebold, Ed.), pp. 340–351. Springer-Verlag, Berlin.

Capitanio, N., Capitanio, G., Boffoli, D., and Papa, S. (2000). **39**, 15454–15461.

Caughey, W. S., Smythe, G. A., O'Keefe, D. H., Maskasky, J. E., and Smith, M. L. (1975). *J. Biol. Chem.* **250**, 7602–7622.

Caughey, W. S., Wallace, W. J., Volpe, J. A., and Yoshikawa, S. (1976). *In* "The Enzymes" (Boyer, P. D., Ed.), 3rd ed., Vol. XIII, pp. 299–344. Academic Press, New York.

Chance, B., Saronio, C., and Leigh, J. S., Jr. (1975). *J. Biol. Chem.* **250**, 9226–9237.

Chothia, C., Levitt, M., and Richardson, D. (1977). *Proc. Natl Acad. Sci. USA* 74, 4130–4134.

Einansdottier, O., and Caughey, W. S. (1984). *Biochem. Biophys. Res. Commun.* **124**, 836–842.

Einansdottier, O., and Caughey, W. S. (1985). *Biochem. Biophys. Res. Commun.* **125**, 840–847.

Errede, B., Haight, G. P., and Kamen, M. D. (1976). *Proc. Natl. Acad. Sci. USA* **73**, 113–117.

Espe, M. P., Hosler, J. P., Ferguson-Miller, S., Babcock, G. T., and McCrachen, J. (1995). *Biochemistry* **34**, 7593–7602.

Fager, L. Y., and Alben, J. O. (1972). *Biochemistry* **11**, 4786–4792.

Fei, M. J., Yamashita, E., Inoue, N., Yao, M., Yamaguchi, H., Tsukihara, T., Shinzawa-Itho, K., Nakashima, R., and Yoshikawa. S. (2000). *Acta. Crystallogr. Sect.* D **56**, 529–535.

Ferguson-Miller, S., and Babcock, G. T. (1995). *Chem. Rev.* **96**, 2889–2907.

Ferguson-Miller, S., Brautigan, D. L., and Margohiash, E. (1976). *J. Biol. Chem.* **251**, 1104–1115.

Garavito, R. M., and Picot, D. (1990). *Methods* **1**, 57–69.

Geles, J., Blair, D. F., and Chan, S. I. (1986). *Biochim. Biophys. Acta* **853**, 205–236.

Gennis, R. B. (1998). *Biochim. Biophys. Acta* **1365**, 241–248.

Gibson, Q. H., and Greenwood, C. (1963). *Biochem. J.* **86**, 541–555.

Gilmour, M. V., Wilson, D. F., and Lemberg, R. (1967). *Biochim. Biophys. Acta* **143**, 487–499.

Greenwood, C., Hill, B. C., Barber, D., Eglinton, D. G., and Thomson, A. J. (1983). *Biochem. J.* **215**, 303–316.

Griffiths, D. E., and Wharton, D. C. (1961). *J. Biol. Chem.* **236**, 1857–1862.

Han, S., Ching, Y.-C., and Rousseau, D. L. (1990). *Proc. Natl. Acad. Sci. USA* **87**, 2491–2495.

Hellwig, P., Rost, B., Kaiser, U., Ostermeire, C., Michel, H., and Mäntele, W. (1996). *FEBS Lett.* **385**, 53–57.

Hellwig, P., Soulimane, T., Buse, G., and Mäntele, W. (1999). *FEBS Lett.* **458**, 83–86.

Hill, B. C. (1994). *J. Biol. Chem.* **269**, 2419–2425.

Hill, B. C., Greenwood, C. and Nichols, P. (1986). *Biochim. Biophys. Acta* **853**, 91–113.

Hosler, J. P., Espe, M. P., Zhen, Y., Babcock, G. T., and Ferguson-Miller, S. (1995). *Biochemistry* **34**, 7586–7592.

Iwase, T., Varotsis, C., Shinzawa-Itho, K., Yoshikawa, S., and Kitagawa, T. (1999). *J. Am. Chem. Soc.* **121**, 1415–1416.

Iwata, S., Ostermeier, C., Ludwig, B., and Michel, H. (1995). *Nature* **376**, 660–669.

Jones, M. G., Bicker, D., Wilson, M. T., Brunori, M., Colosimo, A., and Sarti, P. (1984). *Biochem. J.* **220**, 57–66.

Kadenbach, B., Ungïbauer, M., Jarausch, J., Büge, U., Kuhn-Nentwig, L. (1983). *Trends Biochem. Sci.* **8**, 398–400.

Kannt, B., Pfitzner, U., Ruitenberg, M., Hellwig, P., Ludwig, B., Mäntele, W., Fendler, K., and Michel, H. (1999). *J. Biol. Chem.* **274**, 37974–37981.

Keilin, D., and Hartree, E. F. (1938a). *Nature* **141**, 870.

Keilin, D., and Hartree, E. F. (1938b). *Proc. R. Soc. London Ser. B* **125**, 171–186.

Keilin, D., and Hartree, E. F. (1939). *Proc. R. Soc. London Ser. B Biol. Sci.* **127**, 167–191.

Kitagawa, T., and Ogura, T. (1997). *In* "Progress in Inorganic Chemistry" (Karlin, K. D., Ed.), Vol. 45, pp. 431–479.

Konstantinov, A., Siletsky, S., Mitchel, D., Kaulen, A., and Gennis, R. (1997). *Proc. Natl. Acad. Sci. USA* **94**, 9085–9090.

Kroneck, P. M. H., Antholine, W. E., Kastrau, D. H. W., Buse, G., Steffens, G. C. M., and Zumft, W. G. (1990). *FEBS Lett.* **268**, 274–276.

Kroneck, P. M. H., Antholine, W. E., Riester, J., and Zumft, W. G. (1988). *FEBS Lett.* **242**, 70–74.

Kubota, T., and Yoshikawa, S. (1993). *Biochem. J.* **292**, 519–524.

Landau, E. M., and Rosenbusch, J. P. (1996). *Proc. Natl. Acad. Sci. USA* **93**, 14532–14535.

Larsen, R. W., Pan, L.-P., Musser, S. M., Li, Z., and Chan, S. I. (1992). *Proc. Natl. Acad. Sci. USA* **89**, 723–727.

Lee, H.-M., Das, T. K., Rousseau, D. L., Mills, D., Feguson-Miller, S., and Gennis, R. B. (2000). *Biochemistry* **39**, 2989–2996.

Li, W., and Palmer, G. (1993). *Biochemistry* **32**, 1833–1843.

Lübben, M., and Gerwert, K. (1996). *FEBS Lett.* **397**, 303–307.

Malmström, B. G. (1990). *Chem. Rev.* **90**, 1247–1260.

Malmström, M. G. (1973). *Quart. Rev. Biophys.* **6**, 389–431.

Maxwell, J. C., and Caughey, W. S. (1978). *Methods Enzymol.* **54**, 302–323.

McCoy, S., and Caughey, W. S. (1970). *Biochemistry* **9**, 2387–2393.

Michel, H., and Harrenga, A. (1999). *J. Biol. Chem.* **274**, 33296–33299.

Minnart, K. (1961). *Biochim. Biophys. Acta* **50**, 23–34.

Mochizuki, M., Aoyama, H., Shinzawa-Itho, K., Usui, T., Tsukihara, T., and Yoshikawa, S. (1999). *J. Biol. Chem.* **274**, 33403–33411.

Moody, A. J. (1996). *Biochim. Biophys. Acta* **1276**, 6–20.

Mori, M, and Weil, J. A. (1967). *J. Am. Chem. Soc.* **89**, 3732–3744.

Moss, T. H., Shapiro, E., King, T. E., Beinert, H., and Hartzell, C. (1978). *J. Biol. Chem.* **253**, 8072–8073.

Ogura, T., Takahashi, S., Shinzawa-Itho, K., Yoshikawa, S., and Kitagawa, T. (1990). *J. Am. Chem. Soc.* **112**, 5630–5631.

Ogura, T., Takahashi, S., Hirota, S., Shinzawa-itho, K., Yoshikawa, S., and Kitagawa, T. (1993). *J. Am. Chem. Soc.* **115**, 8527–8536.
Okunuki, K., Sekuzu, I., Yonetani, T., and Takemori, S. (1958). *J. Biochem.* **45**, 847–854.
Ostermeier, C., Harrenga, A., Ermler, U., and Michel, H. (1997). *Proc. Natl. Acad. Sci. USA* **94**, 10547–10533.
Papa, S., Capitanio, N., and Villani, G. (1998). *FEBS Lett.* **439**, 1–8.
Perrin, C. (1989). *Acc. Chem. Res.* **22**, 268–275.
Perrin, C., Lollo, C. P., and Johnston, E. R. (1984). *J. Am. Chem. Soc.* **106**, 2749–2753.
Pfitzner, U., Odenwald, A., Osterman, T., Weingard, T., Ludwig, B., and Richter, O.-M. H. (1998). *J. Bioenerg. Biomembr.* **30**, 89–97.
Proshlyakov, D. A., Ogura, T., Shinzawa-Itho, K., Yoshikawa, S., Appleman, E. H., and Kitagawa, T. (1996). *Biochemistry* **35**, 8580–8586.
Puustinen, A., Bailey, J. A., Dyer, R. B., Mecklenburg, S. L., Wikström, M., and Woodruff, W. H. (1997). *Biochemistry* **36**, 13195–13200.
Ragan, J. J., Ramirey, B. E., Winkler, J. R., Gray, H. B., and Malmström, B. G. (1998). *J. Bioenerg. Biomembr.* **30**, 35–39.
Reinhammer, B., Malkin, R., Jensen, P., Karlsson, B., Andreasson, L.-E., Aasa, R., Vanngard, T., and Malmstrom, B. G. (1980). *J. Biol. Chem.* **255**, 5000–5003.
Rich, P. R. (1995). *Aust. J. Plant. Physiol.* **22**, 479–486.
Robinson, N. C. (1982). *Biochemistry* **21**, 184–188.
Robinson, N. C., Strey, F., and Talbert, L. (1980). *Biochemistry* **19**, 3656–3661.
Schroedl, N. A., and Hartzell, C. R. (1977). *Biochemistry* **16**, 1327–1333.
Sedlák, E., and Robinson, N. C. (1999). *Biochemistry* **38**, 14966–14972.
Shinzawa-Itoh, K., Ueda, H., Yoshikawa, S., Aoyama, H., Yamashita, E., and Tsukihara, T. (1995). *J. Mol. Biol.* **246**, 572–575.
Soulimane, T., Buse, G., Bourenkov, G. P., Bartunik, H. D., Huber, R., and Than, M. E. (2000). *EMBO J.* **19**, 1766–1776.
Speck, S. H., Dye, D., and Margohiash, E. (1984). *Proc. Natl. Acad. Sci. USA* **81**, 347–351.
Stannard, J. N., and Horecker, B. L. (1948). *J. Biol. Chem.* **172**, 599–608.
Steffens, G. C. M., Biewald, R., and Buse, G. (1987). *Eur. J. Biochem.* **164**, 295–300.
Steffens, G. C. M., Soulimane, T., Wolff, G., and Buse, G. (1993). *Eur. J. Biochem.* **213**, 1149–1157.
Takemori (1960). *J. Biochem.* **47**, 382–390.
Tiesjema, R. H., Muijsers, A. O., and Van Gelder, B. F. (1973). *Biochim. Biophys. Acta* **305**, 19–28.
Thomson, A. J., Johnson, M. K., Greenwood, C., and Gooding, P. E. (1981). *Biochem. J.* **193**, 687–697.
Thomson, A. J., Eglinton, D. G., Hill, B. C., and Greenwood, C. (1982). *Biochem. J.* **207**, 167–170.
Tsukihara, T, Aoyama, H., Yamashita, E., Tomizaki, T., Yamaguchi, H., Shinzawa-Itoh, K., Nakashima, R., Yaona, R., and Yoshikawa, S. (1995). *Science* **269**, 1069–1074.
Tsukihara, T., Aoyama, H., Yamashita, E., Tomizaki, T., Yamaguchi, H., Shinzawa-Itoh, K., Nakashima, R., Yaono, R., and Yoshikawa, S. (1996). *Science* **272**, 1136–1144.
Tweedle, M. F., Wilson, L. J., Garcia-Iniguez, L., Babcock, G. T., and Palmer, G. (1978). *J. Biol. Chem.* **253**, 8065–8071.
Tzagoloff, A., and Wharton, D. C. (1965). *J. Biol. Chem.* **240**, 2628–2633.
Van Gelder, B. F. (1966). *Biochim. Biophys. Acta* **118**, 36–46.
Van Gelder, B. F., and Beinert, H. (1969). *Biochim. Biophys. Acta* **189**, 1–24.
Vanneste, W. (1966). *Biochemistry* **5**, 838–848.
Varotsis, C., Woodruff, W. H., and Babcock, G. T. (1990). *J. Am. Chem. Soc.* **112**, 1297.

Verkhovsky, M. I., Jasaitis, A., Verkhovskaya, M. L., Morgan, J. E., and Wikström, M. (1999). *Nature* **400**, 480–483.
Vygodina, T. V., Pecoraro, C., Mitchel, D., Gennis, R., and Konstantinov, A. (1998). *Biochemistry* **37**, 3053–3061.
Warburg, O. (1924). *Biochem. Z.* **152**, 479–494.
Wever, R., Van Drooge, J. H., Muijsers, A. O., Baker, E. P., and Van Gelden, B. F. (1977). *Eur. J. Biochem.* **73**, 149–154.
Wharton, D. C., and Tzagoloff, A. (1964). *J. Biol. Chem.* **239**, 2036–2041.
Wikström, M. (1989). *Nature* **338**, 776–778.
Wikström, M., and Morgan, J. E. (1992). *J. Biol. Chem.* **267**, 10266–10273.
Wikström, M., Bogachev, A., Finel, M., Morgan, J. E., Puustinen, A., Raitio, M., Verkhovskaya, M., and Verkhovsky, M. I. (1994). *Biochim. Biophys. Acta* **1187**, 106–111.
Wikström, M. K. F. (1977). *Nature* **266**, 271–273.
Wilson, D. F., and Miyata, Y. (1977). *Biochim. Biophys. Acta* **461**, 218–230.
Wilson, D. F., and Nelson, D. (1982). *Biochim. Biophys. Acta* **680**, 233–241.
Woodruff, W. H., Einasdóttier, Ó., Dyer, R. B., Bagley, K. A., Palmer, G., Atherton, S. J., Goldbeck, P. A., Dawes, T. D., and Kliger, D. S. (1991). *Proc. Natl. Acad. Sci. USA* **88**, 2588–2592.
Yakushiji, E., and Okunuki, K. (1941). *Proc. Imp. Acad. Jpn.* **17**, 38–40.
Yonetani, T. (1959). *J. Biochem.* **46**, 917–924.
Yonetani, T. (1960). *J. Biol. Chem.* **235**, 845–852.
Yonetani, T. (1961). *J. Biol. Chem.* **236**, 1680–1688.
Yoshikawa, S., and Caughey, W. S. (1982). *J. Biol. Chem.* **257**, 412–420.
Yoshikawa, S., and Caughey, W. S. (1990). *J. Biol. Chem.* **265**, 7945–7958.
Yoshikawa, S., and Caugey, W. S. (1992). *J. Biol. Chem.* **267**, 9757–9766.
Yoshikawa, S., Choc, M. G., O'Toole, M. C., and Caughey, W. S. (1977). *J. Biol. Chem.* **252**, 5498–5508.
Yoshikawa, S., O'Keefe, D. H., and Caughey, W. S. (1985). *J. Biol. Chem.* **260**, 3518–3528.
Yoshikawa, S., Tera, T., Takahashi, Y., Tsukihara, T., and Caughey, W. S. (1988). *Proc. Natl. Acad. Sci. USA* **85**, 1354–1358.
Yoshikawa, S., Mochizuki, M., Zhao, X.-J., and Caughey, W. S. (1995). *J. Biol. Chem.* **270**, 4270–4279.
Yoshikawa, S., Shinzawa-Itho, K., Nakashima, R., Yaono, R., Yamashita, E., Inoue, N., Yao, M., Fei, M. J., Peters, Libeu, C., Mizushima, T., Yamaguchi, H., Tomizaki, T., and Tsukihara, T. (1998). *Science* **280**, 1723–1729.
Yuan, H., Autholine, W. E., and Kroneck, P. M. H. (1998). *J. Inorg. Biochem.* **71**, 99–107.
Zaslansky, D., and Gennis, R. B. (1998). *Biochemistry* **37**, 3062–3067.
Zhao, X.-J., Sampath, V., and Caughey, W. S. (1994). *Biochim. Biophys. Res. Commun.* **204**, 537–543.

# NUCLEAR MAGNETIC RESONANCE SPECTROSCOPY STUDIES ON COPPER PROTEINS

BY LUCIA BANCI,* ROBERTA PIERATTELLI,* AND ALEJANDRO J. VILA†

*CERM, University of Florence, 50019 Sesto Fiorentino, Italy and †Biophysics Section, University of Rosario, 2000 Rosario, Argentina



## I. Introduction

Copper proteins can fill quite different biological roles. In each case, the function is determined by the three-dimensional structure of the biomolecule as well as by the coordination geometry of the metal site, which in turn determines the electronic structure of the metal ion(s) (Bertini *et al.*, 1993c, 1994a; Holm *et al.*, 1996; Solomon *et al.*, 1992).

Nuclear magnetic resonance (NMR) spectroscopy has gained an outstanding role in the characterization of biomolecular structures in solution during the past decade. In the case of metalloproteins in general,



*ADVANCES IN PROTEIN CHEMISTRY, Vol. 60*



0065-3233/02 $35.00

and of copper proteins in particular, NMR spectroscopy has been applied successfully to elucidate the structural, dynamic, and electronic features of the metal sites and as a tool to solve solution structures. Due to the close relationship between these features, NMR is a unique technique for providing information on all these aspects. Both approaches, i.e., NMR as a tool for solving solution structures and as a spectroscopic technique, will be discussed here. However, due to the different electronic and spectroscopic properties of the various classes of copper proteins, as well as their different biological functions and their molecular mass, the NMR studies on the various classes have addressed different aspects of their characterization.

## II. The Influence of the Copper Ion on the NMR Spectra

Copper has two stable magnetically active isotopes, $^{63}$Cu and $^{65}$Cu, and, therefore, copper nuclei are NMR active. However, both nuclei have a nuclear spin quantum number $I = 3/2$ and possess a quadrupole moment. The coupling between the nuclear spin moment and the quadrupole moment in a slow-rotating molecule, such as a protein, enhances the nuclear relaxation rates of the copper nuclei, making the linewidth of the copper NMR signals too broad to be detected (Harris and Mann, 1978). Consequently, NMR can be applied only to the other magnetically active nuclei present in the protein, which are essentially $^{1}$H, $^{13}$C, and $^{15}$N. If the protein is not $^{13}$C and/or $^{15}$N enriched, NMR studies are limited, in general, to $^{1}$H NMR spectra. $^{13}$C and $^{15}$N NMR spectra have been reported for only a few copper proteins. However, $^{13}$C- or $^{15}$N-labeled enzyme inhibitors have been used with unlabeled proteins to characterize their interactions with the paramagnetic copper(II) ion.

Copper in proteins may be present in two oxidation states, copper(I) or copper(II). In a few systems where multicopper centers are present, mixed valence states can be found. As copper proteins are involved in electron transfer processes or catalyze oxidative reactions, both oxidation states are physiologically relevant, and their characterization is crucial to an understanding of the protein function.

The influence of each oxidation state on the NMR spectra is quite different. Copper(I) is a d$^{10}$ metal ion and has no unpaired electrons. Hence, its influence is limited to directly bound ligand atoms whose nuclei can couple to the copper nucleus and experience the influence of its quadrupole moment. This approach has not been exploited yet in proteins. In any case, the relaxation features of nearby nuclei are not greatly affected, and standard NMR experiments can be performed on copper(I) proteins. The presence of the metal needs to be considered essentially only when solution structural studies are performed, and information on the location and coordination geometry of the metal ion is needed. Since NMR spectra cannot provide direct information in this regard, some assumptions should be made.

When the copper ion is in the oxidized state, its presence has dramatic effects on the NMR spectra. Copper(II) is a $d^9$ metal ion with one unpaired electron and consequently is paramagnetic. The magnetic moment associated with this unpaired electron exerts a nonnegligible effect on the magnetic properties of nearby nuclei through magnetic coupling. This coupling between electron and nuclear spins (hyperfine coupling) affects both the chemical shifts and the relaxation rates of nearby nuclei. The extent of the latter effect determines the detectability of the NMR lines of these nuclei, since NMR signal linewidths, $\Delta\nu$, are proportional to the nuclear transverse relaxation rates $R_2$ ($\Delta\nu = \pi R_2$). As a consequence, a strong influence of the electron spin on the nuclear spins can produce broad NMR lines. We will now briefly describe the factors determining this coupling and its influence on the NMR parameters.

### *A. The Electron–Nucleus Coupling*

Hyperfine coupling occurs via two mechanisms: contact (through chemical bonds) and dipolar (through space). The contact coupling reflects the unpaired electron spin density transferred to the resonating nucleus via chemical bonds. The transfer can be due to direct spin delocalization or to spin polarization. In both cases, the coupling constant $a_c$ is proportional to the net spin density $\rho$ at the resonating nucleus, according to

$$a_c = \frac{\mu_0}{3S}\hbar\gamma_I g_e \mu_B \rho, \tag{1}$$

where $\gamma_I$ is the magnetogyric ratio of the resonating nucleus, $g_e$ is the $g$ factor of the free electron, $\mu_B$ is the Bohr magneton, and the other symbols have their usual meanings.[1]

The through-space interaction is a dipolar coupling between the electron and nuclear magnetic moments. When the Zeeman interaction for both the electron and nuclear spins is the dominant term in the spin Hamiltonian of the system, the energy of the dipole–dipole interaction is inversely proportional to the third power of the dipole–dipole distance $r_{IS}$ according to

[1]This equation holds for the free electron. In real systems $g_e$ should be substituted by $g_{av}$, with some approximations.

$$E = \frac{\mu_I \mu_S}{r_{IS}^3}(3\cos^2\theta - 1), \tag{2}$$

where $\theta$ is the angle subtended by the $r_{IS}$ vector and the external magnetic field.[2]

## B. *The Chemical Shifts*

These two coupling mechanisms have effects both on the chemical shifts and on the relaxation rates. The contact contribution to the shift is proportional to the electron spin multiplicity and to the hyperfine contact coupling constant (McConnell and Chesnut, 1958),

$$\delta^{con} = \frac{a_c}{\hbar\gamma_I B_0}\langle S_Z \rangle, \tag{3}$$

where $a_c$ is given by Eq. (1), $\langle S_Z \rangle$ is the expectation value of the $\hat{S}_Z$ operator of an $S$ multiplet, and the other symbols have their usual meanings. As this contribution is operative only through chemical bonds, only nuclei separated by a maximum of five or six bonds from the copper ion are affected by this contribution. The contact shift contribution can directly provide an estimate of the strength of the coordination bond and of the efficiency of the spin density transfer. In blue copper proteins, contact shift values of protons in the copper ligands range from about 1000 ppm for cysteine ligands to less than 100 ppm for histidine ligands (Bertini *et al.*, 1999, 2000).

The dipolar interaction affects the chemical shift as well. A further magnetic field at the resonating nuclei is produced by the electronic magnetic anisotropy averaged on rapid molecular tumbling in solution. It is different from zero when the unpaired electron magnetic moment is anisotropic. Its contribution to the chemical shift is given by (Kurland and McGarvey, 1970)

$$\delta^{pc} = \frac{1}{12\pi r_i^3}\left[\Delta\chi_{ax}(3\cos^2\theta - 1) + \frac{3}{2}\Delta\chi_{rh}\sin^2\theta\cos 2\phi\right], \tag{4}$$

where $r_i$ is the distance of nucleus $i$ from the metal, $\Delta\chi_{ax}$ and $\Delta\chi_{rh}$ are the axial and rhombic components of the anisotropy of the magnetic susceptibility tensor, $\theta$ is the angle between the metal–nucleus vector and the $z$ axis of the $\chi$ tensor, and $\phi$ is the angle between the $x$ axis of the $\chi$ tensor and the projection of the metal–nucleus vector on the $xy$ plane. From Eq. (4) it is apparent that this through-space contribution depends only on the distance of the resonating nucleus from the metal ion (as reciprocal of

[2]Both Eqs. (1) and (2) are valid for isotropic magnetic moments, which is not always valid for the electron magnetic moment.

the third power) and on the orientation of the metal–nucleus vector with respect to the magnetic axes.

In the case of copper(II) ions, the magnetic anisotropy is small, as is easily inferred from the $g$ values obtained from electron paramagnetic resonance (EPR) spectra (Peisach and Blumberg, 1974) that reflect the magnetic anisotropy of the electronic ground level only. As the first excited level is normally not populated at room temperature, its contribution is negligible, and the anisotropy of the squared $g$ values can be taken as a good estimate of the $\chi$ anisotropy (Bertini and Luchinat, 1996).

### C. *The Nuclear Relaxation Rates*

Spin relaxation in a nucleus is induced by random fluctuations of local magnetic fields. These result from time-dependent modulation of the coupling energy of the resonating nuclear spin with nearby nuclear spins, electron spins, quadrupole moments, etc. Any time-dependent phenomenon able to modulate these couplings can contribute to nuclear relaxation. The distribution of the frequencies contained in these time-dependent phenomena is described by a correlation function, characterized by a parameter $\tau_c$, the correlation time. Its reciprocal can be considered as the maximum frequency produced by the fluctuations in the vicinity of the nuclear spin. If more than one process modulates the coupling between the nuclear spin and its surroundings, the reciprocal of the effective correlation time is the sum of the reciprocals of the various contributions

$$\tau_c^{-1} = \tau_r^{-1} + \tau_s^{-1} + \tau_M^{-1}, \tag{5}$$

where $\tau_r$ is the rotational correlation time, $\tau_S$ is the electron relaxation time (when unpaired electrons are present), and $\tau_M$ is the exchange lifetime (if chemical equilibria involving the nuclear spin are present). In most cases, one term dominates, as the contribution of the others becomes irrelevant when they differ by one order of magnitude.

The rotational correlation time, $\tau_r$, depends on the size and the shape of the molecule. In the case of spherical molecules it is isotropic and given by (Einstein, 1956; Stokes, 1956)

$$\tau_r = \frac{4\pi\eta a^3}{3kT}, \tag{6}$$

where $\eta$ is the viscosity of the medium, and $a$ is the radius of the molecule; in the case of anisotropic motions, different correlation times for the principal axes of rotation should be considered.

In the case of paramagnetic systems such as those containing copper(II), $\tau_s$ becomes important. Electron spins change states, i.e., relax, several orders of magnitude faster than nuclear spins. Electron relaxation rates are determined by the electronic structure of the metal ion and so exhibit a range of values depending on the type of metal ion, its donor atoms, and its coordination geometry, as these properties determine the energy of the electronic levels. In order to change its spin state, and consequently to relax to the equilibrium distribution, an electron may access excited electron levels (real or virtual) by coupling with the motions of the ligands. Vibrations of the coordination sphere can modulate the orbit component of the electronic magnetic moment with respect to the spin component. This change in the orbital component is transmitted to the spin component via spin–orbit coupling. In the process of returning to the ground electronic state, the electron can change its spin state and so induce spin relaxation. The efficiency of spin–orbit coupling modulation mechanisms depends on the energy level separation of the excited states with respect to $kT$ and therefore depends on the structure of the chromophore (Banci *et al.*, 1991; Orbach and Stapleton, 1972). The rates of electron relaxation range between $10^{-7}$ s for radicals to $10^{-12}$ s for very-fast-relaxing metal ions, like cobalt(II) or low-spin iron(III) (Banci *et al.*, 1991).

Copper(II) relaxes relatively slowly, with $\tau_s$ values between $10^{-8}$ and $10^{-10}$ s depending on the donor atoms and the coordination geometry (Banci *et al.*, 1991). The slower relaxation rates with respect to other metal ions is a consequence of the small orbital contribution to the electronic moment as well as of the presence of excited states at relatively high energy (Bertini and Luchinat, 1986a). In proteins where the copper ion is coordinated to oxygen or nitrogen donor atoms, the orbital component of the electron magnetic moment is very small (as shown by $g$ values close to 2). Consequently, the electron relaxation time, which dominates the correlation time for slow-rotating systems such as proteins, is on the order of $10^{-8}$–$10^{-9}$ s. In blue copper proteins, copper is strongly bound to a sulfur atom of a cysteinate ligand and has essentially trigonal geometry. In this case, the excited levels are lower in energy and electron relaxation is faster, with times of ca. $10^{-10}$ s.

Nuclear relaxation rates $R_i$ are related to spin transition probabilities. When determined by the paramagnetic center, they depend on the square of the interaction energy between the nuclear and electron spins and on a function of the correlation time $\tau_c$, which describes fluctuations in the interaction energy. The contact contribution is described by (Abragam, 1961; Bloembergen, 1957; Solomon and Bloembergen, 1956)

$$R_1^{cont} = \frac{2}{3}\left(\frac{a_c}{\hbar}\right)^2 S(S+1)\cdot\frac{\tau_c}{1+(\omega_I-\omega_S)^2\tau_c^2} \tag{7}$$

$$R_2^{cont} = \frac{1}{3}\left(\frac{a_c}{\hbar}\right)^2 S(S+1) \cdot \left(\frac{\tau_c}{1+(\omega_I-\omega_S)^2\tau_c^2} + \tau_c\right), \tag{8}$$

where $a_c$ is the same as in Eq. (1), $\omega_I$ and $\omega_S$ are the precession Larmor frequencies for the nuclear and electron spins, respectively, and all other symbols have their usual meanings. The contact coupling is usually modulated by the electron relaxation correlation time $\tau_s$. Chemical exchange between metal-bound and metal-free forms has been found to be relevant in a few cases only (for which $\tau_c^{-1} = \tau_s^{-1} + \tau_M^{-1}$). The contact contribution is seldom dominating in the case of nuclei with large $\gamma$, such as protons, separated from the metal ion by a few chemical bonds. It becomes relevant only when large contact couplings exist, such as for heteronuclei bound directly to the metal ion. Strong contact couplings have been observed for protons in blue copper proteins where nuclear relaxation of the β protons of the copper-bound cysteines is dominated by the contact interaction (*vide infra*).

The dipolar coupling with the unpaired electron is the main source of relaxation for protons and for nuclei of the metal ligands other than those coordinated to the metal ion. This dipolar coupling can be modulated by $\tau_r$ (due to motion of part or whole of the molecule) or by $\tau_s$ (from the change of state of the electron spin). Due to the relatively large molecular weight of protein molecules, electron relaxation is dominant and $\tau_c = \tau_s$. The nuclear relaxation rates are given by (Solomon, 1955)

$$R_1^{dip} = \frac{2}{15}\left(\frac{\mu_0}{4\pi}\right.\frac{^2\gamma_I^2 g_e^2 \mu_B^2 S(S+1)}{r^6}\left(\frac{\tau_c}{1+(\omega_I-\omega_S)^2\tau_c^2} + \frac{3\tau_c}{1+\omega_I^2\tau_c^2} + \frac{6\tau_c}{1+(\omega_I+\omega_S)^2\tau_c^2}\right) \tag{9}$$

$$R_2^{dip} = \frac{1}{15}\left(\frac{\mu_0}{4\pi}\right)^2 \frac{\gamma_I^2 g_e^2 \mu_B^2 S(S+1)}{r^6}\left(4\tau_c + \frac{\tau_c}{1+(\omega_I-\omega_S)^2\tau_c^2} + \frac{3\tau_c}{1+\omega_I^2\tau_c^2} + \frac{6\tau_c}{1+(\omega_I+\omega_S)^2\tau_c^2} + \frac{6\tau_c}{1+\omega_S^2\tau_c^2}\right), \tag{10}$$

where $r$ is the metal–nucleus distance.

From Eq. (10) it is evident that, in the infinite magnetic field limit, the transverse relaxation rate (and therefore the linewidth) is proportional to the correlation time. Consequently the same nucleus at the same distance from the metal would have a linewidth one order of magnitude larger in Type II copper proteins than in blue copper proteins. As an example, a proton at 5 Å from the metal in a system with $\tau_s \approx 10^{-9}$ s (e.g., copper, zinc superoxide dismutase) has a linewidth of 1000 Hz at 800 MHz. The same proton in a system with $\tau_s \approx 10^{-10}$ s (e.g., azurin) has a linewidth of 160 Hz. The difference in detectability is quite significant.

A further contribution to nuclear relaxation, called Curie spin relaxation, originates from the interaction of nuclear spins with the time-averaged static electronic magnetic moment induced by the magnetic field (Gueron, 1975; Vega and Fiat, 1976). This mechanism depends on the square of the external magnetic field and the electron spin multiplicity. In copper proteins of the size that can be studied by NMR, Curie relaxation is never sizable, due to the small $S$ value. Indeed, the increase of the applied magnetic field does not induce significant linewidth broadening (Bertini *et al.*, 1999, 2000). This mechanism is similar in nature to that of nuclear relaxation due to chemical shift anisotropy (Bertini *et al.*, 1993a).

The range of observed values for the $^1$H NMR shifts and for linewidths in the various copper types is summarized in Table I.

### *D. Influence of Polymetallic Centers on the NMR Spectra*

When more than one copper(II) ion is present in the protein, and the metal sites are close enough to interact, further effects are observed in the NMR spectra. Magnetic coupling between the different electron spin moments $S_1, S_2, \ldots, S_n$ of each metal ion gives rise to new energy levels, which are generally relatively close in energy, depending on the coupling constant between the metal ions, $J_n$. In the case of polynuclear metal clusters containing mixed-valence metal ions, complex coupling schemes are needed to describe the new energy levels originating from the magnetic coupling. The occurrence of magnetic coupling has consequences on the behavior of the electronic spins and consequently on the nuclear spins coupled to them.

The electron spin(s) of a given metal ion may relax faster if coupled to another metal ion experiencing more efficient relaxation mechanisms. This electron relaxation enhancement depends on the electron relaxation time of the more rapidly relaxing metal ion and on the magnitude of the coupling constant. In the case of isotropic coupling between two metal ions, Eq. (11) has been proposed, which describes the increase in the electron relaxation rates of the more slowly relaxing metal ion (Banci *et al.*, 1991),

$$\Delta\tau_{S1}^{-1} = \frac{2}{3}\left(\frac{J}{\hbar}\right)^2 S_2(S_2+1)\frac{\tau_{S2}}{1+(\omega_{S1}-\omega_{S2})^2\tau_{S2}^2}, \qquad (11)$$

where $\Delta\tau_{S1}^{-1}$ is the enhancement in the electronic relaxation rate of the slower ion, $\omega_{S1}$ and $\omega_{S2}$ are their respective Larmor frequencies, and $\tau_{S2}$

TABLE I

*Copper Types, Coordination Geometries, Donor Atoms, and NMR Parameters in Copper(II) Proteins*

| Cu(II) center | Idealized geometry | In-plane donors | $^1H\delta$ (ppm) (N donor ligands) | $^1H\delta$ (ppm) (S, O donor ligands) | Weak axial donor(s) | $^1H\delta$ (ppm) axial ligands |
|---|---|---|---|---|---|---|
| Type I | Trigonal | 3 donors (N, N, S) | 20–60 (His ring) | 300–900 (β-$CH_2$Cys) | 0, 1, or 2 donors (S, O) | 0–25 (γH-Met) |
| Type II | Tetragonal | 4 donors ($N_x$, $O_{4-x}$) | 5–20 (His ring) | — | 0 or 1 donors (N, O) | — |
| Type III | Tetragonal | 4 donors (N) | 15–50 (His ring) | — | 0 or 1 donors (N, O) | — |
| $Cu_A$ | Trigonal | 3 donors (N, S, S) | 20–30 (His ring) | 50–500 (β-$CH_2$Cys) | 1 donor (S, O) | — |

is the correlation time for the faster relaxing ion. Equation (11) holds within the Redfield limit, i.e., as long as the $J$ coupling is smaller than the fastest electron relaxation rate. It has been estimated that a $J$ value as low as 0.1 $cm^{-1}$ reduces by a factor of 2 the electron relaxation time of a metal ion with $\tau_s$ of $10^{-9}$ s when coupled to a $S_2 = 3/2$ ion with $\tau_s \approx 10^{-12}$ s. Furthermore, the presence of new spin levels affects the hyperfine coupling as now the unpaired electron(s) has a different spin level distribution.

The presence of magnetic coupling between metal ions also has direct effects on the NMR chemical shifts and nuclear relaxation. As a consequence of the formation of new electron spin levels for the two coupled metal ions, the hyperfine coupling between the resonating nucleus and the unpaired electron(s) is changed. This is due to the fact that the unpaired electron(s) is distributed over spin energy levels characterized by the $S'$ numbers, with a fractional population of the electron spins. This distribution should be taken into account when the hyperfine shifts and the paramagnetic nuclear relaxation rates are evaluated, through coefficients that describe the relative weight of each level. In addition, temperature changes affect the relative population of the different energy levels and so affect the temperature dependence of the shifts and of the nuclear relaxation rates.

Multinuclear metal centers occur naturally in a number of copper proteins that exhibit NMR signals narrower than those of mononuclear centers. Magnetic coupling can be induced purposely to allow detection of signals of nuclei around copper(II). For example, in copper, zinc superoxide dismutase, substitution of paramagnetic cobalt(II) for the native diamagnetic zinc(II) ion allows detection of $^1H$ NMR signals of the copper ligands. These are broadened beyond detection in the native protein (Bertini *et al.*, 1985c) (*vide infra*).

In homo-dinuclear systems, such as two copper(II) ions, no large effects are expected on the electron relaxation rates as the two metal ions relax at the same rate. However, some other relaxation mechanisms are operative, giving rise to faster electron relaxation rates (Clementi and Luchinat, 1998). Consequently, nuclear relaxation is slower than in single copper(II) systems. Several examples from model complexes are available (Brink *et al.*, 1996; Murthy *et al.*, 1997), as well as from a copper(II)-substituted zinc enzyme, the aminopeptidase from *Aeromonas proteolytica* (Holz *et al.*, 1998). In contrast, few NMR studies on native copper proteins containing two coupled copper (II) ions have been reported so far (Bubacco *et al.*, 1999).

In conclusion, the detection of NMR signals even for ligands and for residues in the vicinity of paramagnetic copper(II) ion in proteins is possible. In the case of Type I or Type III copper centers ($\tau_s \approx 10^{-10}$ s),

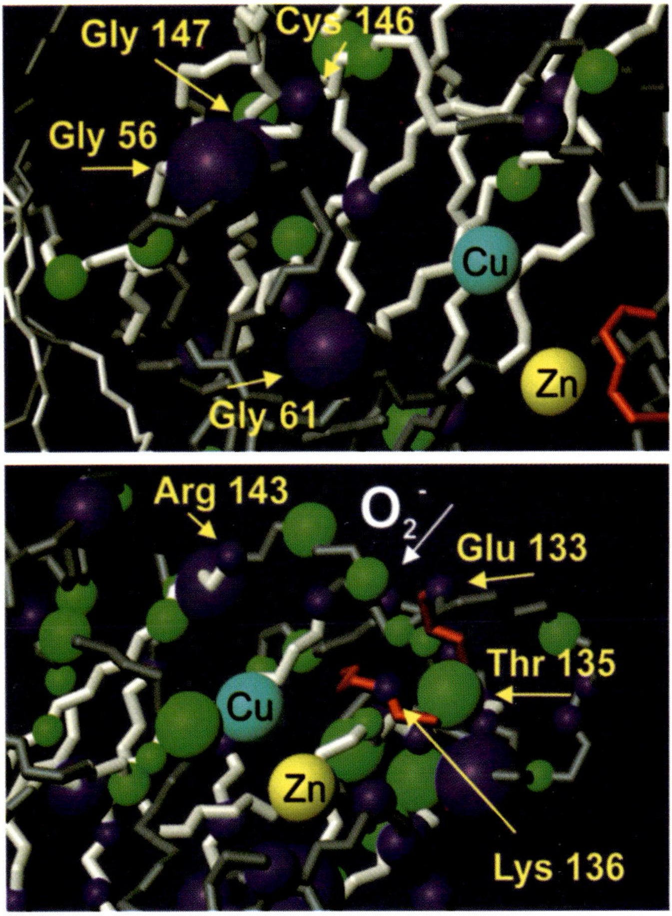


FIG. 7. Display of the differences in mobility in the microsecond–millisecond time range (top) and in the picosecond–nanosecond time range (bottom) between monomeric and dimeric human SOD. Each sphere is centered on the backbone amide nitrogen of the residue considered. The radius of the sphere is proportional to the value of the difference in order parameter, $S^2$, or in the exchange rate, $R_{ex}$. Residues less mobile (i.e., with larger $S^2$ or smaller $R_{ex}$) in the dimer are in purple, whereas residues less mobile in the monomer are in green. The Cu ion is shown in blue, and the Zn ion is in yellow. Elements of secondary structure are highlighted in white (β structure) and red (α structure) (Banci *et al.*, in press).

they can be quite broad but still detectable. The same signals are broadened beyond detection for proteins containing Type II centers.

## III. Additional NMR Tools

### *A. Metal Substitution as a Spectroscopic Probe for Elucidating Active Site Geometry*

The intrinsic difficulty in detecting $^1$H NMR signals in oxidized copper proteins gave rise to numerous NMR studies on other divalent metal derivatives (Bertini and Luchinat, 1986b, 1992; Moratal Mascarell *et al.*, 1993a, c). This approach has been useful for identifying the metal ligands and the active site geometry when the protein structure is not available (Piccioli, 1995; Salgado *et al.*, 1998a; Vila and Fernández, 1996; Vila *et al.*, 1997). These studies can be classified into two categories, depending on whether the coordinated cation is paramagnetic or diamagnetic.

The most commonly used paramagnetic metal ions are Co(II) and Ni(II). Both exhibit fast electron relaxation rates ($10^{-11}/10^{-12}$s) due to the existence of low-lying excited states at room temperature. This allows the detection of sharp signals for nuclei strongly coupled to the paramagnetic center, i.e., for the metal ligands (Banci *et al.*, 1991; Bertini and Luchinat, 1986a). Detection of such hyperfine shifted signals provides information regarding the unpaired electron delocalization on them. In any case, caution should be exercised in transferring the conclusions obtained for a different metal ion to the native Cu(II) proteins.

High-spin Co(II) exhibits a magnetic anisotropy that depends on its coordination geometry. The largest values are observed for five- and six-coordination, with smaller values for tetrahedral coordination (Banci *et al.*, 1991; Bertini and Luchinat, 1986a). As pointed out in the previous section, this magnetic anisotropy gives rise to a pseudocontact contribution to the chemical shifts which, in contrast to Cu(II), may be not negligible. Hence, interpretation of the hyperfine shifts in terms of electron spin density in the metal ligands is not straightforward in these metal derivatives as the two contributions, contact and pseudocontact, should be separated.

When Cu(II) is replaced by diamagnetic metal ions, NMR active metal ions such as Cd(II) or Hg(II) are used. In these cases, all the protein signals can be observed as well as the metal NMR signal itself. In addition, metal–ligand couplings can be detected (Harris, 1986) by heteronuclear 2D experiments such as metal–proton HSQC spectra (Blake *et al.*, 1992; Henehan *et al.*, 1993; Utschig *et al.*, 1995, 1997). These experiments provide cross-peaks originating from the metal nucleus–proton couplings, which are directly related to the MXCH dihedral angle (M, metal ion; X, donor atom). In this way, structural information on the coordination geometry of the metal site can be obtained. These can be particularly precious when paramagnetic broadening prevents acquisition of high-resolution data around the native metal ion.

### *B. Nuclear Magnetic Resonance Dispersion (NMRD)*

The NMRD technique is based on the measurement of nuclear relaxation rates of the protons of the water molecules and has been applied in a few studies on copper proteins to characterize the interaction of the solvent molecules with the copper ion and to determine its electron relaxation rate. It measures proton longitudinal nuclear relaxation rates as a function of the magnetic field. The applied magnetic field is varied typically between $2.3 \times 10^{-4}$ and 1.4 T, corresponding to a range of 0.01–60 MHz for the proton Larmor frequency. Data should be acquired over this relatively large range of magnetic fields, as measurements recorded at a few fields can be misleading in the evaluation of the many parameters that determine relaxation rates. Measurements at low magnetic fields are necessary to measure the larger contributions to relaxation rates,[3] but the sensitivity is quite low. Therefore these measurements can be applied only to solvent protons that are present at high concentration (Banci *et al.*, 1991; Koenig and Brown, 1990).

The relaxation rates of a solvent molecule bound to a protein and in fast exchange with the bulk solvent contain information on the properties of the protein itself. They are the weighted average of the relaxation rates of the free and the bound molecules relaxation rates. As the rates of the former can be measured easily, the latter can be determined. Water molecules can interact with many sites on the protein surface and in its accessible interior, including the metal ion. The former interactions cannot be characterized in a detailed way with this technique but it can be done for water molecules interacting with the metal ion, providing information on the metal coordination properties. By comparing the proton water relaxation rates of oxidized copper(II) with those of reduced copper(I) forms of the protein, the nuclear relaxation rate of the water protons bound to copper(II) can be extracted. These rates contain information on the number of water molecules bound to the copper(II) ion and their distance, according to Eq. (9). Furthermore, as faster electron relaxation rates cannot be measured at room temperature, analysis of the NMRD profiles provides an efficient way of determining $\tau_S$. In the next sections, application of this technique to the characterization of copper-bound water molecules will be also presented.

[3]Equation (9) shows that relaxation rates decrease with increasing magnetic field.

## IV. NMR Studies on Mononuclear Type I Copper Proteins

Type I copper is present at the active site of blue copper proteins (BCP; see chapter by Nersissian and Shipp, this volume) where it is involved in the transfer of a single electron, as well as in multicopper enzymes (Gray *et al.*, 2000; Malmström, 1994; Randall *et al.*, 2000; Sykes, 1991) (see Section V). BCP are single-domain proteins with a β-barrel fold defined by two β-sheets that can contain 6 to 13 strands following a Greek-key motif (Fig. 1) (Adman, 1991; Messerschmidt, 1998; Murphy *et al.*, 1997; Sykes, 1991). These proteins are stable in both the reduced, Cu(I), and the oxidized, Cu(II), forms.

The metal coordination geometry is conserved in most Type I copper sites. The copper ion is bound to a Cys Sγ atom and two His Nδ1 atoms lying approximately in the same plane (Adman, 1991; Sykes, 1991). Different axial ligand-binding motifs, which modulate functional properties, are found in different proteins (Fig. 2). Indeed, the strength of the interaction with the axial ligand is an important factor in tuning the spectral properties of the metal site (Gray *et al.*, 2000; Randall *et al.*, 2000). In most BCP, a weakly bound Met residue is the axial ligand, with Cu–Sδ(Met) bond distances ranging from 2.6 to 3.0 Å (Botuyan *et al.*,

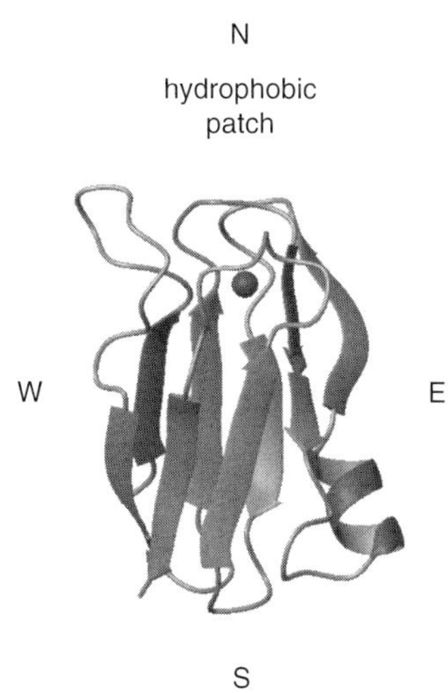


FIG. 1. Ribbon drawing of the cyanobacterium *Synechocystis* sp. PCC 6803 Cu(II)-plastocyanin structure showing the secondary structure elements of the protein. The metal ion is represented as a sphere of arbitrary dimensions (Bertini *et al.*, 2001b). The letters N, E, S, and W refer to the cardinal points. The so-called acidic patch present in the structure of plant plastocyanins is located in the E region.

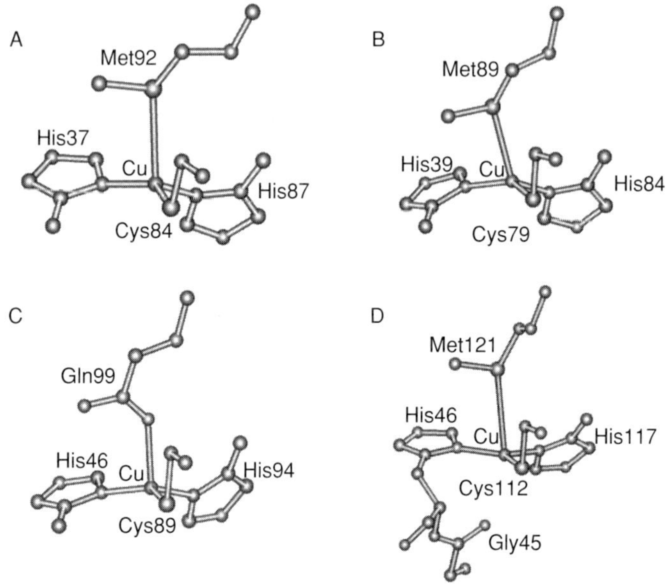


FIG.2. Schematic drawing of the metal coordination sites in typical Type I copper proteins: (A) plastocyanin; (B) rusticyanin; (C) stellacyanin; and (D) azurin.

1996; Guss *et al.*, 1992; Romero *et al.*, 1994). In stellacyanin, a glutamine Oε fulfills this role (Hart *et al.*, 1996; Vila and Fernández, 1996). In azurin, in addition to the axial methionine, a peptide carbonyl of a glycine acts as weak axial ligand located on the opposite side to Met (Baker *et al.*, 1988; Nar *et al.*, 1991) (Fig. 2). The multicopper oxidase laccase possesses a trigonal Type I site, with no axial ligand (Ducros *et al.*, 1998). However, no NMR studies are available for any protein of the latter class and they are not considered further here.

In all BCP, at least one of the His ligands is exposed to solvent, and it may provide a pathway for intramolecular electron transfer (Guss and Freeman, 1983; Van de Kamp *et al.*, 1990). This His residue is usually surrounded by a hydrophobic patch located in the so-called "northern" region of the molecule that has been suggested to be involved in molecular recognition with redox partners (Guss and Freeman, 1983) (Fig. 1). Some BCP, such as plant plastocyanins, also contain a surface acidic patch in the "eastern" region (labeled E in Fig. 1), whose role in intermolecular electron transfer has been matter of debate (see below) (Guss and Freeman, 1983; Ubbink *et al.*, 1998; Ullmann and Kostic, 1996).

### *A. The Diamagnetic Copper(I) State: Spectroscopic Studies and Solution Structures of Blue Copper Proteins*

Early NMR studies on reduced plastocyanins from different sources were aimed at characterizing the metal ligands and their acid–base equilibria before crystal structures were available (Canters *et al.*, 1984; Freeman and Morris, 1978; Hill *et al.*, 1976; Ugurbil and Bersohn, 1977). Residues that bind the metal in their deprotonated forms are tritrable in the apoprotein exclusively. These acid–base equilibria can be followed easily by $^{1}H$ NMR and, due to the relatively small size of BCP (MW 10,000–16,000), signals corresponding to the His ligands were readily identified even at the low magnetic fields available in the early 1980s. Natural abundance $^{13}C$ NMR studies were useful in establishing that the His ligands were coordinated through their Nδ1 atoms in Cu(I) *Pseudomonas aeruginosa* azurin and spinach plastocyanin (Markley *et al.*, 1977; Ugurbil *et al.*, 1977).

Reversible protonation and dissociation of the exposed His ligand have been observed in several BCP in the reduced metal-bound state. Since this protonation renders the proteins inactive, it has been characterized thoroughly (Sykes, 1985, 1991). An active site $pK_a$ of 4.9 was determined by NMR for Cu(I) spinach plastocyanin (Markley *et al.*, 1975). The occurrence of this process was confirmed later by the crystal structure of reduced poplar plastocyanin at low pH (Guss *et al.*, 1986). Similar equilibria have been characterized in *Achromobacter cycloclastes* pseudoazurin ($pK_a$ 4.6) (Dennison *et al.*, 1994b) and in *Thiobacillus versutus* amicyanin ($pK_a$ 6.7) (Lommen *et al.*, 1988). In the latter system a lineshape analysis revealed that this His residue, on protonation and detachment from the copper(I) ion, fluctuates between two conformers (Lommen and Canters, 1990).

Several solution structures of BCP in their reduced state have been determined by NMR, and most of them are available in the RCSB Protein Data Bank (see Table II). The first was that of *Scenedesmus obliquus* Cu(I) plastocyanin, solved by Wright and co-workers, and was one of the first protein structures solved by NMR (Moore *et al.*, 1988). Later, a high-resolution structure of French bean plastocyanin was determined (Moore *et al.*, 1991), using the previously published $^{1}H$ NMR assignments (Chazin and Wright, 1988; Chazin *et al.*, 1988). A number of structures from different sources are now available (Badsberg *et al.*, 1996; Bagby *et al.*, 1994; Babu *et al.*, 1999; Ma *et al.*, 2000) (see Table II), solved by using $^{1}H$ NMR experiments only. The structure of Cu(I) plastocyanin from *Synechocystis* sp. PCC 6803 has been solved from $^{1}H$ and $^{15}N$ assignments (Bertini *et al.*, 2001a). All the crystal and solution structures reveal a similar global fold, with essentially the same backbone structure and hydrogen-bonding patterns. The structures in the β-sandwich

TABLE II
*Structures of Blue Copper Proteins Solved by NMR Spectroscopy*

| PDB ID code | Protein[a] | Source | Reference |
|---|---|---|---|
| — | Plastocyanin | *Scenedesmus obliquus* | Moore *et al.*, 1988 |
| 9PCY | Plastocyanin | *Phaseolus vulgaris* (French bean) | Moore *et al.*, 1991 |
| 1PLA | Plastocyanin | *Petroselinum crispum* (parsley) | Bagby *et al.*, 1994 |
| 1PLB | Plastocyanin | *Pe. crispum* (parsley) | Bagby *et al.*, 1994 |
| 1NIN | Plastocyanin | *Anabaena variabilis* | Badsberg *et al.*, 1996 |
| 1FA4 | Plastocyanin | *An. variabilis* | Ma *et al.*, 2000 |
| 2PCF | Cd(II) plastocyanin (complexed with cytochrome *f*) | *Spinacia oleracea* (spinach) | Ubbink and Bendall, 1997 |
| 1B3I | Plastocyanin (T2S mutant) | *Prochlorothrix hollandica* | Babu *et al.*, 1999 |
| 2B3I | Plastocyanin (T2S mutant) | *Pr. hollandica* | Babu *et al.*, 1999 |
| — | Plastocyanin | *Synechocystis* sp. PCC6803 | Bertini *et al.*, 2001a |
| — | Cu(II) plastocyanin | *Synechocystis* sp. PCC6803 | Bertini *et al.*, 2001b |
| 1CUR | Rusticyanin | *Thiobacillus ferrooxidans* | Botuyan *et al.*, 1996 |
| — | Pseudoazurin | *Paracoccus pantotrophus* | Thompson *et al.*, 2000 |
| — | Amicyanin | *Thiobacillus versutus* | Kalverda *et al.*, 1994 |

[a]In the Cu(I) form, unless stated otherwise.

region are well defined, whereas a certain degree of disorder is observed in the connecting loops.

Full $^{1}H$ and $^{15}N$ assignments for *Ps. aeruginosa* and *Alcaligenes denitrificans* azurins have been reported (Hoitink *et al.*, 1994; Van de Kamp *et al.*, 1992) and a solution structure of the former is now available. It shows some minor structural differences with the crystal structure for those regions involved in intermolecular contacts in the crystal asymmetric unit (G. C. Karlsson, personal communication). The solution structure of Cu(I) *Paracoccus pantotrophus* pseudoazurin (Thompson *et al.*, 2000) is monomeric, contrasting with the formation of a dimeric species in the oxidized state, both in the crystal and in solution (as derived from ultracentrifugation experiments). Despite the different aggregation state, both structures closely resemble one another. The structure of *Thi. versutus* amicyanin was also solved using $^{1}H$ spectra only (Kalverda *et al.*, 1994; Lommen *et al.*, 1991).

*Thiobacillus ferrooxidans* rusticyanin is characterized by high stability at very low pH. Double labeling with $^{13}C$ and $^{15}N$ was helpful for NMR studies of this relatively large protein (MW 16,000). The secondary structure elements of reduced Cu(I) rusticyanin have been identified from their $^{13}C$ shifts (Toy-Palmer *et al.*, 1995), and its solution structure was solved at the same time as its crystal structure (Botuyan *et al.*, 1996; Walter *et al.*, 1996). The metal site is surrounded by hydrophobic aromatic residues that induce sizable ring current shifts on many amide protons surrounding the copper ion. These amide proton resonances are still observed in 99% $^2H_2O$ solution at pH 3.4, indicating that the copper site is protected from the solvent even at low pH (Hunt *et al.*, 1994). The hydrophobic nature of the redox site has been interpreted as a key to both the high protein stability and the high reduction potential.

The occurrence of rigid structures in BCP has been related to the need for minimal structural rearrangement during the redox process to facilitate fast electron transfer. $^{15}N$ relaxation studies on reduced *Ps. aeruginosa* azurin, *Synecchocystis* plastocyanin, and *Pa. pantotropha* pseudoazurin reveal a protein frame that is rigid on the picosecond to nanosecond timescale, consistent with a β-barrel structure (Kalverda *et al.*, 1999; Thompson *et al.*, 2000; Bertini *et al.*, 2001a). Local mobility is restricted to the protein loops connecting the β-strands (Bertini *et al.*, 2001a; Kalverda *et al.*, 1999; Thompson *et al.*, 2000). In azurin, the regions mobile in solution also show larger B factors in the crystal structure (Kalverda *et al.*, 1999). The loop in the northern region (labeled N in Fig. 1) contains three (the Cys and the two His residues) of the four copper ligands. Exchange of backbone NH protons in this loop is extremely slow in azurin, indicating a high level of protection in this region from an extensive pattern of hydrogen bonds. These are thought to constrain the loop conformation to maintain the metal site structure (Van de Kamp *et al.*, 1992). The H117G mutation destroys the characteristic features of the Type I site by eliminating the exposed His ligand and creating a cavity on the protein surface (den Blaauwen *et al.*, 1991; den Blaauwen and Canters, 1993; Hamman *et al.*, 1997). However, the overall protein structure and mobility are not altered (Jeuken *et al.*, 2000), confirming the robustness of the protein fold.

BCP participate in intermolecular electron transfer processes. Hence, once the NMR signals have been assigned, protein–protein interactions may be mapped by NMR. The early availability of the $^1H$ signal assignment of spinach plastocyanin was exploited to map the binding properties of paramagnetic Cr(III) complexes (Armstrong *et al.*, 1986; Cookson *et al.*, 1980; Handford *et al.*, 1980; Jackman *et al.*, 1987). Broadening or disappearance of a cross-peak in 2D spectra was taken as indicative of an interaction with the metal probe (Chazin *et al.*, 1987; Driscoll *et al.*, 1987). These studies provided a basis for analysis of protein–protein interactions between redox partners (Bagby *et al.*, 1990). An earlier study examined the binding of reduced spinach plastocyanin to both the diamagnetic Fe(II) and the paramagnetic Fe(III) forms of turnip cytochrome *c* (used as a model of the physiological partner, cytochrome *f*). Formation of a 1:1 adduct was deduced from changes in the chemical shifts and in the linewidths of the heme signals upon titration (Bagby *et al.*, 1990). Mapping of the chemical shift changes identified regions in spinach plastocyanin that are affected by binding of horse heart cytochrome *c*. More recently, a similar study of the interaction between pea plastocyanin and horse heart cytochrome *c* did not reveal specific paramagnetic effects, suggesting the formation of nonspecific interprotein complexes or a dynamic ensemble of adducts in fast exchange (Ubbink and Bendall, 1997). In contrast, spinach plastocyanin binds to the soluble domain of its physiological partner cytochrome *f* in a single orientation, indicating a short electron transfer path between the metal ions (Ubbink *et al.*, 1998). A low-resolution structural model of the plastocyanin–cytochrome *f* complex was obtained by including paramagnetic constraints (derived from $^{1}$H and $^{15}$N chemical shift differences) in molecular dynamics simulations where the structures of the two partners were kept rigid (Ubbink *et al.*, 1998). The model suggested the formation of an initial electrostatic complex that may later rearrange to a more stable complex in which electron transfer would be mediated through the exposed His-87 copper ligand (Ubbink *et al.*, 1998). The hydrophobic patch surrounding this ligand appears to make van der Waals contact with the heme edge, with an estimated average Cu–Fe distance of ca. 11 Å. Chemical shift mapping experiments in unlabeled mixtures of Cu(I) spinach plastocyanin and with Fe(II) turnip cytochrome *f* helped identify the surface protein regions involved in the complex interface (Ejdeback *et al.*, 2000). Chemical shift changes in the hydrophobic region surrounding the exposed His-87 were larger, suggesting strong interprotein contacts due to hydrophobic interactions. These experiments have been important in assessing the role of the hydrophobic patch and the exposed His residue in intermolecular electron transfer in blue copper proteins.

### *B. Electron Self-Exchange Rates by NMR*

The oxidized and reduced forms of blue copper proteins can coexist in equilibrium in solution. This allows two identical molecules to exchange one electron, giving the so-called electron self-exchange (ESE) reaction (Marcus and Sutin, 1985). The rates of this process usually depend on the pH, temperature, and ionic strength of the solution. When the two forms interconvert at a rate faster than the difference between the Larmor frequencies of two resonances of the same nucleus in each redox species, the system is in a fast exchange regime with respect to the chemical shift. As a consequence, a single NMR signal is observed, which is a weighted average of those of the two species. The presence of a small amount (1–5%) of the oxidized protein in fast exchange with the reduced species can be detected by line broadening of the resonances of the diamagnetic species. This effect was exploited in early NMR studies to identify signals located near the metal site and, in some cases, even to identify the metal ligands (see next section). In addition, the rate of the ESE process may be retrieved from these line-broadening effects. This reaction may not be relevant *in vivo*, where the low protein concentration does not favor the encounter of two identical molecules. However, analysis of ESE rates has been used fruitfully to elucidate the electron transfer process in BCP and it has allowed a better understanding of the electron transfer between actual partners.

Measured ESE rates for BCP range from $10^3$ to $10^6 M^{-1}s^{-1}$ (see Table III). The NMR-derived values are in excellent agreement with those measured by kinetic methods within the frame of the Marcus theory (Marcus and Sutin, 1985). While French bean plastocyanin is characterized by an ESE rate slow on the NMR timescale ($<2 \times 10^4 M^{-1}s^{-1}$) (Beattie *et al.*, 1975), that for *Ps. aeruoginosa* azurin, ca. $10^6 M^{-1}s^{-1}$, is in the fast-exchange regime (Canters *et al.*, 1984). The measured rate for azurin remains fast and fairly constant under different pH and ionic strength conditions (Groeneveld and Canters, 1985). Surprisingly, while His-35 of azurin is located close to the metal site, the conformational rearrangement occurring on its protonation does not affect the ESE rate. Based on this evidence, electron exchange between two azurin molecules was proposed to take place through the hydrophobic region surrounding the exposed His-117 copper ligand. This also represents the closest copper–copper approach possible for electron transfer (ca. 14 Å). The ESE rate decreases by up to three orders of magnitude on the introduction of charged residues into this patch through site-directed mutagenesis. This provides strong support for this hypothesis (Van de Kamp *et al.*, 1990; Van Pouderoyen *et al.*, 1997). The lower self-exchange rate in plastocyanin relative to other BCP may be attributed to the existence of charged patches on the protein surface.

The hydrophobic patch in pseudoazurin is surrounded by several positively charged Lys residues. This array is presumably involved in molecular recognition of its natural partner, nitrite reductase (Williams *et al.*, 1995). The ESE rate of *A. cycloclastes* pseudoazurin is on the order of $10^3 s^{-1}$ at neutral pH, i.e., much lower than those of other BCP, thus suggesting that the presence of these charged residues may hinder protein

TABLE III
*Electron Self-Exchange Rate Constants $k_{ese}$ (298 K) for Blue Copper Proteins Retrieved from $^1H$ NMR Spectra*

| Protein | $k_{ese}(M^{-1}s^{-1})$ | pH | p*I* | Charge[a] | Reference |
|---|---|---|---|---|---|
| *Plastocyanin (Spinacia oleracea)* | $\sim 4.0 \times 10^3$ | 6.0 | 4.2 | −9 | Christensen *et al.*, 1990 |
| *Plastocyanin (Anabaena variabilis)* | $3.2 \times 10^5$ | 7.5 | | +1 | Dennison *et al.*, 1993 |
| *Plastocyanin (Phaseolus vulgaris)* | $< 2.0 \times 10^4$ | 7.4 | | +9 | Beattie *et al.*, 1975 |
| *Amicyanin (Thiobacillus versutus)* | $1.3 \times 10^5$ | 8.6 | 4.7 | −4 | Lommen *et al.*, 1988 |
| *Pseudoazurin (Achromobacter cycloclastes)* | $2.9 \times 10^3$<br>$3.5 \times 10^3$<br>$1.7 \times 10^4$ | 7.5<br>8.2<br>10.9 | 8.4 | +1 | Dennison and Kohzuma, 1999; Dennison *et al.*, 1994a,b |
| *Rusticyanin (Thiobacillus ferrooxidans)* | $1.7 \times 10^4$<br>$1.0 \times 10^4$ | 2.0<br>5.7 | 9.1 | +4 | Kyritsis *et al.*, 1995 |
| *Azurin (Pseudomonas aeruginosa)* | $9.6 \times 10^5$ | 4.5 | 5.4 | −1 | Groeneveld and Canters, 1988 |
| | $7.0 \times 10^5$ | 9.0 | | | |
| *Azurin (Alcigenes denitrificans)* | $4.0 \times 10^5$ | 6.7 | | | Hoitink and Canters, 1992 |
| *Umecyanin (Armoracia laphatifolia)* | $6.1 \times 10^3$ | 7.5 | 5.8 | −4 | Dennison *et al.*, 1996 |

[a]Calculated for proteins containing Cu(II), Asp/Glu with charge −1, Arg/Lys with charge +1, and uncoordinated His with charge 0.

self-association (Dennison *et al.*, 1994a). The rate is consistently increased by one order of magnitude at pH above 10, due to the partial deprotonation of the Lys residues (Dennison and Kohzuma, 1999).

Rusticyanin is stable down to pH 2, and no change in the ESE rate is observed over a broad pH range (Kyritsis *et al.*, 1995). This is consistent with the hydrophobic nature of the residues that surround the active site. The lower ESE rate compared to that of azurin (cf. Table III) is suggested to be related to the longer copper–copper distance due to a more deeply buried metal site (Kyritsis *et al.*, 1995).

### C. *The Paramagnetic Copper(II) State in Blue Copper Proteins*

The first NMR studies on the paramagnetic state of *Ps. aeruginosa* azurin (Hill *et al.*, 1976) and spinach plastocyanin (Beattie *et al.*, 1975; Markley *et al.*, 1975) were reported together with experiments on the corresponding reduced and apo forms. Those studies allowed the prediction that two of the four His residues present in the amino acid sequence were copper ligands (Markley *et al.*, 1977). The approach used was based on measurement of the relaxation enhancement for nuclei close to the paramagnetic Cu(II) ion, as described in Section II, C. Those studies exploited the possibility of resolving the His signals, even at low magnetic fields, because they were shifted outside the diamagnetic envelope.

The slow electron relaxation rates of copper(II) ions were expected to broaden the NMR lines of nearby nuclei beyond detection, and high-resolution NMR studies on blue Cu(II) sites were considered unfeasible for a long time. However, when the $^1$H NMR spectra of oxidized forms from *Thi. versutus* and azurin from *Ps. aeruginosa* were recorded for the first time, relatively well-resolved lines could be observed (Kalverda *et al.*, 1996).

NMR signals of the copper(II) ligands are, in all cases, broad (Fig. 3). The low molecular weight of BCP makes it possible to detect a large number of hyperfine-shifted signals with high sensitivity and resolution, even at very high magnetic fields, since Curie relaxation is negligible. Broadened signals located beneath the diamagnetic envelope can be detected by applying 1D or 2D pulse sequences such as super WEFT (Inubushi and Becker, 1983) or WEFT-NOESY (Chen *et al.*, 1994; Salgado *et al.*, 1997) optimized for the detection of fast-relaxing signals. However, a direct NMR assignment through 1D NOE or 2D experiments is not feasible. Once located, the hyperfine shifted signals may be assigned by 1D or 2D saturation transfer experiments in a sample containing a mixture of the oxidized and reduced forms of the protein. If the two redox species are in the slow-exchange regime, two signals (one for each oxidation state) are present. If the signal of one of the species is saturated by a selective pulse, the saturation is transferred to the same nucleus in the other species. Its signal will experience a reduction in intensity, provided that the ESE rate between the two redox states in equilibrium is on the same order of magnitude as the relaxation rates of the paramagnetic species. If the assignment of one of the species is known (the diamagnetic state, in almost all the cases), it can be transferred to the other species. This approach allowed the assignment of the signals of the His and axial Met ligands in Cu(II) amicyanin and azurin (Kalverda *et al.*, 1996; Salgado *et al.*, 1997, 1998a).

The location of the β-$CH_2$ Cys signals bound to Cu(II) remained elusive until a "blind" saturation transfer experiment was performed (Bertini *et al.*, 1999, 2000). In that experiment, selective saturation was applied over a large spectral range (±2000 ppm) by shifting the position

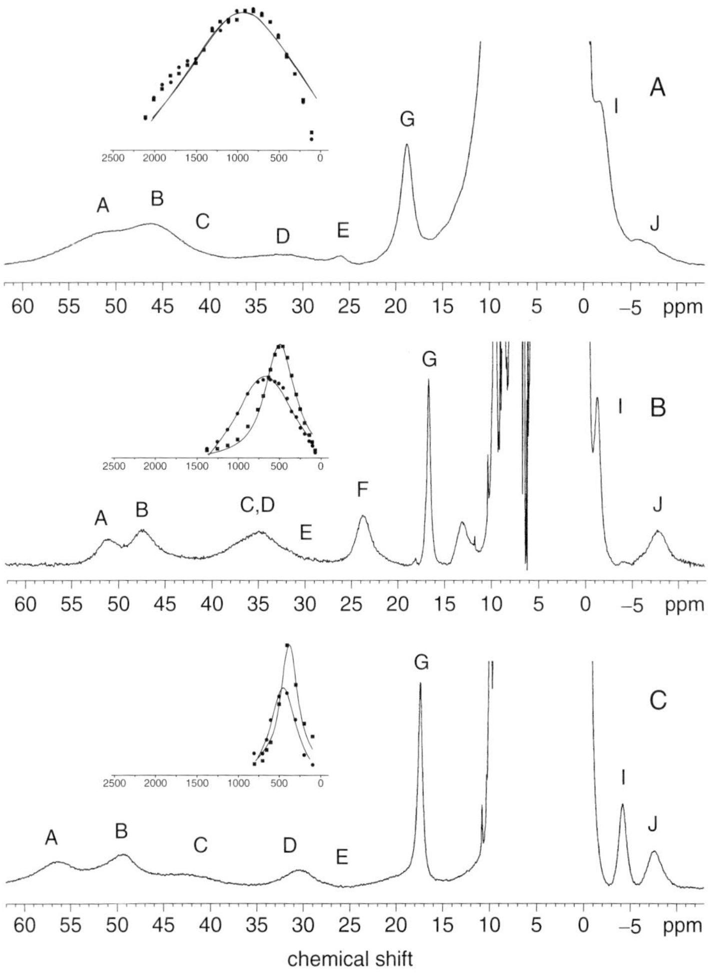


FIG.3. 800-MHz $^1$H NMR spectra of oxidized (A) *Pseudomonas aeruginosa* azurin, (B) spinach plastocyanin, and (C) cucumber stellacyanin recorded in $D_2O$ solution. The letters identify the resonance of the equivalent proton in the three proteins. In the insets the far-downfield regions containing signals not observable in direct detection are shown (Bertini *et al.*, 2000). The positions and the linewidths of the signals of the oxidized species were obtained using saturation transfer experiments over the far-downfield region by measuring the intensity of the exchange connectivity with the corresponding signal in the reduced species (Bertini *et al.*, 1999, 2000).

of the decoupler, although no signal was observed. However, if a signal broadened beyond detectable limits is present in this spectral region, saturation transfer will still be operative and it will be maximized when the decoupler exactly matches the frequency of the broad signal. These experiments allowed reconstruction of the resonances of the β-$CH_2$ Cys protons in plastocyanin, azurin, and stellacyanin (see insets of Fig. 3) from the saturation transfer profile of the signals of the diamagnetic species (Bertini *et al.*, 1999, 2000). In addition, the NH signal of a conserved Asn residue, hydrogen bonded to the Cys Sγ atom, was located in the upfield region (not shown here).

From the assignment of the hyperfine shifted signals of Cu(II) plastocyanin, azurin, and stellacyanin (Bertini *et al.*, 1999, 2000), information on the electron delocalization onto the metal ligands was gained by calculating the contact and pseudocontact contributions to the hyperfine shifts. In any case, since the magnetic anisotropy of the Cu(II) ion is low, the observed shifts can be approximated to the contact contribution, which can be used as an initial criterion to compare the electron spin density on the different nuclei and in the various proteins.

*1. The Cys Ligand*

The large hyperfine coupling constants exhibited by the β-$CH_2$ Cys protons are consistent with a strong Cu(II)–Cys bond with a high degree of covalency that gives rise to the blue color characteristic of these proteins (Randall *et al.*, 2000). The overlap between the sulfur $p_\pi$ orbital and the copper $d_{x^2-y^2}$ orbital containing the unpaired electron clearly favors delocalization of this spin density onto the β protons. In addition, their signals display quite a large range of shift values, from 375 to 850 ppm, and of linewidths among the different proteins examined (Fig. 3). The larger the chemical shifts, the broader the signals, suggesting that the major mechanism contributing to both the hyperfine shifts and the relaxation rates arises from contact coupling between the resonating nucleus and the electron spin density delocalized on the Cys protons. For the Hα Cys proton (signals J in Fig. 3), electron delocalization drops to much lower levels.

The variability of the β-$CH_2$ Cys chemical shifts and linewidths reflect the high degree of sensitivity to changes in copper coordination geometry and in its interaction with the axial ligand. In stellacyanin, the presence of a stronger axial Gln ligand weakens the copper(II)–Cys bond with respect to plastocyanin, where a weaker Met axial ligand is present. In azurin, the Cu(II) ion lies in the equatorial $His_2Cys$ plane, the axial ligands (Met-121 and Gly-45) both being at a longer distance from the metal ion. The presence of two weaker binding axial ligands on both sides of the strong ligands plane forces the metal ion to lie in the plane, thus resulting in the largest Cu(II)–β-$CH_2$ Cys couplings. In summary, a stronger interaction with the axial ligand results in a weakened electron delocalization in the Cys ligand (Bertini *et al.*, 2000). The amide NH of the conserved Asn residue, which is hydrogen bonded to the Cys sulfur and experiences a contact interaction in all the BCP studied, exhibits the same qualitative trend observed for the β-$CH_2$ Cys protons (Bertini *et al.*, 2000).

*2. The Histidine Ligands*

The two His ligands are coordinated to the copper ion through the Nδ1 atom. The signals of the His ring protons usually fall between 30 and 60 ppm. The His signal pattern is conserved in the three proteins studied; the trend in the contact shifts follows the order Hδ2 > Hε1 > Hε2, indicating the existence of a similar delocalization mechanism on the imidazole rings in the different proteins. The hyperfine coupling constants of the Hδ2 protons of both His residues are comparable in the three proteins, the larger shift being experienced by the buried histidine ligand. The His resonances can provide information on the exchange of the His ring NH with the solvent.

*3. The Axial Ligand*

As the axial ligand is weakly bound in BCP (Randall *et al.*, 2000), the spin density delocalized on it is small. Indeed, in azurin the resonances of the axial methionine protons do not experience a significant hyperfine shift contribution. Electron delocalization onto a Hγ of the axial Met has been detected in plastocyanin (signal F in Fig. 3B), suggesting some covalency for the Cu–S(Met) bond. The absence of spin density on the axial Gln ligand in stellacyanin has been attributed to the fact that the γ-$CH_2$ Gln geminal couple is four bonds away from the metal ion, whereas the equivalent protons in a bound Met residue (such as in plastocyanin) are only three bonds away (Bertini *et al.*, 2000).

The recent progress in NMR studies of oxidized BCP has led to the determination of the solution structure of Cu(II) plastocyanin from *Synecchocystis* sp. PCC 6803, which represents the first solution structure of a paramagnetic, oxidized copper protein (Bertini et al., 2001b). This has been achieved on a $^{15}N$-labeled protein by using standard 2D and 3D heteronuclear NMR pulse sequences that were tailored to the fast-relaxing signals of nuclei in the vicinities of the metal site and by applying constraints derived from the paramagnetic center in the structural calculation. The availability of this structure has allowed the comparison with the solution structure of the reduced form (Bertini *et al.*, 2001a), revealing no significant changes in the protein structure on the redox process. Analysis of the $^{1}H$ and $^{15}N$ chemical shift changes in this region indicates that minor conformational rearrangements may occur between the two redox states.

### *D. Metal Substitution in Type I Copper Proteins*

#### *1. Co(II) Substitution*

Co(II) has been the most useful metal probe for the study of BCP. The Co(II) derivatives of *Ps. aeruginosa* azurin (Moratal Mascarell *et al.*, 1993a; Piccioli *et al.*, 1995; Salgado *et al.*, 1995), *Rhus vernicifera* stellacyanin (Fernandez *et al.*, 1997; Vila, 1994; Vila and Fernández, 1996), *Ac. cycloclastes* pseudoazurin (Fernández et al., 2001), *Thi. ferrooxidans* rusticyanin (Donaire *et al.*, 2001), *Thi. versutus* amicyanin (Salgado *et al.*, 1999), several mutants of azurin (Piccioli *et al.*, 1995; Salgado *et al.*, 1996, 1998a; Vila *et al.*, 1997), and the M99Q mutant of amicyanin (Diederix *et al.*, 2000) have been prepared, and their $^{1}$H NMR spectra have been characterized.

The $^{1}$H NMR spectra of Co(II)-substituted BCP are characterized by a large dispersion of signals that, nevertheless, is smaller than that observed in the native Cu(II) proteins. All the hyperfine shifted signals can be detected directly due to the narrower linewidths (Fig. 4). The chemical shift range is not only due to differences in the electron delocalization (contact shifts), but also to the considerable magnetic anisotropy of the Co(II) ion, even when tetracoordinated (Donaire *et al.*, 1998). As a consequence, sizable pseudocontact shifts are induced on nuclei close to the metal ion, whether or not they belong to the metal ligands.

*a. The Cys Ligand.* The β-$CH_2$ Cys resonances can be easily recognized since they are usually the most downfield shifted and the broadest signals (several thousands of hertz) in the spectrum. However, due to these features, they might be not readily detected.

The chemical shifts of these signals are highly sensitive to changes in the electron spin density and in the conformation of the Cys side chain. The shift values of each β-Cys proton are determined by the dihedral angle subtended by the Co(II)–S–C–$H_β$ moiety, and thus changes in the chemical shift separation of the β-$CH_2$ geminal couple may reflect conformational changes in the Co(II)–Cys moiety (Fernandez *et al.*, 1997). Since these protons are located close to the metal site, interpretation of the shifts in terms of electron delocalization in the cobalt derivative can be misleading as the pseudocontact contribution to the overall chemical shift cannot be neglected. This issue has been recently addressed through determination of the magnetic anisotropy tensor in Co(II)-azurin and Co(II)-rusticyanin (Donaire *et al.*, 1998, 2001). The orientation of the χ tensor found in Co(II)-azurin is such that the largest contribution to the chemical shift of the β-$CH_2$ Cys protons arises from the contact interaction and so directly reflects the electron delocalization.

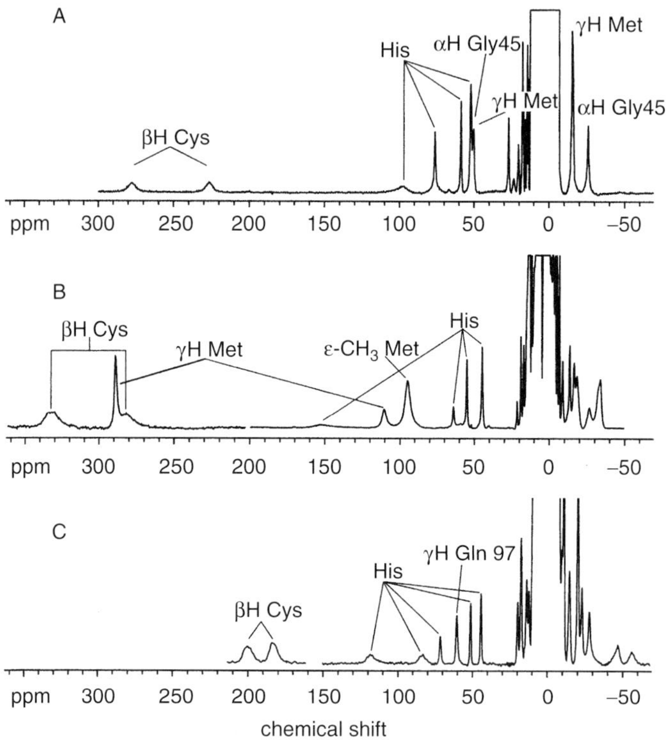


FIG.4. [1]H NMR spectra of cobalt(II)-substituted (A) *Pseudomonas aeruginosa* azurin (Moratal Mascarell *et al.*, 1993b) (B) *Achromobacter cycloclastes* pseudoazurin (Fernández *et al.*, submitted for publication), and (C) *Rhus vernacifera* stellacyanin (Vila, 1994). Spectra (A) and (C) were recorded at 200 MHz at 313 K, whereas spectrum (B) was recorded at 600 MHz and 318 K. All the samples were in 50 mM phosphate buffer at pH 6 in water solution.

When an axial Gln ligand is present (either natural as in stellacyanin or introduced by site-directed mutagenesis), the same orientation of the magnetic anisotropy tensor is maintained, and comparison of the average shift of the two β-$CH_2$ Cys protons in different proteins may prove helpful (Diederix *et al.*, 2000; Fernández *et al.*, 1997; Salgado *et al.*, 1996; Vila and Fernández, 1996).

In rusticyanin and pseudoazurin, the axial Met ligand adopts a different orientation than in azurin, resulting in a stronger Cu(II)-$S_δ$ Met bond and a tetragonal distortion. The magnetic anisotropy tensor is rhombic in the Co(II)-substituted proteins, and the pseudocontact contribution to the β-Cys proton shifts is not negligible (Donaire *et al.*, 2001; Fernández *et al.*, submitted for publication).

*b. The His Ligands.* The signals of the proton His rings are found between 30 and 90 ppm. They are relatively sharp (150–500 Hz), except for the Hε1 imidazole protons, which are located closer to the metal ion. Due to the different degrees of exposure to solvent of the His residues, and the different NH exchange rates, the His resonances can be specifically assigned (Moratal Mascarell *et al.*, 1993a; Vila, 1994).

*c. The Axial Ligand.* The resonances of pairs of geminal protons of axial ligands located close to the metal ion (γ-$CH_2$ in Met and Gln; α-$CH_2$ in Gly) experience quite a large chemical shift separation due to different dihedral angles with the metal ion and, consequently, a different contact shift contribution to the observed chemical shift (Donaire *et al.*, 1998). Furthermore, the pseudocontact shifts are negative for nuclei located in the axial positions. As a consequence, in most cases, this results in a geminal proton couple with one upfield and one downfield resonance (Diederix *et al.*, 2000; Moratal Mascarell *et al.*, 1993a; Piccioli *et al.*, 1995; Salgado *et al.*, 1996; Vila and Fernández, 1996; Vila *et al.*, 1997). When a strongly coordinated Met residue is present, the larger overall contact contribution is such that both geminal protons fall in the downfield region, as in Co(II)-rusticyanin and -pseudoazurin (Donaire *et al.*, 2001; Fernández *et al.*, submitted for publication).

When the axial ligand is a Gln residue, coordination may occur through the oxygen or through the nitrogen atoms of the Gln side chain. Nitrogen-mediated ligation is expected to broaden the Nε2 Gln protons beyond detection. These protons have been detected in the Co(II) derivatives of azurin and amicyanin mutants, suggesting that coordination takes place through the amide oxygen atom (Diederix *et al.*, 2000; Salgado *et al.*, 1996).

*2. Ni(II) Substitution*

Ni(II) in BCP gives rise to smaller pseudocontact shifts, due to a smaller magnetic anisotropy with respect to the cobalt(II) derivatives (Donaire *et al.*, 1998). $^1$H NMR studies are available for Ni(II)-substituted *Ps. aeruginosa* azurin and its M121Q mutant (Blaszak *et al.*, 1982; Moratal Mascarell *et al.*, 1993b,c; Salgado *et al.*, 1996), *R. vernicifera* stellacyanin (Fernández *et al.*, 1998), and *Thi. versutus* amicyanin (Salgado *et al.*, 1999). Ni(II) has been less exploited as a paramagnetic probe than Co(II), even if sharper lines are usually observed in the NMR spectra, as the results are less easily transferred to the native Cu(II) systems due to the different coordination preferences of the two metal ions (Moratal *et al.*, 1995).

### 3. *Cd(II) Substitution*

Substitution of Cu(II) with Cd(II) allows characterization of the coordination binding site through $^{113}$Cd NMR spectroscopy. The $^{113}$Cd NMR shifts are, indeed, quite sensitive to the nature of the donor atoms. NMR studies on Cd(II)-substituted BCP were reported in a pioneering study in 1984 by McMillin and co-workers (Engeseth *et al.*, 1984). The $^{113}$Cd NMR shift for Cd(II) azurin (372 ppm) is close to that of Cd(II)-stellacyanin (380 ppm), with the cadmium ion more shielded than in Cd(II)-plastocyanin (432 ppm) (Engeseth *et al.*, 1984). This similarity of the $^{113}$Cd NMR chemical shift of the first two proteins was attributed to a similar displacement of the Cd(II) ion toward the Gly ligand, as shown in the crystal structure (Blackwell *et al.*, 1994). The $^{113}$Cd longitudinal relaxation times in Cd(II)-substituted BCP are around 100 ms, an order of magnitude shorter than those observed in cadmium derivatives of carbonic anhydrase and superoxide dismutase (Engeseth *et al.*, 1984). The $^{1}$H assignments of Cu(I)- and Cd(II)-pea plastocyanin suggest that small structural changes occur on cadmium substitution (Ubbink *et al.*, 1996).

### 4. *Hg(II) Substitution*

$^{199}$Hg NMR chemical shifts span a range of over 5000 ppm and are highly sensitive to changes in the metal coordination sphere (Oz *et al.*, 1998). This sensitivity has been exploited to probe the metal ligands of BCP. The $^{199}$Hg chemical shifts are −749 ppm for Hg(II)-plastocyanin, −706 ppm for Hg(II)-rusticyanin, and −884 ppm for Hg(II)-azurin (Utschig *et al.*, 1995, 1997). The similarity of the $^{199}$Hg chemical shifts of plastocyanin and rusticyanin reveals a similar ($N_2SS'$) coordination environment. On the other hand, the upfield shift in azurin is consistent with the metal ion being bound to an oxygen atom from the axial Gly residue.

$^{199}$Hg signals in BCP possess $T_1$ values one order of magnitude less than those of $^{113}$Cd in the same site. Faster acquisition rates are then feasible in Hg(II)-substituted BCP, thus balancing the lower sensitivity of $^{199}$Hg and allowing NMR spectra to be recorded in a few hours (Utschig *et al.*, 1995). These fast longitudinal relaxation times did not, however, impede recording of $^{1}$H-$^{199}$Hg HMQC spectra. $^{3}J_{H\text{-}Hg}$ couplings with the Met $\varepsilon$-$CH_3$, imidazole His protons, and Cys $\beta$-$CH_2$ were detected for Hg(II)-plastocyanin and -rusticyanin. No HMQC cross-peaks with the axial Met or Gly protons have been detected in Hg(II)-azurin (Utschig *et al.*, 1997).

## E. *NMRD in Blue Copper Proteins*

Early NMRD studies on Cu(II)-azurin (Koenig and Brown, 1973), recently confirmed and extended to a series of azurin mutants (Kroes *et al.*, 1996), indicate that the electron relaxation times of Type I copper centers are one order of magnitude shorter than those found for Type II copper proteins (Kroes *et al.*, 1996). The NMRD profiles of Cu(II) azurin reveal a small paramagnetic effect, as expected for a metal site not accessible by the solvent. Instead, the water relaxivity is sensibly increased in H46G and H117G azurin in which the mutations create a cavity in the metal site, allowing one and two water molecules, respectively, to bind to the metal ion (Kroes *et al.*, 1996).

## V. NMR Studies on Mononuclear Type II Copper-Containing Proteins

Type II copper(II) sites are present in mononuclear copper enzymes such as dioxygenases, monoxygenases, nitrite reductases, and nonblue oxidases. The high molecular weight of most of these enzymes and the unfavorable electron relaxation time of copper ion in the oxidized form have up to now precluded the application of NMR spectroscopy.

Copper, zinc superoxide dismutases ($Cu_2,Zn_2SOD$) are the one notable exception. $Cu_2,Zn_2SOD$ in eukaryotes is a dimeric enzyme of about 32 kDa (Tainer *et al.*, 1982). Each identical subunit contains a copper and a zinc ion, which are bridged by a histidinato group (Tainer *et al.*, 1982) (Fig. 5). Each subunit is formed by a β-barrel of eight antiparallel β-strands, connected by seven turns and loops. The metal ions are located between loops IV and VII outside the β-barrel (Fig. 6). The latter loop, also called the electrostatic loop, contains several charged residues that are considered to be critical to drive the superoxide anion to the copper ion (Banci *et al.*, 1988a, 1993d; Getzoff *et al.*, 1989, 1992).

In the oxidized enzyme, copper(II) is coordinated by four His ligands (numbered 46, 48, 63, and 120 in the human isoenzyme). His-63, which is deprotonated on both nitrogens, makes the bridge between copper and zinc. The zinc ion is bound to three His ligands (63, 71, and 80) and to Asp-83, which belongs to the β-barrel. A complex network of hydrogen bonds maintains the orientation of the metal ligands. The zinc ion is buried completely within the protein, while the copper ion is accessible to solvent and lies at the bottom of a wide channel, which is about 10 Å deep (Tainer *et al.*, 1982). When the copper ion is reduced, the Cu–His-63 bond is lost, breaking the Cu–His–Zn bridge (see below).

The enzyme catalyzes the disproportionation of superoxide anions to hydrogen peroxide and molecular oxygen. The rate-limiting step of the enzymatic reaction is diffusion of superoxide to the reaction center, under nonsaturating conditions of substrate (Fee and Gaber, 1972; Gralla and Kosman, 1992; Halliwell and Gutteridge, 1989; Klug-Roth *et al.*,

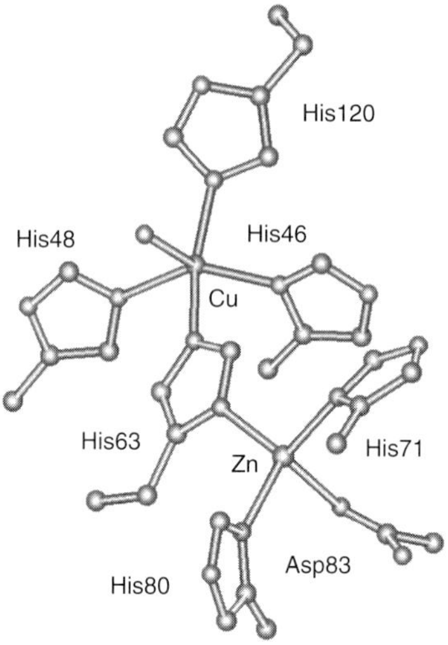


FIG.5. Schematic drawing of the metal site of $Cu_2,Zn_2SOD$ active site.

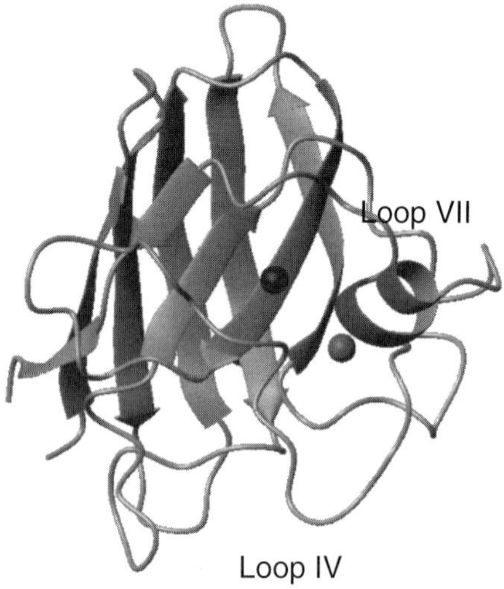


FIG.6. Ribbon drawing of the Cu,ZnSOD monomer (M2SODD133N) showing the secondary structure elements arranged in the Greek-key barrel fold. The metal ions are represented by spheres of arbitrary dimensions (copper, light gray; zinc, dark gray) (Banci *et al.*, 1998).

1973; McCord and Fridovich, 1969). The substrate superoxide anion interacts with the copper ion that either accepts or transfers an electron to it, thus cycling between two oxidation states, copper(I) and copper(II), which are therefore both biologically relevant.

### *A. NMR Structural Studies on Copper(I) Superoxide Dismutase*

The first NMR studies on $Cu_2$,$Zn_2$SOD date back to the 1970s and were performed on the reduced Cu(I) form of the bovine isoenzyme (Burger *et al.*, 1980; Cass *et al.*, 1977; Hill *et al.*, 1980; Lippard *et al.*, 1977). Those works reported the first nonspecific assignment of the active site histidines and their acid/base equilibria (Cass *et al.*, 1977), allowing the authors to establish that, in the bovine isoenzyme, eight histidines are present and that six of them are bound to the metal ions (Fig. 5). Shortly thereafter, the first partial assignment of the His resonances (based on the exchange rate of the NH His protons) and the first NOE studies were reported for this system (Stoesz *et al.*, 1979). Several subsequent NMR studies examined the interaction of the reduced enzyme with anions, but more meaningful results were obtained on paramagnetic derivatives (see next section).

NMR spectra of the reduced form of the human isoenzyme appeared first in 1980, together with a comparative assignment of the yeast and bovine enzymes (Hill *et al.*, 1980). Later results led to the complete assignment of the signals of all the Cu(I) and Zn(II) ligands in human isoenzyme (Bertini *et al.*, 1991). 2D NOESY experiments provided the connectivities between His ring resonances, allowing the differentiation between the NH resonances of histidines bound to a metal through Nδ1 or Nε2 atoms and the identification of intraimidazole cross-peaks and interresidue connectivities. This provided information on the ligand interproton distances (Bertini *et al.*, 1991). Possible ambiguities were removed by using a sample selectively deuterated at the His Hε1 position (Bertini *et al.*, 1991). NOE experiments were also performed with the aim of measuring distances between protons belonging to different histidines. These data were used in a molecular dynamics simulation to refine a first model of the reduced form of the enzyme in solution and to obtain insight regarding possible structural differences between solution and solid state (Banci *et al.*, 1994). The data showed that the bridge connecting the two metal ions is broken in the reduced form of the enzyme, as previously suggested by the pH dependence of the reduction potential (Fee and DiCorleto, 1973) and by the NMR studies on $Cu_2$,$Co_2$SOD derivative (see next section).

A similar approach was used for assigning the active site residues in bovine (Paci *et al.*, 1990) and in prokaryotic isoenzymes (Chen *et al.*, 1995; Sette *et al.*, 2000; Venerini *et al.*, 1999). In the case of the prokaryotic isoenzyme from *Brucella abortus*, the assignment of the relevant resonances was facilitated by the use of a $^{15}$N-enriched sample (Chen *et al.*, 1995).

The complete assignment of the polypeptidic chain was achieved in the late 1990s, when a monomeric analogue of the human isoenzyme became available (Banci *et al.*, 1995b; 1997b; Bertini *et al.*, 1994d). The two subunits of the native dimer are not covalently linked but experience extensive hydrophobic contacts in addition to some hydrophilic interactions that further stabilize the dimeric form (Parge *et al.*, 1986; Tainer *et al.*, 1982). Among the hydrophobic residues, substitution of Phe-50 and Gly-51 with two Glu residues disrupts the quaternary structure of the protein, producing a soluble monomeric form (denoted M2SOD hereafter) (Bertini *et al.*, 1994d). To enhance the enzymatic activity of this monomeric form, which has 10% of the activity of the native species, the Glu-133 present in the electrostatic loop was replaced with Gln (denoted M2SODD133N hereafter) (Banci *et al.*, 1995b), as was previously done in the dimeric enzyme (Getzoff *et al.*, 1992). Also mutations at the subunit interface that maintained the same protein charge as in the native one, such as the placement of two Glu residues (at positions 50 and 51) and two Lys residues (at positions 148 and 151), produced a protein with higher activity than the M2SOD enzyme, although lower than that of the native enzyme (25%) (Banci *et al.*, 1999a).

In order to elucidate the structural and, possibly, the dynamic features that make these artificial monomeric species less active than the native enzyme, selected solution (and crystal) structures have been determined and their mobility has been characterized on various time scales (Banci *et al.*, 1998, 1999a, 2000; Ferraroni *et al.*, 1999).

The reduced molecular weight of the monomeric analogue makes the use of high-resolution NMR amenable, despite the small chemical shift dispersion due to the presence of β-type elements of secondary structure only. In the case of M2SODD133N, the use of $^{13}$C, $^{15}$N triple-resonance NMR experiments allowed complete assignment of the backbone atoms (Banci *et al.*, 1997a) and determination of the three-dimensional structure in solution (Banci *et al.*, 1998). By using distance and dihedral angle constraints, the solution structure, represented by a family of 36 conformers, has been refined up to a backbone root mean square deviation (RMSD) of $0.81 \pm 0.13$ Å over the entire structure (Fig. 6). This structure has been compared with the available X-ray structures of reduced $Cu_2,Zn_2SOD$ as well as with the oxidized form of human and bovine isoenzymes (Banci *et al.*, 1994; Parge *et al.*, 1986; Tainer *et al.*, 1982). The structure of the metal sites and of the backbone are not affected greatly by the monomerization, except in the regions involved in the subunit–subunit interface in the dimeric protein where large disorder is present. Specific structural differences in the active site channel have been found, particularly in the conformation of the electrostatic loop. Furthermore, Arg-143, a catalytically relevant residue (Fisher *et al.*, 1994), moved to a position that was not optimal to drive the superoxide anion toward the copper ion and to place it for efficient electron transfer (Banci *et al.*, 1997a). The very same structural changes at the interface and in the electrostatic loop were found in the M4SOD solution structure (Banci *et al.*, 1999a).

The structure at the copper center is well defined. The bridge between copper and zinc is broken in all reduced isoenzymes examined to date. His-63 is protonated on reduction. This proton is within the van der Waals distance of copper(I) and is strategically located to be involved in the catalytic mechanism, presumably by interacting with $O_2^-$. An H bond between His-63 and $O_2^-$ could be responsible for the attraction of an electron from copper(I) to $O_2^-$, with the subsequent formation of $HO_2^-$.

The same authors performed the backbone assignment of the dimeric form of the enzyme and compared the mobility of the monomeric and the dimeric forms (Banci *et al.*, 2000). The measurement of $^{15}$N $R_1$ and $R_2$ values and of hetero-NOEs for the amide nitrogens provides information on backbone mobility in the picosecond–nanosecond timescale through the so-called order parameter $S^2$ (Lipari and Szabo, 1982). Off-resonance $^{15}$N $R_{1_\rho}$ relaxation data provide information on conformational exchange processes occurring in the microsecond–millisecond timescale and on the rate of this exchange, $R_{ex}$ (Desvaux *et al.*, 1995; Palmer, 1997). The comparisons showed that regions at the subunit interface (residues 50–60, 81–86, and 151–153) are more rigid in the dimer in the picosecond–nanosecond timescale, while the electrostatic loop (residues 131–142) is more rigid in the monomer (Fig. 7, bottom panel, see color insert). Conformational exchange processes were observed in the monomer for the regions around cysteines 57 and 146, which form a disulfide bridge, while they are sizably reduced in the dimer (Fig. 7, top, see color insert). In the dimer, Cys-57 maintains the optimal orientation of the key residue Arg-143 through an H bond with its guanidinium group. This H bond is lost in the monomeric form. The data suggest that the different mobility of the electrostatic loop in the two species in the picosecond–nanosecond time range can have a role in the diffusion of the substrate toward the active site and thus can be relevant in determining the catalytic rates (Banci *et al.*, 2000).

### *B. NMR Studies on Copper(II) Superoxide Dismutase: The "Co Trick"*

In its resting state, SOD contains a paramagnetic oxidized copper(II) ion. As discussed in Section II, a Type II copper(II) ion has a relatively long electron relaxation time, thus producing dramatic broadening in the NMR lines. This results in the disappearance of all signals of the copper ligands and of resonances within 6 Å of the copper iron. Therefore the active site of the oxidized form of this protein cannot be characterized by NMR spectroscopy.

On the other hand, most of the spectroscopic efforts for the characterization of this metalloenzyme were focused on the study of the active site, where the reaction takes place. The paramagnetic ions would in principle act as a probe to shift the signals of the active site resonances outside the diamagnetic envelope. To exploit the effect of a paramagnetic center as a spectroscopic probe in the active site and to overcome the problems produced by the copper ion, the long electron relaxation rates of Cu(II) have been shortened with a very elegant "trick," which makes use of the peculiar properties of the active site. The zinc(II) ion can be replaced by a paramagnetic fast-relaxing metal ion, such as cobalt(II) or nickel(II), which can be magnetically coupled to the copper(II) ion through the histidinato bridge. As already discussed in Section I, magnetic coupling with a fast-relaxing metal ion shortens the electron relaxation time of the slowly relaxing copper ion. NMR signals of nuclei sensing the copper(II) ion can now be easily detected and assigned.

The $Cu_2,Co_2$SOD derivative shows very well resolved, sharp NMR signals for essentially all the protons of the ligands of both metals (Bertini *et al.*, 1985c). This spectrum, reported in Fig. 8A, represented the first high-resolution NMR spectrum reporting the metal site resonances in an oxidized copper protein. Significantly, the signals of the copper ligands are even sharper than those of the cobalt ligands. Magnetic coupling with the fast-relaxing cobalt(II) ion drastically reduces (by about two orders of magnitude) the electronic relaxation time of the copper ion, making it similar to those of the cobalt(II) ion. Furthermore, as copper(II) has only one unpaired electron, while high-spin cobalt(II) has three, the contribution to nuclear relaxation due to the coupling with the unpaired electron(s) is smaller for nuclei coupled to copper than for those coupled to cobalt.

The signals of all the metal ligand protons fall in a chemical shift range of 80 ppm and have been assigned completely. The relative orientation of the ligands was determined through interresidue NOEs (Banci *et al.*, 1989a). These results were confirmed with 2D NOESY spectroscopy (Banci *et al.*, 1993a, b). This spectrum represents a fingerprint of the active site of oxidized SOD and confirms that the bridge is intact in the oxidized form. Spectra of eukaryotic and bacterial SODs are very similar, with minor changes that could be ascribed to small distortions in the coordination spheres of the metal ions (Sette *et al.*, 1995, 2000). Similarly, spectra have been recorded for several mutated forms of the $Cu_2Co_2$ enzyme, in an attempt to understand the role of key amino acids located in the active site channel (Banci *et al.*, 1988a, 1990b; 1993a, 1993c, 1993d, 1995a, 1999b; Bertini *et al.*, 1989; Getzoff *et al.*, 1992).

Anion binding to $Cu_2$,$Co_2$SOD has been extensively studied as they can mimic the interaction of SOD with substrate. $N_3^-$, $CN^-$, $NCO^-$, $NCS^-$, and $F^-$ bind copper(II) in native human and bovine SOD (Bertini *et al.*, 1998, and references therein). $N_3^-$, $CN^-$, and $F^-$ are competitive inhibitors of the enzyme (Rigo *et al.*, 1975, 1977), whereas $NCO^-$ and $NCS^-$ have not been reported to be inhibitors (Ozaki *et al.*, 1988; Rigo *et al.*, 1977; Strothkamp and Lippard, 1981). Except for $CN^-$, exchange between anion-free and anion-bound forms is fast on the NMR timescale. Under these conditions it is possible to determine the anion affinity constant by measuring the variation of the chemical shift of the hyperfine shifted resonances as a function of anion concentration. This information is relevant to an understanding of some aspects of the enzymatic process (Banci *et al.*, 1988b, 1989a, 1989b, 1990a; Bertini *et al.*, 1985c; Ming *et al.*, 1988b).

The resonances of His-48 are most affected by the addition of anions, suggesting a variation of the Cu–N bond length. As all the hyperfine shifted resonances are assigned, structural information on the active site cavity in the presence of anion can be obtained from the analysis of 1D and 2D NOE experiments, even without the need for solving the complete three-dimensional structure. The NMR analysis showed that the position of His-48 is not significantly affected by anion binding and that the variation in the Cu–N (His-48) bond strength is due to a movement of the Cu(II) ion (Banci *et al.*, 1990a). Further studies on the binding of $N_3^-$ by using $^{14}N$ and $^{15}N$ NMR have also provided evidence for an equatorial binding of the anion (Bertini *et al.*, 1994b). This finding has been confirmed by X-ray data (Djinovic *et al.*, 1994a, b).

The use of heteronuclear NMR has also been exploited for the characterization of other anion adducts. The binding of $NCO^-$ and $NCS^-$ to copper has been demonstrated by $^{13}C$, $^{14}N$, and $^{15}N$ NMR spectroscopy (Bertini *et al.*, 1980, 1981), whereas $^{19}F$ NMR has been used for studying the interaction with fluoride (Banci *et al.*, 1989c; Viglino *et al.*, 1979). The affinity of the anions follows the order $NCO^- > NCS^- > F^-$, as does their ability to displace the copper ion. This has been attributed to the competitive effect of some charged residues present in the active site, which can provide a suitable binding site for weakly binding anions (Bertini *et al.*, 1998, and references therein). A similar interaction has also been proposed for phosphate by using $^{31}P$ NMR (Mota De Freitas *et al.*, 1987).

The cyanide-bound and cyanide-free forms of SOD are in slow exchange on the NMR timescale. On addition of increasing amounts of anion, a new set of resonances appears to be due to the CN-bound form. The assignment of the NMR spectrum was achieved through saturation transfer experiments (Paci *et al.*, 1988), which also show that binding of this anion produces structural rearrangements similar to those observed for the other anions.

### *C. Other Metal-Substituted Derivatives of Superoxide Dismutase*

Besides $Cu_2,Co_2SOD$, a number of other metallo-substituted derivatives of SOD have been prepared and studied (Bertini *et al.*, 1994c, and references therein). Ni(II) has been used as a Zn(II) probe in a manner similar to Co(II) (Fig. 8B). Complete assignment of the hyperfine shifted resonances for this derivative has been achieved by 1D and 2D NOE experiments and the results have been compared to those of the better characterized $Cu_2,Co_2SOD$ (Bertini *et al.*, 1992; Ming *et al.*, 1988a). Additionally, Ni(II) has been used as a substitute for the copper ion, as both give rise to square planar coordination geometries, and the $Ni_2,Zn_2SOD$ $^1H$ NMR spectrum has been assigned (Ming and Valentine, 1990). Ag(I) has been used as a probe for Cu(I) in the presence of Ni(II) (Ming *et al.*, 1988a).

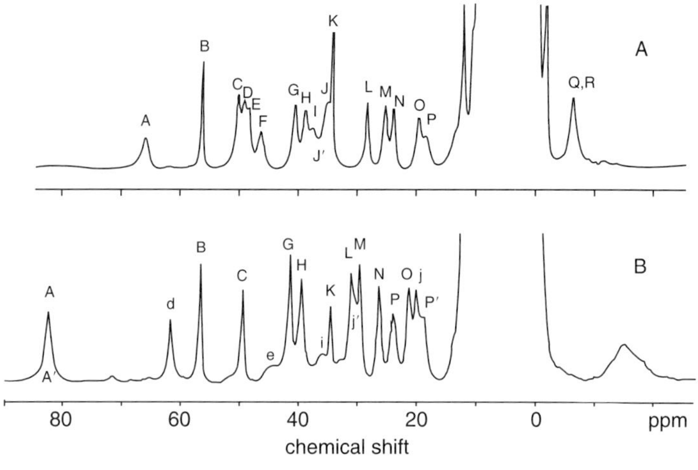


FIG.8. 300-MHz $^1H$ NMR spectra of (A) $Cu_2,Co_2SOD$ and (B) $Cu_2,Ni_2SOD$ in 10 mM acetate buffer at pH 5.5. Signal labeling indicates corresponding signals in the two adducts. Adapted with permission from Ming *et al.* (1988a; Copyright 1988 American Chemical Society).

$^{113}Cd$ NMR was used to investigate the metal-binding sites of SOD (Armitage *et al.*, 1978; Bailey *et al.*, 1980; Kofod *et al.*, 1991). The results indicate that the two subunits are identical and that the coordination of the zinc ion is similar in the absence or in the presence of the copper ion. Both $E_2,Cd_2SOD$ (E denotes empty) and $Cu(I)_2,Cd_2SOD$ show a single $^{113}Cd$ signal at about 330 ppm. Furthermore, the cadmium(II) ion occupies the copper(II)-binding site only when the zinc(II) site is occupied, providing a $Cd_2,Zn_2SOD$ derivative and a resonance at about 185 ppm.

### *D. NMRD Studies on Superoxide Dismutase*

These studies have been essentially devoted to $Cu_2,Zn_2SOD$. The first NMRD profiles were reported for the bovine isoenzyme; they showed a peculiar dispersion that is quite temperature dependent (Gaber *et al.*, 1972). After development of the appropriate theory (Bertini *et al.*, 1985a, 1985b, 1988; Koenig and Brown, 1987), the fitting of the NMRD profiles indicated the presence of one water molecule with the protons ca. 3.4 Å from the copper ion. The latter is characterized by an electron relaxation time of $2.2 \times 10^{-9}$ s at 298 K (Bertini *et al.*, 1988). Typical profiles are reported in Fig. 9. Afterward, this technique was extensively used to characterize the presence or absence of the water molecule close to the copper ion under a variety of experimental conditions, such as in the presence of anions and site-specific mutations (Bertini *et al.*, 1998, and references therein). When anions that act as strong inhibitors and

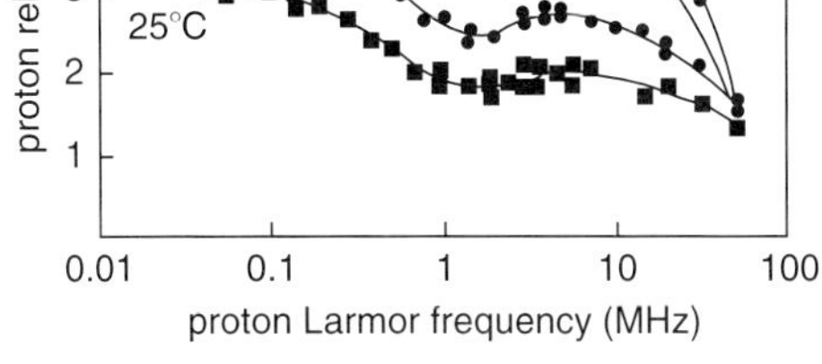


FIG. 9. Proton relaxivity (i.e., water proton $R_1$ for millimolar solutions of protein) as a function of the proton Larmor frequency for $Cu_2,Zn_2SOD$ at different temperatures (Gaber *et al.*, 1972). The lines are best-fit curves with the inclusion of the effect of hyperfine coupling with the metal nucleus (Bertini *et al.*, 1985a).

coordinate to the copper ion are present, the water proton relaxation rates drop to the diamagnetic protein limit, indicating that no water molecule is present close to the copper ion when anions such as azide or cyanide are bound to the copper ion. In the case of weakly bound anions, such as fluoride, an increase in relaxivity was observed, which was interpreted as a consequence of the presence of a water molecule hydrogen bonded to the anion (Banci *et al.*, 1989c). Other isoenzymes show very similar profiles for the native protein as well as similar behavior when anions are added (Bertini *et al.*, 1998).

The NMRD profiles have been investigated for the various mutants on residues in the active site channel. In all the mutants, with the exception of the T137I mutant, a behavior similar to that of the wild-type protein is observed. The interaction with anions is also very similar for these mutants, except for T137I (Banci *et al.*, 1990b). In the latter mutant very low water proton nuclear relaxation rates are observed, indicating that the water molecule present close to the copper ion is missing when threonine 137 is replaced with the bulky, hydrophobic isoleucine residue. Indeed, the coordination geometry of the copper ion is affected by this mutation, becoming more regular toward a tetragonal geometry (Banci *et al.*, 1990b). However, the enzymatic efficiency of this mutant is not greatly affected, thus indicating that the water molecule close to the copper is not involved in the enzymatic process.

## VI. NMR Studies of Proteins Containing Polynuclear Copper Centers

### *A. The $Cu_A$ Center*

This is a binuclear center acting as the primary electron acceptor in terminal oxidases. The electrons are then shuttled to another metal center in the same oxidase (Beinert, 1997; Ferguson-Miller and Babcock, 1996; Ramirez *et al.*, 1995; Randall *et al.*, 2000). In the 16-kDa $Cu_A$-soluble subunit from *Thermus thermophilus* cytochrome *$ba_3$* (Slutter *et al.*, 1996), the copper ions are bridged by the two sulfur atoms of Cys-149 and Cys-153, forming an essentially planar $Cu_2S_2$ rhombic structure with a metal-to-metal distance of 2.5 Å (Williams *et al.*, 1999) (Fig. 10). One of the copper ions is coordinated also to the Nδ1 atom of His-114 and the Sδ atom of Met-160 (at 2.48 Å), whereas the other copper ion is coordinated to His-157 and the backbone carbonyl of Glu-151 (at 2.62 Å). These gross geometrical features are conserved in all structurally characterized $Cu_A$ centers (Iwata *et al.*, 1995; Tsukihara *et al.*, 1995, 1996; Wilmanns *et al.*, 1995). The weak Met and Glu ligands may help maintain the site architecture and regulate its properties.

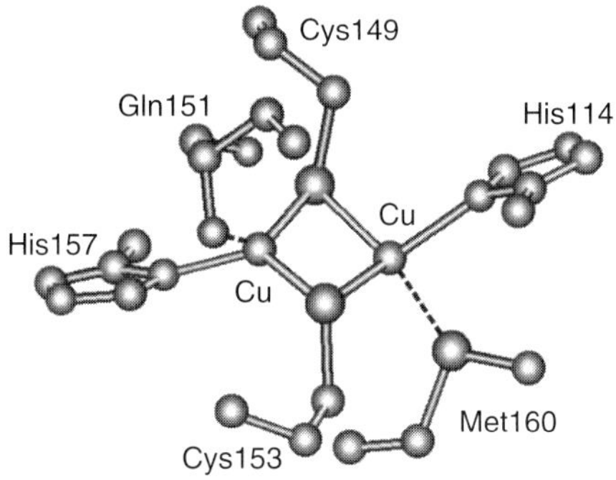


FIG.10. Schematic drawing of the $Cu_A$ copper site.

$Cu_A$ centers exist in two redox states: [Cu(II)Cu(I)] and [Cu(I)Cu(I)]. The oxidized species is a fully delocalized mixed-valence pair (formally two Cu + 1.5 ions), as revealed by EPR spectroscopy (Kroneck *et al.*, 1988, 1990). Despite the similar coordination geometry around copper, these systems display sharper NMR lines than do the BCP due to a shorter electron relaxation time of the paramagnetic center ($\approx 10^{-11}$s) (Clementi and Luchinat, 1998). NMR studies are available for the native $Cu_A$ centers from the soluble fragments of the *The. thermophilus*, *Paracoccus denitrificans*, *Paracoccus versutus*, and *Bacillus subtilis* oxidases (Bertini *et al.*, 1996; Dennison *et al.*, 1995; Luchinat *et al.*, 1997; Salgado *et al.*, 1998a) and *Pseudomonas stutzeri* $N_2O$ reductase (Holz et al., 1999), as well as for engineered $Cu_A$ sites in amicyanin (Dennison *et al.*, 1997) and *Escherichia coli* quinol oxidase (Kolczak *et al.*, 1999).

The four β-$CH_2$ Cys resonances are shifted well downfield, spanning a broad range of chemical shifts from 50 to 450 ppm (Fig. 11). Three are very broad (signals *a–c*) and usually are found between 200 and 450 ppm, whereas a fourth, a sharper one (signal *d*), falls in the range 50–110 ppm. In some cases, line broadening is so drastic that the faster relaxing resonances *a–c* cannot be detected. Full signal identification and assignment are available only for the *The. thermophilus*, *Pa. denitrificans*, and *Pa. versutus* proteins (Bertini *et al.*, 1996; Luchinat *et al.*, 1997; Salgado *et al.*, 1998b). The problem of detecting the signals with large linewidth can be overcome by selective deuteration of the β-$CH_2$ Cys protons and collection of the $^2$H NMR spectrum (see Fig. 11A) (Luchinat *et al.*, 1997). Within a similar range of chemical shifts, different signal patterns for the Cys protons have been found in different $Cu_A$ proteins, reflecting the sensitivity of their shifts to minor changes in the structure of the binuclear unit. Furthermore, the shifts of some of the cysteine protons do not

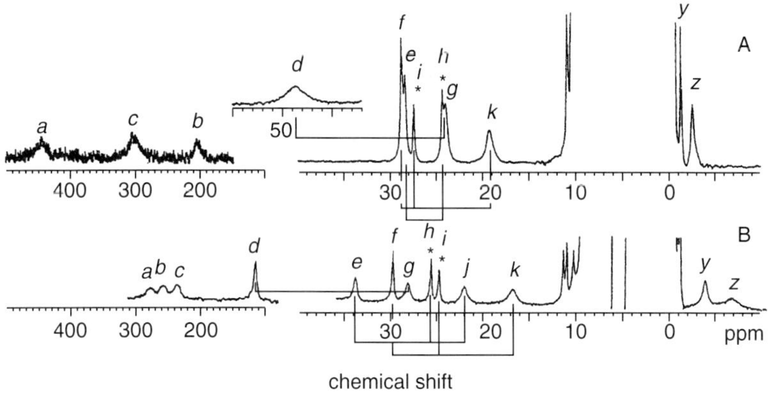


FIG. 11. $^1$H NMR spectra of $H_2O$ solutions of $Cu_A$ domains from (A) *Paracoccus denitrificans*, 800 MHz, pH 5.6, and (B) *Thermus thermophilus*, 600 MHz, pH 4.5. Asterisks denote exchangeable signals. The 100- to 500-ppm region of spectrum (A) is recorded as a $^2$H NMR spectrum. Signals *a* and *c* are not visible in the $^1$H spectrum. NOE connectivities are also shown. Adapted with permission from Luchinat *et al.* (1997; Copyright 1997 American Chemical Society).

follow a normal Curie temperature dependence (Bertini *et al.*, 1996), suggesting the existence of temperature-accessible excited states (Salgado *et al.*, 1998b).

The proton signals of each histidine ring, which are found up to 40 ppm, have been assigned by 1D NOE experiments (Bertini *et al.*, 1996; Luchinat *et al.*, 1997; Salgado *et al.*, 1998b). They can be assigned to specific residues based on the fact that only one of them is solvent exposed, as also occurs in BCP. Electron delocalization onto the two His ligands is slightly different, mainly due to a different orientation of the imidazole planes with respect to the $Cu_2S_2$ rhombus. No electron spin density has been detected in the weakly coordinated Met and Glu ligands.

The overall picture describing the electron delocalization in $Cu_A$ resembles that found in Type I sites: most of the delocalized unpaired spin density is found on the Cys ligands. The electron spin density on each β-$CH_2$ Cys proton is about half of that observed on the equivalent protons in blue copper proteins (Bertini *et al.*, 1996, 1999). The unpaired electron is distributed over the two copper ions and the two Cys ligands. The observed values for the hyperfine shifts are consistent with the fact that the hyperfine couplings found in the β-$CH_2$ Cys protons in Type I sites are twice as large as those observed in $Cu_A$ centers. However, as already discussed, the Cu(II)–Cys covalency in Type I sites can be severely altered by the strength of the Cu(II)–axial ligand interaction. This does not seem to be the case for $Cu_A$. The more rigid structure and the highly delocalized mixed-valence system are essential for providing an efficient binuclear electron transfer unit, with minimal reorganization energy on redox changes.

### B. *Type III Copper Centers*

Type III binuclear copper sites are present in dioxygen carriers, such as hemocyanin, and in oxidases, such as tyrosinase, that appear to be related by evolution (Solomon *et al.*, 1996). These sites are characterized by the presence of two copper ions, each coordinated to three His residues. Different ligands (such as oxygen, hydroxide, and small anions) can bridge the two metal ions (Solomon *et al.*, 1996). These proteins are active in their oxidized Cu(II), Cu(II) state. The cupric ions are antiferromagnetically coupled, giving a diamagnetic ground state that is EPR silent. A temperature-accessible $S = 1$ excited state lies at $J\,\mathrm{cm}^{-1}$ with respect to the ground state, where $J$ is the exchange coupling constant between the two Cu(II) ions (Bubacco *et al.*, 1999). As the $J$ values are in the range 100–300 $\mathrm{cm}^{-1}$, this excited level is partially populated at room temperature and imparts paramagnetism to the system. Canters and coworkers have recorded the $^1$H NMR spectra of Met tyrosinase, as well as of various adducts with chloride and with organic inhibitors (Fig. 12) (Bubacco *et al.*, 1999, 2000). The spectra show well-resolved signals shifted outside the diamagnetic envelope, which have been assigned pairwise to six coordinated His residues. The $J$ values have been estimated from the temperature dependence of the hyperfine shifts. The various adducts can be grouped according to the nature of the inhibitor, which in some cases can bind as a bridging ligand between the two copper ions, thus altering the magnitude of the exchange coupling. This is also reflected in the linewidths of the NMR lines.

## VII. Other Copper-Binding Proteins

Recently, new classes of proteins that are responsible for the homeostasis of copper and for its delivery to specific intracellular targets have been identified. Copper cannot be present as the free ion in solution as its high reactivity leads to the production of radicals. It appears always to be bound to some proteins and is present as copper(I) due to the reducing conditions of the cell environment. Small soluble proteins, called copper chaperones (see chapter by Elam *et al.*, this volume), function to shuttle copper to specific target proteins. Other large, membrane-bound proteins are present to pump copper from one cell compartment to another.

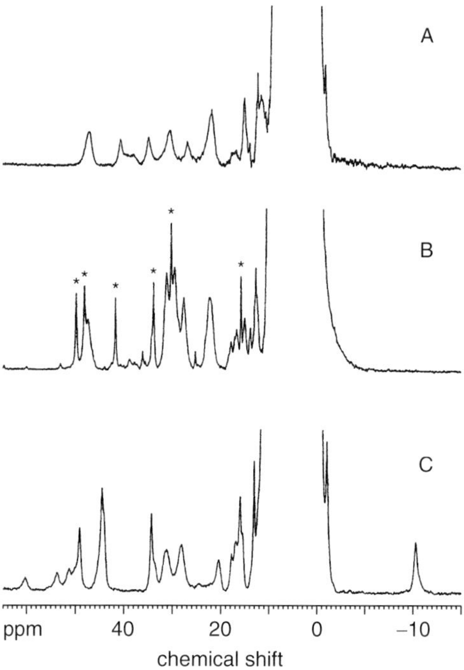


FIG. 12. 600-MHz $^1$H NMR spectra of (A) *Streptomyces antibioticus* Met tyrosinase in 100 mM sodium phosphate, pH 6.8, at 280 K; (B) same as (A) but in the presence of 500 mM NaCl; (C) same as (A) but in the presence of 2 mM 2-hydroxymethyl-5-hydroxy-γ-pyrone (kojic acid). Asterisks indicate the NδH of the six copper-coordinating histidine residues. Reprinted with permission from Bubacco *et al.* (2000).

These contain soluble domains able to bind copper. The first protein demonstrated to be a copper chaperone was Atx1 from yeast, able to bind Cu(I) and to deliver it directly to the soluble domains of Ccc2, a membrane-bound copper ATPase located in the Golgi membrane (Pufahl *et al.*, 1997; Yuan *et al.*, 1997). Another pathway for copper transport in yeast involves the copper chaperone Lys7, which delivers copper to SOD. Lys7 is a large protein, composed of three domains, one of them having the same fold as Atx1. Functional homologues were subsequently found in other organisms (Amaravadi *et al.*, 1997; Himelblau *et al.*, 1998; Klomp *et al.*, 1997; Nishihara *et al.*, 1998; Odermatt and Solioz, 1995; Wakabayashi *et al.*, 1998). Other classes of proteins involved in copper homeostasis have been identified or are being characterized to understand their function.

All copper chaperones, as well as the copper-binding domains of copper ATPases characterized so far, show the typical consensus motif

CXXC (using single-letter code for amino acids and X for any amino acid) that contains the two cysteines that bind copper. The structural environment of the metal site affects the metal-binding affinity and is ultimately responsible for conferring specificity for copper or other metal ions (Steele and Opella, 1997; Veglia *et al.*, 2000).

The copper-bound form of these proteins, as well as the apo state where the two cysteine residues responsible for copper binding oxidize and form a disulfide bridge, is relatively sensitive to oxygen and is not stable for long periods *in vitro*. The chemistry of these proteins is therefore rather different from that of all the other copper-containing proteins studied so far, and the fairly limited application of NMR spectroscopy for their study is specifically devoted to the elucidation of structural features. Solution structures of the native Cu(I) and of the reduced apo form of both Atx1 and the first soluble domain of Ccc2 have been solved recently (Arnesano *et al.*, 2001b; Banci *et al.*, 2000). The NMR structure of the fourth metal-binding domain of the "Menkes disease" protein, the human homologue of Ccc2, is available in both reduced apo and Ag(I)-bound forms (Gitschier *et al.*, 1998). The crystal structures of the oxidized apo form of Atx1 (i.e., with the two cysteines forming a disulfide bridge) and of the Hg(II)-bound form (Rosenzweig *et al.*, 1999) are also available. All these structures share a classical "ferredoxin-like" β1-α1-β2-β3-α2-β4 folding motif (Hubbard *et al.*, 1997). The two cysteines coordinating the copper ion are located between the first loop and the first helix. Comparison between the structures of the Cu(I) and of the apo forms of Atx1 reveals that, on Cu(I) release, the copper-bound cysteines move from a buried location in the bound metal form to a more solvent-exposed conformation in the apo form (Arnesano *et al.*, 2001a). While the structure of Atx1 undergoes changes as a function of copper capture and release, the structure of the Ccc2a domain remains relatively invariant, suggesting that the metal site in Ccc2a is structurally preorganized (Banci *et al.*, 2000). This is one of the key structural difference between the Atx1 metallochaperone family and the homologous metal-binding domains of the copper-transporting P-type ATPases.

NMR structures have been determined also for a few bacterial proteins involved in copper homeostasis (Wimmer *et al.*, 1999; Banci *et al.*, 2001; Banci *et al.*, 2002). The global fold is the same as that of eukariotic proteins. However, while the soluble domain of ATPases, as far as the metal binding region and the hydrophobic interactions are concerned, is quite similar in the two classes of organisms, the small copper transporting proteins have different properties (Arnesano *et al.*, 2002). In particular, in bacterial copper chaperones, like CopZ, the metal binding site is stabilized by conserved hydrophobic interactions between a Met and a Phe/Tyr residues. The latter residue is substituted in eukariotic organisms by a Lys residue which is close to the copper ion in the metalated form while it moves away when the copper ion is released.

## VIII. Perspectives

The characterization of proteins in solution through NMR provides a wealth of information, including the solution structure, protein mobility, modes of interaction with other proteins, factors affecting stability, and the folding–unfolding process. Technological advances in NMR instrumentation, including increasing magnetic field strengths and the development of methodological tools, are extending the size of proteins that can be characterized through NMR spectroscopy. After the successful determination of the solution structure of the copper(II) plastocyanin from *Synechocystis* sp. PCC 6803 (Bertini *et al.*, 2001b), it is apparent that the methods and results typical of diamagnetic proteins can also be obtained for paramagnetic copper(II) proteins. In addition, the presence of the paramagnetic center can be exploited to obtain information on the electronic properties of the system through analysis of contact and pseudocontact shifts, as well as through nuclear relaxation parameters (Bertini *et al.*, 2001c). The very same parameters can be exploited to set structural constraints. Partial molecular orientation of paramagnetic proteins occurring at high magnetic fields can also be exploited for structural constraints (Bertini *et al.*, 2001c). This approach has never been applied to copper-containing systems and it could be the key to obtaining solution structural models of the higher molecular weight copper proteins.

NMR spectroscopy is expected to have a major role in structural genomics research, as it is a fundamental tool for high-throughput structural determination. With the availability of the complete genomes of an increasing number of organisms, new classes of proteins will be identified. An interesting area is represented by those proteins responsible for copper homeostasis, as mentioned before. Little detail is known of the processes that involve copper storage, binding, transport, and regulation. Some of these proteins are membrane bound but feature soluble domains that coordinate copper and then transfer it across the membrane. It can be foreseen that an increasing number of NMR studies in this field will be prompted and, possibly, solid state NMR will become important.

## Note Added in Proof

After this chapter had been submitted, several papers reporting new applications of NMR spectroscopy to copper-containing proteins appeared. In some cases these works address specific points in known systems (see, for example, Hunter *et al.*, 2001; Dennison *et al.*, 2002; Dennison and Lawler, 2002; Sato and Dennison, 2002); in other cases examples of applications to novel systems are given, such as the first NMR characterization of nitrite reductase (Dennison *et al.*, 2000). This is the demonstration that the field is very active and exhibits great potential for the near future.

## Acknowledgments

The stimulating discussions with Professor Claudio Luchinat and Professor Anthony G. Wedd are gratefully acknowledged. A.J.V. is a staff member of CONICET. A.J.V. thanks Fundacion Antorchas and the Fogarty International Center (NIH) for supporting his work on NMR of copper proteins.

## References

Abragam, A. (1961). "The Principles of Nuclear Magnetism." Oxford Univ. Press, Oxford.
Adman, E. T. (1991). *Adv. Protein Chem.* **42**, 144.
Amaravadi, R., Glerum, D. M., and Tzagoloff, A. (1997). *Hum. Genet.* **99**, 329.
Armitage, I. M., Uiterkamp, A. J. M. S., Chlebowski, J. F., and Coleman, J. E. (1978). *J. Magn. Reson.* **29**, 375.
Armstrong, F. A., Driscoll, P. C., Hill, H. A. O., and Redfield, C. (1986). *J. Inorg. Biochem.* **28**, 171.
Arnesano, F., Banci, L., Bertini, I., Cantini, F., Ciofi-Baffoni, S., Huffman, D.L., and O'Halloran, T.V. (2001a). *J. Biol. Chem.* **276**, 41365.
Arnesano, F., Banci, L., Bertini, I., Huffman, D. L., and O'Halloran, T. V. (2001b). *Biochemistry* **40**, 1528.
Arnesano, F., Banci, L., Bertini, I., Ciofi-Baffoni, S., Molteni, E., Huffman, D.L., and O'Halloran, T.V. (2002). *Genome Res.* **12**, 255.
Babu, C. R., Volkman, B. F., and Bullerjahn, G. S. (1999). *Biochemistry* **38**, 4988.
Badsberg, U., Jorgensen, A. M., Gesmar, H., Led, J. J., Hammerstad, J. M., Jespersen, L. L., and Ulstrup, J. (1996). *Biochemistry* **35**, 7021.
Bagby, S., Driscoll, P. C., Goodall, K. G., Redfield, C., and Hill, H. A. (1990). *Eur. J. Biochem.* **188**, 413.
Bagby, S., Driscoll, P. C., Harvey, T. S., and Hill, H. A. (1994). *Biochemistry* **33**, 6611.
Bailey, D. B., Ellis, P. D., and Fee, J. A. (1980). *Biochemistry* **19**, 591.
Baker, E. N., Blundell, T. L., Cutfield, J. F., Dodson, E. J., Dodson, G. G., Hodgkin, D. C., Hubbard, R. E., Isaacs, N. W., Reynolds, C. D., Sakabe, K., Sakabe, N., and Vijayan, N. M. (1988). *Philos. Trans. R. Soc. Ser. B* **319**, 369.
Banci, L., Bencini, A., Bertini, I., Luchinat, C., and Piccioli, M. (1990a). *Inorg. Chem.* **29**, 4867.
Banci, L., Benedetto, M., Bertini, I., Del Conte, R., Piccioli, M., Richert, T., and Viezzoli, M. S. (1997a). *Magn. Reson. Chem.* **35**, 845.
Banci, L., Benedetto, M., Bertini, I., Del Conte, R., Piccioli, M., and Viezzoli, M. S. (1998). *Biochemistry* **37**, 11780.
Banci, L., Bertini, I., Bauer, D., Hallewell, R. A., and Viezzoli, M. S. (1993a). *Biochemistry* **32**, 4384.

Banci, L., Bertini, I., Borsari, M., Viezzoli, M. S., and Hallewell, R. A. (1995a). *Eur. J. Biochem.* **232**, 220.

Banci, L., Bertini, I., Bruni, B., Carloni, P., Luchinat, C., Mangani, S., Orioli, P. L., Piccioli, M., Rypniewski, W., and Wilson, K. (1994). *Biochem. Biophys. Res. Commun.* **202**, 1088.

Banci, L., Bertini, I., Cabelli, D., Hallewell, R. A., Luchinat, C., and Viezzoli, M. S. (1990b). *Inorg. Chem.* **29**, 2398.

Banci, L., Bertini, I., Chiu, C. Y., Mullenbach, G. T., and Viezzoli, M. S. (1995b). *Eur. J. Biochem.* **234**, 855.

Banci, L., Bertini, I., Ciofi-Baffoni, S., Huffman, D.C., and O'Halloran, T. V. (2001). *J. Biol. Chem.* **276**, 8415.

Banci, L., Bertini, I., Ciofi-Baffoni, S., D'Onofrio, M., Gonnelli, L., Marhuenda-Egea, F.C., and Ruiz-Dueñas, F.J., (2002). *J. Mol. Biol.* **317**, 415.

Banci, L., Bertini, I., Cramaro, F., Del Conte, R., Rosato, A., and Viezzoli, M. S. (2000). *Biochemistry* **39**, 9108.

Banci, L., Bertini, I., Del Conte, R., Mangani, S., and Viezzoli, M. S. (1999a). *JBIC* **4**, 795.

Banci, L., Bertini, I., Del Conte, R., Markey, J., and Ruiz- Dueñas, F.J. (2001). *Biochemistry* **40**, 15660.

Banci, L., Bertini, I., Del Conte, R., and Viezzoli, M. S. (1999b). *Biospectroscopy* **5**, 33.

Banci, L., Bertini, I., and Luchinat, C. (1991). "Nuclear and Electron Relaxation: The Magnetic Nucleus–Unpaired Electron Coupling in Solution." VCH, Weinheim.

Banci, L., Bertini, I., Luchinat, C., and Hallewell, R. A. (1988a). *J. Am. Chem. Soc.* **110**, 3629.

Banci, L., Bertini, I., Luchinat, C., Monnanni, R., and Scozzafava, A. (1988b). *Inorg. Chem.* **27**, 107.

Banci, L., Bertini, I., Luchinat, C., Piccioli, M., and Scozzafava, A. (1993b). *Gazz. Chim. Ital.* **123**, 95.

Banci, L., Bertini, I., Luchinat, C., Piccioli, M., Scozzafava, A., and Turano, P. (1989a). *Inorg. Chem.* **28**, 4650.

Banci, L., Bertini, I., Luchinat, C., and Scozzafava, A. (1989b). *J. Biol. Chem.* **264**, 9742.

Banci, L., Bertini, I., Luchinat, C., Scozzafava, A., and Turano, P. (1989c). *Inorg. Chem.* **28**, 2377.

Banci, L., Bertini, I., Luchinat, C., and Viezzoli, M. S. (1993c). *Inorg. Chem.* **32**, 1403.

Banci, L., Bertini, I., Viezzoli, M. S., Argese, E., Orsega, E., Choi Ying Chiu, and Mullenbach, G. T. (1997b). *JBIC* **2**, 295.

Banci, L., Cabelli, D. E., Getzoff, E. D., Hallewell, R. A., and Viezzoli, M. S. (1993d). *J. Inorg. Biochem.* **50**, 89.

Beattie, J. K., Fensom, D. J., Freeman, H. C., Woodcock, E., Hill, H. A., and Stokes, A. M. (1975). *Biochim. Biophys. Acta* **405**, 109.

Beinert, H. (1997). *Eur. J. Biochem.* **245**, 521.

Bertini, I., Banci, L., Brown, R. D., III, Koenig, S. H., and Luchinat, C. (1988). *Inorg. Chem.* **27**, 951.

Bertini, I., Banci, L., Luchinat, C., Bielski, B. H. J., Cabelli, D., Mullenbach, G. T., and Hallewell, R. A. (1989). *J. Am. Chem. Soc.* **111**, 714.

Bertini, I., Borghi, E., Luchinat, C., and Scozzafava, A. (1981). *J. Am. Chem. Soc.* **103**, 7779.

Bertini, I., Bren, K. L., Clemente, A., Fee, J. A., Gray, H. B., Luchinat, C., Malmström, B. G., Richards, J. H., Sanders, D., and Slutter, C. E. (1996). *J. Am. Chem. Soc.* **46**, 11658.

Bertini, I., Briganti, F., Luchinat, C., Mancini, M., and Spina, G. (1985a). *J. Magn. Reson.* **63**, 41.

Bertini, I., Bryant, D. A., Ciurli, S., Dikiy, A., Fernández, C. O., Luchinat, C., Safarov, N., (2001a). *J. Biol. Chem.* Vila, A. J., and Zhao, J. **276**, 47217.

Bertini, I., Capozzi, F., Luchinat, C., Piccioli, M., and Viezzoli, M. S. (1991). *Eur. J. Biochem.* **197**, 691.

Bertini, I., Ciurli, S., Dikiy, A., Fernández, C. O., Luchinat, C., Safarov, N., Shumilin, S., and Vila, A. J. (2001b). *J. Am. Chem. Soc.* **123**, 2405.
Bertini, I., Ciurli, S., Dikiy, A., Gasanov, R., Luchinat, C., Martini, G., and Safarov, N. (1999). *J. Am. Chem. Soc.* **121**, 2037.
Bertini, I., Fernández, C. O., Karlsson, B. G., Leckner, J., Luchinat, C., Malmström, B. G., Nersissian, A. M., Pierattelli, R., Shipp, E., Valentine, J. S., and Vila, A. J. (2000). *J. Am. Chem. Soc.* **122**, 3701.
Bertini, I., Gray, H. B., Lippard, S. J., Valentine, J. S. (Eds.) (1994a). "Bioinorganic Chemistry," University Science Books, Mill Valley, CA.
Bertini, I., Lanini, G., Luchinat, C., Mancini, M., and Spina, G. (1985b). *J. Magn. Reson.* **63**, 56.
Bertini, I., Lanini, G., Luchinat, C., Messori, L., Monnanni, R., and Scozzafava, A. (1985c). *J. Am. Chem. Soc.* **107**, 4391.
Bertini, I., and Luchinat, C. (1986a). "NMR of Paramagnetic Molecules in Biological Systems," Benjamin/Cummings, Menlo Park, CA.
Bertini, I., and Luchinat, C. (1986b). *In* "Biological and Inorganic Copper Chemistry" (Karlin, K. D., and Zubieta, J., Eds.), pp. 23–31, Adenine Press, New York.
Bertini, I., and Luchinat, C. (1992). *In* "Physical Methods for Chemists" (Drago, R. S., Ed.), pp. 500–558, Saunders, Philadelphia.
Bertini, I., and Luchinat, C. (1996). "NMR of Paramagnetic Substances," Coord. Chem. Rev., Vol. 150, Elsevier, Amsterdam.
Bertini, I., Luchinat, C., Macinai, R., Piccioli, M., Scozzafava, A., and Viezzoli, M. S. (1994b). *J. Magn. Reson. Ser. B* **104**, 95.
Bertini, I., Luchinat, C., Ming, L. -J., Piccioli, M., Sola, M., and Valentine, J. S. (1992). *Inorg. Chem.* **31**, 4433.
Bertini, I., Luchinat, C., and Piccioli, M. (1994c). *Progr. NMR Spectrosc.* **26**, 91.
Bertini, I., Luchinat, C., and Piccioli, M. (2001c). *Methods Enzymol.* **339**, 316.
Bertini, I., Luchinat, C., and Scozzafava, A. (1980). *J. Am. Chem. Soc.* **102**, 7349.
Bertini, I., Luchinat, C., and Tarchi, D. (1993a). *Chem. Phys. Lett.* **203**, 445.
Bertini, I., Mangani, S., and Viezzoli, M. S. (1998). *In* "Advanced Inorganic Chemistry" (Sykes, A. G., Ed.), pp. 127–250. Academic Press, San Diego, CA.
Bertini, I., Piccioli, M., Scozzafava, A., and Viezzoli, M. S. (1993b). *Magn. Reson. Chem.* **31**, S17.
Bertini, I., Piccioli, M., Viezzoli, M. S., Chiu, C. Y., and Mullenbach, G. T. (1994d). *Eur. J. Biophys.* **23**, 167.
Bertini, I., Turano, P., and Vila, A. J. (1993c). *Chem. Rev.* **93**, 2833.
Blackwell, K. A., Anderson, B. F., and Backer, E. N. (1994). *Acta Crystallogr. Sect. D* **50**, 263.
Blake, P. R., Lee, B., Summers, M. F., Adams, M. W. W., Park, J. B., Zhou, Z. H., and Bax, A. (1992). *J. Biomol. NMR* **2**, 527.
Blaszak, J. A., Ulrich, E. L., Markley, J. L., and McMillin, D. R. (1982). *Biochemistry* **21**, 6253.
Bloembergen, N. (1957). *J. Chem. Phys.* **27**, 572.
Botuyan, M. V., Toy-Palmer, A., Chung, J., Blake, R. C., Beroza, P., Case, D. A., and Dyson, H. J. (1996). *J. Mol. Biol.* **263**, 752.
Brink, J. M., Rose, R. A., and Holz, R. C. (1996). *Inorg. Chem.* **35**, 2878.
Bubacco, L., Salgado, J., Tepper, A. W., Vijgenboom, E., and Canters, G. W. (1999). *FEBS Lett.* **442**, 215.
Bubacco, L., Vijgenboom, E., Gobin, C., Tepper, A. W., Salgado, J., and Canters, G. W. (2000). *J. Mol. Catal. B* **8**, 27.
Burger, A. R., Lippard, S. J., Pantoliano, M. W., and Valentine, J. S. (1980). *Biochemistry* **19**, 4139.

Canters, G. W., Hill, H. A. O., Kitchen, N. A., and Adman, E. T. (1984). *J. Magn. Reson.* **57**, 1.

Cass, A. E. G., Hill, H. A. O., Smith, B. E., Bannister, J. V., and Bannister, W. H. (1977). *Biochemistry* **16**, 3061.

Chazin, W. J., Rence, M., and Wright, P. E. (1987). *FEBS Lett.* **222**, 109.

Chazin, W. J., Rence, M., and Wright, P. E. (1988). *J. Mol. Biol.* **202**, 603.

Chazin, W. J., and Wright, P. E. (1988). *J. Mol. Biol.* **202**, 626.

Chen, Y. -L., Park, S., Thornburg, R. W., Tabatabai, L. B., and Kintanar, A. (1995). *Biochemistry* **34**, 12265.

Chen, Z. G., de Ropp, J. S., Hernandez, G., and La Mar, G. N. (1994). *J. Am. Chem. Soc.* **116**, 8772.

Christensen, H. E. M., Ulstrup, J., and Sykes, A. G. (1990). *Biochim. Biophys. Acta* **1039**, 94.

Clementi, V., and Luchinat, C. (1998). *Acc. Chem. Res.* **31**, 351.

Cookson, D. J., Hayes, M. T., and Wright, P. E. (1980). *Biochim. Biophys. Acta* **591**, 162.

den Blaauwen, T., and Canters, G. W. (1993). *J. Am. Chem. Soc.* **115**, 1121.

den Blaauwen, T., Vandekamp, M., and Canters, G. W. (1991). *J. Am. Chem. Soc.* **113**, 5050.

Dennison, C., Berg, A., and Canters, G. W. (1997). *Biochemistry* **36**, 3262.

Dennison, C., and Kohzuma, T. (1999). *Inorg. Chem.* **38**, 1491.

Dennison, C., Kohzuma, T., McFarlane, W., Suzuki, S., and Sykes, A. G. (1994a). *Inorg. Chem.* **33**, 3299.

Dennison, C., Kohzuma, T., McFarlane, W., Suzuki, S., and Sykes, A. G. (1994b). *J. Chem. Soc. Chem. Commun.* 581.

Dennison, C., Kyritsis, P., McFarlane, W., and Sykes, A. G. (1993). *J. Chem. Soc. Dalton Trans.* 1959.

Dennison, C., and Lawler, A.T. (2002). *Biochemistry* **40**, 3158.

Dennison, C., Lawler, A.T., and Kohzuma, T. (2002). *Biochemistry* **41**, 552.

Dennison, C., Oda, K., and Kohzuma, T. (2000). *Chem. Commun.*, 751.

Dennison, C., Van Driessche, G., Van Beeumen, J., McFarlane, W., and Sykes, A. G. (1996). *Chem. Eur. J.* **2**, 104.

Dennison, C., Vijgenboom, E., de Vries, S., Van der Oost, J., and Canters, G. W. (1995). *FEBS Lett.* **365**, 92.

Desvaux, H., Birlirakis, N., Wary, C., and Berthault, P. (1995). *Mol. Phys.* **86**, 1059.

Diederix, R. E., Canters, G. W., and Dennison, C. (2000). *Biochemistry* **39**, 9551.

Djinovic, K., Battistoni, A., Carri, M. T., Polticelli, F., Desideri, A., Rotilio, G., Coda, A., and Bolognesi, M. (1994a), *FEBS Lett.* **349**, 93.

Djinovic, K., Polticelli, F., Desideri, A., Rotilio, G., Wilson, K., and Bolognesi, M. (1994b). *J. Mol. Biol.* **240**, 179.

Donaire, A., Jimenez, B., Moratal Mascarell, J. M., Hall, J. F., and Hasnain, S. S. (2001). *Biochemistry*, **40**, 837.

Donaire, A., Salgado, J., and Moratal, J. M. (1998). *Biochemistry* **37**, 8659.

Driscoll, P. C., Hill, A. O., and Redfield, A. G. (1987). *Eur. J. Biochem.* **170**, 279.

Ducros, V., Brzozowski, A. M., Wilson, K. S., Brown, S. H., Ostergaard, P., Schneider, P., Yaver, D. S., Pedersen, A. H., and Davies, G. J. (1998). *Nat. Struct. Biol.* **5**, 310.

Einstein, A. (1956). "Investigations on the Theory of the Brownian Movement." Dover, New York.

Ejdeback, M., Bergkvist, A., Karlsson, B. G., and Ubbink, M. (2000). *Biochemistry* **39**, 5022.

Engeseth, H. R., McMillin, D. R., and Otvos, J. D. (1984). *J. Biol. Chem.* **259**, 4822.

Fee, J., and DiCorleto, P. (1973). *Biochemistry* **12**, 4893.

Fee, J. A., and Gaber, B. P. (1972). *J. Biol. Chem.* **247**, 60.

Ferguson-Miller, S., and Babcock, G. T. (1996). *Chem. Rev.* **96**, 2889.

Fernández, C. O., Sannazzaro, A. I., and Vila, A. J. (1997). *Biochemistry* **36**, 10566.

Fernández, C. O., Niizeki, T., Kohzuma, T., and Vila, A. J. Submitted for publication.
Fernández, C. O., Sannazzaro, A. I., and Vila, A. J. (1998). *Inorg. Chim. Acta* **273**, 367.
Ferraroni, M., Rypniewski, W., Wilson, K. S., Viezzoli, M. S., Banci, L., Bertini, I., and Mangani, S. (1999). *J. Mol. Biol.* **288**, 413.
Fisher, C. L., Cabelli, D. E., Tainer, J. A., Hallewell, R. A., and Getzoff, E. D. (1994). *Proteins Struct. Funct. Genet.* **19**, 24.
Freeman, R., and Morris, G. A. (1978). *J. Chem. Soc. Chem. Commun.* 684.
Gaber, B. P., Brown, R. D., III, Koenig, S. H., and Fee, J. A. (1972). *Biochim. Biophys. Acta* **271**, 1.
Getzoff, E. D., Cabelli, D. E., Fisher, C. L., Parge, H. E., Viezzoli, M. S., Banci, L., and Hallewell, R. A. (1992). *Nature* **358**, 347.
Getzoff, E. D., Tainer, J. A., Stempien, M. M., Bell, G. I., and Hallewell, R. A. (1989). *Proteins Struct. Funct. Genet.* **5**, 322.
Gitschier, J., Moffat, B., Reilly, D., Wood, W. I., and Fairbrother, W. J. (1998). *Nat. Struct. Biol.* **5**, 47.
Gralla, E. B., and Kosman, D. J. (1992). *Adv. Genet.* **30**, 251.
Gray, H. B., Malmström, B. G., and Williams, R. J. (2000). *JBIC* **5**, 551.
Groeneveld, C. M., and Canters, G. W. (1985). *Eur. J. Biochem.* **153**, 559.
Groeneveld, C. M., and Canters, G. W. (1988). *J. Biol. Chem.* **263**, 167.
Gueron, M. (1975). *J. Magn. Reson.* **19**, 58.
Guss, J. M., Bartunik, H. D., and Freeman, H. C. (1992). *Acta Crystallogr. Sect. B* **48**, 790.
Guss, J. M., and Freeman, H. C. (1983). *J. Mol. Biol.* **169**, 521.
Guss, J. M., Harrowell, P. R., Murata, M., Norris, V. A., and Freeman, H. C. (1986). *J. Mol. Biol.* **192**, 361.
Halliwell, B., and Gutteridge, J. M. (1989). *In* "Free Radicals in Biology and Medicine" pp. 22–408, Clarendon, Oxford.
Hamman, C., Van Pouderoyen, G., Nar, H., Ruth, F. X. G., Messerschmidt, A., Huber, R., den Blaauwen, T., and Canters, G. W. (1997). *J. Mol. Biol.* **266**, 357.
Handford, P. M., Hill, H. A. O., Lee, R. W. K., Henderson, R. A., and Sykes, A. G. (1980). *J. Inorg. Biochem.* **13**, 83.
Harris, R. K. (1986). "Nuclear Magnetic Resonance Spectroscopy." Longman, London.
Harris, R. K., and Mann, B. E. (1978). "NMR and the Periodic Table." Academic Press, London.
Hart, J. P., Nersissian, A. M., Herrmann, R. G., Nalbandyan, R. M., Valentine, J. S., and Eisenberg, D. (1996). *Protein Sci.* **5**, 2175.
Henehan, C. J., Pountney, D. L., Zerbe, O., and Vasak, M. (1993). *Protein Sci.* **2**, 1756.
Hill, H. A. O., Lee, W. K., Bannister, J. V., and Bannister, W. H. (1980). *Biochem. J.* **185**, 245.
Hill, H. A. O., Leer, J. C., Smith, B. E., Storm, C. B., and Ambler, R. P. (1976). *Biochem. Biophys. Res. Commun.* **70**, 331.
Himelblau, E., Mira, H., Lin, S. J., Culotta, V. C., Penarrubia, L., and Amasino, R. M. (1998). *Plant Physiol.* **117**, 1227.
Hoitink, C. W., and Canters, G. W. (1992). *J. Biol. Chem.* **267**, 13836.
Hoitink, C. W., Driscoll, P. C., Hill, H. A., and Canters, G. W. (1994). *Biochemistry* **33**, 3560.
Holm, R. H., Kennepohl, P., and Solomon, E. I. (1996). *Chem. Rev.* **96**, 2239.
Holz, R. C., Alvarez, M. L., Zumft, W. G., and Dooley, D. M. (1999). *Biochemistry* **38**, 11164.
Holz, R. C., Bennet, B., Chen, G. J., and Ming, L. J. (1998). *J. Am. Chem. Soc.* **120**, 6329.
Hubbard, T. J. P., Murzin, A. G., Brenner, S. E., and Chothia, C. (1997). *Nucleic Acids Res.* **25**, 236.
Hunt, A. H., Toy-Palmer, A., Assa-Munt, N., Cavanagh, J., Blake, R. C., and Dyson, H. J. (1994). *J. Mol. Biol.* **244**, 370.
Hunter, D.M., McFarlane, W., Sykes, A.G., and Dennison, C. (2001). *Inorg. Chem.* **40**, 354.

Inubushi, T., and Becker, E. D. (1983). *J. Magn. Reson.* **51**, 128.
Iwata, S., Ostermeier, C., Ludwig, B., and Michel, H. (1995). *Nature* **376**, 660.
Jackman, M. P., Sinclairday, J. D., Sisley, M. J., Sykes, A. G., Denys, L. A., and Wright, P. E. (1987). *J. Am. Chem. Soc.* **109**, 6443.
Jeuken, L. J., Ubbink, M., Bitter, J. H., Van Vliet, P., Meyer-klaucke, W., and Canters, G. W. (2000). *J. Mol. Biol.* **299**, 737.
Kalverda, A. P., Salgado, J., Dennison, C., and Canters, G. W. (1996). *Biochemistry* **35**, 3092.
Kalverda, A. P., Ubbink, M., Gilardi, G., Wijmenga, S. S., Crawford, A., Jeuken, L. J., and Canters, G. W. (1999). *Biochemistry* **38**, 12690.
Kalverda, A. P., Wymenga, S. S., Lommen, A., Van de, V., Hilbers, C. W., and Canters, G. W. (1994). *J. Mol. Biol.* **240**, 358.
Klomp, L. W. J., Lin, S. J., Yuan, D., Klausner, R. D., Culotta, V. C., and Gitlin, J. D. (1997). *J. Biol. Chem.* **272**, 9221.
Klug-Roth, D., Fridovich, I., and Rabani, J. (1973). *J. Am. Chem. Soc.* **95**, 2786.
Koenig, S. H., and Brown, R. D. III (1987). *In* "Metal Ions in Biological Systems" (Sigel, H., Ed.), Vol. 21, Dekker, New York.
Koenig, S. H., and Brown, R. D., III (1990). *Progr. NMR Spectrosc.* **22**, 487.
Koenig, S. H., and Brown, R. D., III (1973). *Ann. N. Y. Acad. Sci.* **222**, 752.
Kofod, P., Bauer, R., Danielsen, E., Larsen, E., and Bjerrum, M. J. (1991). *Eur. J. Biochem.* **198**, 607.
Kolczak, U., Salgado, J., Siegal, G., Saraste, M., and Canters, G. W. (1999). *Biospectroscopy* **5**, S19.
Kroes, S. J., Salgado, J., Parigi, G., Luchinat, C., and Canters, G. W. (1996). *JBIC* **1**, 551.
Kroneck, P. M. H., Antholine, W. E., Kastrau, D. H. W., Buse, G., Steffens, G. C. M., and Zumft, W. G. (1990). *FEBS Lett.* **268**, 274.
Kroneck, P. M. H., Antholine, W. E., Riester, J., and Zumft, W. G. (1988). *FEBS Lett.* **242**, 70.
Kurland, R. J., and McGarvey, B. R. (1970). *J. Magn. Reson.* **2**, 286.
Kyritsis, P., Dennison, C., Ingledew, W. J., McFarlane, W., and Sykes, A. G. (1995). *Inorg. Chem.* **34**, 5370.
Lipari, G., and Szabo, A. (1982). *J. Am. Chem. Soc.* **104**, 4546.
Lippard, S. J., Burger, A. R., Ugurbil, K., Pantoliano, M. W., and Valentine, J. S. (1977). *Biochemistry* **16**, 1136.
Lommen, A., and Canters, G. W. (1990). *J. Biol. Chem.* **265**, 2768.
Lommen, A., Canters, G. W., and Van Beeumen, J. (1988). *Eur. J. Biochem.* **176**, 213.
Lommen, A., Wijmenga, S., Hilbers, C. W., and Canters, G. W. (1991). *Eur. J. Biochem.* **201**, 695.
Luchinat, C., Soriano, A., Djinovic-Carugo, K., Saraste, M., Malmström, B. G., and Bertini, I. (1997). *J. Am. Chem. Soc.* **119**, 11023.
Ma, L., Jørgensen, A. M. M., Led, J. J., and Ulstrup, J. (2000). *J. Am. Chem. Soc.* **122**, 9473.
Malmström, B. G. (1994). *Eur. J. Biochem.* **223**, 711.
Marcus, R. A., and Sutin, N. (1985). *Biochim. Biophys. Acta* **811**, 265.
Markley, J. L., Ulrich, E. L., Berg, S. P., and Krogmann, D. W. (1975). *Biochemistry* **14**, 4428.
Markley, J. L., Ulrich, E. L., and Krogmann, D. W. (1977). *Biochem Biophys Res Commun* **78**, 106.
McConnell, H. M., and Chesnut, D. B. (1958). *J. Chem. Phys.* **28**, 107.
McCord, J. M., and Fridovich, I. (1969). *J. Biol. Chem.* **244**, 6049.
Messerschmidt, A. (1998). *Struct. Bonding* **90**, 37.
Ming, L. -J., Banci, L., Luchinat, C., Bertini, I., and Valentine, J. S. (1988a). *Inorg. Chem.* **27**, 4458.
Ming, L. -J., Banci, L., Luchinat, C., Bertini, I., and Valentine, J. S. (1988b). *Inorg. Chem.* **27**, 728.

Ming, L. -J., and Valentine, J. S. (1990). *J. Am. Chem. Soc.* **112**, 6374.
Moore, J. M., Case, D. A., Chazin, W. J., Gippert, G. P., Havel, T. F., Powls, R., and Wright, P. E. (1988). *Science* **240**, 314.
Moore, J. M., Lepre, C. A., Gippert, G. P., Chazin, W. J., Case, D. A., and Wright, P. E. (1991). *J. Mol. Biol.* **221**, 533.
Moratal Mascarell, J. M., Salgado, J., Donaire, A., Jimenez, H. R., and Castells, J. (1993a). *Inorg. Chem.* **32**, 3587.
Moratal Mascarell, J. M., Salgado, J., Donaire, A., Jimenez, H. R., and Castells, J. (1993b). *J. Chem. Soc. Chem. Commun.* 110.
Moratal Mascarell, J. M., Salgado, J., Donaire, A., Jimenez, H. R., Castells, J., and Martinez Ferrer, M. -J. (1993c). *Magn. Reson. Chem.* **31**, S41.
Moratal, J. M., Romero, A., Salgado, J., Perales-Alarcon, A., and Jimenez, H. R. (1995). *Eur. J. Biochem.* **228**, 653.
Mota De Freitas, D., Luchinat, C., Banci, L., Bertini, I., and Valentine, J. S. (1987). *Inorg. Chem.* **26**, 2788.
Murphy, M. E., Lindley, P. F., and Adman, E. T. (1997). *Protein Sci* **6**, 761.
Murthy, N. N., Karlin, K. D., Bertini, I., and Luchinat, C. (1997). *J. Am. Chem. Soc.* **119**, 2156.
Nar, H., Messerschmidt, A., Huber, R., Van de Kamp, M., and Canters, G. W. (1991). *J. Mol. Biol.* **221**, 765.
Nishihara, E., Furuyama, T., Yamashita, S., and Mori, N. (1998). *Mol. Neurosci.* **9**, 3259.
Odermatt, A., and Solioz, M. (1995). *J. Biol. Chem.* **270**, 4349.
Orbach, R., and Stapleton, H. J. (1972). *In* "Electron Paramagnetic Resonance" (Geschwind, S., Ed.), pp. 121–216, Plenum, New York.
Oz, G. L., Pountney, D. L., and Armitage, I. M. (1998). *Biochem. Cell. Biol.* **76**, 223.
Ozaki, S., Hirose, J., and Kidani, Y. (1988). *Inorg. Chem.* **27**, 3746.
Paci, M., Desideri, A., and Rotilio, G. (1988). *J. Biol. Chem.* **263**, 162.
Paci, M., Desideri, A., Sette, M., Cirioli, M. R., and Rotilio, G. (1990). *FEBS Lett.* **263**, 127.
Palmer, A. G., III (1997). *Curr. Opin. Struct. Biol.* **7**, 732.
Parge, H. E., Getzoff, E. D., Scandella, C. S., Hallewell, R. A., and Tainer, J. A. (1986). *J. Biol. Chem.* **261**, 16215.
Peisach, J., and Blumberg, W. E. (1974). *Arch. Biochem. Biophys.* **165**, 691.
Piccioli, M. (1995). *Chem. Tracts* **7**, 99.
Piccioli, M., Luchinat, C., Mizoguchi, T. J., Ramirez, B. E., Gray, H. B., and Richards, J. H. (1995). *Inorg. Chem.* **34**, 737.
Pufahl, R., Singer, C. P., Peariso, K. L., Lin, S. -J., Schmidt, P. J., Fahrni, C. J., Cizewski Culotta, V., Penner-Hahn, J. E., and O'Halloran, T. V. (1997). *Science* **278**, 853.
Ramirez, B. E., Malmström, B. G., Winkler, J. R., and Gray, H. B. (1995). *Proc. Natl. Acad. Sci. USA* **92**, 11949.
Randall, D. W., Gamelin, D. R., LaCroix, L. B., and Solomon, E. I. (2000). *JBIC* **5**, 16.
Rigo, A., Stevanato, R., Viglino, P., and Rotilio, G. (1977). *Biochem. Biophys. Res. Commun.* **79**, 776.
Rigo, A., Viglino, P., and Rotilio, G. (1975). *Anal. Biochem.* **68**, 1.
Romero, A., Nar, H., Huber, R., Messerschmidt, A., Kalverda, A. P., Canters, G. W., Durley, R., and Mathews, F. S. (1994). *J. Mol. Biol.* **236**, 1196.
Rosenzweig, A. C., Huffman, D. L., Hou, M. Y., Wernimont, A. K., Pufahl, R. A., and O'Halloran, T. V. (1999). *Structure* **7**, 605.
Salgado, J., Jimenez, H. R., Donaire, A., and Moratal, J. M. (1995). *Eur. J. Biochem.* **231**(2), 358.
Salgado, J., Jimenez, H. R., Moratal Mascarell, J. M., Kroes, S. J., Warmerdam, G. C. M., and Canters, G. W. (1996). *Biochemistry* **35**, 1810.

Salgado, J., Kalverda, A. P., and Canters, G. W. (1997). *J. Biomol. NMR* **9**, 299.

Salgado, J., Kalverda, A. P., Diederix, R. E., Canters, G. W., Moratal, J. M., Lawler, A. T., and Dennison, C. (1999). *JBIC* **4**, 457.

Salgado, J., Kroes, S. J., Berg, A., Moratal Mascarell, J. M., and Canters, G. W. (1998a). *J. Biol. Chem.* **273**, 177.

Salgado, J., Warmerdam, G., Bubacco, L., and Canters, G. W. (1998b). *Biochemistry* **37**, 7378.

Sato, K., and Dennison, C. (2002). *Biochemistry* **41**, 120.

Sette, M., Bozzi, M., Battistoni, A., Fasano, M., Paci, M., and Rotilio, G. (2000). *FEBS Lett.* **483**, 21.

Sette, M., Paci, M., Desideri, A., and Rotilio, G. (1995). *Eur. J. Biochem.* **227**, 441.

Slutter, C. E., Sanders, D., Wittung, P., Malmström, B. G., Aasa, R., Richards, J. H., Gray, H. B., and Fee, J. A. (1996). *Biochemistry* **35**, 3387.

Solomon, E. I., Baldwin, M. J., and Lowery, M. D. (1992). *Chem. Rev.* **92**, 521.

Solomon, E. I., Sundaram, U. M., and Machonin, T. E. (1996). *Chem. Rev.* **96**, 2563.

Solomon, I. (1955). *Phys. Rev.* **99**, 559.

Solomon, I., and Bloembergen, N. (1956). *J. Chem. Phys.* **25**, 261.

Steele, R. A., and Opella, S. J. (1997). *Biochemistry* **36**, 6885.

Stoesz, J. D., Malinowski, D. P., and Redfield, A. G. (1979). *Biochemistry* **18**, 4669.

Stokes, G. (1956). *Trans. Cambridge Philos. Soc.* **9**, 5.

Strothkamp, K. G., and Lippard, S. J. (1981). *Biochemistry* **20**, 7488.

Sykes, A. G. (1985). *Chem. Soc. Rev.* **14**, 283.

Sykes, A. G. (1991). *Adv. Inorg. Chem.* 377.

Tainer, J. A., Getzoff, E. D., Beem, K. M., Richardson, J. S., and Richardson, D. C. (1982). *J. Mol. Biol.* **160**, 181.

Thompson, G. S., Leung, Y. C., Ferguson, S. J., Radford, S. E., and Redfield, C. (2000). *Protein Sci.* **9**, 846.

Toy-Palmer, A., Prytulla, S., and Dyson, H. J. (1995). *FEBS Lett.* **365**, 35.

Tsukihara, T., Aoyama, H., Yamashita, E., Tomizaki, T., Yamaguchi, H., Shinzawa-Itoh, K., Nakashima, R., Yaono, R., and Yoshikawa, S. (1995). *Science* **269**, 1069.

Tsukihara, T., Aoyama, H., Yamashita, E., Tomizaki, T., Yamaguchi, H., Shinzawa-Itoh, K., Nakashima, R., Yaono, R., and Yoshikawa, S. (1996). *Science* **272**, 1136.

Ubbink, M., and Bendall, D. S. (1997). *Biochemistry* **36**, 6326.

Ubbink, M., Ejdebaeck, M., Karlsson, B. G., and Bendall, D. S. (1998). *Structure* **6**, 323.

Ubbink, M., Lian, L. Y., Modi, S., Evans, P. A., and Bendall, D. S. (1996). *Eur. J. Biochem.* **242**, 132.

Ugurbil, K., and Bersohn, R. (1977). *Biochemistry* **16**, 3016.

Ugurbil, K., Norton, R. S., Allerhand, A., and Bersohn, R. (1977). *Biochemistry* **16**, 886.

Ullmann, G. M., and Kostic, N. M. (1996). *J. Am. Chem. Soc.* **117**, 4766.

Utschig, L. M., Abynard, T., Strong, C., and O'Halloran, T. V. (1997). *Inorg. Chem.* **36**, 2926.

Utschig, L. M., Wright, J. G., Dieckmann, G., Pecoraro, V. L., and O'Halloran, T. V. (1995). *Inorg. Chem.* **34**, 2497.

Van de Kamp, M., Canters, G. W., Wijmenga, S. S., Lommen, A., Hilbers, C. W., Nar, H., and Messerschmidt, A. (1992). *Biochemistry* **31**, 10194.

Van de Kamp, M., Silvestrini, M. C., Brunori, M., Van Beeumen, J., Hali, F. C., and Canters, G. W. (1990). *Eur. J. Biochem.* **194**, 109.

Van Pouderoyen, G., Cigna, G., Rolli, G., Cutruzzola, F., Malatesta, F., Silvestrini, M. C., Brunori, M., and Canters, G. W. (1997). *Eur. J. Biochem.* **247**, 322.

Vega, A. J., and Fiat, D. (1976). *Mol. Phys.* **31**, 347.

Veglia, G., Porcelli, F., DeSilva, T., Prantner, A., and Opella, S. J. (2000). *J. Am. Chem. Soc.* **122**, 2389.

Venerini, F., Sette, M., Stroppolo, M. E., De Martino, A., and Desideri, A. (1999). *Arch. Biochem. Biophys.* **366**, 70.

Viglino, P., Rigo, A., Stevanato, R., Ranieri, G. A., Rotilio, G., and Calabrese, L. (1979). *J. Magn. Reson.* **34**, 265.
Vila, A. J. (1994). *FEBS Lett.* **355**, 15.
Vila, A. J., and Fernández, C. O. (1996). *J. Am. Chem. Soc.* **118**, 7291.
Vila, A. J., Ramirez, B. E., Di Bilio, A. J., Mizoguchi, T. J., Richards, J. H., and Gray, H. B. (1997). *Inorg. Chem.* **36**, 4567.
Wakabayashi, T., Nakamura, N., Sambongi, Y., Wada, Y., Oka, T., and Futai, M. (1998). *FEBS Lett.* **440**, 141.
Walter, R. L., Ealick, S. E., Friedman, A. M., Blake, R. C. I., Proctor, P., and Shoam, M. (1996). *J. Mol. Biol.* **263**, 730.
Williams, P. A., Blackburn, N. J., Sanders, D., Bellamyn, H., Stura, E. A., Fee, J. A., and McRee, D. E. (1999). *Nat. Struct. Biol.* **6**, 509.
Williams, P. A., Fülöp, V., Leung, Y.-C., Moir, J. W. B., Howlett, G., Ferguson, S. J., Radford, S. E., and Hajdu, J. (1995). *Nat. Struct. Biol.* **2**, 975.
Wilmanns, M., Lappalainen, P., Kelly, M., Sauer-Eriksson, E., and Saraste, M. (1995). *Proc. Natl. Acad. Sci. USA* **92**, 11955.
Wimmer, R., Herrmann, T., Solioz, M., and Wüthrich, K. (1999). *J. Biol. Chem.* **274**, 22597.
Yuan, D. S., Dancis, A., and Klausner, R. D. (1997). *J. Biol. Chem.* **272**, 25787.

# AUTHOR INDEX

## B

## C

## D

## G

## H

## I



## J

## K

## M

## N

## O

## P

## Q



## R

## S

## U



## V

## X



## Y

## Z

# SUBJECT INDEX

## A

## I



## K



## L



## M



## N

## X



## Y



## Z

90051
9 780120 342600
ISBN 0-12-034260-X